D. R. Christensen

Soil Organic Matter in Temperate Agroecosystems

Long-Term Experiments in North America

Edited by

E.A. Paul, Ph.D.
Michigan State University
East Lansing, Michigan

K. Paustian, Ph.D.
Colorado State University
Fort Collins, Colorado

E.T. Elliott, Ph.D.
Colorado State University
Fort Collins, Colorado

C.V. Cole, Ph.D.
Colorado State University
Fort Collins, Colorado

CRC Press
Boca Raton New York London Tokyo

LIMITED WARRANTY

Library of Congress Cataloging-in-Publication Data

Soil organic matter in temperate agroecosystems : long-term
experiments in North America / edited by E.A. Paul . . . [et al.].
p. cm.
Includes bibliographical references and index.
ISBN 0-8493-2802-0 (alk. paper)
1. Humus--North America. 2. Soil productivity--North America.
3. Soil ecology--North America. I. Paul, Eldor Alvin.
S592.8.S675 1996
631.4′17′097--dc20 96-23071
CIP

International Standard Book Number 0-8493-2802-0
Library of Congress Card Number 96-23071
Printed in the United States of America 1 2 3 4 5 6 7 8 9 0
Printed on acid-free paper

PREFACE

Soil organic matter (SOM) is an important component of both managed and unmanaged terrestrial ecosystems. It is, therefore, not surprising that it has been strongly implicated as a source-sink in global C calculations that attempt to balance the gains and losses of soil C relative to the predicted changes in atmospheric CO_2. The conversion of the native forests and grasslands of North America to agricultural land made human settlement of this vast and productive area possible. It also constituted the most extensive ecological disturbance that the present Corn Belt and Great Plains areas were subjected to since the end of the last ice age some 10,000–15,000 years ago. There were many economic, environmental, and social consequences of this conversion. A major consequence was the substantial loss of SOM and the resultant release of nutrients that occurred during the first 50 years of cultivation. This in some soils still occurs. These changes made it possible to farm for extended periods without external fertilizers and also indirectly led to N losses to the environment and to increased soil erosion. They also have been a primary source of atmospheric CO_2 in the present day global change scenario.

There is considerable debate as to whether a net loss of SOM is occurring today. Also a concern to both agricultural and environmental scientists is the question of future levels of SOM, one of the world's greatest natural resources. Since SOM levels were generally reduced to levels far below those occurring in the original native ecosystems, there is considerable debate as to whether changes in management techniques (in a climate with increased CO_2 and possibly increased temperatures and altered precipitation) can produce enough change in soil C storage to influence global CO_2 levels.

Many of the agricultural practices that have been traditionally applied in the Great Plains and Corn Belt (e.g., moldboard plowing, fallowing with no plant cover, removal or burning of crop residues, and row crop monoculture cropping) represented "worst-case" scenarios for soil C maintenance. They minimized crop residue inputs and promoted a high degree of SOM decomposition and increased erosion.

Currently, agricultural management in the U.S. and Canada is undergoing rapid change in response to a variety of pressures, including high costs of energy and chemical inputs, environmental concerns about nutrient and pesticide pollution and soil erosion, and food quality demands by consumers. Many of the management techniques designed to address the above issues have also demonstrated a potential for increasing soil C levels above those maintained under past management. Such practices include: (1) reduced or no-till cultivation, (2) reduction in the proportion of fallow relative to crop in semiarid regions, (3) the use of cover crops, green manure and animal manure, (4) the use of more stress-resistant and high-yielding varieties, and (5) in some cases reversion to grassland or forest vegetation conservation under the U.S. Reserve Program and similar programs in Canada. The extent to which soil C levels can be regulated through the application of various combinations of these practices for different cropping systems, soil types, and climatic regions is not clear. However, increased SOM levels are being predicted for a number of management system-soil type combinations.

Increased soil C would have major feedback effects by increasing soil fertility and improving water relations and soil structure. This, in turn, could lead to greater productivity and the return of more residues to the soil. However, a system with higher SOM levels could also lead to environmental contamination if crop uptake of mineralized nutrients is not in synchrony with the microbial activities. The mineralization of N and its accumulation as nitrates in the fall, winter, and early spring, if there were no crop uptake, could lead to leaching to groundwater, and the evolution of gaseous N oxides to the atmosphere.

In addition to changing management practices on agricultural soils, land use changes can have a considerable impact on region-wide C storage potential. Over the past 50 years, significant acreages of cropland have been abandoned or converted back into perennial grasslands and forests. For example, more than 100,000 ha of farmland in northeast Colorado alone were abandoned during the 1930s. In more recent years, 15 million ha of cropland in the U.S. have been converted to grassland under the Conservation Reserve Program. Large decreases in cultivated area have also occurred in the Corn Belt, in particular in the Great Lakes region and in eastern areas bordering on the Appalachians. For example, from 1950 to 1990, the agricultural landbase of Michigan declined by 3.2 million ha. Much of this was previously in corn, soybean and wheat production. To date there has been little systematic evaluation of how these shifts in land-use have affected regional soil C balances or what effect future changes may have. The rigorous testing of models and future concepts concerning SOM-management yield interactions can only be conducted using a first-class database.

Fortunately, the possibility of obtaining the necessary data exists. The part of North America that constitutes the Corn Belt and the Great Plains area has an excellent range of long term field sites. They range in age from the 1886 Morrow plots and the 1888 Sanborn plots to the management studies initiated more recently on the Indian Head Saskatchewan experimental station established in 1985. These plots have produced a wealth of data on crops that include primarily corn, soybeans, and wheat, but also include a variety of forages, both as hay and as cover crops. Also included are sorghum, barley, peas, navy beans, sugar beets, and other crops. Management includes a range of tillage operations, nutrient additions, liming, irrigation, various pesticides, soil leveling, and a range of crop varieties.

Data from the long-term field studies have provided much information about the possible rates and directions of change in soil C. However, our current knowledge base is still fragmented. Different analytical techniques have often been used at different sites and across time. In addition, most of what we know concerning SOM dynamics has been obtained by studying SOM losses. We have much to learn about controls on SOM accumulation and how they vary across soil types, climatic regions, and management regimes. Such information is crucial for estimating the potential for C sequestration in agricultural soils of the Great Plains and Corn Belt regions.

This volume contains: (I) a series of introductory chapters that give a background to the issues surrounding global C questions, and discuss some of the methodology available for measuring SOM dynamics on a regional basis, (II) the site management effects on productivity and SOM content in the Corn Belt, (III) site management effects on productivity and SOM characteristics in the Great Plains, and (IV) soil crop and management databases of long-term experiments in North America. This includes detailed databases on site history, climate, crop yields, and soil properties including C and N levels. These sites are shown in Figure 1.

Support for establishing a long-term experiment site network and assembling the information presented in this book was primarily through an EPA grant entitled "Agroecosystem Carbon Pools and Dynamics (AERL 9101)." Supplemental support was also obtained from the USDA/Agricultural Research Services Global Change program.

Not covered in this volume are some important long-term sites that for various reasons were not included, but which would be valuable additions in the future. Also not covered is an analysis of the data presented in the individual chapters and in Section IV. The EPA grant made it possible to sample many of the cooperating sites discussed in this volume as well as some sites not included in this volume. Analysis of these samples included measurements considered to be sensitive indicators of management effects. In addition to C and N and bulk density, these included particulate organic matter (POM), long term mineralization of C and N, pH and microbial biomass. Discussion of the newly acquired results with the site cooperators, as well as integration and aggregation of the large amount of information on the basis of geography, soil type, cropping systems, nutrient levels input of residues, climate, and tillage still needs to be done. This will allow a detailed evaluation of the role of various management factors in soil C storage in the future. It should also allow the adequate testing and verifications of models that consider the many factors involved and attempt to make reasonably accurate predictions of soil C storage in the future.

E.A. Paul
K. Paustian
E.T. Elliott
C.V. Cole

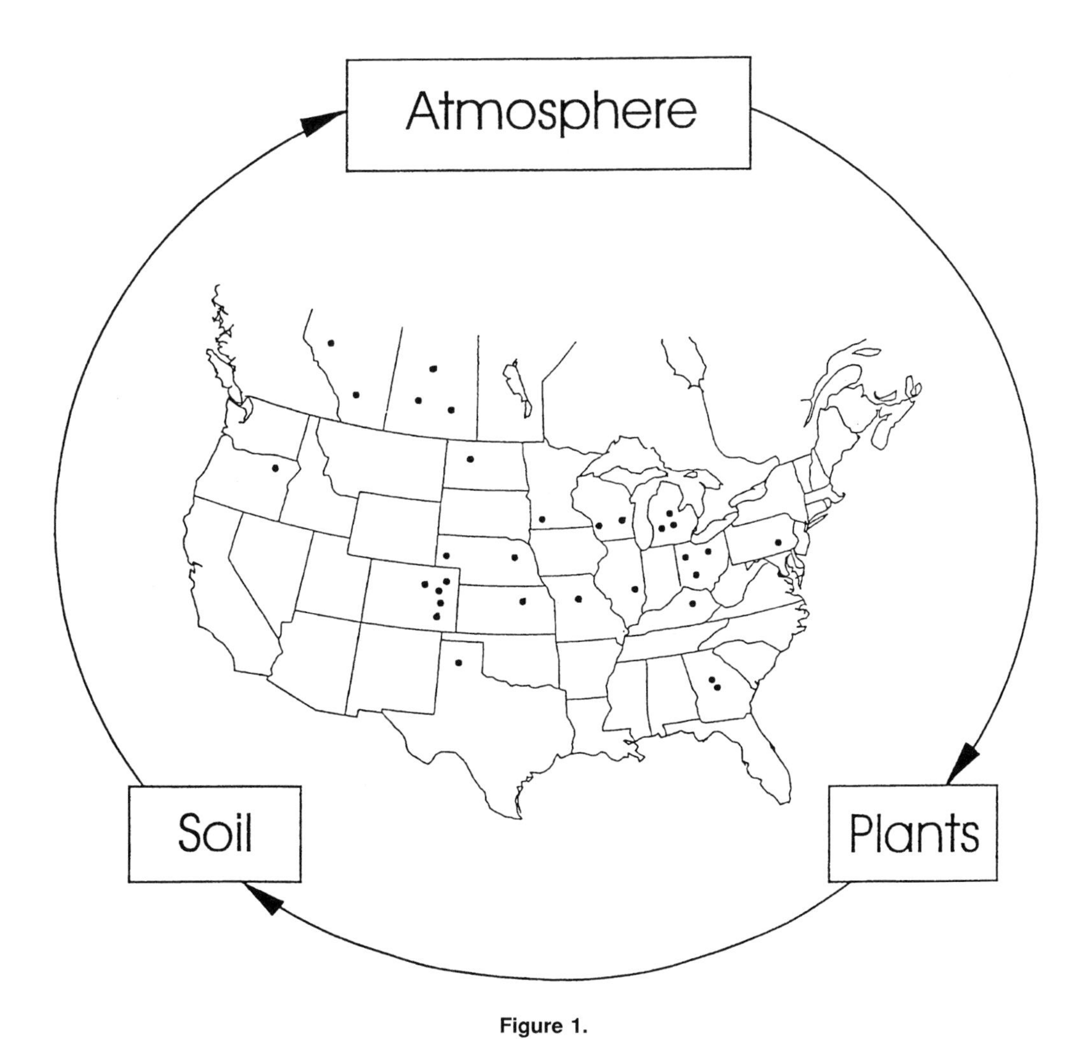

Figure 1.

The Editors

E.A. Paul, Ph.D., Professor, Department of Crop and Soil Sciences, Michigan State University, has served as Chairperson of Crop and Soil Sciences at Michigan State University and as Chair of the Department of Plant and Soil Biology, University of California — Berkeley. Before that he was Professor of Soil Science at the University of Saskatchewan, Saskatoon, Canada. His major fields of interest shown in the extensive papers published have included microbial ecology, soil organic matter dynamics, nutrient cycling, and plant microbial interactions. National and international committees and professional society appointments have helped both his science and his profession. He has taught extensively at the undergraduate and graduate level and advised many successful students both in Canada and the U.S. Books edited include volumes 3, 4, and 5 of *Soil Biochemistry* published by Marcel Dekker. The textbook, *Soil Microbiology and Biochemistry*, co-edited with F.E. Clark and published by Academic Press has extensive readership and is being translated into three languages other than English. He is a fellow of the Canadian Society Soil Science, American Society of Agronomy, Soil Science Society of America, and the American Association for the Advancement of Science and a recipient of the Soil Science Society of America Research Award.

Keith Paustian, Ph.D., is a Senior Research Scientist at the Natural Resource Ecology Laboratory (NREL), Colorado State University. Prior to his appointment at NREL in 1993, he was a Research Assistant Professor at the W.K. Kellogg Biological Station, Michigan State University (1989–1993) and Research Scientist at the Department of Ecology and Environmental Research, Swedish University of Agricultural Sciences, Uppsala, Sweden (1987–1989).

Dr. Paustian's main fields of interest include agroecosystem ecology, soil organic matter, nitrogen cycling, simulation modeling and global change. He is Chair of an Intergovernmental Panel of Climate Change (IPCC) workgroup on developing methodologies for CO_2 emissions from soils in the IPCC Guidelines for National Greenhouse Gas Inventories. He is a member of the IPCC workgroup II on Agricultural Options for Mitigation of Greenhouse Gas Emissions and a member of the IGBP/GCTE Soil Organic Matter Scientific Steering Committee.

Dr. Paustian serves on the Editorial Board for the journal *Applied Soil Ecology*. He is a member of the Soil Science Society of America, Ecological Society of America, International Soil Tillage Research Organization, Phi Kappa Phi and Sigma Xi.

Edward T. Elliott, Ph.D., is the Associate Director of the Natural Resource Ecology Laboratory and holds a joint appointment as Professor in the Soil and Crop Sciences Department at Colorado State University. He is the Director of the University Office of Ecosystem Research and Management. His B.Sc. and M.Sc. in Soil Science were obtained at Colorado State University and his Ph.D. in Ecology is from the University of Georgia, where he studied freshwater ecology.

Dr. Elliott has been an active participant in interdisciplinary teams of ecosystem research scientists throughout his career. His research interests have spanned a wide range of scales in space and time, with a focus on soils within an ecosystem context. Dr. Elliott's studies began with the examination of protozoa in soil, their role in the mineralization of nutrients held in microbial biomass, and the subsequent release and uptake by plants. These research interests naturally led to investigations of soil texture and structure, the protection of microbial biomass from microfaunal grazers, and development of the concept of habitable pore space. Consequently, Dr. Elliott's work shifted toward soil structure and its role in controlling movement of mineralized inorganic nutrients from preferential flow. Subsequent work moved toward soil structure, aggregation and controls on soil organic matter dynamics and the incorporation of this knowledge into conceptual and mathematical models. Dr. Elliott's interest in understanding environmental controls on organic matter dynamics has involved work on site networks and the evaluation of simulation models.

Dr. Elliott's current interests and research projects continue to span a range of topics and has recently included the development of regional projections of soil C storage in soils. He is a Leader for the Soil Organic Matter Task group of the Global Change in Terrestrial Ecosystems core project of the International Geosphere-Biosphere Program.

C. Vernon Cole, Ph.D., is a Senior Research Scientist with the Natural Resource Ecology Laboratory, Colorado State University. He was a Research Soil Scientist with the Agricultural Research Service, USDA from 1950 until his retirement in 1993. He contributed to significant advances in the knowledge of phosphorus chemistry in the first years of his career. A practical soil test developed from this research has been used worldwide and is the basis for identifying phosphorus deficient soils in much of the neutral and slightly weathered soils of the world.

Dr. Cole's research expanded into plant responses to phosphorus and then into broader studies of phosphorus cycling in relation to organic carbon and nitrogen transformations. He was the Principal Investigator of an interdisciplinary project "Organic Matter and Nutrient Cycling in Semiarid Agroecosystems" supported by the Ecosystem Studies branch of the National Science Foundation (NSF), from 1979 to 1985. This was a cooperative effort by Colorado State University and the Agricultural Service, USDA. He was a Principal Investigator with Dr. E.T. Elliott on a subsequent cooperative project, "Organic C, N, S and P Formation and Loss from Great Plains Agroecosystems" also funded by NSF Ecosystems Studies from 1985 to 1989. These interdisciplinary team studies embraced a range of biological and physical-chemical reactions in soils and integrated ecological and pedological principles. This work has led to significant advances in the understanding of management impacts on long-term soil productivity and degradation, and has opened avenues where collaborative research could address production and environmental concerns in a range of agroecosystems.

Dr. Cole served as a member of the Scientific Advisory Committee on Biogeochemical Cycles of the Scientific Committee on Problems of the Environment (SCOPE), and as Chairman of the Scientific Advisory Committee on Global Phosphorus Cycles. He is a member of Working Group II, of the Intergovernmental Panel on Climate Change (IPCC) Second Assessment Report. He is the convening lead author of Chapter 23 "Agricultural Options for Mitigation of Greenhouse Gas Emissions" in a volume *Climate Change 1995 — Impacts, Adaptations and Mitigation of Climate Change: Scientific-Technical Analyses.*

Contributors

T.O. Barnwell, Jr.
National Exposure Research Laboratory
U.S. Environmental Protection Agency
Athens, Georgia

A.L. Black (retired)
Northern Great Plains Research Laboratory
U.S. Department of Agriculture
Agricultural Research Service
Mandan, North Dakota

R.L. Blevins
Department of Agronomy
University of Kentucky
Lexington, Kentucky

J.R. Brown
Department of Soil and Atmospheric Sciences
University of Missouri
Columbia, Missouri

R.E. Brown
Northwest Research and Extension Center
Kansas State University
Manhattan, Kansas

R.R. Bruce (retired)
Southern Piedmont Conservation Research Center
U.S. Department of Agriculture
Agricultural Research Service
Watkinsville, Georgia

L.G. Bundy
Department of Soil Science
University of Wisconsin
Madison, Wisconsin

I.C. Burke
Department of Forest Sciences
Natural Resource Ecology Laboratory
Colorado State University
Fort Collins, Colorado

G.A. Buyanovsky
Department of Soil and Atmospheric Sciences
University of Missouri
Columbia, Missouri

C.A. Campbell
Semiarid Prairie Agriculture Research Station
Agriculture and Agri-Food Canada
Swift Current, SK, Canada

J.M. Carefoot
Research Centre
Agriculture and Agri-Food Canada
Lethbridge, AB, Canada

D.R. Christenson
Department of Crop and Soil Sciences
Michigan State University
East Lansing, Michigan

C. Vernon Cole
Natural Resource Ecology Laboratory
Colorado State University
Fort Collins, Colorado

D.C. Coleman
Institute of Ecology
University of Georgia
Athens, Georgia

H.P. Collins
W.K. Kellogg Biological Station
Michigan State University
Hickory Corners, Michigan

D.A. Crossley, Jr.
Institute of Ecology
University of Georgia
Athens, Georgia

P. Crosson (retired)
Frederick, Maryland

R.G. Darmody
Department of Natural Resources and
Environmental Science
University of Illinois
Urbana, Illinois

L.E. Drinkwater
Rodale Institute Research Center
Kutztown, Pennsylvania

W.A. Dick
School of Natural Resources
Ohio State University
Wooster, Ohio

J.W. Doran
Department of Agronomy
University of Nebraska
Lincoln, Nebraska

W.M. Edwards
North Appalachian Experiment Watershed
U.S. Department of Agriculture
Agricultural Research Service
Coshocton, Ohio

E.T. Elliott
Natural Resource Ecology Laboratory
Colorado State University
Fort Collins, Colorado

K.W. Flach (retired)
Davis, California

M.-C. Fortin
Land Resource Unit
Research Branch
Agriculture Canada
Vancouver, BC, Canada

D.B. Friedman
Department of Agronomy and Range Sciences
University of California
Davis, California

W.W. Frye
Department of Agronomy
University of Kentucky
Lexington, Kentucky

D.J. Fuchs
University of Minnesota
Southwest Experiment Station
Lamberton, Minnesota

A.D. Halvorson
Northern Great Plains Research Laboratory
U.S. Department of Agriculture
Agricultural Research Service
Mandan, North Dakota

G.H. Harris
Department of Crop and Soil Science
University of Georgia
Tifton, Georgia

J.L. Havlin
Department of Agronomy
Kansas State University
Manhattan, Kansas

P.F. Hendrix
Institute of Ecology
University of Georgia
Athens, Georgia

D.R. Huggins
University of Minnesota
Southwest Experiment Station
Lamberton, Minnesota

R.C. Izaurralde
Department of Renewable Resources
University of Alberta
Edmonton, AB, Canada

R.R. Janke
Department of Agronomy
Kansas State University
Manhattan, Kansas

H.H. Janzen
Research Centre
Agriculture and Agri-Food Canada
Lethbridge, AB, Canada

A.M. Johnston
Agriculture and Agri-Food Canada
Melfort, SK, Canada

O.R. Jones
U.S. Department of Agriculture
Agricultural Research Station
Bushland, Texas

N.G. Juma
Department of Renewable Resources
University of Alberta
Edmonton, AB, Canada

D.E. Kissel
Department of Agronomy
University of Georgia
Athens, Georgia

G.P. Lafond
Experimental Farm
Agriculture and Agri-Food Canada
Indian Head, SK, Canada

G.W. Langdale
South Piedmont Conservation
Research Center
U.S. Department of Agriculture
Watkinsville, Georgia

W.K. Lauenroth
Range Science Department
Natural Science Research Laboratory
Colorado State University
Fort Collins, Colorado

G.W. Lesoing
Center for Sustainable Agricultural Systems
University of Nebraska
Lincoln, Nebraska

C.W. Lindwall
Research Centre
Agriculture and Agri-Food Canada
Lethbridge, AB, Canada

R.E. Lucas
Department of Crop and Soil Sciences
Michigan State University
East Lansing, Michigan

D.J. Lyon
Panhandle Research and Extension Center
University of Nebraska
Scottsbluff, Nebraska

E.L. McCoy
School of Natural Resources
Ohio State University
Wooster, Ohio

W.B. McGill
Department of Renewable Resources
University of Alberta
Edmonton, AB, Canada

A.K. Metherell
Agricultural Research
Soil Science Department
Lincoln University
Canterbury, New Zealand

D.G. Milchunas
Rangeland Ecosystem Science Department
and Natural Resource Ecology Laboratory
Colorado State University
Fort Collins, Colorado

C.A. Monz
National Outdoor Leadership School
Lander, Wyoming

A.P. Moulin
Agriculture and Agri-Food Canada
Research Station
Melfort, SK, Canada

B.J. Palik
Jones Ecological Research Center
Newton, Georgia

E.A. Paul
Department of Crop and Soil Science
Michigan State University
East Lansing, Michigan

K. Paustian
Natural Resource Ecology Laboratory
Colorado State University
Fort Collins, Colorado

T.R. Peck
Department of Natural Resources and
Environmental Science
University of Illinois
Urbana, Illinois

S.E. Peters
Seeds of Change, Inc.
Sante Fe, New Mexico

A.E. Peterson
Department of Soil Science
University of Wisconsin
Madison, Wisconsin

G.A. Peterson
Department of Soil and Crop Sciences
Colorado State University
Fort Collins, Colorado

F.J. Pierce
Crop and Soil Sciences Department
Michigan State University
East Lansing, Michigan

K.S. Pregitzer
School of Forestry and Wood Products
Michigan Technological University
Houghton, Michigan

P.E. Rasmussen
Columbia Plateau Conservation Research
U.S. Department of Agriculture
Agricultural Research Service
Pendleton, Oregon

J.A. Robertson
Department of Renewable Resources
University of Alberta
Edmonton, AB, Canada

L.S. Saporito
Department of Agronomy
Pennsylvania State University
University Park, Pennsylvania

G.H. Silva
Department of Crop and Soil Sciences
Michigan State University
East Lansing, Michigan

R.W. Smiley
Columbia Basin Agricultural Research Center
Oregon State University
Pendleton, Oregon

B.A. Stewart
Advances in Soil Science
Amarillo, Texas

D.L. Tanaka
Northern Great Plains Research Laboratory
U.S. Department of Agriculture
Agricultural Research Station
Mandan, North Dakota

L. Townley-Smith
Research Station
Agriculture and Agri-Food Canada
Melfort, SK, Canada

P.W. Unger
Conservation and Produce Research Laboratory
U.S. Department of Agriculture
Agricultural Research Service
Bushland, Texas

M.B. Vanotti
Department of Soil Science
University of Wisconsin
Madison, Wisconsin

M.F. Vigil
Central Great Plains Research Station
U.S. Department of Agriculture
Agricultural Research Service
Akron, Colorado

M.L. Vitosh
Department of Crop and Soil Sciences
Michigan State University
East Lansing, Michigan

G.H. Wagner
Department of Soil and Atmospheric Sciences
University of Missouri
Columbia, Missouri

M.M. Wander
Department of Agronomy
University of Illinois
Urbana, Illinois

D.G. Westfall
Department of Soil and Crop Sciences
Colorado State University
Fort Collins, Colorado

R.P. Zentner
Research Station
Agriculture Canada
Swift Current, SK, Canada

Contents

PART IV: SOIL, CROP, AND MANAGEMENT OF LONG-TERM EXPERIMENTS IN NORTH AMERICA*

---

* This section is contained on the disk located on the back cover.

Part I

The Role of Soil Organic Matter in Agricultural Systems and Global Change

Chapter 1

Impacts of Agriculture on Atmospheric Carbon Dioxide

Klaus W. Flach, Thomas O. Barnwell, Jr., and Pierre Crosson

CONTENTS

I. INTRODUCTION

At some time in the 1950s or 1960s the annual emission of CO_2 from the burning of fossil fuels exceeded, for the first time, the emission of CO_2 associated with the conversion of forests and grasslands to farmland. Although annual emission of CO_2 from fossil fuels has more than doubled since then, and land use-related emissions have temporarily leveled, agriculture remains a significant factor in the world's CO_2 balance. Agriculture-related activities, chiefly deforestation, and biomass burning contribute about one third of the global net CO_2 emission, and it will continue to be a factor if the additional food for the world's rapidly growing population will be produced through the conversion of grasslands and forests to farmland.

In the United States and Canada, most of the land use change took place in the second half of the 19th and the first half of the 20th centuries. During that time farmers used very little fertilizer and returned little biomass to the soil, practices that resulted in a severe loss of soil organic matter (SOM) and a large release of CO_2. In the second half of the 20th century, farmers in these countries have begun to manage the land more intensively and, through higher yields, are producing food for an ever-increasing population. Modern farming practices return more plant residues to the soil, and modeling studies, as well as some field evidence, suggest that the SOM content in some areas has stabilized or may be increasing. Consequently, agriculture in the United States and Canada may have changed from a net source of atmospheric CO_2 to a net sink. This chapter describes major elements of the global balance

0-8493-2802-0/97/$0.00+$.50

of greenhouse gases (chiefly CO_2), the current influence of agriculture on this balance, and technical and economic factors that will influence agriculture's impact of the C balance in the United States in the foreseeable future.

II. WORLDWIDE CARBON POOLS AND FLUXES

Of the world's major pools of carbon (Table 1)[1,2] 60,000,000 Gt (1 Gt - 10^{15} g) reside as carbonates in carbonate rocks and 10,300,000 Gt as organic C in various sedimentary rocks. Fossil fuels are part of the C pool in sedimentary rocks, but only about 4,000 Gt are recoverable. These large pools are relatively stable, but C moves more readily among the much smaller pools in the atmosphere, the oceans, and the terrestrial systems that are of concern to us in this publication. The atmospheric pool (Table 1), of 750 Gt of carbon, is the smallest of the major C pools and reliably estimated to be increasing at a rate of 3 GT C y^{-1}. Many recent studies project that this increase will be a major cause of worldwide climate change. Release of C incident to deforestation (Table 2) and other land use changes (~2 Gt C y^{-1}) and fossil fuel emissions (~4 Gt C y^{-1}) are thought to be the causes of this increase. A mass balance argument indicates that about 4 Gt C y^{-1} of these anthropogenic emissions are recycled back into the oceanic and terrestrial pools. Houghton et al.[5] allocate this recycled C equally between terrestrial and oceanic pools, but Tans et al.[4] in a more recent publication argue that the flux to the terrestrial sink may be as large as 3.4 Gt C y^{-1} and consequently larger than the flux to the oceanic sink. The surface ocean C pool, at 1000 Gt C, is large, and some proposals for active management have been made. The deep ocean C pool, at 34,000 Gt C, is not believed to be amenable to active management.

Table 1 The Major Carbon Pools of the World[1,2]

Reservoir	Carbon Gt (10^{15} g)
Carbonates	60,000,000
Organic carbon	
Sedimentary rocks	10,300,000
Recoverable fossil fuels	4,000
Vegetation	760
Soil	1,400
	15,000,000
Oceans	34,000
Atmosphere	750

Table 2 Fluxes of Carbon to and from the Atmosphere[1,2]

	Gt (10^{15} g)
From Atmosphere to	
Vegetation (photosynthesis)	100–120
Ocean	100–115
Land (silicate weathering)	0.06
To Atmosphere from	
Ocean	100–115
Plant respiration	40–60
Decay of residues	50–60
Fossil fuels	4
Land use change	2

The terrestrial C pool (Table 1) contains just over 2100 Gt C, about 760 Gt C in living vegetation and about 1400 Gt C in SOM. Because this pool is three times larger than the atmospheric pool, and because the annual flux between the two pools is large (Table 2), terrestrial systems are important to the atmospheric pool. It is interesting to examine the chronology of net terrestrial C emissions (Figure 1).[5] For much of the last 100 years, C emissions from land use change (from degradation of biota and soil C stocks) exceeded emissions from fossil fuel combustion. The annual fossil fuel emission of C surpassed that from biota and soils in the 1950s, and has increased dramatically since then. Carbon emissions from land use changes have been increasing less rapidly and were level in the 1960s and 1970s. In the late 19th and early 20th centuries, the settlement of North America was a major source of CO_2, but this source is dwarfed by the effects of modern deforestation in Latin America and tropical Africa (Figure 2).

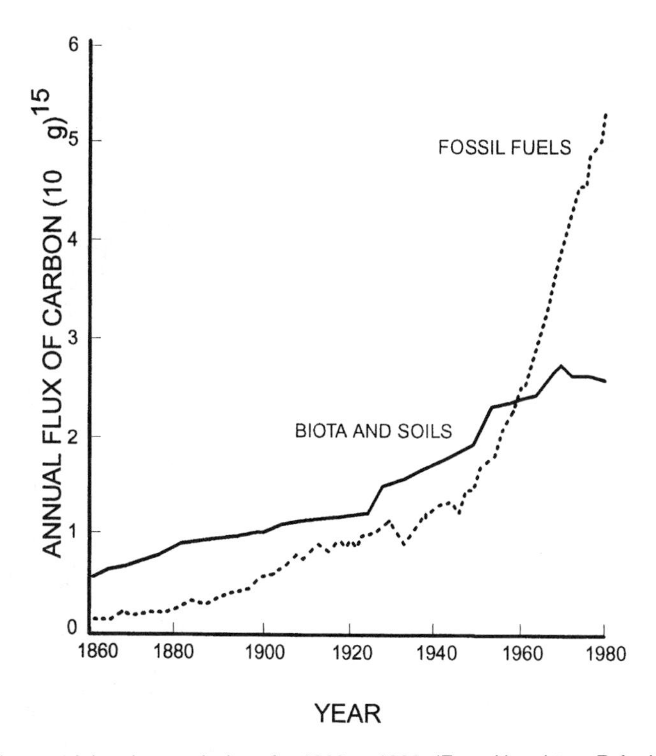

Figure 1 Estimated terrestrial carbon emissions for 1860 to 1980. (From Houghton, R.A., Hobbie, J.E., Melillo, J.M., Moore, B., Peterson, B.J., Shaver, G.R., and Woodwell, G.M., *Ecol. Monogr.* 53(3), 235, 1983. With permission.)

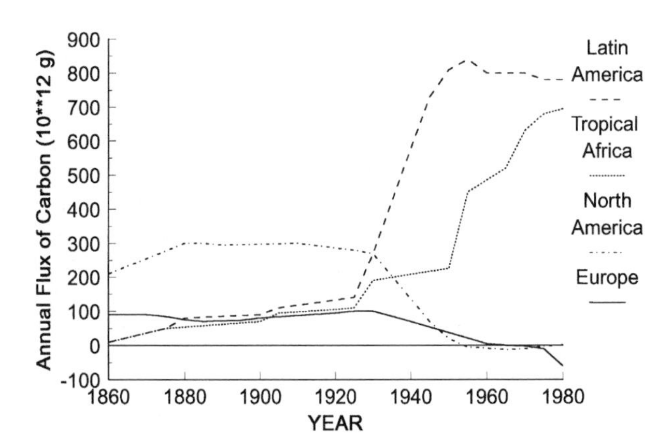

Figure 2 Net flux of carbon to the atmosphere as a result of land-use changes. (From Bouwman, A.F., *Soils and Greenhouse Effect*, Bouwman, A.F., Ed., John Wiley & Sons, New York, 1990. With permission.)

III. OTHER GREENHOUSE GASES

Although this paper, and the publication of which it is a part, deals with the potential role of agriculture in the release and sequestration of CO_2, the relative importance of other greenhouse gases and the role of terragenic systems in influencing their flux should be briefly mentioned. The major greenhouse gases and their potential effect on global warming are listed in Table 3. Only about one half of current global warming can be attributed to CO_2; the other half is attributable to CH_4, O_3, CFC, and N_2O gases that have a much higher radiative absorption potential. Wetlands, including rice paddies, are major terragenic sources of CH_4 and N_2O. The release of both gases is sensitively related to the redox potential of the soil, the soil temperature regime, and the area and thickness of the wetland soil.[6] Since the radiative absorption potentials of CH_4 and N_2O are, respectively, 32 and 150 times larger than that of CO_2, even a relatively small increase in the flux of these gases may increase the total absorption of radiation significantly.

Table 3 Concentration and Absorption Potential of Major Greenhouse Gases (1990 Concentrations)

Type	Residual Time (y)	Annual Rise (%)	1990 Concentration	Radiative Absorption Potential (CO_2 = 1)	Contribution to Global Warming (%)
CO_2	100	0.5	353 ppmv	1	50
CO	0.2	0.6–1.0	90 ppbv	n.a.	n.a.
CH_4	8–12	1	1.72 ppmv	32	19
N_2O	100–200.0	.25	310 ppbv	150	4
O_3	0.1–0.3	2.0	n.a.	2,000	8
CFC	65–110	3.0	26.5 pptv	10,000	15

From Houghton, J.T., Jenkins, G.J., and Ephraums, J.J., *Climate Change; The IPCC Scientific Assessment*, Cambridge University Press, Cambridge, 1990. With permission.

IV. POOLS AND FLUXES OF CARBON IN ECOLOGICAL SYSTEMS

Several attempts have been made to estimate the size of global soil organic C (SOC) pools.[7,8] Figures in Table 4 are largely based on an actual sampling study by Post et al.[9] The global totals of most other estimates are in the same general order of magnitude, but the various estimates use different classification systems, and components of these estimates are not directly comparable. The highest concentrations of SOC (t^{-1}ha) are in the uncultivated soils of the tundra, the boreal and cool temperate forests, the tropical forests, and wetlands. Cultivated soils contain less SOC per unit area, but because of their large extent they contain nearly 12% of the worldwide SOC pool. The distribution of C between the vegetation and the soil varies widely; in general, cool areas have a higher proportion of the C in the soil (Table 4). The precision of estimates, both in terms of land area and their SOC contents, varies greatly. Lack of precision may be relatively unimportant for estimating the impact of human activity on fluxes in boreal and tundra areas, because they are not likely to be converted to farmland. For the long-term prognosis one must, however, consider that these areas may become unstable after global warming and may either release or sequester C, depending on the interplay of rising temperature and rising sea levels. Better data are urgently needed, however, for areas of rapid changes in land use, such as the moist and wet tropical forests, and for wetlands in temperate and tropical climates. Wetlands, in particular, contain more C per unit area than other systems and are probably the largest pool of SOM in the world (Table 4). They may release or sequester large amounts of C under different management systems. The precise size of this pool is uncertain. Bohn,[10,11] for example, reduced his worldwide estimate of C in Histosols form 800 Gt in an earlier publication to 340 Gt in a later publication because he concluded that a bulk density of 0.1 g cm^{-3} was more appropriate than the 0.2 g cm^{-3} he had used before. Houghton et al.[5] attributed the 20% uncertainty in estimates of the world's SOC pool largely to differences in the treatment of Histosols.

Armentano[12] estimated that drainage and cultivation of wetlands release 0.03 Gt C y^{-1} per year worldwide. Because wetlands are thought to have accumulated about 0.14 Gt C y^{-1} before cultivation, drainage of wetlands causes a net loss of 0.17 Gt C y^{-1}, or roughly one tenth of the flux attributed to land use changes (Table 2).

Table 4 Organic Carbon in Soil and Vegetation of Major Life Zones

Life Zone	U.S. Area (10^6 ha)	Soil: Global Area (10^6 ha)	Soil: Global Content (t ha^{-1})	Soil: Global Carbon Stock (10^9 Gt)	Vegetation Global Content (t ha^{-1})	Soil/Vegetation (tt^{-1})
Tundra	50	880	218	191	3	72.67
Boreal desert		200	102	20		
Cool desert		420	99	43		
Warm desert		1400	14	20	20	
Tropical desert brush		120	2	2	3	0.67
Cool temperature steppe		900	3	120	7	0.43
Temperate thorn steppe	242	390	76	30		
Tropical woodland/savanna		2400	54	129		
Boreal forest, moist		420	116	49	90	1.29
Boreal forest, wet		690	193	133		
Temperate forest, cool	265	340	127	43	160	0.79
Temperate forest, warm		860	77	61	135	0.57
Tropical forest, very dry		360	61	22		
Tropical forest, dry		240	99	24		
Tropical forest, moist		530	114	60	160	0.71
Tropical forest, wet		410	191	78	200	0.96

After Post, W.M., Emanuel, W.R., Zinke, P.J., and Stangenberger, A., *Nature*, 298, 156, 1982.

V. FARMING SYSTEMS AND SOIL ORGANIC MATTER

Changes in agricultural management practices can induce dramatic changes in SOC. The classic experiments at Rothamsted (Figure 3)[13] show a threefold increase in soil C storage through annual application of 35 Mg ha^{-1} farmyard manure. In this experiment, soil C is still increasing after 140 years, but manure is applied at nearly twice the rate common in conventional farming. Over the same 140 years, the nonmanured plots lost some SOC and plots fertilized with 144 kg N, 35 kg P, and 90 kg K (about one half of the N and K of the manured plots) gained some SOC until about 1920 and then equilibrated about 10 Mg ha^{-1} higher than the unfertilized plots. Wheat yields, and presumably dry matter production, on the manured plots and the plots receiving mineral fertilizer were about the same and did not change much until the introduction of short-stem varieties of wheat in 1968. Apparently straw was removed from the plots. Predictions from the Rothamsted organic matter turnover model[14] (Figure 3) agree closely with the measured values.

Using data from a 30-year experiment and, in contrast to Rothamsted, starting with a soil that did not have a long history of cultivation, Salter and Green[15] observed different rates of SOC loss for different rotations. From these data, they calculated how "a single year's cropping increased or decreased the organic C content of the soil [in terms of] percentages of the total amount present in the soil." Their values (cropping indices) ranged from –3.12% for corn to +1.36% for hay in a 5-year rotation. They also noticed a high ($r^2 = 0.714$) positive correlation between crop yields and SOC. Flach and Cline[16] used Salter and Green's indices to relate the average SOC of predominantly cash crop and predominantly dairy farming areas with similar climates in central New York. Salter's indices weighted by crop acreages from census data correlated well ($r^2 = 0.88$) with measured SOC for well-drained, medium-textured soils (Figure 4).

VI. A SCENARIO FOR SEQUESTERING CARBON IN A MAJOR AGRICULTURAL PRODUCTION AREA OF THE UNITED STATES

In 1991 the U.S. Environmental Protection Agency (EPA) sponsored an assessment of alternative management practices and policies affecting soil C in agroecosystems of the central United States.[17] The

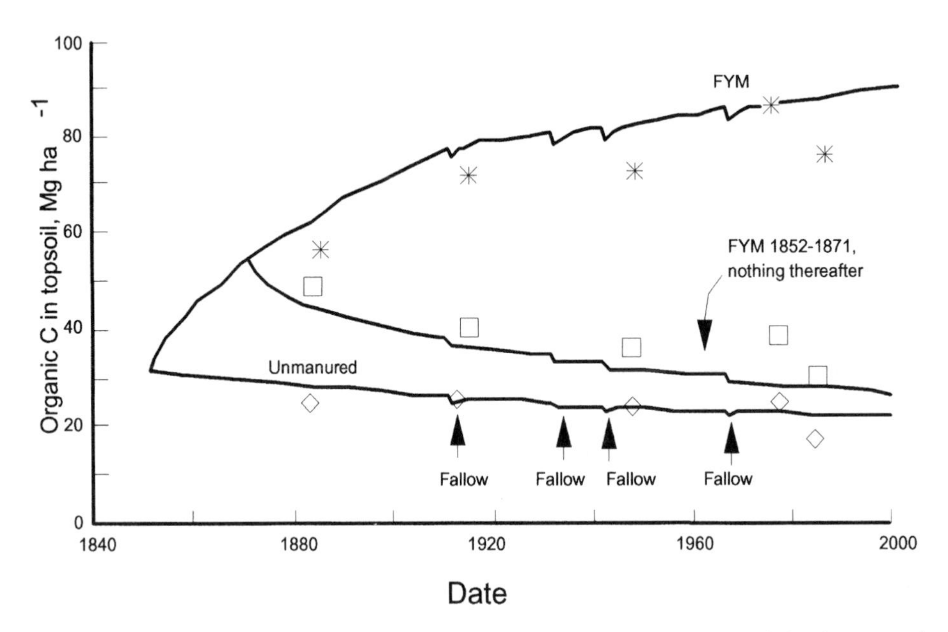

Figure 3 Organic carbon in the topsoil of a plot at Rothamsted, England. The symbols show experimental observations; the solid lines are a model fit. FYM is farm yard manure. (From Jenkinson, D.S., *Agron. J.*, 83, 2, 1991. With permission.)

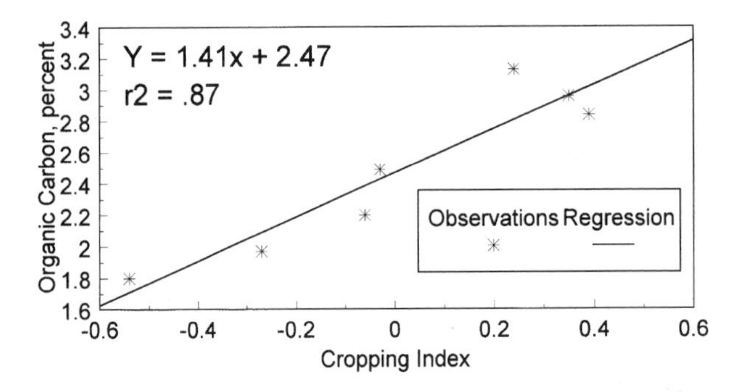

Figure 4 Average SOC in seven New York production areas as predicted by Salter and Green's cropping indices. (From Flach, K.W. and Cline, M.G., *Farm Res.*, 20, 13, 1954. With permission.)

study attempts to predict potential C sequestration in the Corn Belt and Great Lakes regions between 1990 and 2030. The regions encompass 44% of the land area and between 60 and 70% of the agricultural cropland of the coterminous United States. For the study, Donigian et al. selected the Century[18] model from the 14 soil C models they had reviewed because it had been more extensively tested than other models and because it accommodated the range of climates, farming systems, and soils in the study area.

The Century model, like most of the other 13 models that they evaluated, assumes that SOC levels are related to the amount of biomass annually returned to the soil. Consequently, SOC levels, or future C sequestration in soils, will depend on advances in technology that make higher yields and higher biomass production possible and on economic forces that will encourage farmers to take advantage of these possibilities. Donigian et al. assumed increased yields and no major changes in land use during the study period. We analyzed economic and policy factors that are likely to influence agricultural production in the study area. Although we were not comfortable with a projection until 2030, we believe that their assumptions on yields and land use are valid for at least the next 20 years. Neither we nor Donigian consider the impact of climate change.

In a paper prepared for the U.S. EPA, Crosson[19] sketched a scenario of increasing demands for U.S. grain and soybeans from 1987/1990 to 2010 (Table 5[20,21]) because of rising demand in less developed countries. In the same scenario, he assumed that the increase in crop yields over the next 20 years will be 1.4% annually, somewhat less than the 1.8% (USDA[22], pg. 26) of the last 40 years. The projected slowdown reflects concerns that some crops may be approaching maximum potential yields. These projections consider the effect of erosion, which is expected to be small.[23,24] In as much as the projected increase in demand (Table 5) is roughly in line with the expected increase in yields, we expect that the amount of land devoted to the major crops will be about the same in 2010 as in 1987/90. If the rate of yield increase should continue at the 1.8% rate of the last 40 years, some land may actually be released from crop production.

Table 5 Production of Grains and Soybeans 1987/1990 and 2010 (Mt)

	1987/1990	2010	Average Annual Percent Increase
Wheat	59.2	87	1.8
Coarse grain	204.7	264	1.3
Soybeans	49.9	69	1.5
Total	313.8	425	1.4

Sources: 1987/90: grain from USDA;[20] soybeans from USDA;[21] 2010 from Crosson.[19]

The Century model simulates biomass production and conversion to SOM under defined climates, soils, and management practices. The model divides SOM into an active fraction with a turnover time of 1 to 5 years, a slow fraction with a turnover time of 20 to 40 years, and a passive, highly stabilized "recalcitrant" fraction with a turnover time of 200 to 1500 years. Organic matter transformation is affected by soil properties, texture, temperature, and moisture; the lignin content of the biomass; and certain management options such as various tillage practices.

For this study, the model was driven by information on management practices of 27 production areas of the Resource Adjustment Modeling System (RAMS) of the Center for Agricultural and Rural Development. The Century model uses monthly temperature and precipitation data and contains a stochastic weather generator that uses historic mean and skewness data. These values were calculated for 49 Climate Divisions that were superimposed on the production areas and constituted the cells for modeling. Information on soil texture was gained from the EPA's Data Base Analyzer and Parameter Estimator[25] and the Natural Resource Conservation Service's Natural Resources Inventory. The study included a number of policy alternatives, such as incentives to promote conservation tillage and no-till and the Conservation Reserve Program.

To estimate current SOC, the study modeled organic C levels from 1910 to 1990, assuming that SOC was in equilibrium with a native grass vegetation until the land was plowed, and using historic records on management practices and yields. The model (Figure 5) projected future levels for the 1990 to 2030 period, assuming that precipitation and temperatures remained the same as between 1948 and 1988, that land use remained approximately constant, and that yields and biomass production increased at a rate of either 0.5, 1, or 1.5% y^{-1}. The model showed the well-documented decline in SOC between 1910 and 1950 to about one half of the original level, a period of SOC stability until 1970, and then predicted an almost linear increase in SOC at rates of 0.6, 0.9, and 1.2% per year for the three yield scenarios between 1970 and 2030. Allison[26] has identified a period of stable or slightly increasing SOC for the 1950 to 1970 period but there is not, as yet, convincing evidence for any increase in the 1970 to 1990 period. The gains beyond 1990 depend, of course, on the validity of the model and the reality of the projected yield and, more importantly, biomass increase. The continuing linearity of the SOM increase under the high yield assumption to 90% of the SOM of the virgin soil invites scrutiny but then the assumption that yields will continue to increase at the assumed rates for the next 40 years is also open to question. In any case, the projected scenario would result in the sequestration of 0.95, 1.36, and 1.80 Gt C for the study area, assuming yield increase of 0.5, 1.0 and 1.5% y^{-1}, respectively.

Sequestering even 1.8 Gt C over 40 years, or 0.05 Gt y^{-1}, would only mitigate a little more than 0.1% of the current global C loading from the burning of fossil fuels. However, if we were to assume that farmers in the rest of the world were to achieve a 1% annual gain in productivity, and sequestered similar amounts of C (400 kg $ha^{-1}y^{-1}$, accepting the 1.4 Gt C sequestered at the 1% yield increase on 87×10^6 ha) on the world's[5] 1500×10^6 ha of farmland, about 0.6 Gt C, or about one seventh of the current annual emission from fossil fuels, could be sequestered in the soil. We realize, of course, the many pitfalls in this calculation and we offer it only to suggest that further research on the potential role of agriculture in influencing the world's C budget is warranted.

The projections are, however, much less favorable if we consider the C released to the atmosphere by the energy needed for producing fertilizer and other means of production if this energy is produced through burning fossil fuels. The Century model calculated fertilizer needs and arrived at a total N input[27] for the study area of 0.34, 0.39, and 0.44 Gt for the 0.5, 1.0, and 1.5% yield increase scenarios, respectively. According to Pimentel et al.,[28] the production and processing of 1 kg of N fertilizer requires 18,500 kcal, which is equivalent to 1.4 kg C. Consequently, the sequestration of the 0.95, 1.36, and 1.8 Gt C of the three productivity assumptions would cause the release of 0.48, 0.55, and 0.62 Gt C or 53, 40, and 34% of the C the model had predicted as sequestered in the soil. Pimentel et al. calculated that N fertilizers constitute about one half of the energy input for corn production under U.S. conditions. Consequently, some C would be released to provide the energy for the production and distribution of fertilizers other than N and for harvesting operations.

VII. CHANGES IN SOIL ORGANIC CARBON UNDER NO-TILL

There are possibilities to sequester C in soils without using more energy. One such possibility would be more extensive use of conservation tillage and no-till. No-till farming has gained widespread acceptance primarily as a soil conservation measure, but its value in preserving and possibly increasing SOC is just beginning to be recognized. In 1991, 28.1% of the nation's planted cropland was in some form of conservation tillage (no-till, ridge-till, and mulch-till), up from 25.6% in 1989.[29] Of the conservation-tilled land in 1991, mulch-till accounted for 69.9%, no-till for 26.0%, and ridge-till for 4.1%. Corn, small grains, and soybeans accounted, respectively, for 33, 30, and 27% of the land in conservation tillage. Among the USDA's ten producing regions, the Corn Belt, with 35%, has the greatest portion of its cropland in some form of conservation tillage. The Northern Plains region is in second place.

Although the USDA has had programs in place to promote conservation tillage for at least the last 10 years, the practice has spread primarily because of its economic advantages relative to conventional tillage under some conditions of soils and climate.[29] Where soils are well drained, the growing season not too short, and weeds adequately controllable with herbicides, conservations tillage systems suffer little, if any, yield penalty relative to the conventional system. Under these conditions, the labor and energy savings and the soil moisture retention characteristics of conservation tillage give it a competitive edge relative to conventional tillage. The conditions are met throughout much of the Corn Belt and northern Plains, which explains why conservation tillage has been most widely adopted in these regions.

The economic conditions that have favored adoption of conservation tillage over the last 20 years (rising labor and energy costs) are expected to continue indefinitely. In addition, some land currently in conventional tillage has soil and climatic conditions conducive to conservation tillage. There is reason, therefore, to expect continued spread of conservation tillage practices. The principal caveat to this expectation is environmental policy. Conservation tillage systems typically require no less use of herbicides than conventional systems and not infrequently require more. Should concern about environmental impacts of pesticides result in greater restrictions on herbicide use, the continued spread of conservation tillage could be slowed.

Donigian et al.[17] included some evaluation of no-till in their study. In comparing "reduced tillage" and "no-till" with conventional tillage for a corn-soybean rotation, the gain in SOC for the 1990 to 2030 period was between 20 and 80% higher than conventional tillage with spring plowing. The authors concluded, however, that the "degree of the impact [of conservation tillage and no-till] is a function of complex interactions of crops, rotations, climate and soil characteristics" and that "the area warrants further investigation." There is good evidence that disturbing the soil, as in plowing, releases large amounts of CO_2. In an experiment in Minnesota,[30] the cumulative CO_2 flux in the first 19 days after tillage with a plow, disk-harrow, and chisel-plow was 913, 391, and 366 g m^{-2}, respectively. The soil contained 10.4 kg SOC m^{-2} and, assuming that all of the CO_2 came from the SOM, 2.4, 1.0, and 0.96%, respectively, of the SOC were lost. The untreated soil, in contrast, lost only 183 g CO_2 or 0.35% of its SOC in the same 19 days. Evidence that no-till increases SOC also may be drawn from a review of a paper by Kern and Johnson.[31] They reviewed results of 17 field studies (with 28 sites) of SOC of paired plots in which conventional tillage and no-till were compared. The apparent gains in SOC of the no-till plots ranged from 27% for the 0- to 8-cm layer to 16% for the 8- to 15-cm layer and zero for the 15- to 30-cm layer. The duration of the experiments ranges from 3 to 44 years, but only 5 lasted more than

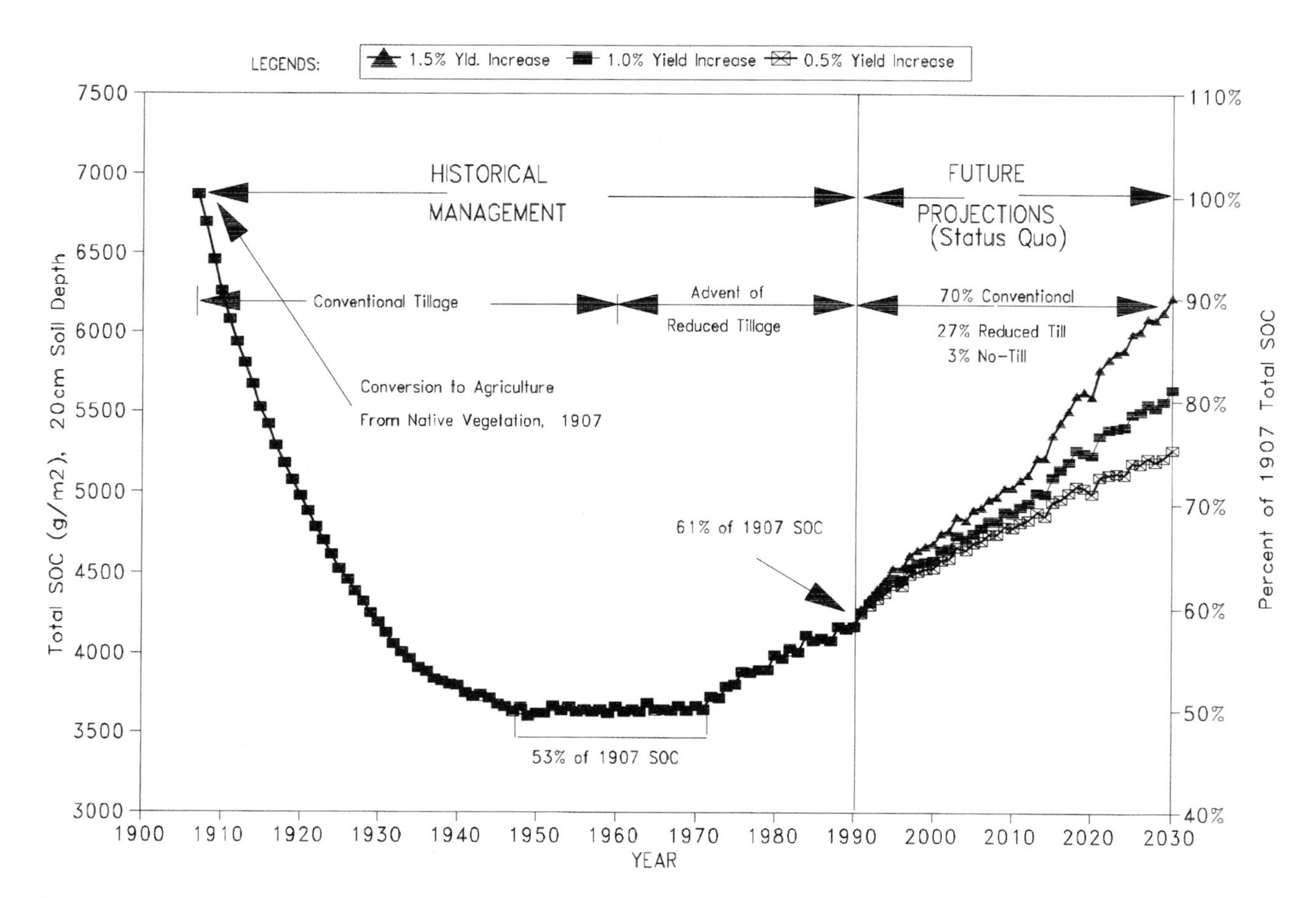

Figure 5 Simulated total carbon levels for three levels of future yield increases. (From Donigian, A.S. et al., Publication No. EPA/600/R-94/064. U.S. EPA, Athens, Georgia, 1994.)

12 years. A review of the source data showed, surprisingly, no significant correlation (r = 0.04) between the duration of the experiment and the gain in SOC. These numbers, uncertain as they may be, suggest nevertheless that no-till increases SOC, but they also suggest that yields and biomass returns as well as management practices must be considered for a quantitative assessment of the potential of no-till for carbon sequestration.

VIII. CARBON IN THE SOIL LANDSCAPE

Although the Century model has the capability to consider loss of SOC by erosion in arriving at estimates of SOC in soils, Donigian et al.[17] chose not to include the effects of erosion. The reasoning is complex. They estimate that erosion losses could reach 0.5 to 1% per year "producing up to 20 to 40% loss of SOC over the 40 years projection period." They argue that much of this loss is likely to be landscape-level movement and that the eroded C is likely to be redeposited in parts of the same field, associated wetlands, and as sediments in bodies of water. Olson et al.[32] have used the C content and rate of deposition of sediments to arrive at a figure of 0.5 Gt for the worldwide sequestration of C "in shallow zones of lakes" and 0.5 Gt "in coastal wetlands, marshes, and bays" for a worldwide total of about 1 Gt in all wetlands. Although they do not identify the source of this C, it can be assumed that a significant portion comes from the erosion of farmland. In addition to sedimentation, the movement of soluble nutrients, particularly the soluble forms of N, may contribute to the sequestration of C in wetlands. Dawson,[33] in a 1950 study, has demonstrated this for the accumulation of C in low moor peats in New York State. Although he considered only N from atmospheric deposition in his calculations, it seems likely that nutrients in the runoff from intensively fertilized farmland may also contribute to the accumulation of C in associated wetlands.

IX. CONCLUSIONS

1. Although currently available data must be considered tentative, there is evidence that future intensification of farming and consequent larger production of biomass will lead to increased sequestration of C in soils. The amounts sequestered may be large enough to be of significance in the world's C cycle.
2. The intensification of farming is expected to continue in the foreseeable future.
3. Significant amounts of C continue to be released to the atmosphere through the conversion of grass and forestlands to cropland. More intensive use of existing farmland rather than opening of new farmland would minimize this source of CO_2.
4. The energy needed for producing fertilizer and for other aspects of intensive farming methods may in part negate the gains in sequestering C made by these methods. Consequently, research on farming practices that increase the sequestration of organic C without requiring additional energy, such as no-till, should be encouraged.
5. Considerable amounts of soil C may be moved through erosion and be deposited in sediments; this movement should be considered in determining the impact of agriculture on the carbon balance.

REFERENCES

1. Post, W.M., T.-H. Peng, W.R. Emanuel, A.W. King, and D.L. DeAngelis, The global carbon cycle, *Am. Sci.*, 78, 310, 1990.
2. Berner, R.A., Atmospheric carbon dioxide levels over Phanerozoic time, *Science*, 249, 1382, 1990.
3. Houghton, J.T., G.J. Jenkins, and J.J. Ephraums, *Climate Change; The IPCC Scientific Assessment*, Cambridge University Press, Cambridge, 1990.
4. Tans, P.P., L.Y. Fung, and T. Takahashi, Observational constraints on the global atmospheric CO_2 budget, *Science*, 247, 1431, 1990.
5. Houghton, R.A., J.E. Hobbie, J.M. Melillo, B. Moore, B.J. Peterson, G.R. Shaver, and G.M. Woodwell, Changes in the carbon content of terrestrial biota and soils between 1860 and 1980: a net releaser of CO_2 to the atmosphere, *Ecol. Monogr.*, 53(3), 235, 1983.
6. Bouwman, A.F., Exchange of greenhouse gases between terrestrial ecosystems and the atmosphere, in *Soils and Greenhouse Effect*, A.F. Bouwman, Ed., John Wiley & Sons, New York, 1990.
7. Buringh, P., Organic carbon in soils of the world, *SCOPE*, 23, 91, 1984.
8. Eswaran, H., E. Van den Berg, and P. Reich, Organic carbon in soils of the world, *Soil Sci. Soc. Am. J.*, 57, 192, 1993.
9. Post, W.M., W.R. Emanuel, P.J. Zinke, and A. Stangenberger, Soil carbon pools and world life zones, *Nature*, 298, 156, 1982.

10. Bohn, H.L., Estimate of organic carbon in world soils, *Soil Sci. Soc. Am. J.*, 46, 1118, 1976.
11. Bohn, H.L., Estimate of organic carbon in world soils. II, *Soil Sci. Soc. Am. J.*, 46, 1118, 1982.
12. Armentano, T.V., Drainage of organic soils as a factor in the world carbon cycle, *Bioscience*, 30, 825, 1980.
13. Jenkinson, D.S., The Rothamsted long-term experiments: are they still of use?, *Agron. J.*, 83, 2, 1991.
14. Jenkinson, D.S., P.B.S. Hart, R.J. Rayner, and L.C. Parry, Modelling the turnover of organic matter in long-term experiments at Rothamsted, *INTECOL Bull.*, 15, 1, 1987.
15. Salter, R.M. and T.C. Green, Factors affecting the accumulation and loss of nitrogen and organic carbon from cropped soils, *J. Am. Soc. Agro.*, 25, 622, 1933.
16. Flach, K.W. and M.G. Cline, Does cropping affect soil organic matter?, *Farm Res.*, 20, 13, 1954.
17. Donigian, A.S., Jr., T.O. Barnwell, R.B. Jackson, A.S. Patwardhan, K.B. Weinreich, A.L. Rowell, R.V. Chinnaswamy, and C.V. Cole, Assessment of alternative management practices and policies affecting soil carbon in agroecosystems of the central United States, Publication No. EPA/600/R-94/067, U.S. Environmental Protection Agency, Athens, GA, 1994.
18. Parton, W.J., J.W.B. Stewart, and C.V. Cole, Dynamics of C, N, P and S in grassland soils: a model, *Biogeochemistry*, 5, 109, 1988.
19. Crosson, P.R., United States agriculture and the environment: perspectives in the next twenty years, U.S. Environmental Protection Agency, unpublished, 1992.
20. U.S. Department of Agriculture, World Grain Situation and Outlook, Foreign Agricultural Service FG8-91, Washington, DC, 1991.
21. U.S. Department of Agriculture, World Oilseed Situation and Outlook, Foreign Agricultural Service FOP 10-91, Washington, DC, 1991.
22. U.S. Department of Agriculture, Economic Indicators of the Farm Sector: Production and Efficiency Statistics 1988, Economic Research Service ECIFS 8-5, Washington DC, 1990.
23. Crosson, P.R., M. Hanthorn, and M. Duffy, The economics of conservation tillage, in *No Tillage and Surface Tillage Agriculture*, M. Sprague and G. Triplett, Eds., John Wiley & Sons, New York, 1986.
24. Pierce, F.J., R. Dowdy, W. Larsen, and W. Graham, Soil productivity in the cornbelt: an assessment of erosion long-term effects, *J. Soil Water Conserv.*, 39, 131, 1984.
25. Imhoff, J.C., R.F. Carsel, J.L. Kittle, Jr., and P.R. Hummel, Database Analyzer and Parameter Estimator (DBAPE) Interactive Program User's Manual, EPA/600/3-89/083, U.S. Environmental Protection Agency, Athens, GA, 1990.
26. Allison, F.E., in *Soil Organic Matter and Its Role in Crop Production*, Elsevier, Amsterdam, 134, 1973.
27. Patwardhan, A.S., personal communication, Aqua-Terra Consultants, 2762 Bayshore Parkway, Mountain View, CA 94043.
28. Pimentel, D., L.E. Hurd, A.C. Bellotti, M.J. Forster, I.N. Oka, O.D. Sholes, R.J. Whitman, Food production and the energy crisis, *Science,* 182, 443, 1973.
29. Conservation Tillage Information Center, National Survey of Conservation Tillage Practices, Executive Summary, NACD's Conservation Technology Center, West Lafayette, IN, 1991.
30. Reicosky, D.C. and J.J. Lindstrom, Effect of fall tillage method on short-term carbon dioxide flux from soil, *J. Soil Water Conserv.*, in press.
31. Kern, J.S. and M.G. Johnson, Conservation tillage impacts on national soil and atmospheric carbon levels, *Soil Sci. Soc. Am. J.*, 57, 200, 1993.
32. Olson, J.S., R.M. Garrels, R.A. Berner, T.V. Armentano, M.I. Dyer, and D.H. Yaalon, The natural carbon cycle, USDE, Office of Energy Research, DOE/ER-0239, 1988.
33. Dawson, J.R., Soil organic matter, in *Agricultural Chemistry*, D.B. Frear, Ed., Van Nostrand, New York, 1950.

Chapter 2

Management Controls on Soil Carbon

Keith Paustian, Harold P. Collins, and Eldor A. Paul

CONTENTS

I. HISTORICAL BACKGROUND

Farmers have long recognized the organic matter content of a soil as a key attribute of soil fertility. It seems likely that this awareness, at least in the intuitive sense of recognizing dark-colored, friable, and earthy smelling soils as favorable sites for growing crops, dates back to the beginnings of agriculture. Yet, the history of agriculture is replete with examples of the depletion of organic matter and the subsequent loss of soil fertility through poor management. In part, this has been due to a lack of knowledge about how various agricultural practices affect the soil. However, it also reflects an inherent conflict between maximizing short-term production at minimal cost vs. providing for the sustained health and long-term productivity of the soil. This tradeoff, between short-term production goals and long-term investment in maintaining soil fertility, characterizes much of the history of agriculture. The development

0-8493-2802-0/97/$0.00+$.50

of agriculture in temperate regions has exhibited recurrent patterns of exploitative soil use followed by the introduction of regenerative practices as soil resources fell to unacceptable levels. In some cases, severe degradation of the soil led to the collapse of societies and a permanent or long-term loss of productive lands.

A. SOIL CARBON AND EUROPEAN AGRICULTURAL DEVELOPMENT

In central and northern Europe, crop production during Neolithic times was based on swidden or "shifting cultivation" techniques where native vegetation was cleared and burned prior to planting.[1] After a few harvests, fields were abandoned as yields declined and new lands were cleared. While these practices were exploitative in nature, periods of cropping were probably short relative to the fallow period and soil organic matter (SOM) was little affected. Also crop production per se played a secondary role to livestock herding and hunting/gathering in the farmer's livelihood.

Beginning with the influx of Bronze age cultures, larger and more sedentary populations developed and crop production became an increasingly important source of food and fiber. Shifting agriculture practices gradually evolved into a permanent agriculture utilizing the same fields year after year. The introduction of wooden, bronze-, and later iron-shod plows allowed greater areas to be tilled and a greater variety of field crops were grown, including wheat, barley, oats, rye, grain legumes, and forages.[2] It seems likely that the development of metal implements and the use of draft animals had an important stabilizing influence on settlement patterns by improving productivity and making it less practical to continually move onto new lands.[3]

A fair amount is known about agricultural techniques in the Mediterranean region during the Greek and Roman eras from Hesiod, Xenophon, Cato, Varro, Pliny, and other classical writers. The limited and seasonal precipitation necessitated the development of dryland farming with the main cereals being autumn-sown barley and wheat. Grain and forage legumes were also widely used and the benefits of legumes on soil fertility were well recognized.[3] Likewise, the value of animal manure and composts of household wastes were understood and green manuring was a recommended practice. The need for a crumbly soil structure as a seed bed was emphasized by contemporary writers and several plowing and cross-plowings per year were the recommended practice. The Roman writer Cato posed the rhetorical question "Wherein does a good system of agriculture consist? In the first place, in thorough ploughing; in the second place, in thorough ploughing; and in the third place, in manuring."[4]

In Europe, north of the Alps, much less is known about the evolution of permanent agriculture. It appears that the development of the open field system may have begun early in the first millennium of the present era[3] and was well established throughout most of Europe by the Middle Ages.[1] The village or open-field system retained its essential structure until the 16th and 17th centuries and even later in parts of Europe.[5]

In this system, a field was typically cropped for 1 to 2 years and then fallowed for 1 or more years. In the classical three-field system, one field was sown to autumn cereals, one to spring-sown cereals, and the third was fallowed.[3] Fallowed land was cultivated or "stirred" as many as four times per year[2] to prevent weeds and perennial grasses from establishing. Stable manure was spread prior to sowing the first crop after fallow. Outlying meadow and woodland commons were used for livestock grazing and fodder production, which provided a transfer of nutrients and organic matter from meadow to field in the form of manure.[6] While this system provided a basis for permanent agriculture, it still represented extensive land use requiring as much as 50 to 70% of total tillable lands to be kept in fallow.[5]

Farming practices on the cultivated fields were generally poor at maintaining SOM levels. Yields on cropped lands were low and most of the straw was removed for thatching, bedding, fuel, or fodder. The straw which remained was grazed by livestock after the harvest and was sometimes burned. Frequent tillage was necessary for weed control. As populations increased, landholdings became more fragmented and crop production encroached on the grazing commons and marginal lands.[1] With decreased grazing and fodder acreage, less manure was available for spreading on cropped areas. Furthermore, under the feudal system, manure was often reserved for fields under the exclusive control of the manor lord and thus was not available to peasant farmers.[2] Under these conditions soil fertility and productivity, and presumably SOM, declined dramatically. Seebohm[2] states that by the 14th century average wheat yields in Britain had fallen from an average of 10 bushels/acre to 6 bushels/acre due to the "increasing exhaustion of the soil". This loss of productivity is believed to have contributed to the breakdown of the manorial system and the beginning of the end of the village farming system.

Beginning in the 17th and 18th centuries, a number of changes in European agriculture led to gradual but dramatic improvements in soil fertility and productivity.[1] These included the consolidation of landholdings into single farm units, the improvement in plows and replacement of oxen by draft horses, and the introduction of new crops and rotations. With both pasture and cropped lands under the control of an individual farmer and with the availability of grass, clover, and later alfalfa seed, it was possible to rotate forage and cereal production on the same fields, leading to the development of the ley farming system. The introduction of root crops as winter livestock feed and improvements in animal breeding led to increased livestock holdings and thus greater availability of manure. In addition, the practice of composting, utilizing a variety of farm, household, and industrial organic wastes, developed to a high degree.[2] These improvements enabled the use of more sophisticated rotations, with sown hay crops (leys) and legumes alternated with cereals and root crops, providing a basis for continuous production from a given field.

The growth in prosperity and population in Europe, from the 17th century onward, is testament to the increased productivity of these improved farming practices. As an example, one of the early advocates of the new agriculture, the famous French chemist Lavoisier, was able to obtain wheat yields of 20 to 24 bushels/acre (about double the norm in the surrounding district) and maintain five times the normal number of livestock.[3] By the early 19th century improved farming had spread through much of Europe. The use of leys, legumes, and manure additions was well suited for the maintenance of SOM insuring a stable supply of plant nutrients from mineralization. These practices characterized European farming up until this century, when mechanization and the widespread use of chemical fertilizer and pesticides again revolutionized agriculture.

B. SOIL CARBON AND NORTH AMERICAN AGRICULTURAL DEVELOPMENT

Agriculture was an important part of the economy for many of the native North American tribes. Swidden agriculture, similar in form to that practiced in Europe in the Neolithic era, was a predominant practice in the eastern deciduous forests.[7] Trees were girdled in the fall or winter, a practice which was effective for killing the root systems and prevented girdled trees from sprouting.[8] Maize was planted in mounds, with other crops such as beans, pumpkins, squash, and sunflowers interspersed between the maize. Production levels were primarily maintained by periodically abandoning fields and clearing new forest stands, although fish were also used to fertilize maize in coastal regions.[7] As in early Europe, the sparse population and abundant land area meant that agricultural practices per se probably had little impact on SOM. In contrast, the use of fire by native Americans may have significantly influenced prairie ecosystems, and thereby soil development, in the midwest and Great Plains regions of North America (see Chapter 5).

In addition to swidden agriculture, more permanent agricultural fields existed along river systems and in the irrigated agriculture of the Southwestern U.S. Interestingly, investigations of prehistoric fields on runoff terraces in New Mexico by Sandor et al.[9] found that organic matter declines attributable to cultivation were around 50% in the A horizon. This implies that, where more sedentary agriculture was practiced, losses of organic matter and soil degradation due to cultivation could be similar to those which have occurred under industrial-era agriculture.

Following European colonization, agriculture in North America was characterized by a prolonged period of exploitation. Early colonial farming was mainly subsistence based and utilized techniques (i.e., shifting cultivation) and crops — corn, squash, beans, tobacco — adopted from the native peoples. The most important food crop introduced from Europe was wheat. Tillage implements were largely the hoe and planting stick.

In the early European colonies of the Chesapeake Bay region, the typical practice entailed growing one or two crops of tobacco (as a cash crop) followed by several years of maize and then wheat.[8] When wheat yields became unacceptably low, fields were abandoned for 20 to 30 years before again being cleared for a new cropping cycle. While these practices led to soil degradation, particularly during the summer fallow preceding wheat crops,[8] sufficient land areas and a long forest fallow allowed the system to be maintained for many years. However, by the early 1800's population pressures led to an elimination or shortening of the forest fallow period resulting in severe land degradation due to erosion and organic matter loss. A contemporary observer lamented that the low land prices had "greatly contributed to accelerate among our land killers, the exhaustion of our soil."[8] As the unsustainability of existing practices became increasingly apparent, management reforms were introduced. These included the use of grass and clover in rotation with tobacco and grain crops, lime and guano fertilizer applications, and increased livestock holding.[10] However, these reforms were largely restricted to the more densely settled regions and, as population spread westward, the pattern of extensive and exploitative agriculture was repeated.

Social and political forces, as well as technological developments such as the iron and steel plow developed by John Deere in 1837, sparked the rapid agricultural development of the midwest and western U.S. in the 19th century and the Canadian prairies during the late 19th and early 20th centuries. With the expansion of agriculture into the prairie regions, the organic matter-rich soils provided, initially, abundant nutrients and high yields. Organic matter contents of these soils, however, declined in many cases precipitously during several years of continuous cropping.[11] By the end of the 19th century, the frontier was largely closed and a greater emphasis was again placed on the reform of farming practices, primarily directed at rebuilding and maintaining SOM and soil fertility. This is evidenced in part by the establishment of several long-term agricultural field experiments (e.g., References 11 and 12) around the turn of the century.

C. SOIL CARBON AND MODERN AGRICULTURE

During the 20th century, changes in agriculture throughout the temperature regions of the world have been driven by the introduction of fossil-fuel powered machinery, the development of agrochemicals, including inorganic fertilizer and pesticides, and advanced crop breeding. These developments have had important implications on SOM. The replacement of the horse by the tractor, together with a shift toward grain-fed livestock production, sharply reduced the need for forage-producing areas. With the availability of inorganic fertilizers, agriculture moved away from its emphasis on organic matter management as the key to maintaining soil fertility — consequently the use of legumes and green manures to provide nutrients to the crop declined. In Europe and North America, farming trends were toward monocultures or simpler rotation dominated by grain production, reduced ley acreages, residue burning to reduce disease and promote clean tillage, and increases in the number and intensity of tillage operations.[6] These developments were generally detrimental to the maintenance of SOM However, in areas which had historically been farmed for cash grain production, as in much of the North American prairies, increased productivity from fertilizers and improved crop varieties and the introduction of combine harvesters increased crop residue inputs, partially offsetting the negative effects of increased tillage intensity.[13]

While the new agriculture was able to produce unprecedented gains in agricultural productivity, there were early concerns that continuous fertilizer use and minimum residue return would lead to organic matter depletion and deterioration in soil tilth and quality. In this context, a number of field experiments investigating the balance of organic matter in soils were established in the 1950s and 1960s.[14–17] This renewed interest in SOM has continued to intensify to the present day and organic matter management is a major consideration in efforts to develop sustainable farming systems that minimize soil erosion and nutrient leakage and produce high-quality products.

The role of agricultural soils in the production and consumption of greenhouse trace gases is of more recent scientific interest.[18–21] While current contributions of land use to atmospheric CO_2 increase (approximately 25%[22]) are mainly from tropical regions, the rise in CO_2 during the 19th and early 20th centuries was largely associated with the rapid expansion of temperate zone agriculture. As much as 110 Pg (10^{15} g) of C may have been released from temperate zone ecosystems, contributing substantially to increased CO_2 prior to the mid-1900s.[23] Currently, the C balance in temperate agricultural soils appears to have stabilized[24], but there is considerable interest in evaluating the potential for these soils to regain some of their lost C and help ameliorate the continued increase in atmospheric CO_2.[24–26] A thorough knowledge of how soil C responds to various management practices is essential for these assessments. A prime source of this information is long-term field experiments of the kind described in this volume.

II. A CONCEPTUAL FRAMEWORK FOR ANALYZING MANAGEMENT EFFECTS ON SOIL C

A number of excellent reviews have been written about organic matter in agriculture soils and its response to management (e.g., References 27 to 31). Our purpose in this review is to organize knowledge and experimental results according to a systems framework where the focus will be on management factors as *process controls* on inputs and outputs of C from the soil.

A. PROCESSES DETERMINING SOIL C BALANCE

There is a tremendous store of empirical knowledge about the influence of management on SOM, and the practices which promote SOM formation and maintenance are well known. These practices include grass and legume production (in rotation or as permanent pasture), manure application, elimination of bare fallow, increased residue return, and reduced tillage. However, agricultural management is seldom, if ever, dictated by a primary objective of increasing SOM. For a given management system, the combination of tillage, cropping, fertilization, and manuring practices used is likely to have both positive and negative effects on SOM. Moreover, management effects on organic matter will vary for different soil types, climatic zones, and past management histories. Therefore, it is necessary to examine the fundamental processes which determine organic matter formation and decomposition and then analyze how specific management practices affect these processes.

In the simplest terms, the level of soil organic carbon in a soil will be governed by the difference between inputs of organic matter and outputs through mineralization, erosion, and leaching. Erosion can significantly affect the SOM content at a specific location, through the physical removal and transport of particulate organic matter. However, the extent to which erosion impacts the balance of C and other nutrients within the larger landscape will depend on the depositional pattern and subsequent fate of the eroded material. Ignoring for the moment the influence of soil erosion, the balance of soil carbon (the dominant constituent of SOM) is depicted in Figure 1.

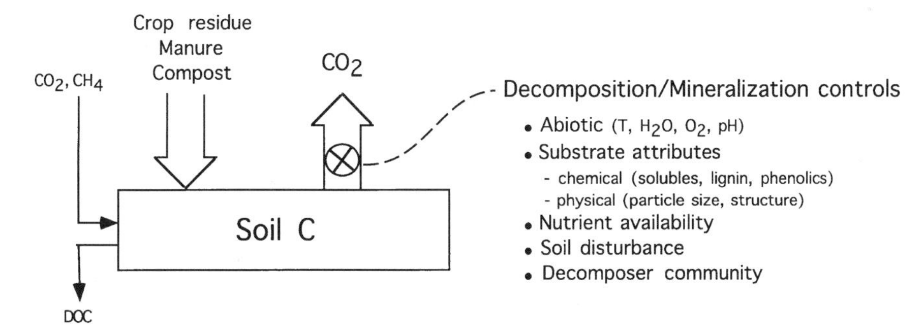

Figure 1 Inputs and outputs of C from organic matter in soils. The main controls on decomposition are shown as regulating CO_2 evolution by soil heterotrophs. (DOC = dissolved organic C).

The input of organic C to the soil consists of crop residues (including roots), animal manure, compost, and in some cases industrial-derived products in the form of sewage sludge and other wastes. Small amounts of carbon (as CO_2) are also incorporated in the soil by autotrophic microorganisms. Losses of C occur through the decomposition and mineralization of organic compounds by soil heterotrophs with CO_2 being the dominant product. Under reducing conditions (e.g., flooded soils), significant amounts of CH_4 and other hydrocarbons may be produced but in most well-drained agricultural soils there will be a net consumption of these gaseous compounds.[19] Carbon-containing compounds may also be leached from the soil profile as dissolved organic carbon (DOC). Except in acid soils, soil C losses as DOC are probably small.[32] However, there is relatively little information on this process in agricultural soil and some speculate that C losses via this pathway might be significant over the long term.[33,34]

The controls on decomposition processes are, in most respects, more complex and less easily manipulated by management than are inputs of organic matter. These controls include abiotic factors (soil temperature, water, aeration, pH), the physiochemical nature of the organic matter (i.e., chemical composition, structure, and particle size), its physical exposure to decomposers, the availability of mineral nutrients required for microbial growth and metabolism, and, finally, the nature and composition of the decomposer community itself (Figure 1).

B. INTERACTION OF MANAGEMENT CONTROLS

While the influence of individual control factors may be fairly well understood, the interactions and aggregate effects of multiple controls on decomposition and C inputs are more difficult to predict. Various components of agricultural management affect all of these process controls (Table 1). Moreover, a specific management practice (such as fertilization) may affect more than one control on decomposition or input and the multiple effects of a practice may act synergistically or in opposition.

There are many examples of interacting effects of specific management practices. Addition of N fertilizer might enhance the decomposability of crop residue (tending to reduce soil C) and at the same time increase crop residue production (tending to increase soil C). Mulching increases surface soil moisture and decreases soil temperature — however, the net effect on decomposition rates may be different for soils in a cool, wet climate as opposed to one that is warm and dry or for soils of different

Table 1 Summary of Management Practices Affecting Controls on Soil C Turnover

Process Control	Management Practice
C Inputs	Crop productivity (crop type, water and nutrient availability, pest control)
	Fallow frequency
	Residue return/removal
	Amendments (e.g., manure, compost)
	Cover crop production
Decomposition and mineralization	
Residue composition and particle size	Crop type, age, nutrient content
	Residue management (stubble retention, burned vs. unburned, chopping)
	Composting treatments
Temperature/moisture/aeration	Crop type and density (i.e., canopy shading)
	Tillage
	Mulching
	Irrigation
	Drainage
	Crop type (H_2O use efficiency)
Nutrient availability/soil reaction	Fertilization
	Liming
	Manure addition
	N-fixing crops and green manures
Soil disturbance/aggregation	Tillage
	Residue management
Decomposer community	Microorganism inoculation
	Earthworm addition

textures. An interaction between crop productivity and decomposition exists via the soil water balance, in that transpiration accounts for the bulk of water loss from most well-drained soils. Thus, management practices that increase productivity may tend to decrease soil moisture, thereby favoring a buildup of C through both increased C inputs and decreased decomposition.

To fully evaluate management influences on the soil C balance the processes must be analyzed within an ecosystem context.[35] A "causal loop" model of a number of soil properties and processes relating to the effects of soil tillage illustrates this point (Figure 2). The direct effects of tillage include the mixing of surface residues, roots, and other organic matter within the plow layer as well as the disturbance of the mineral soil matrix. These primary effects then spawn a series of secondary effects and feedbacks involving a multitude of physical, chemical, and biological processes and soil properties which are of consequence for the soil C balance. These effects vary temporally as well. For example, tillage could increase porosity and aeration of a soil initially, but lead to subsequent declines in these properties, relative to untilled soils, due to a decrease in aggregate stability.

Given the extreme complexity of the ecosystem interactions outlined above, one might ask whether it will ever be possible to make reliable quantitative predictions about the effects of a specific management system on C levels for a particular soil. Over the past several years, simulation models have begun to address some of the processes and interactions described here with a degree of success.[36–39] As more information is gathered, particularly results from long-term field studies such as those documented in this volume, and with accompanying advances in theory, the predictive ability of these models will continue to improve. Moreover, while we have focused considerable attention on the complexity of ecosystem responses to management, we would suggest that the effects of management are most predominant in two areas, (1) in controlling the amount and kind of organic matter returned to the soil and (2) in determining the degree of soil disturbance through tillage. Thus we will attempt to generalize the effects of management, focusing on tillage, organic matter addition, and residue management, crop rotation, and fertilization, based primarily on field studies conducted in temperate-zone agricultural soils. Before doing so, we will give an overview of the main crop-management zones in temperate North America and discuss constraints to agricultural production within the different regions. This will provide a context for subsequent discussion of how different management alternatives are likely to affect soil C levels in a particular region.

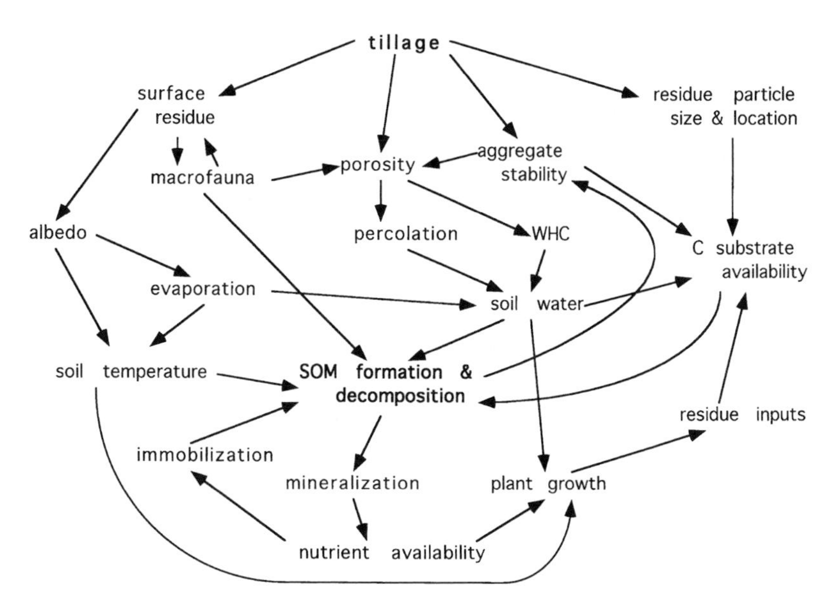

Figure 2 Causal loop model showing interactions and feedbacks between a management practice (i.e., tillage) and soil factors affecting organic C levels.

III. TEMPERATE CROPPING SYSTEMS AND MANAGEMENT CONSTRAINTS

North America, by virtue of its large size and the variety of its climate and soils, includes examples of the major types of cropping systems in current use in temperate regions worldwide. We recognize that there are crop species and management practices in other temperate regions which differ from those found in North America. However, we contend that in such cases reasonable analogs exist in North America for which the general principles of management and soil C interactions will be similar. Therefore, we will restrict our discussion to cropping systems and management constraints in U.S. and Canadian agriculture with the expectation that these can be compared to temperature zone agriculture globally.

Major land resource areas (MLRAs) for the principle crop growing regions of the U.S., as delineated by USDA/Soil Conservation Service,[40] are shown in Figure 3, with descriptions of major crops, average temperature and precipitation, dominant soil types, and land area coverage (Table 2). We have excluded MLRAs for which agriculture is of less importance, such as in the Rocky Mountain, Great Basin, and Pacific Coast regions of the West and the upper Northeast, which are dominated by forest or rangeland cover. We also neglect the MLRAs for central California and central and southern Florida. While these latter two regions are extremely important in terms of agricultural production, particularly of products for direct human consumption, they are more limited in area relative to most other crop-growing regions. In addition, their high diversity in crops and management practices makes it difficult to make meaningful generalizations on a regional basis.

The dominant control on the distribution of cropping systems is climate. As Jenny[41] pointed out, the Great Plains region of the central U.S. and Canada is characterized by regular, and roughly orthogonal, gradients of temperature and precipitation. Major cropping systems regions are arrayed according to these gradients (Figure 3). Moving eastward from the Rockies, precipitation is the major control of the transition from rangeland to cereal cropping with summer fallowing, to continuous cereal cropping, and then to row crop agriculture. The north-south temperature gradient also influences which systems are viable, by determining the length of the growing season and overwintering capacity of crops [e.g., accounting for the dominance of spring wheat in the Northern Great Plains (F) vs. winter wheat in the Central Great Plains (H)].

In the midwestern and eastern U.S., differences in precipitation are less important and the main determinant of large-scale cropping regions is temperature. In northern regions (e.g., Northern Lake States; K), cool moist conditions are most suited to perennial forage crops and short-season annuals. To the south are the main feed-grain-producing areas in the Corn Belt (M) and Appalachians and Ozarks (N), dominated by corn and soybean production. Farther south in the Atlantic and Gulf Slope and Coastal

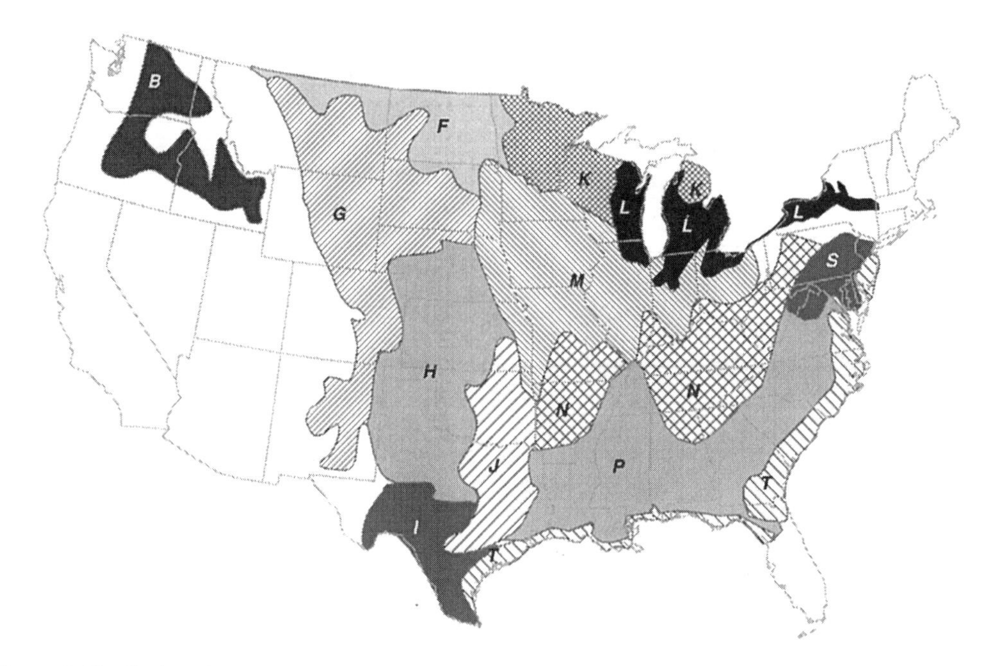

Figure 3 Distribution of the largest agriculturally important land areas in the continental U.S., based on major land resource area (MLRA) designations by the USDA, Soil Conservation Service.[40] See Table 2 for area codes and descriptions. (From USDA, Land Resource Regions and Major Land Resource Areas of the United States, Agriculture Handbook 296, 1981.)

regions (P,T) additional crops including cotton, tobacco, peanuts, etc., become important and the potential for double cropping and winter crops increases.

While not depicted in Figure 3, management systems in the major agricultural areas in Canada are broadly similar to those shown along the U.S.-Canadian border. For example, the spring wheat growing region (F) extends into southern Alberta, Saskatchewan, and Manitoba and the feed-grain and mixed cropping zones (M and L) extend into southern Ontario and Quebec. Further north, there is a decline in growing season length and in soil water deficit, resulting in progressively greater use of perennial forage crops, both in eastern and western Canada.

Soil C levels across the central U.S. (Figure 4) coincide to some degree with climate patterns,[42] although there is great variability locally associated with differences in parent material, topography, and drainage. In the central U.S., the precipitation gradient across the Great Plains influences productivity and rooting depth to a relatively greater degree than it does decomposition rates, giving a general trend of increasing soil C as precipitation increases west to east.[43] Within the semiarid and subhumid zones, SOM tends to increase south to north, due to lower temperatures (decreasing decomposition rates) and lower water deficits (enhancing productivity). The eastern U.S. lacks a strong regional gradient in precipitation and there is less of a regional gradient in soil C. Soil C tends to be greater in cooler regions and high C soils occur in floodplains and coastal areas due to the deposition of organic matter-rich illuvium and the presence of wetland ecosystems.

Climate and soil differences constrain management options and thereby determine which practices are most important in affecting soil C. For example, in the semi-arid regions of the Great Plains and Pacific Northwest, where wheat is the dominant crop, moisture limitations to productivity are of overriding concern. Therefore, fallow frequency and tillage intensity and their effects on water utilization are key management factors impacting soil C. The potential for increasing soil C levels will largely rest on the success in minimizing both fallow frequency and tillage (see Chapter 28). In the more mesic regions to the east, management controls on productivity levels (e.g., fertilization, crop type), as well as tillage practices and cover cropping, are key factors. In both cases, management practices that increase C inputs to soil and reduce decomposition rates associated with intensive tillage will promote maintenance and buildup of soil C.

Table 2 Characteristics of Agriculturally Important MLRAs in the Continental United States

MLRA	Precipitation (mm)	MAT (°C)	Dominant Soils	Principle Land Use and Crops	Area (km^2)
B. Northwest wheat and range region	250–570	7–10	Xerolls, Borolls, Ochrepts	Wheat-fallow; continuous grain and field peas where precipitation > 350 mm	240,000
F. Northern Great Plains spring wheat region	250–550	4–9	Borolls, Aquolls	Spring wheat-fallow Other spring grains, flax, hay	350,000
G. Western Great Plains range and irrigated region	275–600	7–16	Ustolls, Agrids	Range; irrigated feed grains; wheat-fallow on eastern edge	570,000
H. Central Great Plains winter wheat and range region	500–750	10–18	Ustolls	Winter wheat; irrigated corn, alfalfa and (in south) cotton	575,000
I. SW plateaus and plains range and cotton region	500–700	16–22	Argids, Othids (W) Ustalfs, Ustolls, Usterts (E)	Range; wheat and sorghum on favorable sites; irrigated cotton	175,000
J. SW prairies cotton and forage region	625–1150	15–22	Ustalfs, Ustolls, Usterts	Cotton, wheat, sorghum, other feed grains, hay	145,000
K. Northern Lake States forest and forage region	500–825	2–7	Boralfs	Forage and feed grains for dairy production	280,000
L. Lake States fruit, truck and dairy region	675–925	6–11	Udalfs, Boralfs	Dairy, corn, winter wheat, beans, sugar beets; fruit (on eastern edges of Great Lakes)	195,000
M. Central feed grains and livestock region (Corn Belt)	625–900	6–13	Udolls, Udalfs Aualfs, Aquolls	Corn, soybeans, oats, other field grains	725,000
N. East and Central farming and forest region (Appalachian and Ozarks)	1025–1275	9–17	Udalfs, Udults	Corn, soybeans, small grains, hay	610,000
P (&O). South Atlantic and Gulf slope (plus Mississippi Delta MLRA O)	1025–1525	16–20	Udults, Udalfs Aquents, Fluvents, Ochrepts in flood plains	Cotton, soybeans, small grains, corn; tobacco and peanuts locally	800,000
S. Northern Atlantic Slope diversified farming region	900–1275	8–14	Udalfs, Udults, Ochrepts	Truck crops, forage, soybean, feed grain	105,000
T. Atlantic and Gulf Coast lowland	1025–1525	13–21	Aquults, Aqualfs, Udults, Aquepts (E)	Rice (in west), corn, soybeans	250,000
			Uderts, Aqualfs, Aquolls, Aquepts (W)	Tobacco, peanuts, sorghum (locally)	

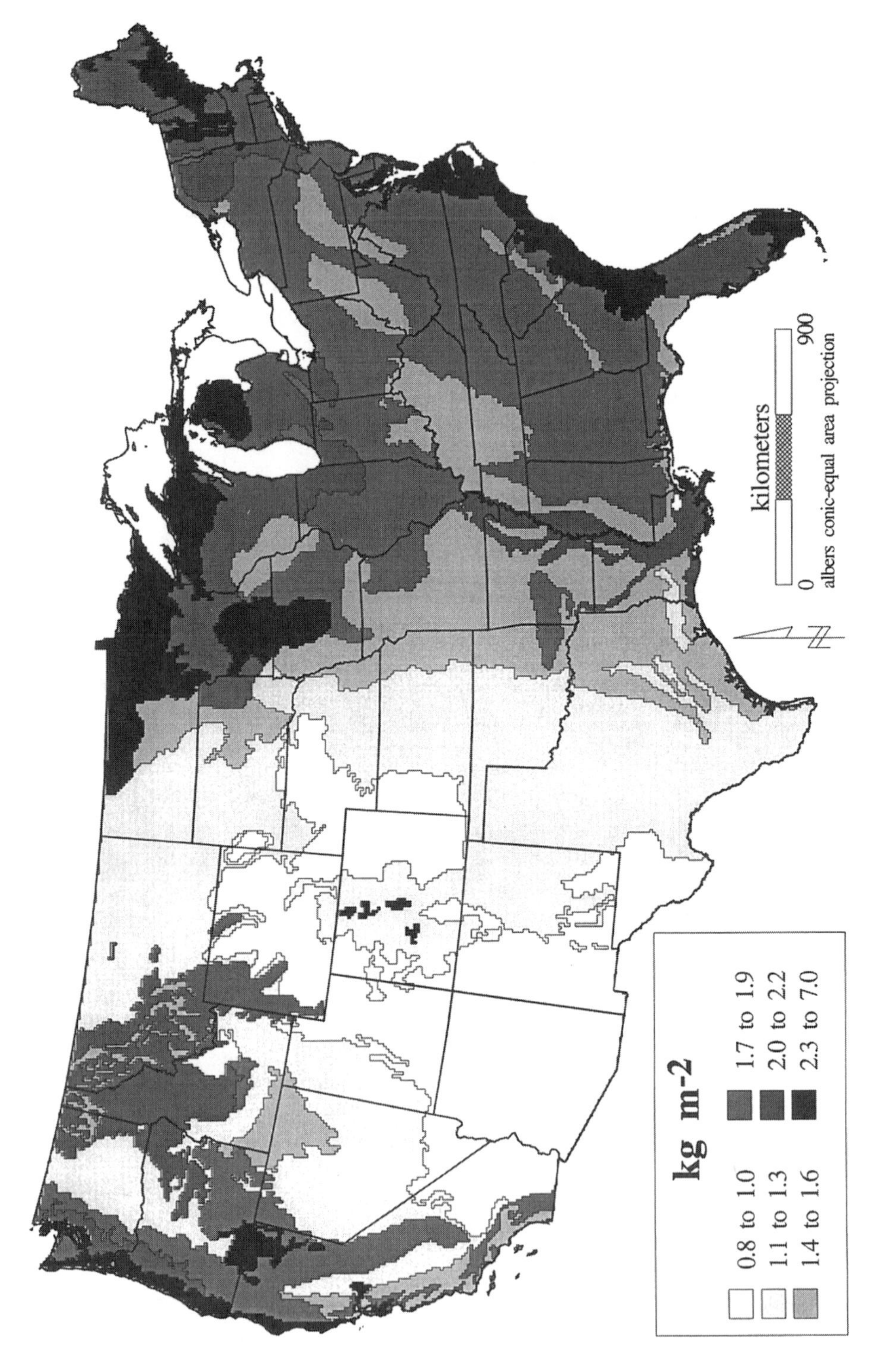

Figure 4 Spatial patterns of mean agricultural soil organic C for the continental U.S. (Figure courtesy of J. Kern, US-EPA, Corvallis.)

IV. MANAGEMENT EFFECTS ON SOIL C

A. CULTIVATION OF NATIVE SOILS

The most dramatic effects of agricultural land use on soil C are associated with the initial cultivation of native soils. Typically, C concentrations in the plow layer decline rapidly for several years following cultivation and eventually stabilize only after many years (Figure 5).

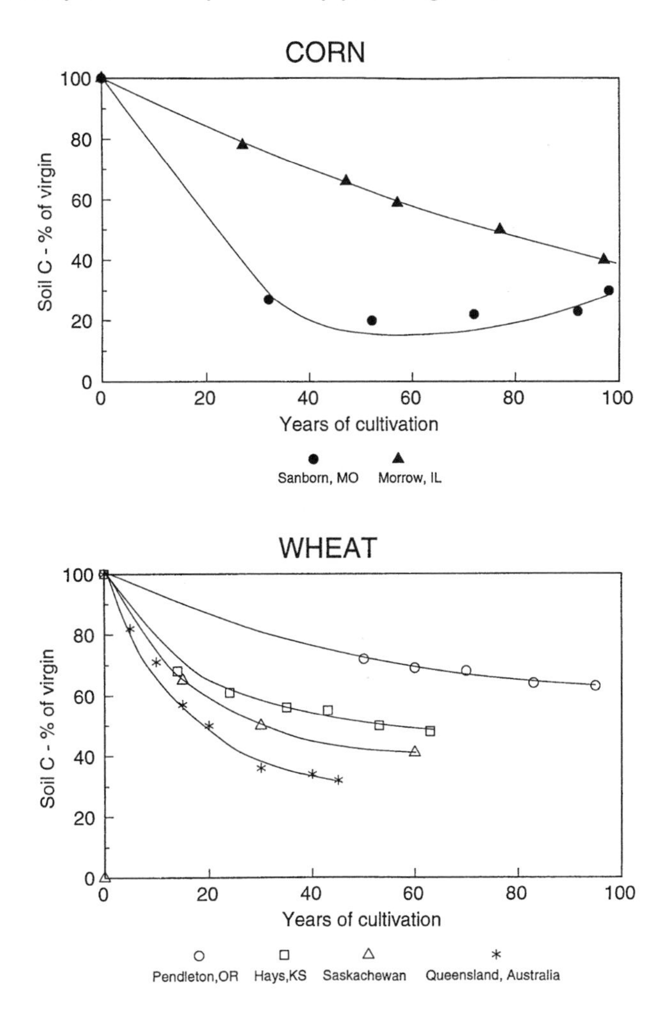

Figure 5 Declines in soil C with time since initial cultivation, grouped by corn- and wheat-dominated cropping systems. Data from Saskatchewan,[44] Hays, Kansas,[11] Pendleton, Oregon,[45] Queensland, Australia,[46,47] Morrow plots, Illinois,[48] and Sanborn plots, Missouri.[49]

The magnitude of change due to cultivation can vary substantially between locations. For example, Haas et al.[11] estimated decreases in C concentrations of 28 to 59% following 30 to 43 years of cropping for 11 sites in the Great Plains. Bowman et al.[50] reported a very high loss of C, 62% in the top 15 cm and 32% from the 15- to 30-cm depth, for a sandy soil in eastern Colorado after 60 years of cultivation. Mann[51] summarized much of the available literature of paired comparisons between virgin and cultivated soils in the U.S. and calculated regression estimates of C loss stratified by soil type and length of cultivation period. Estimates of maximum loss on a percent C basis in the upper 15 cm ranged from 11% for Psamments to 77% for Ustolls. However, *amounts* of C lost, calculated on a per area basis, were generally lower, as were estimates based on samples taken to 30 cm, i.e., below normal plow depth. She estimated average C losses for all soils to 30-cm depth to be 23% and concluded overall that *relative* C losses increased as a function of initial C content. Using a smaller but more detailed data set, Davidson and Ackerman[52] found that cultivation losses of surface soil C were generally between 20 and 40%. In contrast to the conclusion of Mann,[51] they found no effect of initial C content or soil texture on relative

C loss, where the data were calculated on a soil horizon basis rather than on the basis of fixed sampling depths. They, and others[53] point out that changes in bulk density or A horizon depth associated with cultivation can bias interpretations of C loss when comparing cultivated vs. uncultivated soils.

A number of factors may contribute to losses in soil C upon cultivation including reduced C inputs, physical soil disturbance, a more favorable decomposition environment, and increased soil erosion. The predominance of any one factor is highly site specific and thus broad generalizations are difficult. Furthermore, as management systems evolve and change, so does the soil C status relative to the original native condition. Specific effects of individual management practices are discussed later in the paper. In this section we are primarily concerned with the impacts of the initial disturbance of native soils.

The degree of change in C inputs depends on the production system which replaces the native vegetation. Even with harvest removal of grain, high input agricultural systems could return as much or more C to the soil as the native systems they replaced. Conversely, more extensive production systems including, for example, use of summer fallow, as well as extensive residue removal, may greatly reduce C return relative to native ecosystems. In addition, the distribution of C inputs may be altered; for example, lower proportions of C are added below ground in annual crops vs. perennial grassland.[54]

There are relatively few comparisons of native vs. cultivated ecosystems which specifically address differences in plant residue inputs. Buyanovsky et al.[55] made a detailed study of C budgets for winter wheat and native grassland on similar sites in Missouri. Although net primary production in winter wheat was about 10% higher than that in the native system, the amount of C allocated to SOM was 80% greater in the grassland. They found that decomposition rates of fresh residues were also higher for winter wheat, due to tillage disturbance and incorporation of residues and higher soil moisture contents during summer. Consequently, the grassland was able to maintain higher SOM levels due both to higher C inputs as well as slower residue decomposition rates compared with cultivated winter wheat. van Veen and Paul[56] compiled data for southern Saskatchewan and reported that residue production from grassland (2000 kg C ha^{-1} yr^{-1}) was greater than for corresponding wheat (1425 kg C ha^{-1} yr^{-1}) systems.

Cultivation invariably affects soil structure and conversion from grassland or pastures to cultivated land yields substantial reductions in aggregate stability within a few years.[57–61] The breakdown of stable macroaggregates is thought to promote decomposition of previously physically-protected organic matter through greater exposure to soil organisms and improved aeration.[62–64] Differences in cultivation response as a function of soil texture may be associated with the vulnerability of a soil to physical disturbance. Several authors have reported higher relative C losses following cultivation in coarse-textured compared with medium- and fine-textured soils.[65–68] In coarser-textured soils, the structural stability of the soil depends mainly on binding agents such as roots and fungal hyphae,[60] which are disrupted and rapidly decomposed upon disturbance. Higher clay contents promote organo-mineral complexes which may allow smaller aggregates to persist, protecting some of the physically stabilized SOM from decomposing following disturbance.[69]

Improved abiotic conditions for decomposition following cultivation could include more optimal temperature or moisture regimes, increased aeration, improved pH, and increased nutrient availability. In temperate regions, soil temperatures are below optimal for the microflora during a major portion of the year. The lower albedo of bare soil surfaces and higher incident radiation during spring and fall tend to increase soil temperatures in annual crops. Increases in soil moisture following conversion of grassland or forest to cultivated land could occur as a consequence of lower evapotranspiration rates, particularly with the practice of summer fallowing. Thus well-drained soils under cultivation are likely to have temperature and moisture regimes that are more favorable for decomposition. At the other extreme, drainage of wet soils improves aeration which also speeds decomposition. Changes in soil reaction and nutrient availability with cultivation are probably of less importance as *direct* controls on SOM levels in agricultural soils. However, nutrient and pH status have significant indirect influences on soil C turnover, primarily through effects on productivity and residue inputs, as discussed later.

Once native soils have been cultivated they are much more susceptible to erosion. Reduction in the C content of surface soils as a result of erosion is due both to the loss of topsoil and to a dilution effect from subsoil mixing. On a landscape basis, the net effect of erosion on C balance will depend on the deposition pattern of the eroded soil and changes in productivity and decomposition rates of the eroded soil. Depending on these factors, net soil C storage on a landscape basis could be either reduced or increased. If eroded soil is deposited in upland sites, decomposition rates of the C in the depositional soil may not be greatly altered. However, if soil is transported to basins or wetland areas, where decomposition rates are reduced, then net soil C storage on a landscape could conceivably be increased by erosion. Conversely, if productivity of the eroded soil is significantly reduced, net C storage in soil would be decreased due to decreased C inputs. However, if productivity was largely maintained (e.g., through the use of fertilizers) then subsoil mixed into the root zone through tillage could become enriched in organic matter and net soil C storage, minus erosional losses, could increase.

Most long-term field experiments were established on relatively level ground in order to minimize water erosion and soil variability. Nevertheless, several studies have shown that water erosion can be significant even with slight slopes and in many cases erosional losses by both water and wind are a confounding factor in estimating C turnover from long-term studies. For the Sanborn plots, Gantzer et al.[70] calculated that topsoil thickness had been reduced by 56% under continuous corn and by 30% under corn rotated with oats, wheat, and perennial crops, compared with permanent timothy grass. In long-term plots at Wooster, Ohio, Dick et al.[71] estimated that conventionally tilled plots lost about 3.7 cm more soil than no-tilled plots over 18 years, an amount equivalent to about 500 g C m^{-2}. Slater and Carleton[72] showed that decreases in soil C content were closely correlated to cumulative soil loss for corn plots and corn rotated with cereals and hay, in two sets of erosion plots in Iowa. They found that C losses from erosion *exceeded* that accounted for by reduced C content of the top soil. From this it can be inferred that either SOM formation exceeded decomposition losses — the difference making up the additional C lost to erosion — or that subsoil mixed into the surface horizon was sufficiently high in C to dampen the apparent losses due to erosion. In a toposequence of soils under cultivation for varying lengths of time, Gregorich and Anderson[73] quantified C losses due to erosion vs. C loss through mineralization, for Boroll soils on the Canadian prairies. They found that, for upslope positions, mineralization accounted for most of the carbon loss (70%) during the first years following cultivation but that erosion accounted for 70 to 75% of C loss for the two soils that had been cultivated more than 50 years. Carbon loss in the lower slope (depositional) area was more than double the C lost by erosion upslope, suggesting that the decomposition rate of the deposited material was at least as high as it would have been if it had remained in place and not been eroded.

B. CARBON INPUTS AND RESIDUE MANAGEMENT

One side of the soil C balance equation is the amount of C entering the soil, the other side being the amount decomposed. Plant residues are the major source of C inputs in all terrestrial ecosystems. In agricultural systems an additional "source" of organic matter is available in the form of manure, sewage sludge, and other organic by-products. However, depending on the scale considered, these inputs might be more properly viewed as a redistribution, since animal and human waste, food processing waste, etc., originate mainly from harvested plants.

For a given climatic region and soil type, the rate of C input is an important factor determining the amount of C which can be maintained in the soil. This fact has been recognized for some time and numerous calculations have been made of the amount of residue return needed to maintain organic matter levels at a particular site.[15,74–77] Residue inputs are under a high degree of control by the farmer, via crop selection, productivity levels (as influenced by fertilizers), residue management, and use of manure and other external additions. Thus, in our subsequent evaluations, the role of specific management practices (e.g., crop rotation, fertilizer use, etc.) on C inputs will be an important focus.

If organic matter decomposition is viewed as a series of first-order processes (i.e., constant *fractions* of organic matter decomposed per unit time) as most current theories espouse, then it is easily seen that the amount of SOM that can be potentially maintained in soil (i.e., the steady-state level, C_t^*) is directly proportional to the rate of C inputs (I). This holds true regardless of the number of different fractions considered as comprising SOM[47,78] as shown in the equation below. The rate of change of total C can be expressed as a function of C inputs and the specific decomposition rate (k) of soil C fractions, i.e.,

$$dC_t/dt = d(C_1 + C_2 + \ldots + C_n)/dt = I - k_1C_1 - k_2C_2 - \ldots - k_nC_n \tag{1}$$

Since at steady-state each individual pool would make up a constant fraction (f) of total C ($C_1 = f_1C_t^*$), then,

$$I = k_1(f_1C_t) + k_2(f_2C_t) + \ldots + f_n(k_nC_t) \text{ ,}$$

$$I = (k_1f_1 + k_2f_2 + \ldots + k_nf_n)\, C_t^*$$

$$I = k_tC_t^* \tag{2}$$

A number of long-term field experiments, where residue additions have been carefully controlled, support the theoretical proportionality between inputs and soil C levels (Figure 6). While the soil C data shown here do not necessarily reflect differences at equilibrium, the linearity between differences in soil C levels and C input rates (normalized as mean annual C change over the experimental period) conform to the predictions of the multiple first-order model. The degree of change in soil C (i.e., given by the slope coefficient) depends on climatic and edaphic factors affecting decomposition rates at a particular site. Also, since the instantaneous rate of change of SOM decreases over time following a change in inputs, short-duration experiments would be expected to show higher *mean* changes in soil C, as appears to be the case for experiments at Lind, WA (17 y) and Culbertson, MT (7 y).

Most other studies of residue additions or removals generally show similar effects on organic matter,[77,81–84] although the magnitude of the response varies considerably depending on the length of the experiment, climate and soil type, and management system. Short- and medium-term changes in SOM are often difficult to detect against background soil variability. Saffigna et al.[85] used initial values of soil C as a covariate in analyzing effects of sorghum residue retention and showed significantly higher soil C levels, compared to where residues were removed after only 6 years.

A few studies have shown surprising little response of soil C to differences in carbon inputs. Referring back to Equation 2, it is unreasonable that equilibrium SOM can increase indefinitely with ever-increasing C inputs (plus there are practical limits to the amount of residue which can be added to soil). This implies an upper limit to the amount of C which can be sequestered in mineral soils, independent of input rate. This is suggested by the C balance analyses by Campbell et al.[86] for a high organic matter soil at Melfort, Saskatchewan, which showed no effect of varying C inputs on SOM levels (Figure 7). One hypothesis is that there is a maximum amount of C which can be stabilized in organomineral complexes that are resistant to decomposition and that, once this capacity is saturated, additional residues remain accessible to rapid microbial decomposition and add little to the total soil C storage.

Residue return to the soil can be reduced by burning. Burning of crop residues, to control diseases and to remove impediments to tillage or furrow irrigation, has been practiced in many cereal-growing regions. Burning over many years has been observed to decrease soil C levels,[82,87–89] although in some instances the effects appear to be less than would be expected considering the amount of C volatilized during burning.[76,90] The creation of more resistant C fractions (e.g., charcoal) by burning and variation in the amount of residue remaining after burning (25 to 70% for wheat straw[88]) may account for differences in soil C response.[91]

Of the commonly used organic amendments, farmyard manure is the most recognized in its ability to maintain and build SOM as shown in numerous long-term field experiments.[38,80,83,92–99] Having passed through the digestive tracts of animals, manure is enriched in more refractory compounds and is therefore stabilized to a relatively higher degree than fresh plant residues. Thus, effects of manure applications on SOM content can persist for many years after applications have ceased.[83,96,100] Particle-size fractionation studies show that manure applications increase the relative amounts of C in clay and silt size fractions, which have longer turnover times compared to the sand fraction.[101,102]

The effects of green manures on soil C are more variable than those for farmyard manure. In their review, MacRae and Mehuys[103] cite nine experiments, of which five showed net increases, two showed decreases, and two showed no change in SOM over the study period. However, in the two studies which showed no change (in the Woburn plots, U.K.), the original paper by Chater and Gasser[92] reported that organic matter declined in the absence of green manuring (i.e., with straw return alone) and that green manure was as effective as farmyard manure in *maintaining* soil C levels. In contrast, green manure crops in several long-term plots in the Central Great Plains were generally not effective at reducing C or N losses compared with rotations with only cereals.[11,104] In these studies, low yields of green manure crops and the negative effects of their water use on following grain crops did not increase (or may have decreased) C inputs compared with rotations with cereals only (e.g., Krall et al.[104] cite several years in which green manure crops failed). Under these conditions soil C contents would not be expected to increase.

The effect of different kinds of residue (including manure) on C stabilization efficiency has long been of interest to farmers and soil scientists. The conventional wisdom has been that animal manures and composts are more effective at building SOM than harvest residue, which in turn is more effective than fresh plant material such as green manures. However, it has been pointed out that, if the stabilization efficiency of manure-derived C is based on the amount of original plant material it represents, then the humification efficiency for manure is not necessarily higher than for the original plant material.[28,83] More important perhaps is the degree to which stabilization efficiency can be related to general characteristics

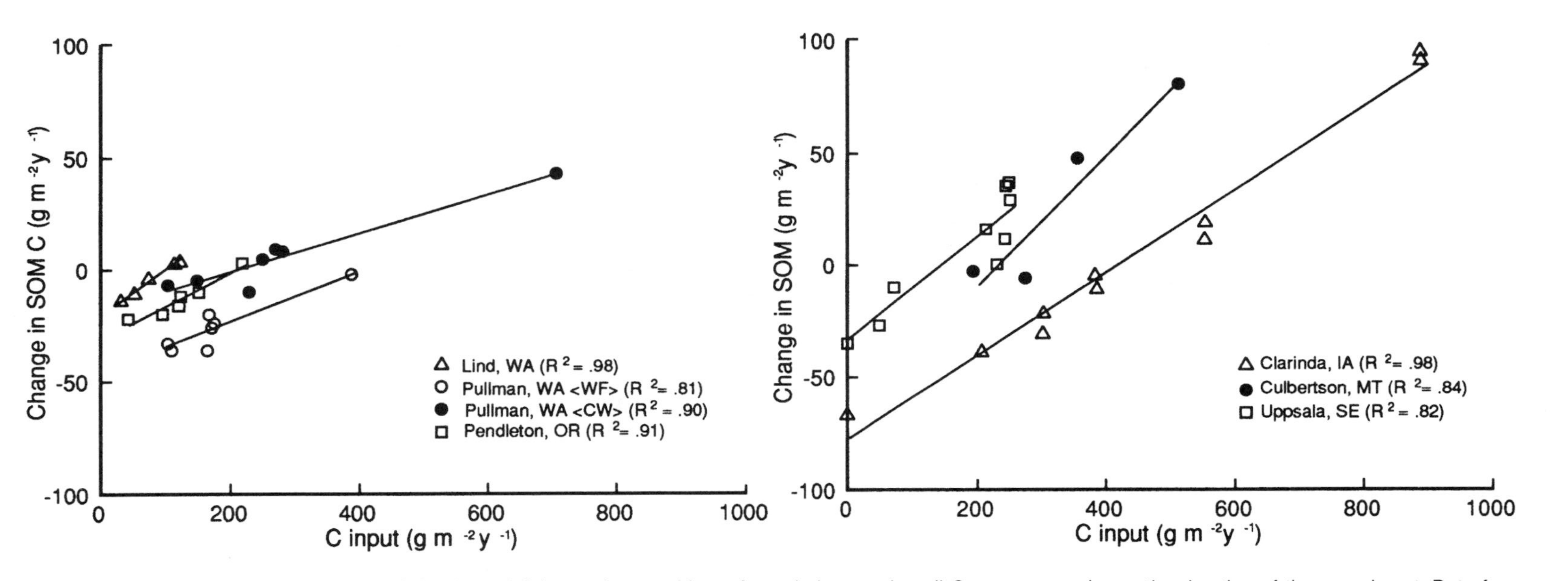

Figure 6 Linearity between soil organic C levels and C inputs from residues. Annual changes in soil C are averaged over the duration of the experiment. Data from Black,[79] Horner et al.,[80] Larson et al.,[15] Paustian et al.,[38] and Rasmussen et al.[76] (From Paustian K., Robertson, G.P., and Elliott, E.T., *Soil Management and Greenhouse Effect*, Lal, R., Kimble, J., Levine, E., and Stewart, B.A., Eds., *Advances in Soil Science*, Lewis Publishers, Boca Raton, FL, 1995. With permission.)

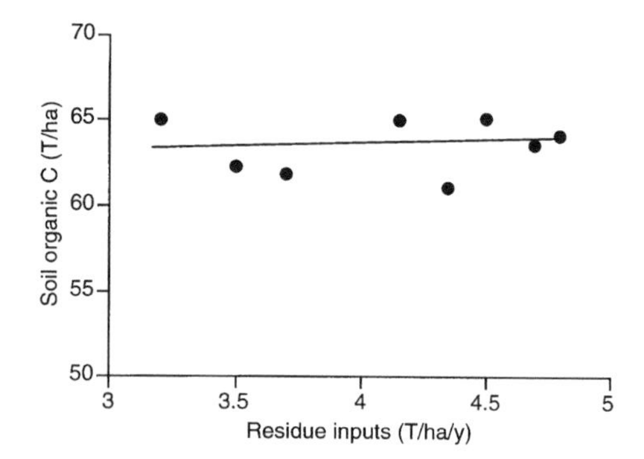

Figure 7 Soil C levels as a function of residue inputs, showing no response to input levels, for a 30-year-old experiment at Melfort, Saskatchewan. (Modified from Campbell, C.A., Bowren, K.E., Schnitzer, M., Zentner, R.P., and Townley-Smith, L., *Can. J. Soil Sci.*, 71, 377, 1991c.)

of residues. Paustian et al.[38] found that differences in soil C levels for various organic amendments in field plots in Sweden were well described by the lignin contents of the residues (Figure 8). Similarly, small plot experiments in Canada showed significant influences of residue quality on changes in soil C levels, where C input rates were held constant (Figure 8). The lignin and N content of residues also helped account for differences in SOM in long-term plots receiving straw, pea vine, or manure residues at Pendleton, Oregon.[39]

In addition to their role as the primary source of C inputs, crop residues and the way in which they are managed have significant impacts on soil physical properties including bulk density, water infiltration, pore size distribution, and aggregate stability.[106,107] With respect to soil C maintenance, residue effects on aggregation and aggregate stability are of particular importance since aggregation is viewed as a key mechanism in promoting C stabilization.[63,69,108] Retention of residues generally increases the number and stability of soil aggregates.[107,109–111] Morachan et al.[112] found that aggregate stability increased in proportion to residue addition rates. Smika and Greb[113] reported that the proportion of aggregates greater than 0.84 mm (classified as nonerodible by wind) increased with increasing rates of wheat straw retention. Halstead and Sowden[114] reported greater aggregate stabilities in all residue amendments compared with unamended soil and found that stability was greatest with straw, manure, or sludge amendments and less for alfalfa, deciduous leaves, or ryegrass litter. By enhancing aggregate stability, increasing residue inputs may have a synergistic effect by providing more raw material for humification as well as increasing the stabilization potential of that material through increased aggregation.[106]

C. TILLAGE

The idea of tillage is fundamental to most people's view of agriculture and the invention of the plow and its subsequent refinement over several thousand years has been a major force in human development. Jethro Tull, whose writings and inventions were highly influential in the preindustrial European agricultural revolution, energetically promoted the benefits of intensive tillage as a means of breaking down the soil into minute particles which could then be absorbed by plants.[3] Although his theory of plant nutrition was subsequently refuted, Tull's prescription contained an element of truth in that part of the effectiveness of intensive tillage is to speed the release of nutrients contained in SOM. However, over the long term, intensive tillage has caused or contributed to soil degradation in many regions. Reducing erosion and organic matter losses in cultivated soils has been the primary reason for the development of less-intensive tillage practices.[115–117]

Tillage affects the processes determining soil C balance in two fundamental ways, (1) through the physical disturbance and mixing of soil and the exposure of soil aggregates to disruptive forces and (2) through controlling the incorporation and distribution of plant residues in the soil. Important secondary effects of tillage on the soil microclimate include the influence of surface residue coverage on soil temperature, water interception and infiltration, and the effects of tillage-induced changes in porosity and soil structure on soil aeration and water relations. Conventional tillage, based on the moldboard plow (CT) and no-till (NT), represent opposite ends of the tillage spectrum and would be expected to show the greatest differences in soil C response. Therefore, our discussion will largely focus on comparisons of these two tillage systems.

The degree of soil disturbance caused by tillage is difficult to define and quantify but a commonly used measure is aggregate stability. By increasing the effective soil surface area and continually exposing new soil to wetting/drying and freeze/thaw cycles at the surface, tillage makes aggregates more susceptible to disruption and physically protected interaggregate organic material becomes more available for

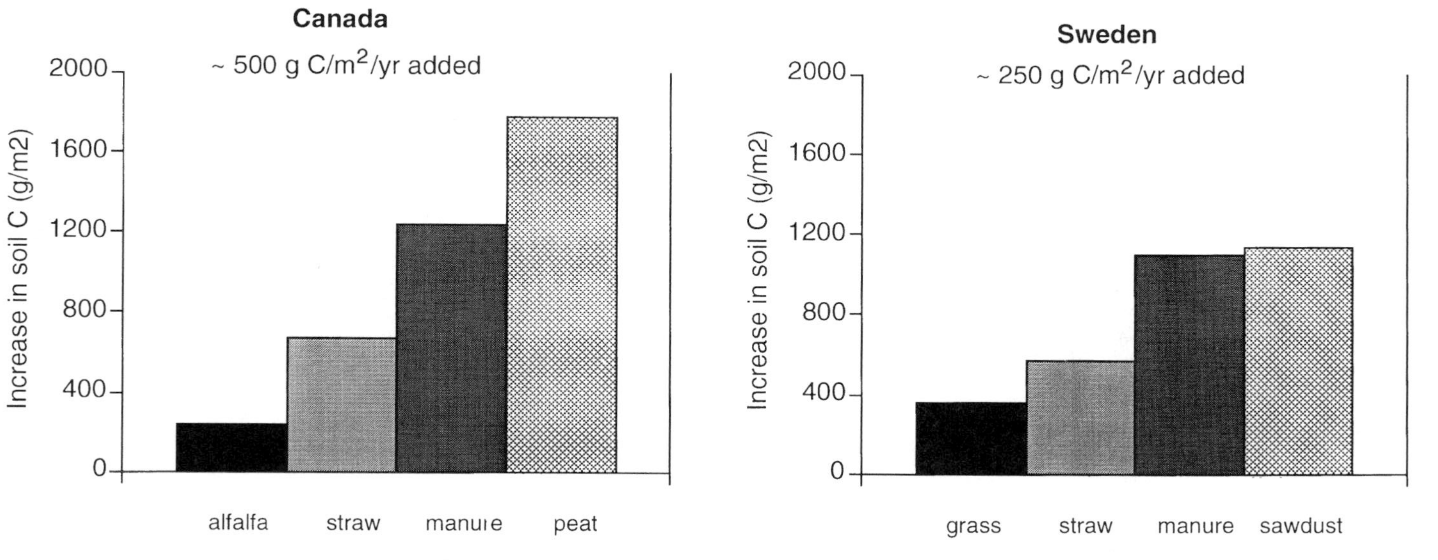

Figure 8 Influence of residue quality on soil C, showing net increment in soil C over the duration of the experiment, where constant amounts of residues were added each year. Data from Sowden and Atkinson[105] for a 20-year experiment with addition of ~500 g C m^{-2} y^{-1} to a sandy soil in Canada and from Paustian et al.[38] for a 31-year experiment with ~250 g C m^{-2} y^{-1} added to a sandy clay loam in Sweden. (From Paustian, K., Robertson, G.P., and Elliott, E.T., *Soil Management and Greenhouse Effect*, Lal, R., Kimble, J., Levine, E., and Stewart, B.A., Eds., *Advances in Soil Science*, Lewis Publishers, Boca Raton, FL, 1995. With permission.)

decomposition.[63,64] Numerous field studies show increases in macroaggregate stability with reduced tillage, with no-till generally showing the highest degree of aggregate stability compared with conventional tillage employing moldboard plowing.[118–122] While differences between various forms of intermediate tillage are less clear, aggregate stability tends to decrease with increasing tillage intensity.[123]

Tillage practices which promote the greatest degree of soil disturbance also tend to be the most effective at burying crop residues. Estimates of residue burial rates for a variety of primary and secondary tillage implements are given in Table 3. The degree of residue incorporation has a major effect on the initial rate of decomposition. In temperate environments, decomposition rates of residues are generally slower when left on the surface than when buried in soil.[124–127] Drier conditions and reduced mineral nutrient availability are probably the main reasons for reduced microbial activity in surface residues. Under some conditions, temperature extremes in surface residues may also be detrimental to microbial activity. For example, temperatures as high as 60°C have been recorded in unshaded residues under wind-still conditions in Australia[128] and the western U.S.[129]

Table 3 Percent of Crop Residue Remaining on Soil Surface Following Tillage Operations

	Residue Remaining (%)		
Tillage Implement	**Good and Smika**	**Tindall and Crabtree**	**Stott**
Moldboard plow	NI	0-5	2–4
Tandem disk	25	50	30–60
One-way disk	50	50–60	NI
Chisel plow	90	75	50–75[a]
Sweep plow	90	85	85–90
Rod weeder	85	85–95	80–85
No-till drill	NI	NI	90–95[b]

Note: Values by Good and Smika[130] and Tindall and Crabtree[131] are for wheat residues. Ranges reported by Stott[132] are from calculations using a general model for residue management which is parameterized for 21 major crop species. NI denotes not included in the cited reports.

[a] Chisel with straight shank.

[b] No-till drill with smooth coulters.

The combination of reduced litter decomposition rates and less soil disturbance usually results in greater amounts of soil C in no-till vs. conventionally tilled systems. Differences in C content between no-till and conventional till are most extreme near the surface, primarily due to differences in the distribution of C inputs.[133] Consequently, comparisons of tillage effects based on sampling only the top few centimeters of soil can give a misleading and overly positive impression of the C buildup under no-till. However, comparisons based on deeper sampling where C levels are summed to below depth of plowing usually show higher overall levels in no-till (Figure 9). These data are from a number of long-term field studies, with paired conventional and no-till treatments. Carbon content and bulk density data were used to calculate C on a square-meter basis to below depth of plowing (generally 30 cm) and expressed on an equivalent soil mass basis to correct for differences in bulk density due to tillage.[134] A number of other studies,[135–141] most of which report increased C concentrations under no-till, were not included because of a shallower sampling depth or a lack of information on bulk density.

The average increase in soil C under NT was about 300 g m^{-2} with a few site/treatments showing increases as high as 1 kg C m^{-2} (Figure 9a). On a relative basis, most sites showed 5 to 20% increases in soil C under NT vs. CT (Figure 9b). The range in absolute differences in C varied little as a function of total soil C (as measured under CT) and thus the relative difference tended to decrease with increasing soil C. It should be noted that the surface mulch which builds up on many NT soils may not be fully accounted for in comparisons of plowed and NT systems, if sampling has been restricted to the mineral soil. In such cases the effect of NT on total C storage would be underestimated.

There is no apparent pattern related to soil texture, with the exception of two instances showing substantially lower C under NT (e.g., Figure 9a). Both of these comparisons are for poorly drained clay soils with cropping systems producing lower residue yields (i.e., continuous soybeans[148] and corn-soybean[144]). The negative results may be associated with reduced yields and residue inputs with NT on

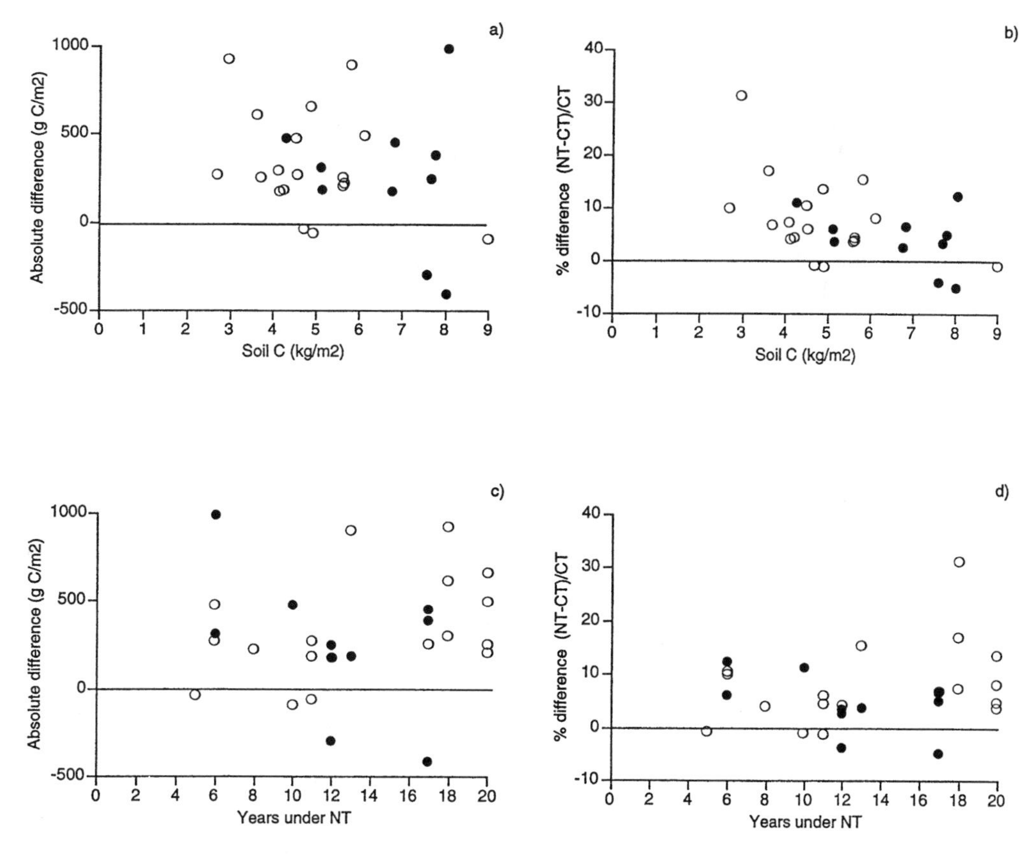

Figure 9 Soil carbon levels in pair-wise comparisons of no-till (NT) and conventional tillage (CT), with moldboard plow, from several long-term experiments. Shown are (a) absolute difference (NT-CT) and (b) relative difference ((NT-CT)/CT) as a function of soil C (under CT) and (c) absolute and (d) relative differences as a function of time under NT. Values are for total organic C to depths at or below depth of plowing and adjusted for differences in bulk density for comparison on an equivalent soil mass basis (see description in text). Filled circles are for clay and clay loam soils, all other textures shown as open circles. Each site/treatment is only represented once, by the most recent published value. Data from Powlson and Jenkinson,[134] Dick,[142] Groffman,[143] Dick et al.,[144–145] Doran,[133] Dalal,[146] Balesdent,[147] Havlin et al.,[148] Chan et al.,[89] and Ismail et al.[149]

the wet clay soils and perhaps less moisture limitation on the decomposition of surface residues compared with better-drained soils. Overall, fine-textured soils tended to have the highest organic matter contents and hence show smaller relative differences between NT and CT.

There was no clear trend in tillage-induced changes as a function of time under no-till (Figure 9cd). There appears to be an early rapid gain in C following conversion to NT followed by a stabilization or much slower rate of C increase. However, site-specific and other management differences likely confound the interpretation so that generalizations regarding temporal dynamics cannot be made on the basis of cross-site comparisons. Unfortunately, relatively few studies of NT have been in existence for sufficient time to generate reliable time-series from repeated measurements. Plots at Lexington, Kentucky, and at Hoytville and Wooster, Ohio (see Chapter 12) have been sampled repeatedly over a 20- to 30-year period. At Lexington, there was an increase in C with time under NT, such that, after 20 years, C contents exceed that of the original bluegrass sod.[149] However, there was a similar although lower increase under CT, reflecting the influence of the high residue inputs under both tillage treatments. In 25-year-old tillage experiments, Dick et al.[71] reported that the most rapid changes in C levels under NT occurred during the first 10 years. Repeated sampling of NT plots in Saskatchewan over a 13-year period showed an increase of 15% in soil C (for 0- to 15-cm depth) after 6 years, whereas C increased by only an additional 4% during the subsequent 7 years (Campbell, C.A., unpublished data). In a study of a NT chronosequence at Coshocton, Ohio, Staley et al.[150] found that total C was positively correlated to years under NT and that microbial biomass-C responded quickly to NT management and reached a new equilibrium value after about 10 years.

One reason that soil C responses to tillage are difficult to generalize is that tillage practices can have indirect effects by influencing crop productivity and thereby crop residue inputs. Lower soil temperatures under NT may delay germination and early plant development in regions with short growing seasons.[116] Lower yields with NT under such conditions has been cited as limiting its applicability in cool regions.[151,152] However, major yield reductions in cool climates are not universal. Riley et al.,[153] in summarizing field experiments for Denmark, Norway, and Sweden, found that NT yields averaged between 91 and 99% of yields obtained with plow tillage. Interestingly, Knight and Lewis[154] reported *higher* spring soil temperatures and comparable yields in NT vs. CT in central Alaska; this was attributed to the insulating effect of greater snow retention due to more residue coverage under NT.

No-till often gives lower yields in poorly drained soils, which are subject to compaction in the absence of tillage, causing restricted root growth and greater susceptibility to disease than under CT.[155–157] The retention of surface residues and lack of soil disturbance associated with NT have been found to increase the severity of a variety of wheat diseases, resulting in reduced yields.[158] However, crop rotation and use of disease-resistant cultivars can reduce disease-induced yield declines in NT.[156] In Indiana, yields under NT were lower for poorly drained soils with high organic matter contents, but yields were greater with NT on low organic matter soils with either poor or good drainage.[121,159] Reduced tillage systems (chisel plow, ridge tillage) showed yield responses intermediate to plow tillage and NT.

Many other studies have shown no or variable differences in crop yield response to tillage[160–162] and higher production under no-till has been shown for well-drained soils, particularly where water use efficiency is improved.[116,163] Prasad and Power[91] summarized the expected differences in yield between no-till and conventional till as follows: (1) little difference under conditions of adequate soil water, good drainage, and adequate available N; (2) increased yields under no-till where there is limited precipitation and soil water and adequate weed control and fertilization; and (3) reduced yields under no-till in areas with excessive precipitation, low temperatures, poor drainage, poor weed control, or low fertility levels.

Effects of tillage on soil physical parameters such as bulk density, porosity, pore size distribution, and pore continuity provide additional indirect controls on decomposition and soil C levels. Increased bulk density and reduced porosity in upper soil layers can occur under NT,[7,164–165] accompanied by a reduction in the number of large pores.[165,166] For the root zone as a whole, however, the effect of surface compaction under no-till may be compensated for by lower bulk densities and increased porosity deeper in the profile.[160,161,166,167] Plow pans, which can impede water movement and root penetration, may not form so readily with reduced tillage.[166,168] In other cases, NT or other reduced tillage practices have had no effect on or have decreased bulk density.[122,167,169,170] In most cases, irrespective of changes in overall porosity, no-till soils have equal or increased water infiltration rates,[122,160,161,164,166] probably due to greater continuity of macropores, such as earthworm burrows and root channels. The net effect of tillage-induced differences on soil water will vary for individual soils. However, the greater water capture and lower evapotranspiration of reduced tillage systems will generally increase soil moisture which may favor higher decomposition rates of root residues within the soil, in contrast to the lower decomposition rates of residues on the soil surface, as discussed earlier.

To summarize, it seems clear that reduced tillage, and especially no-till, is generally effective in increasing soil C, provided that yields and residue production are not adversely affected. Whether increases in soil C continue over long periods of time and how closely no-till soils can be made to mimic the characteristics and C levels of soils under native perennial vegetation remains to be seen. Because of the greater below-ground allocation in native perennial systems and the removal of a substantial portion of the net primary productivity in agricultural systems, it is unlikely that use of no-till practices alone can achieve the soil C levels of native ecosystems. However, if C inputs to soil can be increased to a similar magnitude as in native systems, by increasing nutrient and water supply and using highly productive crop species, then a restoration of soil C to precultivation levels (or higher) may be possible under no-till management.

D. CROP ROTATION

The selection of crops to be grown is the most basic management decision faced by the farmer and, as discussed in the earlier section on land resource areas, this decision is constrained by both climatic and economic factors as well as land suitability. It is recognized that the selection of crops grown may be closely linked to other management factors such as tillage (e.g., annual vs. perennial crops) and fertility (e.g., legumes vs. non-N-fixing crops). However, in this instance we focus mainly on how crop type, residue yield, and fallow frequency influence C inputs and the decomposition environment.

Overall comparisons of soil C contents under rotational systems vs. continuous corn, or continuous wheat, reveal some general patterns (Figure 10). Where corn is the dominant annual crop, only rotations which contain grass or forage crops show consistently higher C levels compared to continuous corn. For the most part, corn-soybean rotations show lower values, with the notable exception being the Morrow plot corn/soybean (earlier corn/oats) rotation (far right, Figure 10). The Morrow plot results, however, are atypical, as discussed below. In the wheat-dominated systems, there are few examples of rotations which yield higher C levels than continuous wheat except for a few treatments including green manures or legume crops. Rotations including 1 or 2 years of hay as well as rotations including other cereal crops show little effect on C content compared with continuous wheat. Wheat-fallow systems, however, show as much as 40% lower C compared to continuous wheat (Figure 10).

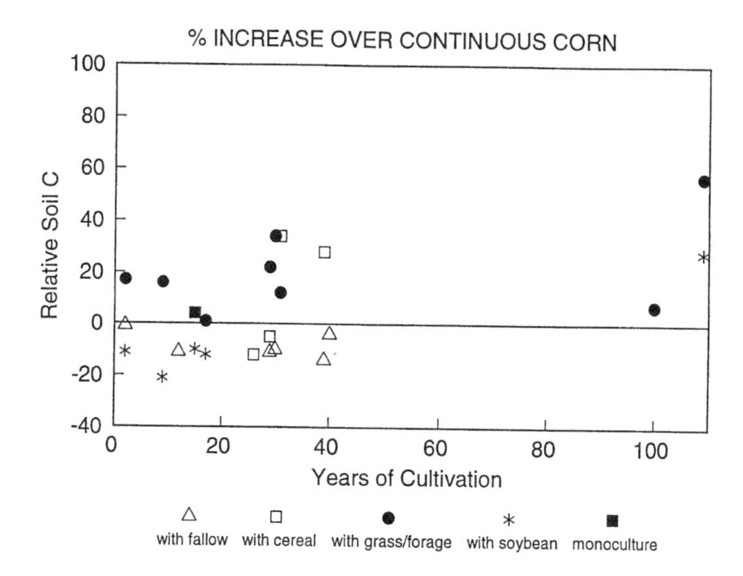

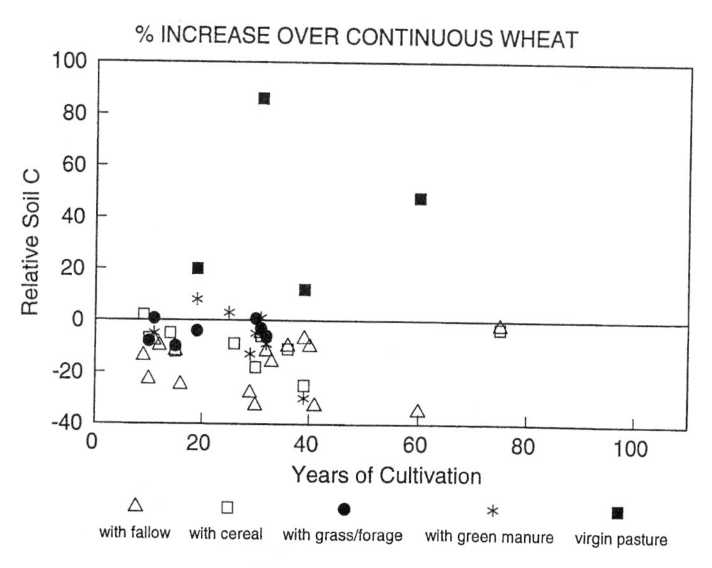

Figure 10 Summary of rotation influence on soil C levels, shown as percent of soil C under continuous corn or continuous wheat treatments in the same experiment, plotted as a function of experiment duration. Symbols depict different classes of rotations for corn- and wheat-based cropping systems. Data from Haas et al.,[11] Anderson and Peterson,[93] Barber,[16] Hooker et al.,[171] Odell et al.,[48] Upchurch et al.,[172] Dick et al.,[144,145] Soon and Broersma,[173] Insam et al.,[174] Rasmussen et al.,[45] Campbell et al.,[86,175] and Monreal and Janzen.[176]

1. Rotations with Annual Crops

The effect of rotations of annual crops on SOM may be attributable mainly to the amount of residues produced and returned to the soil. Representative yields of major field crops and estimates of associated above-ground crop residues for different regions in the United States are summarized in Table 4. Typically, highest residue levels for row crops are produced by C_4 feed grains such as corn and sorghum. Soybeans, mostly grown in rotation with corn, produce less than half as much residue. Residue production from most small-grain cereals is intermediate. It is more difficult to generalize about below-ground residue production, due to the limited number of measurements and differences in methodologies to determine below-ground productivity. However, for most annual crops, C inputs from root production usually account for 20 to 40% of total dry matter production.[177–179]

Table 4 Approximate Crop Residue Yields (above Ground Only) for Different Crops Based on Average Yields for the U.S.[180]

Crop	Yield (bushels/ha)	Yield (kg/ha)	Harvest Index (%)	Crop Residue (kg/ha)
Soybean	34	2000	50–60	1300–2000
Oats	51	1800	40–50	1800–2700
Wheat	34	2300	35–45	2800–4300
Barley	55	3000	45–50	3000–3700
Sorghum	59	3300	35	6100
Corn	108	6800	40–50	6800–10200
Sorghum (silage)	na	9070	na	na
Corn (silage)	na	11,790	na	na

Note: Estimates of harvest index values of most crops based on Anderson and Vasilas,[181] Brinkman and Rho,[182] Cox et al.,[183] Donald and Hamblin,[184] Meyers et al.,[185] Russell,[186] and Walker and Fioritto.[187] Harvest index for sorghum based on silage yields as an estimate of total above-ground dry matter production.

Other species characteristics, such as residue quality, are also important factors affecting rotation responses. Cereals such as wheat and barley appear to have somewhat higher lignin contents (e.g., 16 to 24%) compared to corn (11 to 16%)[188,189] which can retard decomposition rates and increase C stabilization efficiency as discussed previously.

The influence of crop rotation on residue production and soil C changes has been demonstrated in several long-term field experiments. Zielke and Christenson[190] found that changes in soil C for six rotations including corn, sugar beet, navy bean, oats, and alfalfa were closely correlated with the amount of residue returned. Carbon levels increased with the frequency of corn in the rotation. Similarly, Havlin et al.[148] reported that rotation effects on soil C were directly related to the amount of residues produced, where continuous sorghum > sorghum-soybean > continuous soybean at two sites and continuous corn > corn-soybean > continuous soybean at another site. Campbell and Zentner[191] found that changes in SOM reflected the residue production of each rotation and its susceptibility to erosion. Organic matter in several rotations increased during a 15-year period of high crop yields and then decreased with lower production during a subsequent dry period.

At first glance, results from the Morrow plots (Figure 10, far right for "corn") appear to contradict this pattern with the corn-oat (later corn-soybean) rotation having higher values than continuous corn, despite the lower residue yields expected from oats and soybean. However, as Guernsey et al.[192] report, corn yields in the corn-oat rotation were nearly twice as high as in continuous corn through the first 80 years of the experiment. Moreover, the corn-oat rotation included a legume catch crop following oats, which provided additional organic matter and probably accounted for some of the higher productivity in the following corn crop. Based on yield information it appears likely that past residue inputs in the corn-oat rotation exceeded that in continuous corn. At two sites in Minnesota, Crookston et al.[193] reported an average of 10% higher corn yields in rotation with soybean as compared to continuous corn. Thus, if corn production is increased by rotation this can help offset lower residue inputs from other crops in the rotation.

In semiarid agricultural systems, where summer fallowing is routinely practiced, decreases in SOM with increasing fallow frequency have been well documented.[45,110,191,194–197] Data from three Canadian sites show a roughly linear decrease in soil C with increasing proportion of fallow in the rotation

(Figure 11). Summer fallowing is conducive to increased rates of organic matter decomposition for several reasons, including increased soil moisture, increased soil temperatures, increased soil disturbance associated with mechanical weed control during the fallow period, and greater susceptibility to erosion.

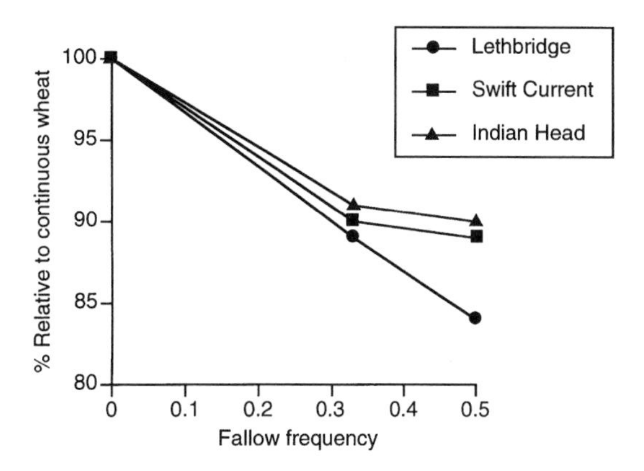

Figure 11 Soil C response to fallow frequency, as a percentage of C content under continuous cropping, for three long-term experiments with fallow-wheat, fallow-wheat-wheat, and continuous wheat rotations in Saskatchewan. Data from Campbell et al.[86,175] and Janzen.[194] Rotations at Indian Head and Swift Current were N+P fertilized and those at Lethbridge were unfertilized.

Summer fallowing also affects total C inputs over the course of the rotation. In unfertilized systems, wheat yields following fallow may be nearly double those with continuous wheat[11,80] and if residues are assumed to be proportional to yields then average C inputs might not be greatly different. However, moisture stress often results in a higher proportion of C allocation to crop roots[198–200] so that the ratio between total residue production (including roots) and yield could be greater in continuous wheat compared to wheat-fallow systems. The impact of fallow frequency on average residue return rates can also vary with N availability. Where N fertilizer was applied, Horner et al.[80] found that annual wheat yields with continuous cropping were about 70% of those in wheat-fallow, which would give greater residue inputs with continuous cropping over the course of the rotation.

2. Rotations with Perennial Crops

Inclusion of perennial crops into the rotation has long been recognized as an effective means of increasing SOM. Interestingly, Johnston[201] attributes the relatively high organic matter content of many English arable soils, which have been under cultivation for several hundred years, to periodic reversions to pasture during times of economic depression. As discussed in the first part of this paper, the development of ley cropping played an important role in the increased productivity and soil fertility of late pre-industrial agriculture.

The efficiency of rotations that include hay crops (i.e., leys) in maintaining or increasing soil C tends to be greater in more humid regions. Clement and Williams[202] reported an average increase of 15% in total soil C after 4 years of pasture whereas C decreased in annually cropped treatments. Grazed pastures had greater increases than where hay was mowed and removed. Long-term rotations at Woburn (U.K.) had approximately 25% greater C levels in 5-year rotations with 3 years of ley compared with 5 years of only annual crops.[201] In experiments on both old grassland soil and old arable soil at Rothamsted, 6-year rotations with 3 years of fertilized grass or grass-clover ley increased soil C by 10 to 15% compared with rotations with annual crops only. In contrast, where alfalfa was used as the hay crop, C levels were no higher than in the annual crop rotation.[201] Increases in total C levels under ley cropping have been reported in a number of other European studies.[17,94,203–205] In Indiana, 7 years of continuous alfalfa or bromegrass yielded up to 25% more soil C than continuous corn.[206] However, 4 years of pasture crops followed by 3 years of corn resulted in similar C levels as found in continuous corn. In a study comparing continuous cotton with lespedeza, a perennial forage crop, Davidson et al.[207] found dramatic increases in soil C under lespedeza (to 30-cm depth), with C levels nearly double that in cotton in the top 10 cm.

Use of perennial crops in rotation in dry climates is constrained by moisture limitations and cereal crop harvests following perennial crops are often reduced.[11,104] In summarizing results from 17 long-term sites in the Great Plains, Haas et al.[11] concluded that biennials and perennials (e.g., sweet clover, ryegrass) grown for 1 year as green manure crops were not effective at reducing soil C declines in 2-, 3- and 4-year cereal-based rotations. However, other studies under semiarid conditions in the U.S. have shown less decline in soil C using hay crops in rotation with cereals compared to cereal-only rotations.[208–210] In several studies for the subhumid regions of the Canadian prairies, summarized by Campbell et al.,[211] rotations with forage crops gave roughly similar C levels to those in continuous wheat. In Australia, use of forage crops in wheat rotations has become an important management practice for dryland agriculture.[212] As an example, Drover[197] reported that fallow-wheat-lupin-lupin rotations increased C levels by 12% compared to continuous wheat, in a sandy loam brown soil, and 33% in a lateritic sand. On these nutrient-poor soils, productivity and residue production of the wheat crop was greatly enhanced by the increased N supplied by the legume. Grace et al.[213] cite several other examples of soil C enhancement of dryland wheat systems in Australia by the incorporation of 1- or 2-year legume pastures into the rotation.

Changes in the distribution and amount of C inputs and the absence of tillage are key components of the ability of perennial crops to sequester C. Rapid increases in total C following initiation of ley cropping are largely due to increases in particulate organic matter (POM), comprised of partially decomposed root and leaf material. Garwood et al.[214] reported that POM (>0.25 mm) was twice as high under grass-legume leys as in arable soils and that about half of the increase in total carbon during a 4-year ley cycle was in this fraction. Comparing soils under pasture (2.7% C) and wheat-fallow (1% C), Oades and Turchenek[215] reported that the pasture soil was enriched mainly in the POM fraction and in silt-sized fractions which were thought to represent microbial debris. Tyson et al.[203] found that POM in a 30-year-old ley was four times that in the annually cropped treatment and POM comprised 15 to 20% of the total C in the pasture. Similarly, budgets of continuous barley, grass, and alfalfa leys reported by Paustian et al.[205] showed a total C increase of ~1.5% in the two leys the 2nd year after ley establishment, due to increases in POM and litter and standing root biomass.

Table 5 Influence of Previous Cropping History on Aggregate Stability, Expressed as Mean Weight Diameter (MWD, i.e., aggregate stability increases with increasing MWD) and Organic Carbon in a Lismore Silt Loam in New Zealand

Cropping History	Aggregate Stability (MWD)	Organic C (%)
10 year arable	1.0	2.0
4 year arable	1.2	2.4
1 year arable	1.3	2.4
1 year pasture	2.0	2.4
4 year pasture	2.5	2.5
10 year pasture	2.7	3.2

Note: Cropping histories for arable systems indicate years under annual cropping after coming out of pasture and conversely, for pastures, years in pasture after coming out of annual cropping.

Data from Haynes, R. J., Swift, R. S., and Stephen, R. C., *Soil Tillage Res.*, 19, 77, 1991.

Thus, the maintenance of higher soil C under perennial crops will be associated with a relative increase in POM and an increase in the physical protection of SOM. Aggregate stability increases rapidly after perennials are established (Table 5), both due to a lack of tillage disturbance and the characteristics of the below-ground system of most perennial grasses. These characteristics include a dense, fibrous root system, which helps to form and bind soil aggregates, and the large production of fungal hyphae and microbially derived polysaccharides and gums, which are effective in binding mineral and organic matter particles. Over extended time periods, POM under perennial crops can be maintained at higher levels and gradually more resistant humic substances are formed. However, the potential for long-term C stabilization may not be realized until after several cycles of aggregate dissolution and reformation have led to the encapsulation and protection of organic matter within stable soil aggregates. Thus, while rapid losses of physically stabilized C can occur with cultivation of native grassland or old pastures, restocking this stable C pool through conversion to grassland or pastures may occur more slowly.[217] Accordingly, increases in soil C under rotations of alternating annual and perennial crops are likely to be transient[206] unless inclusion of the ley substantially increases C inputs over the course of the rotation, as appears to be the case in many of the European experiments.

E. FERTILIZATION

Many of the oldest field experiments in Europe (e.g., Rothamsted, Askov, Halle/Saale, Limberhof, Grossenzerdorf) were set up specifically to study crop production responses to the then newly developed mineral fertilizers. Treatments were designed to compare a variety of fertilizer additions with traditional fertility management using animal or green manure additions and controls with no added nutrients. A secondary objective was to monitor changes in soil conditions, including organic matter, as affected by mineral fertilizers. Numerous long-term studies involving fertilizers have subsequently been established; thus there is a wealth of field-based information available. Comparisons of N additions are most common, although many studies have included varying P and K additions in their designs. In this discussion we will focus on the effects of N fertilizer additions on soil C.

Nitrogen availability can influence soil C levels in a variety of ways. It is clear that by increasing crop production, and thereby residue inputs, N fertilization can contribute to increased SOM contents. By increasing plant growth, fertilization can also lead to increased transpiration, drier soils, and decreased decomposition rates.[126] Results from many long-term studies show a general tendency of increases in soil C with substantive additions of N, compared to zero or low N additions (Table 6). The response across levels of N is less clear, although several sites show a roughly monotonic increase in C levels as N inputs increase. An exception is the site at Melfort, Saskatchewan, which shows no response to N addition in this very high organic matter soil (approximately 6% C). As discussed earlier, Campbell et al.[86] reported that soil C contents at this site were also unaffected by different levels of C input, which suggests that the soil C holding capacity is essentially saturated. Thus, a lack of response to N, at least as it affects C inputs, is not surprising.

Application of ammonium-based fertilizer in the absence of liming can promote soil acidification, resulting in decreased decomposition rates, as shown for the unlimed, NH_4-N fertilized plots in the Parkgrass experiment at Rothamsted.[222]

Nitrogen additions can affect decomposition rates and C stabilization efficiency in other ways that contribute to higher SOM levels. Fog[223] reviewed 60 papers which reported zero or negative effects of N addition on decomposition rates. He offered several possible explanations for negative effects of N additions on decomposition, including the repression of lignolytic enzymes by ammonium and an increase in the amount of amino compounds which can act as precursors in the formation of recalcitrant humic compounds. At the microbial level, insufficient N can lead to lower yield efficiencies (i.e., more CO_2 respired per unit C assimilated[224]). Under such conditions addition of N could increase growth efficiency resulting in a higher proportion of C inputs retained in SOM.

In an analysis of long-term plot studies with constant above-ground C, with and without N fertilization, Paustian et al.[38] found that increased root residue inputs could not account for observed increases in soil C in the N fertilized treatments (Figure 12). They also found that soil C:N ratios were lower in the unfertilized treatments, where straw or sawdust were added, suggesting that N limitation may have reduced C stabilization efficiency. In a study by Campbell et al.,[225] soil C was found to be similar in fertilized plots where straw was removed compared with fertilized plots where straw was retained. This was despite the fact that C inputs were estimated to be greater in the treatment with straw retention. One interpretation offered by the authors was that roots, rather than straw, were the primary source of C to build stable organic matter.

In most field experiments it is difficult, if not impossible, to partition the interacting, and potentially conflicting, effects of N addition on soil C. However, when viewed in a broad (if somewhat tautological) sense, it is reasonable that, since C and N are the major constituents of SOM and their proportionality (i.e., C:N ratio) is relatively constant across a range of agricultural soils, then an adequate supply of N is required to build SOM. If inputs of these two elements are too much out of balance then the efficiency of soil C sequestration will be reduced.

Table 6 Summary of Long-Term Field Experiments Showing Soil C Responses to Differential Levels of N Fertilizer Application (N Levels are Shown in Bold Type)

Site	Years	Treatment	N Levels (kg ha⁻¹) Soil C Change (% of Control)			
Mandan, MT[209]	7		**0**	**30**	**60**	
		Sandy loam — continuous wheat	100	109	—	
		Sandy loam — crested wheatgrass	100	99	100	
		Silt loam — continuous wheat	100	107	—	
		Silt loam — crested wheatgrass	—	100	102	
Pendleton, OR[218]	44		**11**	**22**	**50**	
		Wheat/fallow — moldboard plow	100	98	102	
		Wheat/fallow — disk	100	94	108	
		Wheat/fallow — sweep	100	103	108	
Swift Current, Saskatchewan[191]	24		**0**	**29**		
		Fallow/wheat/wheat	100	104		
		Continuous wheat	100	109		
Indianhead, Saskatchewan[175]	29		**0**	**24**		
		Fallow/wheat	100	104		
		Fallow/wheat/wheat	100	106		
		Continuous wheat	100	106		
Melfort, Saskatchewan[86]	30		**0**	**52**		
		Fallow/wheat/wheat	100	100		
		Continuous wheat	100	100		
		Fallow/wheat/hay rotation	100	97		
Queensland, Australia[146]	13	Cereals — conventional tillage	**0**	**23**	**69**	
		Straw burned	100	101	106	
		Straw retained	100	102	105	
		Cereals — no-till				
		Straw burned	100	94	98	
		Straw retained	100	112	113	
Eastern Kansas[148]	8		**0**	**252**		
		Continuous soybean	100	95		
		Corn/soybean	100	101		
		Continuous corn	100	102		
Purdue, IN[16]	12		**0**	**67**	**200**	
		Continuous corn	100	106	107	
Lamberton, MN[77]	19		**0**	**45**	**90**	**180**
		Continuous corn	100	104	103	105
Lexington, KY[149]	20		**0**	**84**	**168**	**336**
		Continuous corn — plow tillage	100	115	115	126
		Continuous corn — no-till	100	105	106	120
Askov, Denmark[94]	78		**0**	**35**	**70**	**105**
		Cereal/root crop rotation — loam	100	106	109	111
		Cereal/root crop rotation — sand	100	111	121	—
Southern Sweden[17]	15		**0**	**50**	**100**	**150**
		Mixed cropping	100	107	108	111
Uppsala, Sweden[38]	30	Cereal/root crops	**0**	**80**		
		Straw removed	100	118		
		Straw added	100	116		
		Sawdust added	100	115		
Ås, Norway[84]	20	Cereals	**0**	**34**	**68**	**136**
		Straw added	100	98	101	103
		Straw removed	100	101	104	103
Øsaker, Norway[84]	20	Cereals	**0**	**34**	**68**	**136**
		Straw added	100	102	100	104
		Straw removed	100	100	99	107

Table 6 (continued) Summary of Long-Term Field Experiments Showing Soil C Responses to Differential Levels of N Fertilizer Application (N Levels are Shown in Bold Type)

Site	Years	Treatment	N Levels (kg ha^{-1}) Soil C Change (% of Control)			
Ås, Norway[84]	31		**60**	**120**		
		Cereals	100	102		
		Cereal + row crops	100	102		
		2 ley + 4 arable	100	108		
		4 ley + 2 arable	100	104		
Halle/Saale, Germany[219]	80		**0**	**40**		
		Continuous rye	100	108		
Gottingen, Germany[220]	81		**0**	**30–50**		
		Mixed rotation	100	111		
Riverside, CA[221]	28	Citrus grove	**0**	**310**		
		Urea N	100	105		
		NH_4-N	100	125		
		NO_3-N	100	123		
Rothamsted, U.K.[222]	120	Pasture (parkgrass plots)	**0**	**48**	**96**	**144**
		Unlimed				
		NH_4 — fertilizer	100	100	138	153
		NO_3 — fertilizer	100	108	97	—
		Limed				
		NH_4 — fertilizer	100	95	101	105
		NO_3 — fertilizer	100	99	100	—

Note: Soil C levels represent the total change over the duration of the experiment, given as percent of the unfertilized (or least fertilized) treatment.

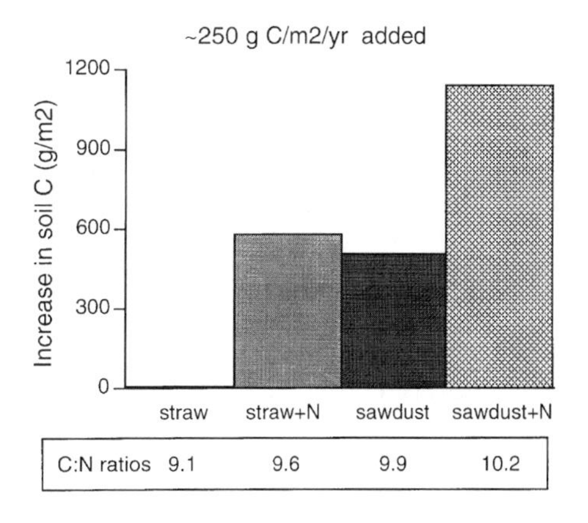

Figure 12 Nitrogen addition effects on the net change in soil C over the duration of a 31-year experiment with constant (~250 g C m^{-2} y^{-1}) of straw or sawdust addition. Above-ground crop residues were removed from the plots. N fertilized plots received rates equivalent to 80 kg N ha^{-1} y^{-1} as $Ca(NO_3)_2$.

V. CONCLUDING REMARKS

Agricultural practices and SOM dynamics are intimately linked and, as we have discussed, virtually all facets of management impact the amount of C which can be maintained in soil. However, because the amounts of C in soils are large and change comparatively slowly, the implications of a particular management system on the soil may be apparent only after several years to decades. We are fortunate that a number of far-sighted individuals initiated and subsequently maintained long-term field experiments, providing us with a unique legacy of agricultural and ecological information.

The intelligent management of soil resources, including organic matter, is of critical importance, not only for productivity and sustainability of the farmer's field, but also for the health and sustainability of our global environment. The practical knowledge as well as theoretical insights which have been derived from long-term field experiments provide us with some essential tools to improve the sustainability and environmental quality of agroecosystems.

ACKNOWLEDGMENTS

The authors wish to thank Dr. C.A. Campbell and two anonymous referees for reviewing earlier versions of the manuscript.

REFERENCES

1. Grigg, D., *The Dynamics of Agriculture Change: The Historical Experience*, Hutchinson & Co., London, 1983.
2. Seebohm, M.E., *The Evolution of the English Farm*, 2nd ed., E.P. Publishing, 356 pp., 1976.
3. Fussell, G.E., *Farming Technique from Prehistoric to Modern Times*, Pergamon Press, Oxford, 1965.
4. Whitney, M., The development of man on the earth and the beginnings of organized agriculture, in *Soil and Civilization: A Modern Concept of the Soil and the Historical Development of Agriculture*, D. Van Nostrand Company, New York, 176, 1925.
5. Olsson, G., Nutrient use and productivity for different cropping systems in south Sweden during the 18th century, in *The Cultural Landscape: Past, Present and Future*, Birks, H.H., Birks, H.J.B., Kaland, P.E., and Moe, D., Eds., Cambridge University Press, Cambridge, 1988, 123.
6. Steen, E., Agricultural outlook, in *Ecology of Arable Land — Organisms, Carbon and Nitrogen Cycling*, Andrén, O., Lindberg, T., Paustian, K., and Rosswall, T., Eds., Chapter 8, Ecological Bulletins, Copenhagen, 1989, 40.
7. Curwen, E.C. and Hatt, G., *Plough and Pasture: The Early History of Farming*, Collier Books, New York, 1961.
8. Percy, D.O., Ax or plow?: Significant colonial landscape alteration rates in the Maryland and Virginia tidewater. *Agric. Hist.*, 66, 66, 1992.
9. Sandor, J.A., Gersper, P.L., and Hawley, J.W., Soils at prehistoric agricultural terracing sites in New Mexico. II. Organic matter and bulk density changes, *Soil Sci. Soc. Am. J.*, 50, 173, 1986.
10. Cochrane, W.W., *The Development of American Agriculture*, University of Minnesota Press, Minneapolis, 1979.
11. Haas, H.J., Evans, C.E., and Miles, E.F., Nitrogen and Carbon Changes in Great Plains Soils as Influenced by Cropping and Soil Treatments, Technical Bulletin No. 1164 USDA, State Agricultural Experiment Stations, 111 pp., 1957.
12. Mitchell, C.C., Westerman, R.L., Brown, J.R., and Peck, T.R., Overview of long-term agronomic research, *Agron. J.*, 83, 24, 1991.
13. Cole, C.V., Stewart, J.W.B., Ojima, D.S., Parton, W.J., and Schimel, D.S., Modelling land use effects of soil organic matter dynamics in the North American Great Plains, in *Ecology of Arable Land*, Clarholm, M. and Bergström, L., Eds., Kluwer Academic, Dordrecht, 89, 1989.
14. Persson, J., Detailed Investigation of the Soil Organic Matter in a Long Term Frame Trial, *Department of Soil Sciences Report 128*, Swedish University of Agricultural Sciences, 1980.
15. Larson, W.E., Clapp, C.E., Pierre, W.H., and Morachan, Y.B., Effects of increasing amounts of organic residues on continuous corn. II. Organic carbon, nitrogen, phosphorus, and sulfur, *Agron. J.*, 64, 204, 1972.
16. Barber, S.A., Corn residue management and soil organic matter, *Agron. J.*, 71, 625, 1979.
17. Jansson, S.L., Bördighetsstudier för Markvård. Försök i Malmöhus län 1957–74. (Long-term soil fertility studies. Experiments in Malmöhus County 1957–74), Supplement 10, Stockholm, 1975.
18. Bouwman, A.F., Ed., *Soils and the Greenhouse Effect*, John Wiley & Sons, Chichester, England, 1990.
19. Duxbury, J.M., Harper, L.A., and Mosier, A.R., Contributions of agroecosystems to global climate change, in *Agricultural Ecosystem Effects on Trace Gases and Global Climate Change*, ASA Special Publication Number 55, 1993, 1.
20. Wisniewski, J. and Lugo, A.E., Eds., *Natural Sinks of CO_2*, Kluwer Academic, Dordrecht, 1992.
21. Wisniewski, J. and Sampson, R.N., Eds., *Terrestial Biospheric Carbon Fluxes: Quantification of Sinks and Sources of CO_2*, Kluwer Academic, Dordrecht, 1993.
22. Houghton, R.A. and Skole, D.L., Carbon, in *The Earth as Transformed by Human Action*, Turner, B.L., II, Clark, W.C., Kates, R.W., Richards, J.F., Mathews, J.T., and Meyer, W.B., Eds., Cambridge University Press, Cambridge, 1990, chap. 23.
23. Wilson, A.T., Pioneer agriculture explosion and CO_2 levels in the atmosphere, *Nature*, 273, 40, 1978.
24. Cole, C.V., Paustian, K., Elliott, E.T., Metherell, A.K., Ojima, D.S., and Parton, W.J., Analysis of agroecosystem carbon pools, *Water, Air, Soil Pollut.*, 70, 357, 1993a.
25. Barnwell, T.O., Jackson, R.B., Elliott, E.T., Burke, I.C., Cole, C.V., Paustian, K., Paul, E.A., Donigian, A., Patwardhan, A., Rowell, A., and Weinrich, K. An approach to assessment of management impacts on agricultural soil carbon, *Water, Air, Soil Pollut.*, 64, 423, 1992
26. Cole, C.V., Flach, K., Lee, J., Sauerbeck, D. and Stewart, B., Agricultural sources and sinks of carbon. *Water, Air, Soil Pollut.*, 70, 111, 1993b.
27. Kononova, M.M., *Soil Organic Matter. Its Nature, Its Role in Soil Formation and in Soil Fertility*, 2nd ed., Nowakowski, T.Z. and Newman, A.C.D., (Translators) Pergamon Press, Oxford, 1966.
28. Allison, F.E., *Soil Organic Matter and Its Role in Crop Production*, (Developments in Soil Science 3), Elsevier, Amsterdam, 1973.

29. Campbell, C.A., Soil organic carbon, nitrogen and fertility, in *Soil Organic Matter*, Schnitzer, M. and Khan, S.U., Eds., Elsevier, Amsterdam, 1978.
30. Tate, R.L., *Soil Organic Matter. Biological and Ecological Effects*, John Wiley & Sons, New York, 1987, 291.
31. Rasmussen, P.E. and Collins, H.P., Long-term impacts of tillage fertilizer, and crop residue on soil organic matter in temperate semiarid regions, *Adv. Agron.*, 45, 93, 1991.
32. McKaig, N., Jr. and Roller, E.M., The effects of organic matter added to lysimeters containing Norfolk coarse sand, *Soil Sci. Soc. Proc.*, 2, 195, 1938.
33. Mann, L.K., A regional comparison of carbon in cultivated and uncultivated Alfisols and Mollisols in the central United States, *Geoderma*, 36, 241, 1985.
34. Meints, V.M. and Peterson, G.A., The influence of cultivation on the distribution of nitrogen in soils of the Ustoll suborder, *Soil Sci.*, 124, 334, 1977.
35. Elliott, E.T., Janzen, H.H., Campbell, C.A., Cole, C.V., and Myers, R.J.K., An ecosystem approach to integrated nutrient management for sustainable land use, in *Sustainable Land Management in the 21st Century*, Vol. 2, Wood, R.C. and Dumanski, J., Eds., Lethbridge, Saskatchewan, 1994, 35.
36. Jenkinson, D.S., Hart, P.B.S., Rayner, J.H., and Parry, L.C., Modelling the turnover of organic matter in long-term experiments at Rothamsted, *INTECOL Bull.*, 15, 1, 1987.
37. Parton, W.J., Schimel, D.S., Cole, C.V., and Ojima, D.S., Analysis of factors controlling soil organic matter levels in Great Plains grasslands, *Soil Sci. Soc. Am. J.*, 51, 1173, 1987.
38. Paustian, K., Parton, W.J., and Persson, J., Modeling soil organic matter in organic-amended and nitrogen-fertilized long-term plots, *Soil Sci. Soc. Am. J.*, 56, 476, 1992.
39. Parton, W.J. and Rasmussen, P.E., Long-term effects of crop management in wheat-fallow. II. CENTURY model simulations, *Soil Sci. Soc. Am. J.*, 58, 530, 1994.
40. USDA, SCS, Land Resource Regions and Major Land Resource Areas of the United States, Agriculture Handbook 296, 1981.
41. Jenny, H., *Factors of Soil Formation*, McGraw-Hill, London, 1941.
42. Kern, J.S. and Johnson, M.G., Conservation tillage impacts on national soil and atmospheric carbon levels, *Soil Sci. Soc. Am. J.*, 57, 200, 1993.
43. Anderson, D.W., Pedogenesis in the grassland and adjacent forests of the Great Plains, *Adv. Soil Sci.*, 7, 53, 1987.
44. Martel, Y.A. and Paul, E.A., Effects of cultivation on the organic matter of grassland soils as determined by fractionation and radiocarbon dating, *Can. J. Soil Sci.*, 54, 419, 1974.
45. Rasmussen, P.E., Collins, H.P., and Smiley, R.W., Long-Term Management Effects on Soil Productivity and Crop Yield in Semi-Arid Regions of Eastern Oregon, Station Bulletin 675, USDA-ARS and Agricultural Experiment Station, Oregon State University, Pendleton, 1989.
46. Dalal, R.C. and Mayer, R.J., Long-term trends in fertility of soils under continuous cultivation and cereal cropping in southern Queensland. I. Overall changes in soil properties and trends in winter cereal yields, *Aust. J. Soil Res.*, 24, 265, 1986a.
47. Dalal, R.C. and Mayer, R.J., Long-term trends in fertility of soils under continuous cultivation and cereal cropping in southern Queensland. II. Total organic carbon and its rate of loss from the soil profile, *Aust. J. Soil Res.* 24, 281, 1986b.
48. Odell, R.T., Melsted, S.W., and Walker, W.M., Changes in organic carbon and nitrogen of Morrow plot soils under different treatments 1904–1973, *Soil Sci.*, 137, 160, 1984.
49. Balesdent, J., Wagner, G.H., and Mariotti, A., Soil organic matter in long-term field experiments as revealed by carbon-13 natural abundance, *Soil Sci. Soc. Am. J.*, 52, 118, 1988.
50. Bowman, R.A., Reeder, J.D., and Lober, R.W., Changes in soil properties in a Central Plains rangeland soil after 3, 20, and 60 years of cultivation, *Soil Sci.*, 150, 851, 1990.
51. Mann, L.K., Changes in soil carbon storage after cultivation, *Soil Sci.*, 142, 279, 1986.
52. Davidson, E.A. and Ackerman, I.L., Changes in soil carbon inventories following cultivation of previously untilled soils, *Biogeochemistry*, 20, 161, 1993.
53. Tiessen, H., Stewart, J.W.B., and Bettany, J.R., Cultivation effects on the amounts and concentration of carbon, nitrogen, and phosphorous in grassland soils, *Agron. J.*, 74, 831, 1982.
54. Anderson, D.W. and Coleman, D.C., The dynamics of organic matter in grassland soils, *J. Soil Water Conserv.*, 40, 211, 1985.
55. Buyanovsky, G.A., Kucera, C.L., and Wagner, G.H. Comparative analyses of carbon dynamics in native and cultivated ecosystems, *Ecology*, 68, 2023, 1987.
56. van Veen, J.A. and Paul, E.A., Organic carbon dynamics in grassland soils. I. Background information and computer simulation, *Can. J. Soil Sci.*, 61, 185, 1981.
57. Angers, D.A., Pesant, A., and Vigneux, J., Early cropping-induced changes in soil aggregation, organic matter, and microbial biomass, *Soil Sci. Soc. Am. J.*, 56, 115, 1992.
58. Elustondo, J., Angers, D.A., Laverdière, M.R., and N'Dayegamiye, A., Étude comparative de l'agrégation et de la matière organique associée aux fractions granulométriques de sept sols sous culture de mais ou en prairie, *Can. J. Soil Sci.*, 70, 395, 1990.
59. Elliott, E.T., Aggregate structure and carbon, nitrogen, and phosphorus in native and cultivated soils, *Soil Sci. Soc. Am. J.*, 50, 627, 1986.

60. Tisdall, J.M. and Oades, J.M., Organic matter and water-stable aggregates in soils, *J. Soil Sci.*, 33, 141, 1982.
61. Low, A.J., The effect of cultivation on the structure and other physical characteristics of grassland and arable soils, *J. Soil Sci.*, 23, 363, 1972.
62. Rovira, A.D. and Greacen, E.L., The effect of aggregate disruption on the activity of microorganisms in the soil, *Aust. J. Agric. Res.*, 8, 659, 1957.
63. Elliott, E.T. and Coleman, D.C., Let the soil work for us, *Ecol. Bull.*, 39, 23, 1988.
64. Beare, M.H., Cabrera, M.L., Hendrix, P.F., and Coleman, D.C., Aggregate-protected and unprotected organic matter pools in conventional- and no-tillage soils, *Soil Sci. Soc. Am. J.*, 58, 787, 1994.
65. Aguilar, R., Kelly, E.F., and Heil, R.D., Effects of cultivation on soils in Northern Great Plains rangeland, *Soil Sci. Soc. Am. J.*, 52, 1081, 1988.
66. Burke, I.C., Yonker, C.M., Parton, W.J., Cole, C.V., Flach, K., and Schimel, D.S., Texture, climate and cultivation effects on soil organic matter content in U.S. grassland soils, *Soil Sci. Soc. Am. J.*, 53, 800, 1989.
67. Campbell, C.A. and Souster, W., Loss of organic matter and potentially mineralizable nitrogen from Saskatchewan soils due to cropping, *Can. J. Soil Sci.*, 58, 331, 1982.
68. Doughty, J.L., Cook, F.D., and Warder, F.G., Effect of cultivation on the organic matter and nitrogen of brown soils, *Can. J. Agric. Sci.*, 34, 406, 1954.
69. Tiessen, H., Stewart, J.W.B., and Hunt, W.H., Concepts of soil organic matter transformations in relation to organo-mineral particle size fractions, *Plant Soil*, 76, 287, 1984.
70. Gantzer, C.J., Anderson, S.H., Thompson, A.L., and Brown, J.R., Evaluation of soil loss after 100 years of soil and crop management, *Agron. J.*, 83, 74, 1991.
71. Dick, W.A., McCoy, E.L., Edwards, W.M., and Lal, R., Continuous application of no-tillage to Ohio soils, *Agron. J.*, 83, 65, 1991.
72. Slater, C.S. and Carleton, E.A., The effect of erosion on losses of soil organic matter, *Soil Sci. Soc. Proc.*, 2, 123, 1938.
73. Gregorich, E.G. and Anderson, D.W., Effects of cultivation and erosion on soils of four toposequences in the Canadian prairies, *Geoderma*, 36, 343, 1985.
74. Lucas, R.E., Holtman, J.B., and Connor, L.J., Soil carbon dynamics and cropping practices, *Agric. Energy*, 1977.
75. Kruglov, L.V. and Proshlyakov, A.A., Humus replenishment in the plowed soils of the non-Chernozem belt, *Soviet Soil Sci.*, 11, 313, 1979.
76. Rasmussen, P.E., Allmaras, R.R., Rohde, C.R., and Roager, N.C., Jr., Crop residue influences on soil carbon and nitrogen in a wheat-fallow system, *Soil Sci. Soc. Am. J.*, 44, 596, 1980.
77. Bloom, P.R., Schum, W.M., Malzer, G.L., Nelson, W.W., and Evans, S.D., Effect of N fertilizer management of organic matter in Minnesota mollisols, *Agron. J.*, 74, 161, 1982.
78. Paustian, K., Robertson, G.P., and Elliott, E.T., Management impacts on carbon storage and gas fluxes (CO_2, CH_4) in mid-latitude cropland and grassland ecosystems, in *Soil Management and Greenhouse Effect*, Lal, R., Kimble, J., Levine, E., and Stewart, B.A., Eds, (Advances in Soil Science), Lewis Publishers, Boca Raton, FL, in press.
79. Black, A.L., Soil property changes associated with crop residue management in a wheat-fallow rotation, *Soil Sci. Soc. Am. Proc.*, 37, 943, 1973.
80. Horner, G.M., Oveson, M.M., Baker, G.O., and Pawson, W.W., Effect of Cropping Practices on Yield, Soil Organic Matter and Erosion in the Pacific Northwest Wheat Region, Bulletin 1, Agricultural Experiment Stations of Idaho, Oregon and Washington, USDA-ARS, July 1960.
81. Uhlen, G., Effect of nitrogen, phosphorous and potassium fertilizers and farm manure in long-term experiments with rotation crops in Norway, *Ann. Agron.*, 27, 547, 1976.
82. Hooker, M.L., Herron, G.M., and Penas, P., Effects of residue burning, removal, and incorporation on irrigated cereal crop yields and soil chemical properties, *Soil Sci. Soc. Am. J.*, 46, 122, 1982.
83. Sauerbeck, D.R., Influence of crop rotation, manurial treatment and soil tillage on the organic matter content of German soils, in *Soil Degradation*, Boels, D., Davies, D.B., and Johnston, A.E., Eds., Proceedings of the Land Use Seminar on Soil Degradation, Wageningen, October 13–15, 1980, A.A. Balkema, Rotterdam, 1982, 163.
84. Uhlen, G., Long-term effects of fertilizers, manure, straw and crop rotation on total-N and total-C in soil, *Acta Agric. Scand.*, 41, 119, 1991.
85. Saffigna, P.G., Powlson, D.S., Brookes, P.C., and Thomas, G.A., Influence of sorghum residues and tillage on soil organic matter and soil microbial biomass in an Australian vertisol, *Soil Biol. Biochem.*, 21, 759, 1989.
86. Campbell, C.A., Bowren, K.E., Schnitzer, M., Zentner, R.P., and Townley-Smith, L., Effect of crop rotations and fertilization on soil biochemical properties in a thick Black Chernozem, *Can. J. Soil Sci.*, 71, 377, 1991c.
87. Dormaar, J.F., Pittman, U.J., and Spratt, E.D., Burning crop residues: effect on selected soil characteristics and long-term wheat yields, *Can. J. Soil Sci.*, 59, 79, 1979.
88. Biederbeck, V.O., Campbell, C.A., Bowren, K.E., Schnitzer, M., and McIver, R.N., Effect of burning cereal straw on soil properties and grain yields in Saskatchewan, *Soil Sci. Soc. Am. J.*, 44, 103, 1980.
89. Chan, K.Y., Roberts, W.P., and Heenan, D.P., Organic carbon and associated soil properties of a red earth after 10 years of rotation under different stubble and tillage practices, *Aust. J. Soil Res.*, 30, 71, 1992.
90. Pikul, J.L., Jr. and Allmaras, R.R., Physical and chemical properties of a Haploxeroll after fifty years of residue management, *Soil Sci. Soc. Am. J.*, 50, 214, 1986.
91. Prasad, R. and Power, J.F., Crop residue management, *Adv. Soil Sci.*, 15, 205, 1991.

92. Chater, M. and Gasser, J.K., Effects of green manuring, farmyard manure, and straw on the organic matter of soil and of green manuring on available nitrogen, *J. Soil Sci.*, 21, 127, 1970.
93. Anderson, F.N. and Peterson, G.A., Effects of continuous corn (*Zea mays* L.), manuring, and nitrogen fertilization on yield and protein content of the grain and on the soil nitrogen content, *Agron. J.*, 65, 697, 1973.
94. Kofoed, A.D. and Nemming, O., Fertilizers and manure on sandy and loamy soils, *Ann. Agron.*, 27, 583, 1976.
95. Steineck, O. and Ruckenbauer, P., Results of a 70 years long-term rotation and fertilization experiment in the main cereal growing area of Australia, *Ann. Agron.*, 27, 803, 1976.
96. Jenkinson, D.S. and Rayner, J.H., The turnover of soil organic matter in some of the Rothamsted classical experiments, *Soil Sci.*, 123, 298, 1977.
97. Sommerfeldt, T.G., Chang, C., and Entz, T., Long-term annual manure applications increase soil organic matter and nitrogen, and decrease carbon to nitrogen ratio, *Soil Sci. Soc. Am. J.*, 52, 1668, 1988.
98. N'Dayegamiye, A. and Cote, D., Effect of long-term pig slurry and soil cattle manure application on soil chemical and biological properties, *Can. J. Soil Sci.*, 69, 39, 1989.
99. Liang, B.C. and Mackenzie, A.F., Changes in soil organic carbon and nitrogen after six years of corn production, *Soil Sci.*, 153, 307, 1992.
100. Meek, B., Graham, L., and Donovan, T., Long-term effects of manure on soil nitrogen, phosphorous, potassium, sodium, organic matter, and water infiltration rate, *Soil Sci. Soc. Am. J.*, 46, 1014, 1982.
101. Christenson, B.T., Effects of animal manure and mineral fertilizer on the total carbon and nitrogen contents of soil size fractions, *Biol. Fertil. Soils*, 5, 304, 1988.
102. Angers, D.A. and N'Dayegamiye, A.N., Effects of manure application on carbon, nitrogen, and carbohydrate contents of a silt loam and its particle-size fractions, *Biol. Fertil. Soils*, 11, 79, 1991.
103. MacRae, R.J. and Mehuys, G.R., The effect of green manuring on the physical properties of temperate-area soils, *Adv. Soil Sci.*, 3, 70, 1985.
104. Krall, J.L., Army, T.J., Post, A.H., and Seamans, A.E., A summary of dryland rotations and tillage experiments at Havre, Huntley and Moccasin, Montana, *Montana Agric. Exp. Stn. Bull.*, 599, 1965.
105. Sowden, F.J. and Atkinson, H.J., Effect of long-term additions of various organic amendments on the organic matter of a clay and a sand, *Can. J. Soil Sci.*, 48, 323, 1968.
106. Boyle, M., Frankenberger, W.T., Jr., and Stolzy, L.H., The influence of organic matter on soil aggregation and water infiltration, *J. Prod. Agric.*, 2, 290, 1989.
107. Elliott, L.F. and Papendick, R.I., Crop residue management for improved soil productivity, *Biol. Agric. Hortic.*, 3, 131, 1986.
108. Campbell, C.A., Moulin, A.P., Curtin, D., Lafond, G.P., and Townley-Smith, L., Soil aggregation as influenced by cultural practices in Saskatchewan. I. Black Chernozemic soils, *Can. J. Soil Sci.*, 73, 579, 1993.
109. Adem, H.H. and Tisdall, J.M., Management of tillage and crop residues for double-cropping in fragile soils of south-eastern Australia, *Soil Tillage Res.*, 4, 577, 1984.
110. Dormaar, J.F. and Pittman, U.J., Decomposition of organic residues as affected by various dryland spring wheat-fallow rotations, *Can. J. Soil Sci.*, 60, 97, 1980.
111. Emmond, G.S., Effect of rotations, tillage treatments and fertilizers on the aggregation of a clay soil, *Can. J. Soil Sci.*, 51, 235, 1971.
112. Morachan, Y.B., Moldenhauer, W.C., and Larson, W.E., Effects of increasing amounts of organic residues on continuous corn. I. Yields and soil physical properties, *Agron. J.*, 64, 199, 1972.
113. Smika, D.E. and Greb, B.W., Nonerodible aggregates and concentration of fats, waxes, and oils in soils as related to wheat straw mulch, *Soil Sci. Soc. Am. Proc.*, 39, 104, 1975.
114. Halstead, R.L. and Sowden, F.J., Effect of long-term additions of organic matter on crop yields and soil properties, *Can. J. Soil Sci.*, 48, 341, 1968.
115. Baeumer, K. and Bakermans, W.A.P., Zero-tillage, *Adv. Agron.*, 25, 77, 1973.
116. Phillips, R.E., Blevins, R.L., Thomas, G.W., Frye, W.W., and Phillips, S.H., No-tillage agriculture, *Science*, 208, 1108, 1980.
117. Gebhardt, M.R., Daniel, T.C., Schweizer, E.E., and Allmaras, R.R., Conservation tillage, *Science*, 230, 625, 1985.
118. Chaney, K., Hodgson, D.R., and Braim, M.A., The effects of direct drilling, shallow cultivation and ploughing on some soil physical properties in a long-term experiment on spring barley, *J. Agric. Sci., Cambridge,* 104, 125, 1985.
119. Douglas, J.T. and Goss, M.J., Stability and organic matter content of surface soil aggregates under different methods of cultivation and in grassland, *Soil Tillage Res.*, 2, 155, 1982.
120. Hamblin, A.P., Changes in aggregate stability and associated organic matter properties after direct drilling and ploughing on some Australian soils, *Aust. J. Soil Res.*, 18, 27, 1980.
121. Kladivko, E.J., Griffith, D.R., and Mannering, J.V., Conservation tillage effects on soil properties and yield of corn and soya beans in Indiana, *Soil Tillage Res.*, 8, 277, 1986.
122. Prove, B.G., Loch, R.J., Foley, J.L., Anderson, V.J., and Younger, D.R., Improvements in aggregation and infiltration characteristics of a krasnozem under maize with direct drill and stubble retention, *Aust. J. Soil Res.*, 28, 577, 1990.
123. Unger, P.W., Surface soil properties after 36 years of cropping to winter wheat, *Soil Sci. Soc. Am. J.*, 46, 796, 1982.
124. Parker, D.T., Decomposition in the field of buried and surface-applied cornstalk residue, *Soil Sci. Soc. Am. Proc.*, 26, 559, 1962.

125. Brown, P.L. and Dickey, D.D., Losses of wheat straw residue under simulated field conditions, *Soil Sci. Soc. Am. Proc.*, 34, 118, 1970.
126. Andrén, O., Decomposition in the field of shoots and roots of barley, lucerne and meadow fescue, *Swed. J. Agric. Res.*, 17, 113, 1987.
127. Beare, M.H., Parmelee, R.W., Hendrix, P.F., Coleman, D.C., Crossley, D.A., Jr., and Cheng, W., Microbial and faunal interactions and effects on litter nitrogen and decomposition in agroecosystems, *Ecol. Monogr.*, 62, 569, 1992.
128. Bristow, K.L., The role of mulch and its architecture in modifying soil temperature, *Aust. J. Soil Res.*, 26, 269, 1988.
129. Aiken, R., Flerchinger, G., Nielsen, D., Alonso, C., and Rojas, K., Instrumented measure and simulated solutions for the soil energy balance under residues, *Agron. Abstr.*, 8, 1993.
130. Good, L.G. and Smika, D.E., Chemical fallow for soil and water conservation in the Great Plains, *J. Soil Water Conserv.*, 33, 89, 1978.
131. Tindall, T.A. and Crabtree, R.J., Crop Residue Management in Livestock Production and Conservation Systems. II. Agronomic Considerations of Crop Residue Removal, *Research Report P-796*, Oklahoma State University, 1980.
132. Stott, D.E., A tool for soil conservation education, *J. Soil Water Conserv.*, 46, 332, 1991.
133. Doran, J.W., Microbial biomass and mineralizable nitrogen distributions in no-tillage and plowed soils, *Biol. Fertil. Soils*, 5, 68, 1987.
134. Powlson, D.S. and Jenkinson, D.S., A comparison of the organic matter, biomass, adenosine triphosphate and mineralizable nitrogen contents of ploughed and direct-drilled soils, *J. Agric. Sci.*, 97, 713, 1981.
135. Granatstein, D.M., Bezdicek, D.F., Cochran, V.L., Elliott, L.F., and Hammel, J., Long-term tillage and rotation effects on soil microbial biomass, carbon and nitrogen, *Biol. Fertil. Soils*, 5, 265, 1987.
136. Gallaher, R.N. and Ferrer, M.B., Effect of no-tillage vs. conventional tillage on soil organic matter and nitrogen contents, *Commun. Soil Sci. Plant Anal.*, 18, 1061, 1987.
137. Fleige, H. and Baeumer, K., Effect of zero-tillage on organic carbon and total nitrogen content, and their distribution in different N-fractions in Loessial soils, *Agro-Ecosystems*, 1, 19, 1974.
138. Hargrove, W.L., Reid, J.T., Touchton, J.T., and Gallaher, R.N., Influence of tillage practices on the fertility status of an acid soil double-cropped to wheat and soybeans, *Agron. J.*, 74, 684, 1982.
139. Agenbag, G.A. and Maree, P.C.J., The effect of tillage on soil carbon, nitrogen and soil strength of simulated surface crusts in two cropping systems for wheat (*Triticum aestivum*), *Soil Tillage Res.*, 14, 53, 1989.
140. Arshad, M.A., Schnitzer, M., Angers, D.A., and Ripmeester, J.A., Effects of till vs no-till on the quality of soil organic matter, *Soil Biol. Biochem.*, 22, 595, 1990.
141. Campbell, C.A., Biederbeck, V.O., Schnitzer, M., Selles, F., and Zentner, R.P., Effect of 6 years of zero tillage and N fertilizer management on changes in soil quality of an orthic brown Chernozem in southwestern Saskatchewan, *Soil Tillage Res.*, 14, 39, 1989.
142. Dick, W.A., Organic carbon, nitrogen and phosphorus concentrations and pH in soil profiles as affected by tillage intensity, *Soil Sci. Soc. Am. J.*, 47, 102, 1983.
143. Groffman, P.M., Nitrification and denitrification in conventional and no-tillage soils, *Soil Sci. Soc. Am. J.*, 49, 329, 1984.
144. Dick, W.A., Van Doren, D.M., Jr., Triplett, G.B., Jr., and Henry, J.E., Influence of Long-Term Tillage and Rotation Combinations on Crop Yields and Selected Soil Parameters. I. Results Obtained for a Mollic Ochraqualf Soil, Research Bulletin 1180, The Ohio State University, Ohio Agricultural Research and Development Center, Wooster, OH, December 1986a.
145. Dick, W.A., Van Doren, D.M., Jr., Triplett, G.B., Jr., and Henry, J.E., Influence of Long-Term Tillage and Rotation Combinations on Crop Yields and Selected Soil Parameters. II. Results Obtained for a Typic Fragiudalf soil, Research Bulletin 1181, The Ohio State University, Ohio Agricultural Research and Development Center, Wooster, OH, December 1986b.
146. Dalal, R.C., Long-term effects of no-tillage, crop residue, and nitrogen application on properties of a vertisol, *Soil Sci. Soc. Am. J.*, 53, 1511, 1989.
147. Balesdent, J., Mariotti, A., and Boisgontier, D., Effect of tillage on soil organic carbon mineralization estimated from ^{13}C abundance in maize fields, *J. Soil Sci.*, 41, 587, 1990.
148. Havlin, J.L., Kissel, D.E., Maddux, L.D., Claassen, M.M., and Long, J.H., Crop rotation and tillage effects on soil organic carbon and nitrogen, *Soil Sci. Soc. Am. J.*, 54, 448, 1990.
149. Ismail, I., Blevins, R.L., and Frye, W.W., Long-term no-tillage effects on soil properties and continuous corn yields. *Soil Sci. Soc. Am. J.*, 58, 193, 1994.
150. Staley, T.E., Edwards, W.M., Scott, C.L., and Owens, L.B., Soil microbial biomass and organic component alterations in a no-tillage chronosequence, *Soil Sci. Soc. Am. J.*, 52, 998, 1988.
151. Amemiya, M., Conservation tillage in the western Corn Belt, *J. Soil Water Conserv.*, 32, 29, 1977.
152. Karlen, D.L., Conservation tillage research needs, *J. Soil Water Conserv.*, 45, 365, 1990.
153. Riley, H., Børresen, T., Ekeberg, E., and Rydberg, T., Trends in reduced tillage research and practice in Scandinavia, in *Conservation Tillage in Temperate Regions*, Lewis Publishers, Chelsea, MI, 1992.
154. Knight, C.W. and Lewis, C.E., Conservation tillage in the subarctic, *Soil Tillage Res.*, 7, 341, 1986.
155. Boone, F.R., Slager, S., Miedema, R., and Eleveld, R., Some influences of zero-tillage on the structure and stability of a fine-textured river levee soil, *Neth. J. Agric. Sci.*, 24, 105, 1976.

156. Dick, W.A. and Van Doren, D.M., Jr., Continuous tillage and rotation combinations effects on corn, soybean, and oat yields, *Agron. J.*, 77, 459, 1985.
157. Hughes, K.A., Horne, D.J., Ross, C.W., and Julain, J.F., A 10-year maize/oats rotation under three tillage systems. II. Plant population, root distribution and forage yields, *Soil Tillage Res.*, 22, 145, 1992
158. Cook, R.J. and Haglund, W.A., Wheat yield depression associated with conservation tillage caused by root pathogens in the soil not phytotoxins from the straw, *Soil Biol. Biochem.*, 23, 1125, 1991.
159. Griffith, D.R., Kladivko, E.J., Mannering, J.V., West, T.D., and Parsons, S.D., Long-term tillage and rotation effects on corn growth and yield on high and low organic matter, poorly drained soils, *Agron. J.*, 80, 599, 1988.
160. Francis, G.S., Cameron, K.C., and Swift, R.S., Soil physical conditions after six years of direct drilling or conventional cultivation on a silt loam soil in New Zealand, *Aust. J. Soil Res.*, 25, 517, 1987.
161. Coote, D.R. and Malcolm-McGovern, C.A., Effects of conventional and no-till corn grown in rotation on three soils in eastern Ontario, Canada, *Soil Tillage Res.*, 14, 67, 1989.
162. White, P.F., The influence of alternative tillage systems on the distribution of nutrients and organic carbon in some common Western Australian wheatbelt soils, *Aust. J. Soil Res.*, 28, 95, 1990.
163. Carefoot, J.M., Nyborg, M., and Lindwall, C.W., Tillage-induced soil changes and related grain yield in a semi-arid region, *Can. J. Soil Sci.*, 70, 203, 1990.
164. Gantzer, C.J. and Blake, G.R., Physical characteristics of Le Sueur clay loam soil following no-till and conventional tillage, *Agron. J.*, 70, 853, 1978.
165. Hamblin, A.P. and Tennant, D., Effect of tillage on soil water behaviour, *Soil Sci.*, 132, 233, 1980.
166. Tollner, E.W., Hargrove, W.L., and Langdale, G.W., Influence of conventional and no-till practices on soil physical properties in the southern Piedmont, *J. Soil Water Conserv.*, 39, 73, 1984.
167. Edwards, J.H., Wood, C.W., Thurlow, D.L., and Ruf, M.E., Tillage and crop rotation effects on fertility status of a Hapludult soil, *Soil Sci. Soc. Am. J.,* 56, 1577, 1992.
168. Tanchandrphongs, S. and Davidson, J.M., Bulk density, aggregate stability, and organic matter content as influenced by two wheatland soil management practices, *Soil Sci. Soc. Am. Proc.*, 34, 302, 1970.
169. Carter, M.R. and Kunelius, H.T., Comparison of tillage and direct drilling for Italian ryegrass on the properties of a fine sandy loam soil, *Can. J. Soil Sci.*, 66, 197, 1986.
170. Blevins, R.L., Smith, M.S., Thomas, G.W., and Frye, W.W., Influence of conservation tillage on soil properties, *J. Soil Water Conserv.*, 38, 301, 1983.
171. Hooker, M.L., Gwin, R.E., Herron, G.M., and Gallagher, P., Effects of long-term, annual applications of N and P on corn grain yields and soil chemical properties, *Agron. J.*, 75, 94, 1983.
172. Upchurch, W.J., Kinder, R.J., Brown, J.R., and Wagner, G.H., Sanborn Field, Research Bulletin 1054, University of Missouri, Columbia, 1985.
173. Soon, Y.K. and Broersma, K., A study of cropping systems on gray and dark gray luvisols in the Peace River Region of Alberta, Agriculture Canada, Research Station Beaverlodge, Bulletin 86-94, 1986.
174. Insam, H., Parkinson, D., and Domsch, K.H., Influence of macroclimate on soil microbial biomass, *Soil Biol. Biochem.*, 21, 211, 1989.
175. Campbell, C.A., Biederbeck, V.O, Zentner, R.P., and Lafond, G.P., Effect of crop rotations and cultural practices on soil organic matter, microbial biomass and respiration in a thin Black Chernozem, *Can. J. Soil Sci.*, 71, 363, 1991b.
176. Monreal, C.M. and Janzen, H.H., Soil organic-carbon dynamics after 80 years of cropping a Dark Brown Chernozem, *Can. J. Soil Sci.*, 73, 133, 1993.
177. Buyanovsky, G.A. and Wagner, G.H., Post-harvest residue input to cropland, *Plant Soil*, 93, 57, 1986.
178. Paustian, K., Bergström, Jansson, P.-E., and Johnson, H., Ecosystem dynamics, in *Ecology of Arable Land — Organisms, Carbon and Nitrogen Cycling*, Andrén, O., Lindberg, T., Paustian, K., and Rosswall, T., Eds., Chapter 7, Ecological Bulletins, Copenhagen, 40, 153, 1989.
179. van Veen, J.A., Merckx, R., and Van de Geijn, S.C., Plant- and soil-related controls of the flow of carbon from roots through the soil microbial biomass, in *Ecology of Arable Land*, Clarholm, M. and Bergström, L., Eds., Kluwer Academic, Dordrecht, 1989, 43.
180. USDA, Agricultural Statistics 1992, United States Department of Agriculture, National Agricultural Statistics Service, U.S. Government Printing Office, Washington, DC, 1992.
181. Anderson, L.R. and Vasilas, B.L., Effects of planting date on two soybean cultivars: seasonal dry matter accumulation and seed yield, *Crop Sci.*, 25, 999, 1985.
182. Brinkman, M.A. and Rho, Y.D., Response of three oat cultivars to N fertilizer, *Crop Sci.*, 24, 973, 1984.
183. Cox, T.S., Shroyer, J.P., Ben-Hui, L., Sears, R.G., and Martin, T.J., Genetic improvement in agronomic traits of hard red winter wheat cultivars from 1919 to 1987, *Crop Sci.*, 28, 756, 1988.
184. Donald, C.M. and Hamblin, J., The biological yield and harvest index of cereals as agronomic and plant breeding criteria, *Adv. Agron.*, 28, 361, 1976.
185. Meyers, K.B., Simmons, S.R., and Stuthman, D.D., Agronomic comparison of dwarf and conventional height oat genotypes, *Crop Sci.*, 25, 964, 1985.
186. Russell, W.A., Genetic improvement of maize yields, *Adv. Agron.*, 46, 245, 1991.
187. Walker, A.K. and Fioritto, R.J., Effect of cultivar and planting pattern on yield and apparent harvest index in soybean, *Crop Sci.*, 24, 154, 1984.

188. Fox, D.G., Sniffen, C.J., O'Connor, J.D., Russell, J.B., and Van Soest, P.J., The Cornell Net Carbohydrate and Protein System for Evaluating Cattle Diets. I. A Model for Predicting Cattle Requirements and Feedstuff Utilization. Search:Agriculture, Ithaca, NY, Cornell University Agricultural Experiment Station No. 34, 1990.
189. Theander, O. and Åman, P., Anatomical and chemical characteristics, in *Straw and Other Fibrous By-Products as Feed*, Sundstol, F. and Owen, E., Eds., (Developments in Animal and Veterinary Sciences, 14), Elsevier, Amsterdam, 1984, chap. 4.
190. Zielke, R.C. and Christenson, D.R., Organic carbon and nitrogen changes in soil under selected cropping systems, *Soil Sci. Soc. Am. J.,* 50, 363, 1986.
191. Campbell, C.A. and Zentner, R.P., Soil organic matter as influenced by crop rotations and fertilization, *Soil Sci. Soc. Am. J.*, 57, 1034, 1993.
192. Guernsey, C.W., Fehrenbacher, J.B., Ray, B.W., and Miller, L.B., Corn yields, root volumes, and soil changes on the Morrow plots, *J. Soil Water Conserv.*, 24, 101, 1969.
193. Crookston, R.K., Kurle, J.E., Copeland, P.J., Ford, J.H., and Lueschen, W.E., Rotational cropping sequence affects yield of corn and soybean, *Agron. J.*, 83, 108, 1991.
194. Janzen, H.H., Soil organic matter characteristics after long-term cropping to various spring wheat rotations, *Can. J. Soil Sci.*, 67, 845, 1987.
195. Biederbeck, V.O., Campbell, C.A., and Zentner, R.P., Effect of crop rotation and fertilization on some biological properties of a loam in southwestern Saskatchewan, *Can. J. Soil Sci.*, 64, 355, 1984.
196. Ridley, A.O. and Hedlin, R.A., Soil organic matter and crop yields as influenced by the frequency of summer fallowing, *Can. J. Soil Sci.*, 48, 315, 1968.
197. Drover, D.P., The influence of various rotations on coarse-textured soils at Chapman and Wongan Hills Research Stations, western Australia. I. Nitrogen and organic-carbon contents of the soils, *J. Soil Sci.*, 7, 219, 1956.
198. Silvius, J.E., Johnson, R.R., and Peters, D.B., Effect of water stress on carbon assimilation and distribution in soybean plants at different stages of development, *Crop Sci.*, 17, 713, 1977.
199. Sharp, R.E. and Davies, W.J., Solute regulation and growth by roots and shoots of water-stressed maize plants, *Planta*, 147, 43, 1979.
200. Hall, M.H., Sheaffer, C.C., and Heichel, G.H., Partitioning and mobilization of photoassimilate in alfalfa subjected to water deficits, *Crop Sci.*, 28, 964, 1988.
201. Johnston, A.E., Soil organic matter, effects on soils and crops, *Soil Use Manage.*, 2, 97, 1986.
202. Clement, C.R. and Williams, T.E., Leys and soil organic matter. I. The accumulation of organic carbon in soils under different leys, *J. Agric. Sci.*, 63, 377, 1964.
203. Tyson, K.C., Roberts, D.H., Clement, C.R., and Garwood, E.A., Comparison of crop yields and soil conditions during 30 years under tillage or grazed pasture, *J. Agric. Sci., Cambridge* 115, 29, 1990.
204. Kooistra, M.J., Lebbink G., and Brussaard, L., The Dutch programme on soil ecology of arable farming systems. II. Geogenesis, agricultural history, field site characteristics and present farming systems at the Lovinkhoeve Experimental Farm, *Agric. Ecosyst. Environ.*, 27, 361, 1989.
205. Paustian, K., Andrén, O., Clarholm, M, Hansson, A.-C., Johansson, G., Lagerlöf, J., Lindberg, T., Pettersson, R., and Sohlenius, B., Carbon and nitrogen budgets of four agroecosystems with annual and perennial crops, with and without N fertilization, *J. Appl. Ecol.*, 27, 60, 1990.
206. Gupta, U.C. and Reuszer, H.W., Effect of plant species on the amino acid content and nitrification of soil organic matter, *Soil Sci.*, 104, 395, 1967.
207. Davidson, J.M., Gray, F., and Pinson, D.I., Changes in organic matter and bulk density with depth under two cropping systems, *Agron. J.*, 59, 375, 1967.
208. Hobbs, J.A. and Brown, P.L., Effects of Cropping and Management on Nitrogen and Organic Carbon Contents of a Western Kansas Soil, Technical Bulletin 144, Agricultural Experiment Station, Kansas State University of Agriculture and Applied Science, Manhattan, 1965.
209. Haas, H.J., Power, J.F., and Reichman, G.A., Effect of Crops and Fertilizer on Soil Nitrogen, Carbon, and Water Content, and on Succeeding Wheat Yields and Quality, ARS-NC-38, 1976.
210. Granatstein, D.M., Long-Term Tillage and Rotation Effects on Soil Microbial Biomass, Carbon, and Nitrogen, Masters Thesis, Department of Agronomy and Soils, Washington State University, Pullman, 1986.
211. Campbell, C.A., Zentner, R.P., Janzen, H.H., and Bowren, K.E., *Crop Rotation Studies on the Canadian Prairies*, Canadian Publishing Center, Ottawa, 1990, 29.
212. Russell, J.S., Soil fertility changes in long-term experimental plots at Kybybolite, South Australia. I. Changes in pH, total nitrogen, organic carbon, and bulk density, *Aust. J. Agric. Res.*, 11, 902, 1960.
213. Grace, P.R., Ladd, J.N., and Skjemstad, J.O., The effect of management practices on soil organic matter dynamics, in *Soil Biota: Management in Sustainable Farming Systems*, Pankhurst, C.E., Doube, B.M., Gupta, V.V.S.R., and Grace, P.R., Eds., CSIRO, Melborne, 1994, 162.
214. Garwood, E.A., Clement, C.R., and Williams, T.E., Leys and soil organic matter. III. The accumulation of macro-organic matter in the soil under different swards, *J. Agric. Sci.,* 78, 333, 1972.
215. Oades, J.M. and Turchenek, L.W., Accretion of organic carbon, nitrogen and phosphorus in sand and silt fractions of a red-brown earth under pasture, *Aust. J. Soil Res.* 16, 351, 1978.
216. Haynes, R.J., Swift, R.S., and Stephen, R.C., Influence of mixed cropping rotations (pasture-arable) on organic matter content, water stable aggregation and clod porosity in a group of soils, *Soil Tillage Res.*, 19, 77, 1991.

217. Dormaar, J.F. and Smoliak, S., Recovery of vegetative cover and soil organic matter during revegetation of abandoned farmland in a semiarid climate, *J. Range Manage.*, 38, 487, 1985.
218. Rasmussen, P.E. and Rohde, C.R., Long-term tillage and nitrogen fertilization effects on organic nitrogen and carbon in a semiarid soil, *Soil Sci. Soc. Am. J.*, 52, 1114, 1988.
219. Welte, E. and Timmermann, F., Fertilite du sol et bilan de l'azote dans l'essai permanent de fumure < Ewinger Roggenbau > (culture continue de seigle) a Halle/Saale, *Ann. Agron.*, 27, 721, 1976.
220. Timmermann, F. and Welte, E., Effects de la fumure sur les rendements et l'absorption des éléments minéraux par les cultures dans l'essai a long terme du champ < E > de Gottingen, *Ann. Agron.*, 27, 703, 1976.
221. Pratt, P.F., Goulben, B., and Harding, R.B., Changes in organic carbon and nitrogen in an irrigated soil during 28 years of differential fertilization, *Soil Sci. Soc. Proc.*, 21, 215, 1957.
222. Thurston, J.M., Williams, E.D., and Johnston, A.E., Modern developments in an experiment on permanent grassland started in 1856: effects of fertilizers and lime on botanical composition and crop and soil analyses, *Ann. Agron.*, 27, 1043, 1976.
223. Fog, K., The effect of added nitrogen on the rate of decomposition of organic matter, *Biol. Rev.*, 63, 433, 1988.
224. Tempest, D.W. and Neijssel, O.M., The status of Y_{ATP} and maintenance energy as biologically interpretable phenomena, *Ann. Rev. Microbiol.*, 38, 459, 1984.
225. Campbell, C.A., Lafond, G.P., Zentner, R.P., and Biederbeck, V.O., Influence of fertilizer and straw baling on soil organic matter in a thin Black Chernozem in western Canada, *Soil Biol. Biochem.*, 23, 443, 1991a.

Chapter 3

Characterization of Soil Organic Carbon Relative to Its Stability and Turnover

H.P. Collins, E.A. Paul, K. Paustian, and E.T. Elliott

CONTENTS

I. INTRODUCTION

A wide variety of techniques have been applied to the measurement and characterization of soil organic matter (SOM) in relation to its dynamics under various management conditions. The selection of methods

0-8493-2802-0/97/$0.00+$.50

to describe SOM dynamics depends upon the purpose of the study, be it for chemical characterization and identification of specific SOM components, the description of SOM pools important in the cycling and release of plant nutrients, or the quantification of ecosystem carbon (C) budgets. Classical approaches have combined chemical extractions with identification of specific chemical compounds. Functional approaches have attempted to provide a description of SOM pool dynamics by incorporating radio- and stable isotopes as tracers or using ^{14}C dating techniques to identify specific fractions that are biologically active. Stevenson and Elliott[1] stressed that methods used to evaluate SOM dynamics should be related to fractions or pools that have biological significance if they are to relate to the potential for soils to provide nutrients to plants. In the following we discuss methods that permit functional descriptions of SOM transformations relative to ecosystem functioning, biodegradation, soil fertility, and global change.

II. TOTAL SOIL ORGANIC MATTER: ORGANIC CARBON

Estimates of SOM are derived primarily from the determination of total organic C since it comprises 48 to 58% of SOM mass.[2] The most commonly used analytical procedures involve dry combustion or wet digestion. In dry combustion, total C is determined by burning soil in a stream of pure O_2 in a resistance or induction furnace; CO_2 is determined by titration or thermal conductivity. In wet combustion, soil is digested in the strong oxidant $K_2Cr_2O_7$ and a 3:1 mixture of H_2SO_4 and H_3PO_4. Oxidizable C is determined by either titration of excess $K_2Cr_2O_7$ or the measurement of liberated CO_2 trapped in 1 *N* NaOH, which is then titrated with HCl after addition of $BaCl_2$.[3] For a complete description of the materials and methods involved in total C determination see the review of Nelson and Sommers.[2]

Recent improvements in automated instruments such as the Carlo Erba CHN* (Carlo Erba Strumentazione, Milan), LECO CHN-600 (Laboratory Equipment Corp., St. Joseph, MO), or Perkin-Elmer total C and N analyzers have increased the precision and accuracy of determining total soil C. Although the initial cost of this type of equipment is high, large numbers of samples can be run in a shorter period of time, compared to wet oxidation.[4] Sheldrick,[5] in comparing the LECO CHN-600 to wet oxidation, concluded that the automated method was preferable because it provided more reliable and precise data and was more efficient since both total C and N could be determined faster than C alone in a single analysis. The small sample size often associated with automated techniques requires very fine grinding of soil. Fine grinding of soil to particle sizes ranging to below 0.08 mm has been reported using hammer-, ball-, or roller-mill grinders.[6–8] Multisample conveyor-belt grinders that can be built with normal laboratory equipment and supplies have been described by Smith and Um[9] and Kelly.[10]

III. ISOLATION AND FRACTIONATION OF SOIL ORGANIC MATTER

A. CHEMICAL EXTRACTION AND PURIFICATION

Classical chemical fractionation of SOM yields three major fractions: humic acids (HA), fulvic acids (FA), and humin.[11] This technique is based upon differences in solubilities of humic substances in alkaline and acid solutions. The technique involves extraction with an alkaline reagent, usually NaOH or $Na_4P_2O_7$, separation of the soluble extract from the soil residue, following acidification of the extract to pH 2 with HCl (Figure 1A). FAs are soluble in both alkali and acid, HAs are soluble in alkali but precipitated by acid, and humin is insoluble in both alkali and acid. The fractions should not be considered distinct or discrete compounds since each can be further purified to reduce heterogeneity.[12,13] Common techniques used to purify humic extracts include electrophoresis, electro-ultra filtration, ion exchange, gel chromatography, salting out, changes in pH, and further use of differences in solubility. Humic substances differ in molecular weight, elemental composition, acidity, and cation exchange capacity (Figure 1B).

Fulvic extracts are typically composed of a variety of phenolic and benzene carboxylic acids that are held together by hydrogen bonds to form stable polymeric structures.[14] These are associated with polysaccharides that are easily separated by adsorption on charcoal or gel and resin chromatography. The low molecular weight FAs have higher oxygen but lower C contents than HAs and contain more acidic functional groups, particularly COOH.[15] The HA fraction consists of hydroxyphenols, hydroxybenzoic acids, and other aromatic structures with linked peptides, amino compounds, and fatty acids.

* Trade names and company names are included for the benefit of the reader and do not imply any endorsement or preferential treatment of the product listed.

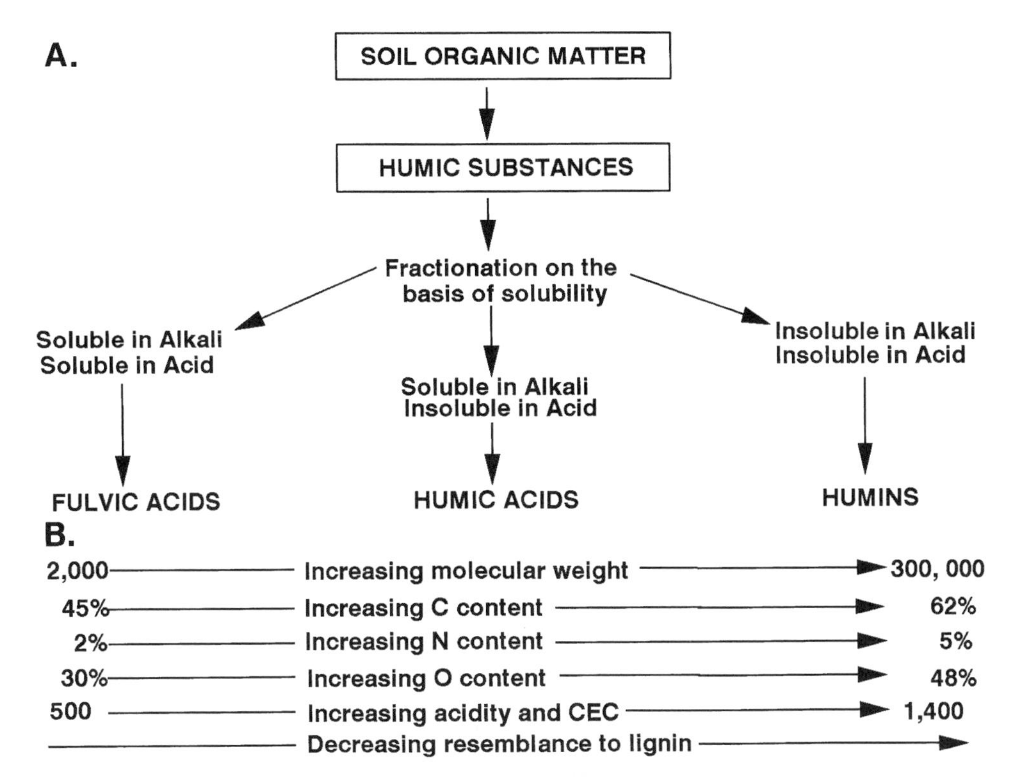

Figure 1 (A) Classical fractionation scheme of SOM and (B) characteristic properties of humic substances.[1,11,13]

Humic extracts are generally contaminated with silicates that can be removed with HF. Proteins and carbohydrates can be removed using 6 *M* HCl.

Soil humins are considered to be nonextractable humic type polymers that form strong associations with the mineral fraction and are not as easily separated by the usual alkaline reagents.[14] Hatcher et al.[16] suggested that the humin fraction consisted of highly condensed HAs, fungal melanins, and paraffinic structures. Almendros and Gonzalez-Vila[17] found a high proportion of polymethylene compounds (fatty acids) that seem to be inherited from the waxes of higher plants and further suggested that, during biodegradation, the lipid polymers are altered and incorporated into the humin fraction. Although HAs and humins constitute the majority of organic C in a system, they contribute only a small portion to the annual cycling of soil C because of their extremely slow turnover rate. The aromatic HAs acids are considered to be very stable in soil,[18] corresponding to the chemical stabilized organic matter postulated by Jenkinson and Rayner[19] and van Veen and Paul.[20] FAs have turnover times of hundreds of years, whereas those HAs and humins approach several thousand years.[21]

An alternative to NaOH or $Na_4P_2O_7$ extraction involves acid hydrolysis. This simple rapid technique provides an acid-soluble fraction ranging from 25 to 50% of the soil C and a nonhydrolyzable fraction constituting 50 to 75% of the C. Experiments with ^{14}C involving both C dating and enriched samples show that there are some artifacts, e.g., modern lignin ends up in the non-hydrolyzable plant fraction. However, there still is usually a difference of 1000 years in the C age of the soluble and much older acid-soluble fractions giving an estimate of the size and tracer age of the resistant old fractions.

B. HUMUS FORMATION

The prevailing theory of humus formation is based upon a multistage process involving (1) decomposition of plant material to simple C compounds; (2) assimilation and repeated cycling of C through the microbial biomass with formation of new cells; and (3) concurrent polymerization of microbial synthesized polyphenols (quinones) and alteration of plant-derived lignin to form high molecular weight polymers (Figure 2).[15,22,23]

Plant material is the primary source of material for SOM formation. Decomposition products are incorporated into various SOM fractions. In general, water-soluble C compounds (simple sugars, proteins) degrade first, followed by structural polysaccharides (cellulose and hemicelluloses), and then

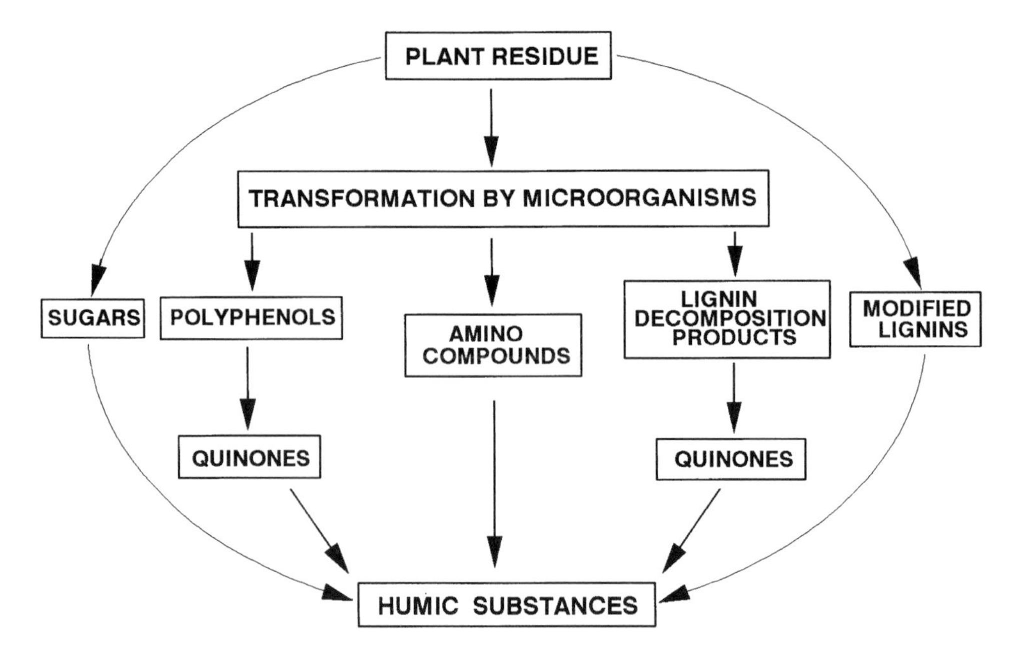

Figure 2 Transformation processes during the formation of humic substances.[11,13]

lignin, which decomposes at a much slower rate. Initially, between 40 and 60% of the plant C is assimilated by the microbial biomass, which itself is subject to biodegradation and transformation. For many crop residues in temperate climates, 50 to 70% of the original C is lost as CO_2 after 1 year. Of the residual soil C 10 to 20% is microbial biomass and the remainder has become incorporated into new SOM.[24] Approximately 20% of the residual C is associated with the HA fraction in the form of peptides and polysaccharides. The major portion of residual plant lignin C is associated with aromatic complexes.[25] Stott et al.[26] found that the majority of polysaccharide and protein C in wheat (*Triticum aestevum*) residues became associated with the FA fraction of new organic matter. From 36 to 54% of the C derived from wheat straw lignin was found in the HA fraction. Corn (*Zea mays)* residues underwent similar incorporation into new humus fractions.

C. WHOLE SOIL ANALYSIS

A difficulty with the chemical extractions of SOM and fractionation in HA, FA, and humin is that they are tedious and labor intensive and not suitable for large numbers of samples.[27] New approaches that include ^{13}C-nuclear magnetic resonance (NMR) spectroscopy, solid-state cross-polarization with magic angle spinning (CP/MAS-NMR), and pyrolysis-soft field ionization mass spectrometry (Py-FIMS) have been successfully applied to the study of in situ SOM in a number of soils.[27–30] The ^{13}C-NMR spectrum provides specific information on the chemical structures involving ^{13}C atoms within a molecule. The C skeleton of humic materials is observed rather than the adjacent protons, allowing the functional groups to be detected. Carbon nuclei are spread over a wide range of chemical shifts that effectively separate signals even when carbons have only small differences in diverse structural environments.[31] Carbon structures are determined in relative terms from the chemical shifts that occur when energy is absorbed by a molecule spinning in a magnetic field.[31] The types of C which can be detected by ^{13}C-NMR spectroscopy are presented in Table 1. The chemical shift is expressed as parts per million (ppm). The intensity of the signal detected and the spectral quality of that signal (signal:noise ratio) are dependent upon the amount of ^{13}C present in the sample and the concentration of the ^{13}C nucleus present among the wide range of C structures that are found in soil.[31,32] The presence of paramagnetic ions also reduces the efficiency of signal acquisition. The major paramagnetic ion in soil is Fe^{3+} but other transition metal cations may cause signal reduction if present in high concentrations. Baldock et al.[32] reported that the concentration of the ^{13}C nuclei could also be increased by incubating soil with ^{13}C-labeled glucose and following changes in the chemical structure of the substrate carbon followed as it was decomposed by the soil microflora and incorporated into SOM. Oades et al.[33] found that fractionating soils on the basis of particle size and density concentrated organic C and improved the signal quality of solid-state

^{13}C-NMR spectra. Solid state ^{13}C-NMR spectra represents the sum of all organic structures in the sample but cannot differentiate among litter, partly humified, and humified organic matter. It is also not possible, using unfractionated soil samples, to distinguish between structures that are associated to some degree with inorganic components in the soil and those that are not.

Table 1 Types of Carbon which Can Be Detected by ^{13}C-NMR Spectroscopy in Humic Materials and the Corresponding Chemical Shifts

	(ppm)
Aliphatic C (alkanes + fatty acids)	0–40
Protein C, peptide C, amino acid C, C in OCH3	41–60
Carbohydrate C	61–105
Aromatic C	106–150
Phenolic C	151–170
Carboxylic C (total acidity)	171–190
Carbonyl C	210–230
Aliphaticity (%)	$\frac{(0\text{–}105) \times 100}{(0\text{–}170)}$
Aromaticity (%)	$\frac{(106\text{–}170) \times 100}{(0\text{-}170)}$

From Wilson, M.A., *NMR Techniques and Applications in Geochemistry and Soil Chemistry*, Pergamon Press, Oxford, 1987. With permission.

In applications of Py-FIMS, the sample is pyrolyzed under vacuum directly in the ion source of the mass spectrometer, and the volatile components identified by field ionization (FI) mass spectra. The advantages of Py-FIMS and descriptions of the methodology have been summarized by Schulten.[34] Schnitzer and Schulten[35] reported that the dominant fractions obtained from whole soil analysis include carbohydrates, phenols, lignin monomers and dimers, and, to a lesser extent, *n*-fatty acids. Minor components included *n*-alkyl monoesters and diesters, *n*-alkenes, *n*-alkylbenzenes, and N-compounds. It is noteworthy that both Py-FIMS and ^{13}C-NMR provide similar results. In the near future, development of quantitative analyses will greatly improve the application of the information gathered by these techniques to changes in soil C resulting from land management. These methods applied to whole soils avoid structural alterations which occur during extractive methods. They have given useful empirical estimates of the degree of humification as well as suggesting some fundamental aspects of organic matter structure.

IV. PHYSICAL FRACTIONATION OF SOIL ORGANIC MATTER

Although chemical extractions have been routinely used to characterize the chemical structure and composition of SOM, they have not been very useful in identifying specific SOM pools that diminish upon intensive management. The physical occlusion or "protection" of organic materials, restricting the accessibility of microorganisms and enzymes, is believed to be an important mechanism controlling SOM turnover.[18,36,37] Physical fractionation of soil has been more useful in making these distinctions.[38–40] Physical separations address this issue by yielding information on the distribution and concentration of SOM within different parts of the soil matrix. The distribution of organic matter within physical fractions of soil can be measured by disrupting soil structure followed by the separation of fractions by either particle size or density gradients (Table 2).

A. METHODS OF DISRUPTION

Mechanical disruption of soil particles is commonly achieved using either sonication or shaking.[41,42] The principle behind sonic dispersion is the transmission of vibrating sound waves that create microscopic bubbles, which upon collapse produce a high energy of cavitation.[43,44] At this time there are no standard protocols for sonic dispersion. The duration of ultrasonic treatment required to achieve complete dispersion will vary among soil types, soil/water ratio, soil-suspension volume as well as probe dimensions and power output intensity. Christensen[42] stressed that the optimum treatment time should be determined for each soil under study to ensure that dispersion was maximized. The minimum energy at which

Table 2 Comparison of Methods for Physical Disruption and Separation of SOM

Method	Principles	Size Fractions	Advantages	Disadvantages
		Disruption		
Sonication	Vibrating sound waves create microscopic bubbles that upon collapse produce a high energy of cavitation that disrupts the bonds of soil aggregates	Particle size distribution	Minimizes chemical transformations	Potential for redistribution of organic matter among size/density fractions; currently no standard protocol
Shaking	Used principally with chemical extraction of whole soil; apparatus include end-over-end, reciprocal, wrist action, and rotary shakers	Particle size distribution	Simple, a wide range of disrupting energies can be obtained	Soil samples having high sand contents increase the rate of disruption compared to heavier textured soils
Chemical	Treatment is commonly used prior to disruption; some chemical dispersants can selectively solubilize or oxidize various OM binding aggregates, e.g., periodate-tetraborate specific for carbohydrates	Particle size distribution	Useful in understanding function of various fractions of organic matter	Some components of organic matter are modified through chemical transformations, solubilization, or oxidation
		Separation		
Dry Sieving	Separates soil particles based strictly on size and used primarily in aggregate separations of nondisrupted soil	Particles >50 μm		Sieve abrades aggregates resulting in an average aggregate size below that of the field
Wet Sieving	Separation of aggregates into various sizes through a nest of sieves under water	<2 mm	Useful index of aggregate stability	
Sedimentation	Separates soil particles based on equivalent spherical diameter and used most often with a disruption pretreatment	<1 mm		Major limitation is that aggregates >1 mm in diameter settle too rapidly to be measured accurately
Density	Separates particles based upon the weight per unit volume, independent of size and shape and is used to separate light and heavy fractions; requires centrifugation when working with fine particles	Particle size-fractions with density gradients from 1.6 to 2.2 mg m^{-3}		

cavitation occurs should be used if minimal disruption of soil structure is desired. Low energy levels are difficult to control and require careful calibration. Although seldom done, the sonicator should be calibrated so that results among investigators can be compared. The energy emitted by the probe can be estimated by measuring the downward energy of the probe on a sensitive balance. Christensen[45] showed that a sonication time of 15 minutes and 695 J ml^{-1} (300 W) was generally sufficient for maximal dispersion in three sandy soils (Figure 3). Clay yield increased and silt decreased with increased sonication time, indicating the progressive disruption of silt-sized particles. One of the greatest problems with the use of sonication in SOM studies is the potential for redistribution of organic matter among size/density fractions. Increasing the intensity of sonication results in the recovery of increasing amounts of organic matter in the fine silt and clay fractions. However, low levels of sonication provides incomplete dispersion of soils, causing microaggregates of smaller size particles to be included in silt and sand size separates.

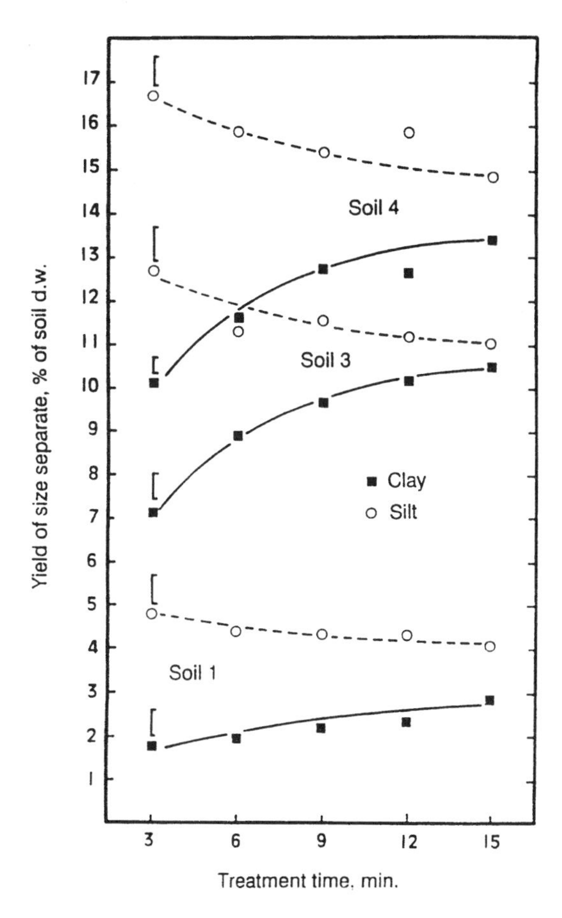

Figure 3 Influence of ultrasonic treatment on yields of clay (<2 μm) and silt (2 to 20 μm) from three Danish soils. (From Christensen, B.T., *Adv. Soil Sci.*, 20, 2, 1992. With permission.)

Shaking is a less intense alternative dispersion method to sonication. Typical apparatus used are end-over-end, reciprocal, wrist action, and rotary shakers.[42] The duration of shaking and addition of chemical dispersants contribute to the uniformity with which soil is disrupted. Introducing glass beads or agate balls enhances the dispersion effect during shaking. The use of chemical dispersants during physical disruption may create changes in some SOM components through chemical transformations, solubilization, or oxidation.

B. METHODS OF SEPARATION

The physical fractionation of soil is generally accomplished using dry or wet sieving, sedimentation, and/or density separations. Dry sieving separates soil particles based upon size and has been used primarily in separating large aggregates of relatively undisturbed soil samples. Dry sieving is often used for wind erosion studies to determine particle sizes susceptible to wind erosion.[46] The duration of sieving, abrasion, and impact forces influences the final size distribution of aggregates. An indication of how the sieving procedure has affected the aggregate size distribution can be obtained by running the soil through the sieve a second time. Comparison of the amounts of soil in the various aggregate sizes after the first and second sieving provides an estimate of how much the abrasion has changed the aggregate distribution. Wet sieving is used more often than dry to determine particle size distribution and stability of soil aggregates. For primary particle size separations, soil is generally dispersed using sodium hexametaphosphate and NaOH to raise the pH and complex bridging ions to further enhance dispersion of fine particles. Sand fractions are separated on sieves, with silts and clays separated by sedimentation.

Fractionation using sedimentation separates particles that can vary in size, shape, or density based upon their equivalent spherical diameter. Sedimentation dynamics are described by Stokes Law which assumes (1) that terminal velocity of each particle is reached as soon as settling begins, (2) settling and resistance are due to the viscosity of the liquid, (3) particles have similar density and are spherical and smooth, and (4) there is no interaction between settling particles.[43] Density fractionation has historically relied on organic liquids, but aqueous solutions of inorganic salts have become increasingly popular. The use of inorganic liquids is essential where the organic matter content of particle size fractions needs to be determined. The most frequently used organic liquids are tetrabromoethane (2.96 g cm^{-3}), bromoform (2.88 g cm^{-3}), and tetrachloromethane (1.59 g cm^{-3}). Inorganic liquids include polytungstate, sodium iodide (NaI), and sodium bromide (Na_2Br).

Edwards and Bremner[47] proposed that a large portion of primary soil particles (sand, silt, and clay) and organic matter are contained in microaggregates consisting of clay and humified matter linked by polyvalent metals. Tisdall and Oades[38] and Oades[48] present a conceptual model of the association of SOM and soil structure based upon a description of three physical fractions that exist in mineral soils: free primary particles, microaggregates, and macroaggregates. Aggregates are bound together by three types of cementing agents: (1) transient, comprising microbial and plant derived polysaccharides; (2) temporary, which include roots and fungal hyphae; and (3) persistent, aromatic humic material associated with Fe and Al, and other polyvalent metal cations. The metals act as clay-organic matter and organic matter-organic matter bridges. Persistent organic matter binds primary minerals into microaggregates (0.002 to 0.020 mm) which are stable to sonication. These, in turn, are bound by additional persistent organic matter to form microaggregates 0.02 to 0.25 mm in size which disintegrate with sonication but remain intact over long periods of cultivation. Tisdall and Oades[38] further suggested that microaggregates are bound together into macroaggregates (0.25 to 2.0 mm) by transient and temporary organic matter.

Elliott[39] and Elliott and Coleman[49] modified the micro-/macroaggregate model by suggesting that the matter held within macroaggregates has an intermediate turnover time and it is this pool which makes up the bulk of the organic matter lost due to cultivation. As physical disruption energy increases, particle distributions shift to contain smaller particles.[39,50,51] The organic matter associated with the sand fraction released during sonication and sedimentation is composed mostly of the light fraction, i.e., undecomposed organic matter of plant origin.[36,52–54]

Organic matter associated with different primary particles has been shown to have varying rates of turnover.[36,52,55] There is a large body of literature supporting both the quantitative importance and the resistant nature of the SOM associated with the silt/coarse clay fraction and sometimes faster turnover rate of the fine clay separates obtained from sonication and sedimentation.[52,56–60] This, however, may be somewhat an artifact of the fractionation technique in which labile microbial constituents and other compounds become adsorbed upon the fine clays during fractionation. In several studies, Tiessen et al.[61,62] found that the greatest loss of C following 60 years of cultivation was associated with the sand fraction (light fraction). Some of the loss may have been the consequence of the disintegration of aggregates resulting in an accumulation of SOM in finer size fractions. The proportion of SOM in smaller particle-size fractions increased with years of cultivation.

Buyanovsky et al.[63] tracked the distribution of ^{14}C originating from labeled soybean [*Glycine max* (L.) merr.] residues, among vegetative fragments (2 to 0.2 mm and 0.2 to 0.053), the fine silt (25 to 2 μm), clay (<2 μm) fractions, and within natural aggregates (2 to 1, 1 to 0.5, 0.5 to 0.25, and <0.25 mm) over a 4-year period on Sanborn Field, Columbia, MO. Following a combination of wet sieving and ultrasonication, Buyanovsky et al.[63] compared residence times of each fraction to values of the C pools presented in C cycling models of Jenkinson and Rayner[19] and Parton et al.[64] (Table 3). Theoretical pools I and II are characterized by the quality of plant material, and were suggested to be associated with the vegetative fractions >0.053 mm, with turnover times ranging between 0.5 to 2 years. Pool III represents the soil microbial biomass with a turnover time ranging from 1 to 10 years, which encompassed aggregate sizes ranging from nonaggregated to 2 mm with a turnover spanning 1 to 7 years. Turnover times for pools IV and V were obtained from estimates derived by Balesdent et al.[59] and Hsieh.[81,82] These results provide a link between the physical soil fractions and accepted conceptual models describing soil C dynamics.

Table 3 Comparison of Mean Residence Times (Years) of C in Theoretical Pools of SOM and in Soil Physical Fractions

Pool	Jenkinson and Rayner[19]	Parton et al.[64]	Physical Fractions; Buyanovsky et al.[63]
I	Decomposable plant material, 0.24	Metabolic plant residues, 0.5	Vegetative fragments (2–0.2 mm), 0.5–1
II	Resistant plant material, 3.33	Structural plant residues, 3.0	Vegetative fragments (>0.053), 1–2
			Vegetative fragments (0.053–0.025 mm), 2–3
			Macroaggregates (2–1 mm), 1–4
III	Soil biomass, 2.44	Active soil C, 1.5–10	Aggregates (1–0.5 mm), 2–10
			Aggregates (0.5–0.1 mm), 3–10
			Nonaggregated soil, 7
IV	Physically stabilized, 72	Slow soil C, 25–50	Fine silt (internal), 40
V	Chemically stabilized, 2857	Passive soil C, 1000–1500	Fine clay (internal), 1000

From Buyanovsky, G.A., Aslam, M., and Wagner, G.H., *Soil Sci. Soc. Am. J.*, 58, 1167, 1994. With permission.

Christensen and Sorensen[65] reported that after 5 to 6 years of cultivation the highest concentration of C was associated with the clay fraction (66 to 84%), whereas silt accounted for 4 to 19%, and sand less than 2%. The decrease in labile but physically protected organic matter, resulting from aggregate breakdown, reduces the ability of the SOM to supply plant nutrients via mineralization.[66] Fine clay mineral colloids are important in protecting labile SOM, thereby slowing its transformation to recalcitrant forms or loss as CO_2 through decomposition.[18,56] Physical protection of organic matter results from surface adsorption on a molecular scale or by occlusion within microaggregates.[37,67]

Elliott[40] compared aggregate size distributions of native and cultivated soils and found that, as a result of cultivation, the organic matter which was readily lost was that which normally bound organic matter-poor microaggregates into organic matter-rich macroaggregates. He proposed that, if macroaggregates are composed of microaggregates that contain more humified organic matter, plus interaggregate material that has less humic character, then there should be differences in organic matter chemistry between micro- and macroaggregates and that an important pool of organic matter is the glue that cements microaggregates into macroaggregates. Higher concentrations of organic C and N in macro- compared to microaggregates have been reported.[50,51,68,69] Dormaar[51] found greater concentrations of recalcitrant organic matter (resin extractable C), higher HA:FA ratios, and less polysaccharide, polyuronide, and phenol in microaggregates than macroaggregates in cultivated soil.

C. DENSITY FRACTIONS OF ORGANIC MATTER

Based on floatation in high specific gravity liquids, SOM can be divided into two broad categories: (1) a light fraction (LF) consisting of relatively mineral-free, incompletely decomposed plant and animal debris, and associated microorganisms, and (2) a heavy fraction (HF) consisting of organic matter adsorbed to mineral surfaces or contained within organomineral microaggregates.[70]

The LF of soil is usually defined as having a density ranging from 1.5 to 2.0 g cm^{-3}, a relatively wide C:N ratio, and is biologically active with a rapid turnover rate. The HF has a narrow C:N ratio but decomposes more slowly than the LF.[21,43,71] The relatively rapid mineralization rate of LF may be related to the labile nature of its constituents and to the lack of protection by soil colloids. Janzen et al.,[72] in evaluating the LF as a measure of labile SOM, studied three long-term rotations in western Canada. The LF comprised 2 to 17% of the soil organic C and was highest where soils were in perennial forages or continuously cropped and lowest where bare fallow was part of the management. The respiration rate and microbial biomass N were highly correlated to the LF content. N-mineralization varied somewhat because of some immobilization during decomposition.

Cambardella and Elliott[42] isolated a particulate organic matter (POM) fraction that they suggest corresponds to the characteristics of the intermediate or slow SOM pools described in SOM simulation models.[20,73,74] POM can be isolated by shaking 10 g of soil in 30 ml of 0.5% sodium hexametaphosphate for 15 minutes. The dispersed soil is passed through a 53-µm sieve and rinsed several times with water. The POM fraction retained on the sieve and mineral-associated organic matter passing through the sieve are collected, then dried in a forced air-oven at 50°C overnight, and total organic C from each determined using either dry or wet oxidation techniques. The difference between the C value of the material passing through the screen and that determined from a nondispersed sample is equal to the C retained on the sieve, i.e., POM. Since the sand fraction is removed from the mineral-associated fraction during dispersion, a correction factor is used to obtain a per gram soil percentage.

Cambardella and Elliott[42] compared 20 years of bare fallow, stubble mulch and no-till management with native prairie. Wheat cropping reduced the POM-C fraction from 40 to 18% for plowed wheat, 19% for stubble mulch, and 25% for no-till. Stable C isotope composition of the POM isolated from the no-till and stubble-mulch treatments showed that 87% of the POM-C was derived from the original grassland with only 13% derived from the wheat crop in the plowed treatment. The percent of C derived from wheat increased to 31% under no-till management. Scanning electron microscopy indicated that the POM fraction was composed mostly of root fragments in various stages of decomposition.

D. PROTECTION OF SOIL ORGANIC MATTER WITHIN AGGREGATES

SOM is derived either directly from plant material or from microbial products derived secondarily from the processing of plant material. These materials may be physically protected from microbial access within soil aggregates, be they labile or recalcitrant. In either case the SOM mineralization rate may be reduced. If labile substrates are released from physical protection, mineralization may be rapid, whereas mineralization may be slow where recalcitrant SOM is released from the disruption of aggregates. For these reasons, physical fractionation followed by chemical or biological assays of lability is a useful approach to the study of SOM dynamics. A modified version of the hierarchical aggregate hypothesis of Tisdall and Oades[38] can explain many aspects controlling the rate of SOM turnover in many soils from temperate regions.[75]

POM is a significant component of macroaggregate SOM.[41] Analysis of differences in C and N content and C/N of macroaggregates from different tillage treatments suggests a strong relationship between the stability of the aggregates and the presence of POM.[69] In addition, it was suggested that a greater proportion of wheat-derived POM was found in more intensively cultivated soils. Beare et al.[76] found that POM-N comprised a greater proportion of aggregate N in no-till compared with conventional cultivations and values increased with increasing aggregate size. They suggested that large microaggregates (106 to 250 µm) were formed in conjunction with POM at the center of macroaggregates, as proposed earlier by Oades.[48]

Crushing aggregates releases SOM that is otherwise less available as substrate for microorganisms.[40,77] It is unclear whether this released SOM is of microbial or plant origin but both may be physically protected. Disruption of aggregates from no-till increased mineralization by almost 20% but had much less effect (5 to 10%) in conventionally tilled soils.[78] The effect of crushing on mineralization may be reduced by storage of air-dried soils.

An enriched labile fraction was isolated using a combination of wet sieving to obtain macroaggregates, gentle sonication, particle size separation (2 to 20 µm), and densitometry (2.07 to 2.22 g cm^{-3}).[79] It comprised a significant portion of the total SOM (e.g., 25% of total N in no-till soil) and was considerably more labile than the aggregates from which it was isolated. This observation supports the suggestion that organic matter between microaggregates within macroaggregates is labile but physically protected.[40] Interestingly, soil containing the lowest proportion of this fraction was from native grassland, where a much larger proportion of total SOM was isolated in the POM fraction.

V. TRACER USE IN SOIL ORGANIC MATTER STUDIES

A. CARBON DATING

The bombardment of ^{14}N by cosmic radiation in the atmosphere produces a low level of ^{14}C which when incorporated into plant tissue by photosynthesis provides a long-term tracer for SOM dating (^{14}C half-life is 5568 years). Radiocarbon dating using naturally occurring ^{14}C has become an important tool in describing SOM dynamics by providing the opportunity to study the distribution of naturally occurring ^{14}C in soils.[80–82] It is particularly useful in characterizing the age of the more resistant fractions of SOM.[83] Above-ground testing of thermonuclear devices in the 1950s and 1960s resulted in the temporary enrichment of atmospheric radiocarbon (Figure 4). These explosions released neutrons that reacted with atmospheric N_2 to produce ^{14}C which also entered the terrestrial carbon cycle.[19,84] Although atmospheric testing was suspended in the 1970s the biosphere maintains a ^{14}C activity about 30% above 1950 values but continues to fall, producing a measurable signal that allows determination of turnover rates in the important 30- to 60-year time interval to be measured[85] (Figure 4).

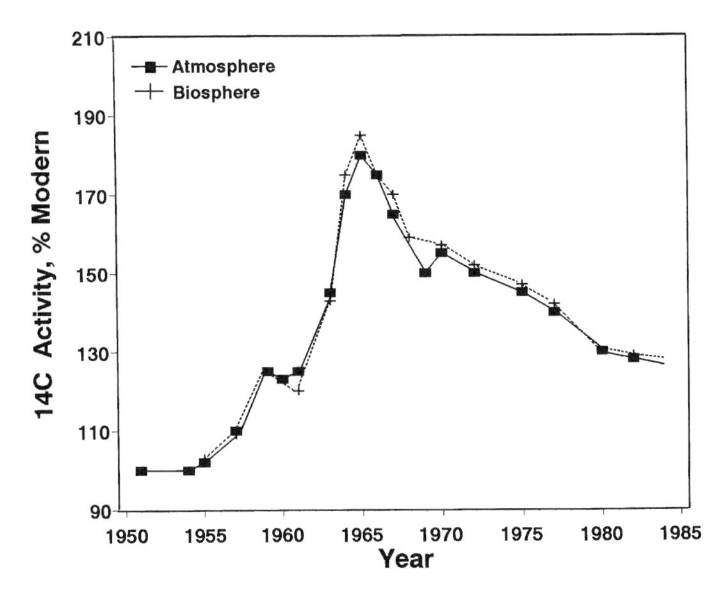

Figure 4 The mean annual specific ^{14}C activity in the atmosphere and biosphere of the Northern Hemisphere, 1950 to 1981. (From Hsieh, Y.-P., *Soil Sci. Soc. Am. J.*, 57, 1020, 1993. With permission.)

Following incorporation of recent plant material into the soil, ^{14}C is diluted by older soil C so that the enriched ^{14}C present in old SOM comprises a smaller fraction of the total C than in the plant material. Soil organic C is a mixture of recently added plant debris and the associated biota and very much older humic materials associated with mineral particles. Therefore, absolute ages of SOM of present-day soils are difficult to determine. The terms radio carbon age or mean residence time (MRT) are most often used. Table 4 provides representative MRTs of total SOM and organic matter fractions for several soils of the Northern Great Plains. Paul et al.[86] found half-lives of 250 to 1900 years for SOM of western Canadian soils. Estimates of the MRT of organic matter in Mollisols of the Central Great Plains range from 2000 to 7000 years.[87]

Martel and Paul[88] measured the ^{14}C content of field soils and compared them to the ^{14}C distribution in a soil incubated with ^{14}C-acetate. Carbon dating showed the MRTs for the surface horizon was 350 years but varied from modern to 1910 years for different chemical fractions. Their work also indicated that SOM in surface horizons turned over more rapidly than lower horizons. Jenkinson and Rayner[19] constructed a model describing the turnover of SOM from soils under long-term management at the Rothamsted Experiment Station in Harpenden, U.K. Using a combination of ^{14}C-labeled plant material and radiocarbon dating of SOM they were able to follow the flux of C through five SOM pools (decomposable plant material, resistant plant material, soil microbial biomass, physically protected, and chemically stabilized organic matter). Ages of these pools ranged from several months for plant materials to several thousand years for chemically and physically stabilized SOM fractions. Campbell et al.[89] estimated compartments for the labile and stable C pools to have turnover rates of 53 and 1429 years, respectively.

Hsieh[81] proposed that the turnover of soil organic C could be described by two major pools, an active pool with a turnover time of less than a few decades and a stable pool exhibiting a considerably longer turnover time. This was based on the assumptions that cultivation causes significant loss in the active but not stable, and that bomb effects apply primarily to active, and not to stable SOM pools. Comparisons between the Morrow plots (Urbana, IL) and the Sanborn Field (Columbia, MO) showed the mean age

Table 4 Representative Mean Residence Times of SOM and Organic Matter Fractions

Site Description	Mean Residence Time[a] (years)	Ref.
Saskatchewan, Canada — Mollisol	1000	86
Saskatchewan, Canada		89
Mollisol	870 ± 50	
Alfisol	250 ± 60	
Saskatchewan, Canada — Catena		
Crest to depression	545 to modern	88
B horizon	700–4000	
Akron, CO, U.S.	2000–7000	87
Soil Fractions		
Unfractionated soil	870 ± 50	89
Fulvic acid	495 ± 60	
Humic acid (total)	1235 ± 60	
Humin (total)	1140 ± 50	
Decalcified soil	1450	19
Acid hydrolysate	515	
Residue from hydrolysis	2560	
Humin	1240	

[a] MRT = 18,500 $\log_{10}(A_o/A)$, where A_o is the activity of the modern standard (NBS oxalic acid) and A is the activity of the unknown sample.

of the stable SOM pool at both sites was much greater than a few hundred years at 2973 and >600 years, respectively. These data confirmed that the stable SOM pool is very resistant to biodegradation. Hsieh[82] also demonstrated that the MRT of the active SOM pool in surface soils could be determined using bomb effect ^{14}C if the yearly variation of ^{14}C level in the atmosphere was known.

B. NATURAL ^{13}C ABUNDANCE

The difference in assimilation of ^{13}C between plants having C_3 vs. C_4 photosynthetic pathways provides an additional approach for assessing the long-term stability of SOM.[90] The stable isotope ^{13}C occurs naturally in the atmosphere at a concentration of 1.1% ($\delta^{13}C = -7‰$).[91] During photosynthesis C_3 plants incorporate less $^{13}CO_2$ than do C_4 plants.[92,93] The ratio of ^{13}C and ^{12}C of plants or SOM are reported as $\delta^{13}C$ values measured relative to a standard:

$$\delta^{13}C‰ = \frac{(^{13}C/^{12}C\ \text{sample} - {}^{13}C/^{12}C\ \text{standard})}{^{13}C/^{12}C\ \text{standard}} \times 10^3 \tag{1}$$

The standard was initially a limestone fossil (*Belemnitella americana*) from the Cretaceous Pee Dee formation of South Carolina.[94] This PDB standard is no longer available; however, other standards calibrated against PDB can be obtained from the National Institute of Standards and Technology (NIST) or the International Atomic Energy Agency (IAEA).

The $\delta^{13}C$ values of C_3 plants range from –23 to –40‰, with many occurring at about –26‰ (Figure 5). The C_4 plants have a $\delta^{13}C$ range from –9 to –19‰ and most often are found to occur at –13‰.[95–97] Plants with the Crassulacean acid metabolism (CAM) pathway often have a wide range between the two major types of vegetation. These however, are restricted to desert vegetation. Since the range of C_3 plants does not overlap the range of C_4 plants, differences in isotope ratios can be used to quantify the contribution of each photosynthetic pathway to SOM in mixed plant communities.[98,99] Since there is little further discrimination during microbial attack and humification, SOM has a C isotopic composition comparable to that of the source plant material prior to humification. With each change in vegetation between C_3 and C_4 plants, there is a corresponding change in the ^{13}C value of SOM.[100–102] This allows fairly short-term (<5 years) changes in SOM accumulation and turnover to be measured. The ^{13}C technique can now be conducted by readily available automated mass spectrometers. This equipment is cheaper and more reliable than the dating techniques necessary for ^{14}C analyses.

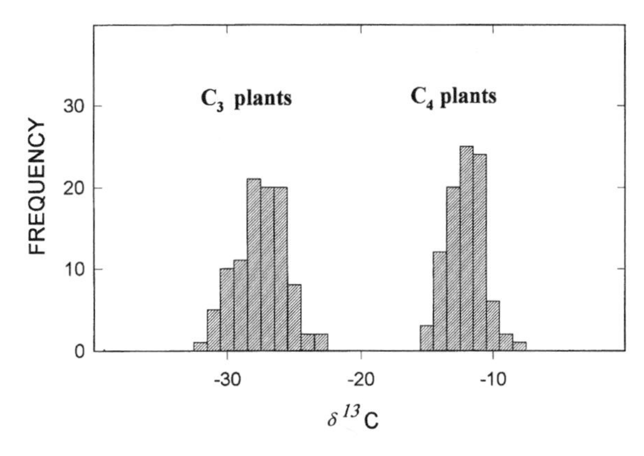

Figure 5 Frequency distribution for a collection of C_3 and C_4 plants depicting the variation in δ ^{13}C composition of photosynthetically fixed carbon. (Adapted from Deines, P., *Handbook of Environmental Isotope Geochemistry,* Vol. 1, Fritz, P. and Fontes, J.C., Eds., Elsevier, Amsterdam, 1980, 329.)

Natural ^{13}C abundance has been used to follow SOM transformations associated with land management.[59,90,103–106] Changes in the natural abundance of different organomineral fractions (micro- and macroaggregates) from soil surface horizons can be used to identify changes in SOM resulting from intensive management. Balesdent et al.[59] measured SOM turnover for a forest soil interplanted to continuous corn for 23 years. They reported that the cropping of a C_4 plant on soils that had been historically dominated by C_3 plants (Hardwood forest) could be used as in situ labeling of SOM. Balesdent et al.[90] demonstrated that the turnover time of the stable SOM pool of the surface soil of the Sanborn Field was >600 years which remained virtually unchanged after decades of cultivation. Since forests yield a C_3 signal, the growth of corn or sorghum (C_4) in any of the Eastern North American sites provides a usable signal. Grasslands vary in their ^{13}C content from the –26‰ of the C_3 grasslands (cool grasslands) to –14‰ from areas such as Kansas where C_4 grasses (warm season) predominate. Most Corn Belt cultivated soils have a label of approximately 20‰. With the accuracy of present-day mass spectrometers it should be possible to follow SOM dynamics from 2 to 4 years of continuous corn (C_4) following C_3 crops such as wheat or soybeans or vice versa in most soils of the central United States.[107]

Carbon dating with ^{14}C, ^{13}C analyses, and extended incubation of soils from historical plots can be combined to determine pool sizes and fluxes required in C balance calculations. Paul et al.[107] found that 70% of the total soil C in 85-year cultivated wheat plots at Akron, CO, was derived from the native prairie vegetation. Carbon dating showed the residue of acid hydrolysis to constitute 56% of total C and to date over 2600 years.

The power of the ^{13}C natural abundance technique is demonstrated using some preliminary data from a never-tilled grassland and corn-soybean plots in southwestern Michigan (Paul, E.A., unpublished). The present day grassland, cleared from a woodlot circa 1950, shows an oak-hickory forest and C_3 grass heritage with a ^{13}C content of –25.5‰ (Table 5). The 80 to 100-year cultivated field with 60% as much total C has a ^{13}C content of –20.9‰. On the basis of a residue return of two parts corn to one part soybeans plus wheat (–17‰), one can calculate that 42% of the total C remaining in the cultivated soil is derived from precultivation SOM. However, 67% of the residue of acid hydrolysis was derived from the precultivation SOM. Carbon dating shows the grassland to reflect bomb ^{14}C inputs at 1.07 ^{14}C pm C. The residue of hydrolysis at 0.98 pm C has an overall MRT of 170 years. The corn-soybean soil shows an overall C dating age of 546 years. The 56% of the C that is nonhydrolyzable dates 1435 years. Reviews of the methods and protocols describing the use of carbon isotopes can be found in Carbon Isotope Techniques edited by Coleman and Fry[108] and Theory and Application of Tracers by Schimel.[109]

VI. FUNCTIONAL DESCRIPTIONS OF SOIL ORGANIC MATTER

SOM is made up of a range of materials that degrade at variable rates due to their chemical composition as well as the degree of protection by absorption to minerals or entrapment within aggregates. Since SOM transformations are controlled by the soil biota, emphasis in this section will be given to analytical approaches that identify changes in active SOM pools or in pools defined by their C-mineralization rates. Microbial biomass has a unique position within the active SOM pool since it is both a source and sink

Table 5 Carbon Dynamics of Soil from a Perennial-Grassland and Corn-Soybean Field on the W.K. Kellogg Biological Station, Long-Term Ecological Research Site (Michigan State University) as Determined by ^{13}C Analyses, C Dating, and Acid Hydrolysis (E.A. Paul, unpublished)

	Grassland		Corn-Soybeans	
	Total	Hydr. Res.[a]	Total	Hydr. Res.[a]
C μg^{-1}	15000	7950 (53%)	9000	5040 (56%)
δ ^{13}C‰	–25.5	–27.8	–20.9	–24.4
%C from native	—	—	42	67
14C pmC	1.07	0.98	0.93	0.084
MRT[b] (y)	—	170	546	1435

[a] Fraction resistant to 6 *N* HCl acid hydrolysis.
[b] Mean residence time.

of soil C, as well as other important nutrients.[110,111] Microbial biomass and its activity are usually positively correlated with SOM due to a dependence on both the quantity and quality of degradable C sources.[112,113] Biomass C typically comprises 1 to 4% of the SOM and represents a significant part of the active SOM pool.[111,114] The size of this pool and its rate of turnover has been related to shifts in agricultural management,[113,115] differences in soil chemical and physical characteristics,[116] as well as climatic variables.[117]

The energy flux through the microbial biomass drives decomposition of residues and detritus and indicates whether the system is building or depleting the SOM pool.[118] If the amount decomposed exceeds C inputs, SOM will decline and C and N will be lost from the system. The change in the SOM pool is dependent upon the quality and quantity of C inputs, the fluctuation of microbial biomass, and the rate of decomposition.

A. DETERMINATION OF MICROBIAL BIOMASS C

Of the three current most frequently used chemical methods to determine microbial biomass, two are based upon the technique of Jenkinson and Powlson,[110] in which soil is subjected to chloroform fumigation then either (1) incubated for a 10- to 20-day period at which point respired CO_2 is determined or (2) following fumigation, soil is immediately extracted with potassium sulfate.[119] The third method is based upon the addition of saturating quantities of a readily available substrate, such as glucose, that causes a large immediate increase in CO_2 respiration.[120] Table 6 presents a brief comparison of these methods.

1. Chloroform Fumigation-Incubation (CFI)

Beakers of moist sieved soil are placed in a desiccator with a beaker containing 50-ml ethanol-free chloroform. The desiccator is evacuated until the chloroform boils and then placed in the dark for 24 hours, at room temperature. Following the fumigation the beaker of chloroform is removed and the desiccator evacuated to remove chloroform vapors. Fumigated samples and nonfumigated control samples are inoculated with 1 ml of an untreated soil-water suspension, then placed in a 500-ml wide-mouthed jar containing a vial of water and a 5-ml vial of 1 *N* NaOH to trap respired CO_2. The original procedure of Jenkinson and Powlson[110] used 150 g of soil; since then the amount of soil used in the analysis has continued to decline. Our laboratory routinely uses a 25-g soil sample and a 7-day preincubation. The preincubation period reduces the CO_2 flush resulting from sample handling and sieving.

During incubation microbial cells lysed by chloroform fumigation are mineralized at a fairly constant rate that represents about two fifths of the biomass-C in the sample. Biomass is calculated from the expression

$$C_M = (CO_{2(F)} - CO_{2(NF)})/k_C \quad (2)$$

where k_C is the proportion of microbial-C mineralized to CO_2, usually taken as 0.41,[120] $CO_{2(F)}$ is the quantity of CO_2 evolved from the fumigated sample in the 10 days following incubation, and $CO_{2(NF)}$ the amount produced from the nonfumigated sample for the same period. Disturbance caused by sieving of the soil prior to incubation often stimulates microbial respiration resulting in CO_2 values of the control

Table 6 Comparison of Methods Used for Determination of Microbial Biomass in Soil

Method	Principles	C Determination	Advantages	Disadvantages
CFI[a]	20–50 g of soil is fumigated in ethanol-free chloroform atmosphere for 1–5 d, incubated for 10 d; nonfumigated controls are incubated for 10–20 d	Headspace CO_2 determined by gas chromatography or infrared gas analyzer	Reference method	Use of appropriate control, interference from nonmicrobial labile C
			Simple and equipment readily available	
		Trapped CO_2 determined by titration, $BaCl_2$, or dual endpoint acid-base		
	Liberated CO_2 is measured directly in the headspace or trapped in NaOH			
CFE	Similar to CFIM, except following fumigation samples are extracted with 0.5 M K_2SO_4 (5:1 extractant to soil)	C determined on suitable automated C-analyzer	Low interference from non-microbial labile C	Poor correlation to CFIM
		Wet-combustion, either dichromate or persulfate digestion, CO_2 determined by titration	Shorter analysis time	Chemical disposal
	Total soluble C determined in both fumigated and non-fumigated soil extracts			
SIR	Glucose amendments are added to soil to determine the lowest concentration that will give the maximum respiratory response	Headspace CO_2 determined by gas chromatography or infrared gas analyzer	Simple, short analysis period (1–3 h)	
			Needs to be calibrated to other methods	
	Hourly CO_2 evolution is measured to encompass increases, decreases or lags in CO_2 flux			

[a] CFI — chloroform fumigation-incubation method;[110] CFE — chloroform fumigation-extraction method,[119] SIR — substrate induced respiration.[120]

greater than that of the fumigated soil. Therefore, the CO_2 flush may need to be calculated using CO_2 respired between 10 and 20 days of the unfumigated soil. Calculations without using a control may give values that are too high but these are highly correlated to biomass estimates determined by direct microscopy.

2. Chloroform Fumigation-Extraction (CFE)

The CFE method is based upon the fumigation technique but microbial constituents released by fumigation are extracted directly. Soil samples are prepared as described above in containers suitable for fumigation and extraction. The length of fumigation ranges from 1 to 5 days dependent upon soil texture. Fumigation efficiency varies with clay content.[121] Therefore, Horwath and Paul[122] recommend up to 5-day fumigation for soils with a high clay content. Extraction of the fumigated and nonfumigated control soil is done with 0.5 *M* K_2SO_4. After shaking the suspension is filtered and the filtrate stored frozen until analyzed. Extracted C can be determined either by wet combustion using potassium dichromate or persulfate digestion procedure or on automated soluble C analyzers.[123] Liberated CO_2 is trapped in 1 *M* NaOH and titrated with HCl.[3] Total N rendered extractable by fumigation can also be determined using the autoclave/persulfate digestion method of Cabrera and Beare[124] on the same extracts used for extractable-C determinations. Biomass C is calculated from the equation

$$C_M = (C_F - C_{NF})/k_{ec} \qquad (3)$$

where, C_F and C_{NF} are µg C determined in fumigated and nonfumigated samples, respectively, and k_{ec} is the proportion of microbial-C extracted from soil, usually taken as 0.35.[121]

3. Substrate-Induced Respiration (SIR)

The SIR method is based upon the addition of saturating quantities of a readily available substrate, such as glucose, that causes a large immediate increase in CO_2 respiration.[120] The CO_2 flush prior to new synthesis of biomass is used as an index of soil microbial biomass. The first step is to determine the lowest glucose level that will give the maximal CO_2 production. The range of glucose amendments is typically 25 to 400 μM g^{-1} soil. Soils high in organic matter will require greater concentrations of glucose addition. In the original procedure glucose was added in a dry form; however, West and Sparling[125] suggested using glucose in solution to enhance substrate dispersion and reduce water limitations during incubation. Samples are amended in gas tight bottles and incubated at 22°C for up to 3 hours. Respired CO_2 is determined using a gas chromatography or an infrared gas analyzer. Results are commonly expressed as CO_2 h^{-1} g^{-1} soil dry weight. Biomass C is determined from the expression of Anderson and Domsch;[120]

$$C_m = 40.04y + 0.37 \tag{4}$$

where, C_m is the microbial biomass and y is the maximal rate of CO_2 respiration. This equation is valid for SIR incubations done at 22°C. To obtain further descriptions of these and other methods used in determining soil microbial biomass, readers are directed to the reviews of Smith and Paul[118] and Horwath and Paul.[122]

B. C-MINERALIZATION POTENTIALS AND POOL SIZE DETERMINATIONS

Mineralization of SOM plays a fundamental role in soil fertility through the release of nutrients and subsequent influence on net primary productivity. The measurement of CO_2 evolution from soil has been widely used to determine the effect of environmental variables on the oxidation of SOM. The C-mineralization coefficient, i.e., the percentage of total organic C evolved as CO_2, has been used to compare soils under varying management.[115,126,127]

Mineralizable soil C can be measured by incubating soil samples in gas-tight containers. At periodic intervals headspace CO_2 can be determined using a variety of techniques, e.g., infrared gas analyzer, gas chromatograph, or NaOH traps.[3] In the latter case trapped CO_2 is precipitated by the addition of $BaCl_2$ and back titrated with HCl. This technique also provides samples for ^{14}C studies. The analysis of CO_2-C has also been automated using continuous flow gas chromatography.[128,129] Zibilske[3] provides an excellent review of the various methods and apparatus used in both field and laboratory settings.

Long-term incubations (>200 days) of soil with measurements of the CO_2 evolved have been widely used to differentiate functional C pools in soil. This method constitutes a biological fractionation of organic matter, whereby the most labile fractions are the most rapidly depleted by the soil microorganisms and subsequent soil C are more slowly mineralized. By analyzing the CO_2 release rates, a variety of mathematical models can be fit to derive estimates for functional C pool sizes and their turnover rates.

The most commonly used models are based on the assumption of first-order kinetics, i.e., where the rate of C mineralization is directly proportional to the amount of C in the organic matter pool. When integrated over time this produces an exponential decay curve. However, in most cases the cumulative CO_2 release curve is analyzed. Models (Table 7) that have been used to analyze incubation data often include two or more first-order components.[107,130–132] Alternatively, "mixed order" models, which include a zero-order (i.e., a constant rate of C mineralization independent of pool size) term for a recalcitrant soil fraction and first-order terms for more labile pools, have been proposed.[131–133]

Most incubations of unamended fresh soil show an initial rapid release of CO_2 as labile organic materials are mineralized. The degree of disturbance involved in sample preparation (e.g., sieving, wetting of dried soils) can influence the initial mineralization flush and therefore differences in sample preparation need to be considered in comparing studies.[134,135] Microbial biomass, which typically declines over the course of the incubation, appears to contribute a substantial portion (20 to 40%) of the C (and N) released.[133,136–138] The initial mineralization flush is often well described by the first-order model while subsequent C release generally follows a linear or slower exponential pattern.

An example of the use of incubation data to compare C pool sizes and turnover rates in soils with different management histories is given in Table 8. Soil from never-tilled grassland, a corn-soybean field, and a 4-year-old reversion plot (abandoned from cropping) were incubated for 200 days at 25°C. The CO_2 released over the 200-day incubation was described as the sum of three first-order rate reactions. The size of the old resistant pool was shown by acid hydrolysis to comprise 53% of the total C in the grassland and 56% in the corn-soybean rotation. Since the residue of acid hydrolysis dated greater than

Table 7 Commonly Used Statistical Models to Estimate Pool Sizes and Rate Constants from Cumulative CO_2 Data

Designation	Integral Form	Parameters (and Units)
Single first order	$C_m = C_0(1-e^{-kt})$	C_0 — mineralizable C pool (mass or concentration)
		k — specific rate of mineralization (time^{-1})
Double first order	$C_m = C_1(1-e^{-k_1t}) + (1-e^{-k_2t})$	C_1 — labile C pool (mass or concentration)
		C_2 — resistant C pool (mass or concentration)
		k_1 — specific rate of labile C mineralization (time^{-1})
		k_2 — specific rate of resistant C mineralization (time^{-1})
Mixed order	$C_m = C_1(1-e^{-k_1t}) + at$	C_1 — labile C pool (mass or concentration)
		k_1 — specific rate of labile C mineralization (time^{-1})
		a — mineralization rate from nonlabile C (mass time^{-1})

Note: In all models, C_m is cumulative CO_2-C and t is time.

150 years, it was assumed that negligible amounts of the CO_2 evolved in 200 days was derived from the resistant pool. This assumption made it possible to analyze the CO_2 data as the sum of two first-order rate reactions. The C mineralized represented 7.3% of total C in the grassland soil which had approximately 50% more organic C and the highest amount of C mineralized. More significant is the mineralization response of the reversion plot which showed a much greater labile C pool compared to the corn-soybean field, despite having the same total soil C. The greater C_1 pool and the higher rate of mineralization of the intermediate (C_2) pool suggest that organic matter is accumulating in the reversion plots. This accumulation was detected in the more labile fractions, but not in the total C values. The 200-day incubation and kinetic analyses indicate that the MRT of the C_2 pool for the three cases were different, ranging from 5 to 10 years.

Table 8 Kinetic Analysis of CO_2 Mineralization Curves for Three Treatments of the W.K. Kellogg Biological Station, Long-Term Ecological Research Site (Michigan State University)

	Grassland	Reversion	Corn/Soybeans
Total C (C_T) (μg g^{-1})	15,000	9500	9500
C mineralized (C_m) (μg C g^{-1} 200 d^{-1})	1100	930	560
C_m/C_T (%)	7.3	9.8	5.9
Microbial C (μg g^{-1})	345 (51)	251 (38)	141 (22)
Microbial C/C_T (%)	2.3	2.6	1.5
Pool size (C_1) (μg g^{-1})	623 (103)	404 (42)	170 (24)
Mineralization kinetics			
C_1^a/C_T (%)	4.1	4.2	1.8
k_1 d^{-1}	0.022	0.029	0.024
k_2 d^{-1}	0.00035	0.00052	0.00029
k_1 MRT (d)	45	34	41
k_2 MRT (y)	7.8	5.3	9.5
R^2	0.997	0.999	0.999

[a] C_1 corrected for microbial growth.
Note: Standard error of the mean shown in parenthesis.

From Paul, E.A., Horwath, W.R., Harris, D. et al., in *Soils and Global Change*, Lal, R., Kimble, T., Levine, E., and Stewart, B.A., Eds., Lewis Publishers, Boca Raton, FL, 1995, 297.

Although laboratory incubations provide for controlled conditions with uniform samples, they present problems with interpretation and extrapolations to in situ mineralization rates. The length of incubation used can substantially affect the parameter estimates for a given model and there is a high inverse autocorrelation of the rate constant (k) and pool size (C_o) using the first-order model.[135] Thus, pool size estimates are highly sensitive to relatively small deviations in the rate constant. To avoid these difficulties in model interpretation, Campbell et al.[139] proposed combining the two parameters (i.e., k*C_o) into a single index of mineralization potential. The calculation involved comparing treatment effects on N mineralization. In addition, manipulation of soil (i.e., sieving) changes characteristics of gas exchange, soil water, and makes available to microorganisms a portion of the organic matter which is less available under field conditions. Consequently, laboratory incubations overestimate C mineralization rates in the field. Despite these limitations, incubation procedures provide one of the most widely used and functionally meaningful measurements of the bioavailability of SOM.

VII. SUMMARY

A wide variety of techniques have been applied to the measurement of soil organic C (SOC). These generally have not been standardized, creating difficulties in comparing and interpreting SOC data across diverse soil types or management practices. In this review we have described a suite of methods that encompass both classical approaches of extraction and identification of specific chemical compounds to approaches that provide functional descriptions of SOC pools and their dynamics. Physical separations yield information on the distribution and concentration of SOM within different fractions of the soil matrix, which combined with functional approaches provide reliable information for estimating both labile and passive SOC pool sizes and turnover. The wealth of information generated by the combination of approaches will improve our understanding of how land management influences the quantity and quality of various SOM pools important in the cycling and release of plant nutrients or the quantification of ecosystem carbon budgets.

REFERENCES

1. Stevenson, F.J. and Elliott, E.T., Methodologies for assessing the quality and quantity of soil organic matter, in *Tropical Soil Organic Matter*, Coleman, D.C., Oades, J.M., and Uehara, G., Eds., University of Hawaii Press, 1989, 173.
2. Nelson, D.W. and Sommers, L.E., Total Carbon, organic carbon, and organic matter, in *Methods of Soil Analysis Part 2. Chemical and Microbiological Properties*, 2nd ed., Page, A.L., Ed., American Society of Agronomists, Madison, WI, 1982.
3. Zibilske, L.M., Carbon mineralization, in *Methods of Soil Analysis Part 2, Microbiological and Biochemical Properties*, Weaver, R.W., Angle, J.S., and Bottomley, P.S., Eds., American Society of Agronomists, Madison, WI, 1994, chap. 38.
4. Tabatabai, M.A. and Bremner, J.M., Automated instruments for determination of total carbon, nitrogen, and sulfur in soils by combustion techniques, in *Soil Analysis: Instrumental Techniques and Related Procedures*, Smith, K.A., Ed., Marcel Dekker, New York, 1983, 171.
5. Sheldrick, B.H., Test of the Leco CHN-600 determinator for soil carbon and nitrogen analysis, *Can. J. Soil Sci.*, 66, 543, 1986.
6. Craswell, E.T. and Eskew, D.L., Nitrogen and nitrogen-15 analysis using automated mass and emission spectrometers, *Soil Soc. Sci. Am. J.*, 55, 750, 1991.
7. Jensen, E.S., Evaluation of automated analysis of ^{15}N and total N in plant material and soil, *Plant Soil*, 133, 83, 1991.
8. Harris, D. and Paul, E.A., Automated analysis of ^{15}N and ^{14}C in biological samples, *Commun. Soil Sci. Plant. Anal.*, 20, 935, 1989.
9. Smith, J.L. and UM, M.H., Rapid procedures for preparing soil and KCl extracts for ^{15}N analysis, *Commun. Soil Sci. Plant Anal.*, 21, 2173, 1990.
10. Kelly, K.R., Conveyor-belt apparatus for fine grinding of soil and plant materials, *Soil Sci. Soc. Am. J.*, 58, 144, 1994.
11. Hayes, M.H.B., Extraction of humic substances from soil, in *Humic Substances in Soil, Sediment, and Water*, Aiken, G.R., Mcknight, D.M., and Wershaw, R.L., Eds., John Wiley & Sons, New York, 1985, 329.
12. Swift, R.S., Fractionation of soil humic substances, in *Humic Substances in Soil, Sediment, and Water*, Aiken, G.R., Mcknight, D.M., and Wershaw, R.L., Eds., John Wiley & Sons, New York, 1985, 387.
13. Hayes, M.H.B., Concepts of the origins, composition, and structure of humic substances, in *Advances in Soil Organic Matter Research: The Impact on Agriculture and the Environment*, Wilson, W.S., Ed., Royal Society of Chemists, Redwood Press, Melksham, Wiltshire, 1991, 3.
14. Schnitzer, M. and Kahn, S.U., *Soil Organic Matter. Developments in Soil Science 8*, Elsevier, Amsterdam, 1978.
15. Stevensen, F.J., Geochemistry of soil humic substances, in *Humic Substances in Soil, Sediment, and Water,* Aiken, G.R., Mcknight, D.M., and Wershaw, R.L., Eds., John Wiley & Sons, New York, 1985, 13.
16. Hatcher, P.G., Schnitzer, M., Dennis, L.W., and Maciel, G.E., Aromaticity of humic substances in soil, *Soil Sci. Soc. Am. J.*, 45, 1089, 1985.
17. Almendros, G. and Gonzalez-Villa, F.G., Degradative studies on a soil humin fraction — sequential degradation of inherited humin, *Soil Biol. Biochem.*, 19, 513, 1987.
18. Anderson, D.W., Process of humus formation and transformation in soils of the Canadian Great Plains, *J. Soil Sci.*, 30, 77, 1979.

19. Jenkinson, D.S. and Rayner, J.H., The turnover of soil organic matter in some of the Rothamsted classical experiments, *Soil Sci.*, 123, 298, 1977.
20. van Veen, J.A. and Paul, E.A., Organic carbon dynamics in grassland soils. I. Background information and computer simulation, *Can. J. Soil Sci.,* 61, 185, 1981.
21. Paul, E.A. and van Veen, J.A., The use of tracers to determine the dynamic nature of organic matter, in *Proc. 11th Int. Congr. Soil Sci.*, Vol. 3, International Soil Science Society, Edmonton, 1978, 1.
22. Kononova, M.M., *Soil Organic Matter*, Pergamon Press, Elmsford, NY, 1966.
23. Flaig, W., Beutelspacher, H., and Rietz, E., Chemical composition and physical properties of humic substances, in *Soil Components*, Vol. I, Gieseking, J.E., Ed., Springer-Verlag, Berlin, 1975, 1.
24. Martin J.P. and Haider, K., Microbial degradation and stabilization of ^{14}C-labeled lignins, phenols, and phenolic polymers in relation to soil humus formation, in *Lignin Biodegradation: Microbiology, Chemistry, and Potential Applications*, Vol. 2, Kirk, T.K., Higuchi, T., and Chang, H.M., Eds., CRC Press, Boca Raton, Florida, 1980, 77.
25. Martin, J.P. and Stott, D.E., Characteristics of organic matter decomposition and retention in arid soils, in 6th Annu. Int. Symp. Environ. Biogeochem., Skujins, J.J., Ed., UNEP, Paris, 1984.
26. Stott, D.E, Kassim, G., Jarrell, W.M., Martin, J.P., and Haider, K., Stabilization and incorporation into biomass of specific plant carbons during biodegradation in soil, *Plant Soil*, 70,15, 1983.
27. Schnitzer, M. and Schulten, H.R., The analysis of soil organic matter by pyrolysis-field ionization mass spectrometry, *Soil Sci. Soc. Am. J.*, 56, 1811, 1992.
28. Wilson, M.A., Pugmire, R.J., Zilm, R.W., Goh, K.M., Heng, S. and Grant, D.M., Cross polarization 13 C-NMR spectroscopy with "magic angle" spinning characterizes organic matter in whole soil, *Nature*, 29, 648, 1984.
29. Skjemstad, J.O. and Dalal, R.C., Spectroscopic and chemical differences in organic matter of two vertisols subjected to long periods of cultivation, *Aust. J. Soil Res.*, 25, 323, 1987.
30. Skjemstad, J.O., Dalal, R.C., and Barron, B.F., Spectroscopic investigations of cultivation effects on organic matter of vertisols, *Soil Sci. Soc. Am. J.*, 50, 354, 1986.
31. Wilson, M.A., *NMR Techniques and Applications in Geochemistry and Soil Chemistry*, Pergamon Press, Oxford, 1987.
32. Baldock, J.A., Currie, G.J., and Oades, J.M., Organic matter as seen by solid state ^{13}C NMR and Pyrolysis Tandem Mass Spectrometry, in *Advances in Soil Organic Matter Research: The impact of Agriculture and the Environment*, Wilson, W.S., Ed., The Royal Society of Chemists, Thomas Graham House, Science Park, Cambridge, 1991, 45.
33. Oades, J.M., Waters, A.G., Vassallo, A.M., Wilson, M.A., and Jones, G.P., Influence of management on the composition of organic matter in a red-brown earth as shown by ^{13}C nuclear magnetic resonance, *Aust. J. Soil Res.*, 26, 289, 1988.
34. Schulten, H.R., Pyrolysis and soft ionization mass spectrometry of aquatic/terrestrial humic substances and soils, *J. Anal. Appl. Pyrolysis*, 12, 149, 1987.
35. Schnitzer, M. and Schulten, H.R., The analysis of soil organic matter by pyrolysis-field ionization mass spectrometry, *Soil Sci. Soc. Am. J.*, 56, 1811, 1992.
36. Turchenek, I.W. and Oades, J.M., Fractionation of organic-mineral complexes by sedimentation and density techniques, *Geoderma*, 21, 311, 1979.
37. Elliott, E.T., Anderson, R.V., Coleman, D.C., and Cole, C.V., Habitable porespace and microbial trophic interactions, *Oikos*, 35, 327, 1980.
38. Tisdall, J.M. and Oades, J.M., Organic matter and water-stable aggregates in soil, *J. Soil Sci.*, 33, 141, 1982.
39. Elliott, E.T., Aggregate structure and carbon, nitrogen and phosphorus in native and cultivated soils, *Soil Sci. Soc. Am. J.*, 50:627, 1986.
40. Cambardella, C.A. and Elliott, E.T., Particulate soil organic matter changes across a grassland and cultivation sequence, *Soil Sci. Soc. Am. J.*, 56, 777, 1992.
41. Elliott, E.T. and Cambardella, C.A., Physical separation of soil organic matter, *Agric. Ecosys. Environ.*, 34, 407, 1991.
42. Christensen, B.T., Physical fractionation of soil and organic matter in primary particle size and density separates, *Adv. Soil Sci.*, 20, 2, 1992.
43. Gee, G.W. and Bauder, J.W., Particle size analysis, in *Methods of Soil Analysis, Part I, Physical and Mineralogical Methods*, 2nd ed., Klute, A., Ed., American Society of Agronomists, Madison, WI, 1986, chap. 15.
44. Gregorich, E.G., Kachanoski, R.G., and Voroney, R.P., Carbon mineralization in soil size fractions after various amounts of aggregate disruption, *J. Soil Sci.*, 40, 649, 1989.
45. Christensen, B.T., Carbon and nitrogen in particle size fractions isolated from Danish arable soils by ultrasonic dispersion and gravity-sedimentation, *Acta Agric. Scand.*, 35, 175, 1985.
46. Kemper, W.D. and Rosenau, R.C., Aggregate stability and size distribution, in *Methods of Soil Analysis, Part I, Physical and Mineralogical Methods*, 2nd ed., Klute, A., Ed., American Society of Agronomists, Madison, WI., 1986, chap. 17.
47. Edwards, A.P. and Bremner, J.M., Microaggregates in soil, *J. Soil Sci.*, 18, 64, 1967.
48. Oades, J.M., Soil organic matter and structural stability: mechanisms and implications for management, *Plant Soil*, 76, 319, 1984.
49. Elliott, E.T. and Coleman, D.C., Let the soil work for us, *Ecol. Bull.*, 39, 23, 1988.
50. Tisdall, J.M. and Oades, J.M., The effect of crop rotation on aggregation in a red-brown earth, *Aust. J. Soil Res.*, 18, 423, 1980.

51. Dormaar, J.F., Chemical properties of soil and water-stable aggregates after sixty-seven years of cropping to spring wheat, *Plant Soil*, 75, 51, 1983.
52. Tiessen, H. and Stewart, J.W.B., Particle-size fractions and their use in studies of soil organic matter. II. Cultivation effects on organic matter composition in size fractions, *Soil Sci. Soc. Am. J.*, 47, 509, 1983.
53. Christensen, B.T., Decomposability of organic matter in particle-size fractions from field soils with straw incorporation, *Soil Biol. Biochem.*, 19, 429, 1987.
54. Oades, J.M. and Turchenek, L.W., Accretion of organic carbon, nitrogen and phosphorus in sand and silt fractions of a red-brown earth under pasture, *Aust. J. Soil Res.*, 16, 351, 1978.
55. Amato, M. and Ladd, J.N., Studies of nitrogen mineralization in calcareous soils. V. Formation and distribution of isotope-labelled biomass during decomposition of 14C and 15N labelled plant material, *Soil Biol. Biochem.*, 12, 405, 1980.
56. Anderson, D.W., Saggar, S., Bettany, J.R., and Stewart, J.W.B., Particle size fractions and their use in studies of soil organic matter. I. The nature and distribution of forms of carbon, nitrogen, and sulfur, *Soil Sci. Soc. Am. J.*, 45, 767, 1981.
57. Anderson, D.W. and Paul, E.A., Organo-mineral complexes and their study by radio-carbon dating, *Soil Sci. Soc. Am. J.*, 48, 298, 1984.
58. Tiessen, H., Stewart, J.W.B., and Hunt, H.W., Concepts of soil organic matter transformations in relation to organo-mineral particle size fractions, *Plant Soil*, 76, 287, 1984.
59. Balesdent, J., Wagner, G.H., and Mariotti, A., Soil organic matter turnover in long-term field experiments as revealed by carbon-13 natural abundance, *Soil Sci. Soc. Am. J.*, 52, 118, 1988.
60. Christensen, B.T. and Sorensen, L.H., Nitrogen in particle size fractions of soils incubated for five years with ^{15}N-ammonium and ^{14}C-hemicellulose, *J. Soil Sci.*, 37, 241, 1986.
61. Tiessen, H.J., Stewart, J.W.B., and Bettany, J.R., Cultivation effects on the amounts and concentration of carbon, nitrogen and phosphorus in grassland soils, *Agron. J.*, 74, 831, 1982.
62. Tiessen, H., Stewart, J.W.B., and Moir, J.O., Changes in organic and inorganic phosphorus composition of two grassland soils and their particle size fractions during 60–90 years of cultivation, *J. Soil Sci.*, 34, 815, 1983.
63. Buyanovsky, G.A., Aslam, M., and Wagner, G.H., Carbon turnover in soil physical fractions, *Soil Sci. Soc. Am. J.*, 58, 1167, 1994.
64. Parton, W.J., Stewart, J.W.B., and Cole, C.V., Dynamics of C, N, P, and S in grassland soil: a model, *Biogeochemistry*, 5, 109, 1988.
65. Christensen, B.T. and Sorensen, L.H., The distribution of native and labelled carbon between soil particle size fractions isolated from long-term incubation experiments, *J. Soil Sci.*, 36, 219, 1985.
66. Tate, R.L., III, *Soil Organic Matter, Biological and Ecological Effects*, John Wiley & Sons, New York, 1987.
67. Young, J.L. and Spycher, G., Water-dispersible soil organic-mineral particles. I. Carbon and nitrogen distribution, *Soil Sci. Soc. Am. J.*, 43, 324, 1979.
68. Christensen, B.T., Straw incorporation and soil organic matter in macro-aggregates and particle size separates, *J. Soil Sci.*, 37, 125, 1986.
69. Cambardella, C.A. and Elliott, E.T., Carbon and nitrogen distribution in aggregates from cultivated and native grassland soils, *Soil Sci. Soc. Am. J.*, 57, 1071, 1993.
70. Sollins, P., Spycher, G., and Glassman, C.A., Net mineralization from light- and heavy-fraction forest soil organic matter, *Soil Biol. Biochem.*, 16, 31, 1984.
71. Ford, G.W., Greenland, D.J., and Oades, J.M., Separation of the light fraction from soils by ultrasonic dispersion using halogenated hydrocarbons containing a surfactant, *J. Soil Sci.*, 20, 291, 1969.
72. Janzen, H.H., Campbell, C.A., Brandt, C.A., Lafond, G.P., and Townley-Smith, L., Light fraction organic matter in soils from long-term crop rotations, *Soil Sci. Soc. Am. J.*, 56, 1799, 1992.
73. Parton, W.J., Schimel, D.S., Cole, C.V., and Ojima, D.S., Analysis of factors controlling soil organic matter levels in Great Plains grasslands, *Soil Sci. Soc. Am. J.*, 51, 1173, 1987.
74. Paul, E.A., Dynamics of organic matter in soils, *Plant Soil*, 76, 275, 1984.
75. Elliott, E.T., Cambardella, C.A., and Gupta, V.V.S.R., Characteristics of biologically available organics, in *Ecosystems Research Report 1. Decomposition and Accumulation of Organic Matter in Terrestrial Ecosystems: Research Priorities and Approaches*, van Breemen, N, Ed., European Science Foundation, Doorwerth, The Netherlands, 1991.
76. Beare, M.H., Hendrix, P.F., and Coleman, D.C., Water-stable aggregates and organic matter fractions in conventional- and no-tillage soils, *Soil Sci. Soc. Am. J.*, 58, 777, 1994.
77. Gupta, V.V. and Germida, J.J., Populations of predatory protozoa in field soils after 5 years of elemental S fertilizer application, *Soil Biol. Biochem.*, 20, 787, 1988.
78. Beare, M.H., Cabrera, M.L., Hendrix, P.F., and Coleman, D.C., Aggregate-protected and unprotected organic matter pools in conventional- and no-tillage soils, *Soil Sci. Soc. Am. J.*, 58, 787, 1994.
79. Cambardella, C.A. and Elliott, E.T., Carbon and nitrogen dynamics of soil organic matter fractions from cultivated grassland soils, *Soil Sci. Soc. Am. J.*, 58, 123, 1994.
80. O'Brien, B.J., Soil organic carbon fluxes and turnover rates estimated from radiocarbon enrichments, *Soil Biol. Biochem.*, 16, 115, 1984.
81. Hsieh, Y.-P., Pool size and mean age of stable soil organic carbon in cropland, *Soil Sci. Soc. Am. J.*, 56, 460, 1992.

82. Hsieh, Y.-P., Radiocarbon signatures of turnover rates in active soil organic carbon pools, *Soil Sci. Soc. Am. J.*, 57, 1020, 1993.
83. Paul, E.A. and Clarke, F.E., *Soil Microbiology and Biochemistry*, Academic Press, San Diego, CA, 1993.
84. Goh, K.M., Stout, J.D., and Rafter, T.A., Radiocarbon enrichment of soil organic matter fractions in New Zealand soils, *Soil Sci.*, 123, 385, 1977.
85. O'Brien, B.J. and Stout, J.D., Movement and turnover of soil organic matter as indicated by carbon isotope measurements, *Soil Biol. Biochem.*, 10, 309, 1978.
86. Paul, E.A., Campbell, C.A., Rennie, D.A., and McCallum, K.J., Investigations of the Dynamics of Soil Humus Utilizing Carbon Dating Techniques, Trans. 8th Int. Cong. Soil Sci., Romania, 1964, 201.
87. Paul, E.A., Follett, R.F., Leavitt, S.W., Halvorson, A., Peterson, G.A., and Lyon, D.I., Determination of soil organic matter, pool sizes and dynamics: use of radiocarbon dating for Great Plains soil, *Soil Sci. Soc. Am. J.*, in press.
88. Martel, Y.A. and Paul, E.A., Effects of cultivation on the organic matter of grassland soils as determined by fractionation and radiocarbon dating, *Can. J. Soil Sci.*, 54, 419, 1974.
89. Campbell, C.A., Paul, E.A., Rennie, D.A., and McCallum, K.J., Applicability of the carbon-dating method of analysis to soil humus studies, *Soil Sci.*, 104, 217, 1967.
90. Balesdent, J., Mariotti, A., and Guillet, B., Natural 13C abundance as a tracer for studies of soil organic matter dynamics, *Soil Biol. Biochem.*, 19, 25, 1987.
91. Keeling, C.D., Mook, W.G., and Tans, P.P., Recent trends in the $^{13}C/^{12}C$ ratio of atmospheric carbon dioxide, *Nature*, 277, 121, 1979.
92. Vogel, J.C., *Fractionation of the Carbon Isotopes During Photosynthesis*, Springer-Verlag, Berlin, 1980.
93. O'Leary, M., Carbon isotope fractionation in plants, *Phytochemistry*, 20, 553, 1981.
94. Boutton, T.W., Stable isotope ratios of natural materials. I. Sample preparation and mass spectrometric analysis, in *Carbon Isotope Techniques*, Coleman, D.C. and Fry, B., Eds., Academic Press, San Diego, CA, 1991, 155.
95. Smith, B.N. and Epstein, S., Two categories of $^{13}C/^{12}C$ ratios for higher plants, *Plant Physiol.*, 47, 380, 1971.
96. Delens E., Lerman, J.C., Nato, A., and Moyse, A., Carbon isotope discrimination by the carboxylating reactions in C_3, C_4, and CAM plants, in *Proceedings of the Third International Congress on Photosynthesis*, Avron, M., Ed., Elsevier, Amsterdam, 1974, 1267.
97. Deines, P. The isotopic composition of reduced organic carbon, in *Handbook of Environmental Isotope Geochemistry*, Vol 1, Fritz, P. and Fontes, J.C., Eds., Elsevier, Amsterdam, 1980, 329.
98. Ludlow, M., Troughton, J., and Jones, R., A technique for determining the proportion of C_3 and C_4 species in plant samples using stable isotopes of carbon, *J. Agric. Sci.*, 87, 625, 1976.
99. Ode, D.J., Tiessen, L.L., and Lerman, J.C., The seasonal contribution of C_3 and C_4 plant species to primary production in a mixed prairie, *Ecology*, 61, 1304, 1980.
100. Dzurec, R.S., Boutton, T.W., Caldwell, M.M., and Smith, B.N., Carbon isotope ratios of soil organic matter and their use in assessing community composition changes in Curlew Valley, Utah, *Oecologia*, 66, 17, 1985.
101. Schwartz, D., Mariotti, A., Lanfranchi, R., and Gulliet, B., $^{13}C/^{12}C$ ratios of soil organic matter as indicators of vegetation changes in the Congo, *Geoderma*, 39, 97, 1986.
102. Martin, A., Mariotti, A., Balesdent, J., Lavelle, P., and Vuattoux, R., Estimate of organic matter turnover rate in a savanna soil by ^{13}C natural abundance measurements, *Soil Biol. Biochem.*, 22, 517, 1990.
103. Nissenbaum, A., and Shallinger, K.M., The distribution of the stable isotope ($^{13}C/^{12}C$) in fractions of soil organic matter, *Geoderma*, 11, 137, 1974.
104. Barnes, P.W., Tiessen, L.L., and Oade, D.J., Distribution, production, and diversity of C_3 and C_4 dominated communities in a mixed prairie, *Can. J. Bot.*, 61, 741, 1983.
105. Cerri, C.C., Balesdent, J., Feller, C., Victoria, R., and Plenecassagne, A., Application du tracage isotopique naturel en ^{13}C a l'etude de la dynamique de la matiere organique dans les sols, *C.R. Acad. Sci.*, 9, 423, 1985.
106. Vitorello, V.A., Cerri, C.C., Andreux, F., Feller, C., and Victoria, R.L., Organic matter and natural carbon-13 distribution in forested and cultivated oxisols, *Soil Sci. Soc. Am. J.*, 53, 773, 1989.
107. Paul, E.A., Horwath, W.R., Harris, D., Follet, R., Levitt, S.W., Kimball, B.A., and Pregitzer, K., Establishing the pool sizes and fluxes in CO_2 emissions from soil organic matter turnover, in *Soils and Global Change*, Lal, R., Kimble, T., Levine, E., and Stewart, B.A., Eds., Lewis Publishers, Boca Raton, FL, 1995, 297.
108. Coleman, D.C. and Fry, B., *Carbon Isotope Techniques*, Academic Press, San Diego, CA, 1991.
109. Schimel, D.S., *Theory and Application of Tracers*, Academic Press, San Diego, CA, 1993.
110. Jenkinson, D.S. and Powlson, D.S., The effect of biocidal treatments on metabolism in soil. V. A method of measuring soil biomass, *Soil Biol. Biochem.*, 8, 209, 1976.
111. Schnürer, L., Clarholm, M., and Rosswall, T., Microbial biomass and activity in an agricultural soil with different organic matter contents, *Soil Biol. Biochem.*, 17, 611, 1985.
112. Kassim, G., Martin, J.P., and Haider, K., Incorporation of a wide variety of organic substrate carbons into soil biomass as estimated by the fumigation procedure, *Soil Sci. Soc. Am. J.*, 45, 1106, 1981.
113. Adams, T. McM., and Laughlin, R.J., The effects of agronomy on the carbon and nitrogen contained in the soil biomass, *J. Agric. Sci.*, 97, 319, 1981.
114. Anderson, T.H. and Domsch, K.H., Ratios of microbial biomass carbon to total organic carbon in arable soils, *Soil Biol. Biochem.*, 21, 471, 1989.

115. Collins, H.P., Rasmussen, P.E., and Douglas, C.L., Jr., Crop rotation and residue management effects on soil carbon and microbial dynamics, *Soil Sci. Soc. Am. J.*, 56, 783, 1992.
116. Gregorich, E.G. and Anderson, D.W., Effects of cultivation and erosion on four toposequences in the Canadian prairies, *Geoderma*, 36, 343, 1985.
117. Insam, H., Parkinson, D., and Domsch, K.H., Influence of macroclimate on soil microbial biomass, *Soil Biol. Biochem.*, 21, 211, 1989.
118. Smith J. L. and Paul, E.A., The significance of soil microbial biomass estimations, in *Soil Biochemistry*, Vol. 6, Bollag, J.M. and Stotzky, G., Eds., Marcel Dekker, New York, 1990, 357.
119. Brookes, P.C., Landman, A., Pruden, G., and Jenkinson, D.S., Chloroform fumigation and the release of soil nitrogen: a rapid direct extraction method to measure microbial biomass nitrogen in soil, *Soil Biol. Biochem.*, 17, 837, 1985.
120. Anderson, J.P.E. and Domsch, K.H., A physiological method for the quantitative measurement of microbial biomass in soils, *Soil Biol. Biochem.*, 10, 215, 1978.
121. Voroney, R.P. and Paul, E.A., Determination of Kc and Kn in situ for calibration of the chloroform fumigation-incubation method, *Soil Biol. Biochem.*, 16, 9, 1984.
122. Horwath, W.R. and Paul, E.A., *Microbial biomass, in Methods of Soil Analysis, Part 2, Microbiological and Biochemical Properties*, Weaver, R.W., Angle, J.S., and Bottomley, P.S., Eds., American Society of Agronomists, Madison, WI, 1994, chap. 36.
123. Tate, K.R., Ross, D.J., and Feltham, C.W., A direct extraction method to estimate soil microbial biomass C: effects of experimental variables and some different calibration procedures, *Soil Biol. Biochem.*, 20, 329, 1988.
124. Cabrera, M.L. and Beare, M.H., Alkaline persulfate oxidation for determining total nitrogen in microbial biomass extracts, *Soil Sci. Soc. Am. J.*, 57, 1007, 1993.
125. West, A.W. and Sparling, G.P., Correlation between four methods to estimate total microbial biomass in stored, air-dried and glucose amended soils, *Soil Biol. Biochem.*, 18, 569, 1986.
126. Carter, M.R. and Rennie, D.A., Changes in soil quality under zero tillage farming systems: distribution of microbial biomass and mineralizable C and N potentials, *Can. J. Soil Sci.*, 62, 587, 1982.
127. Campbell, C.A., Biederbeck, V.O., Schnitzer, M., Selles, F., and Zentner, R.P., Effect of 6 years of zero tillage and N fertilizer management on changes in soil quality of an Orthic Brown Chernozem in southwestern Saskatchewan, *Soil Till. Res.*, 14, 39, 1989.
128. Hendricks, C.W., Paul, E.A., and Brooks, P.D., Growth measurements of terrestrial microbial species by a continuous flow technique, *Plant Soil*, 101, 189, 1987.
129. Nordgren, A., Apparatus for the continuous, long-term monitoring of soil respiration rate in large numbers of samples, *Soil Biol. Biochem.*, 20, 955, 1988.
130. Molina, J.A.E., Clapp, E.E., and Larson, W.E., Potentially mineralizable nitrogen in soil: the simple exponential model does not seem to apply for the first 12 weeks of incubation, *Soil Sci. Soc. Am. J.*, 44, 442, 1980.
131. Bonde, T.A. and Rosswall, T., Seasonal variation of potentially mineralizable nitrogen in four cropping systems, *Soil Sci. Soc. Am. J.*, 51, 1508, 1987.
132. Bonde, T.A. and Lindberg, T., Nitrogen mineralization kinetics in soil during long-term laboratory incubations: a case study, *J. Env. Qual.*, 17, 414, 1988.
133. Blet-Charaudeau, C., Muller, J., and Laudelot, H., Kinetics of carbon dioxide evolution in relation to microbial biomass and temperature, *Soil Sci. Soc. Am. J.*, 54, 1324, 1990.
134. Stanford, G. and Smith, S.J., Nitrogen mineralization potentials of soils, *Soil Sci. Soc. Am. Proc.*, 36, 465, 1972.
135. Paustian, K. and Bonde, T.A., Interpreting incubation data on nitrogen mineralization from soil organic matter, *INTECOL Bull.*, 15, 101, 1987.
136. Brookes, P.C., Powlson, D.S., and Jenkinson, D.S., The microbial biomass in soil, in *Ecological Interactions in Soil*, Fitter, A.H., Atkinson, D., Read, D.J., and Usher, M.B., Eds., Spec. Publ. No. 4, British Ecology Society, Oxford, 1985, 123.
137. Bonde, T.A., Schnürer, J., and Rosswall, T., Microbial biomass as a fraction of potentially mineralizable nitrogen in soils from long-term field experiments, *Soil Biol. Biochem.*, 20, 447, 1988.
138. Robertson, K., Schnürer, J., Clarholm, M., Bonde, T.A., and Rosswall, T., Microbial biomass in relation to C and N mineralization during laboratory incubations, *Soil Biol. Biochem.*, 20, 281, 1987.
139. Campbell, C.A., Lafond, G.P., Leyshon, A.J., Zentner, R.P., and Janzen, H.H., Effect of cropping practices on the initial potential rate of N mineralization in a thin Black Chernozem, *Can J. Soil Sci.*, 57, 43, 1991.

Chapter 4

Crop Residue Input to Soil Organic Matter on Sanborn Field

G.A. Buyanovsky and G.H. Wagner

CONTENTS

I. INTRODUCTION

The value of soil organic matter (SOM) has been long recognized. From earliest times, its level in soil was used as a general indicator of soil productivity. Maintaining or increasing SOM was the goal of many early research projects. When J.W. Sanborn established the plots that bear his name, interest in rotations and the practice of manuring were part of a management philosophy which sought to find methods for increasing SOM and the nitrogen (N) contained therein in order to increase yields.[1] Soils research at Missouri in the early years of this century, however, did not achieve practical success in this matter. After some 35 years of accumulating data on Sanborn and other experimental fields, nearly all plots, including heavily manured ones, showed N losses. With a continued view toward improving management that recognizes the value of SOM, there was a philosophical shift at that time to a consideration of the turnover of SOM and toward N turnover in relation to productivity of the soil.[2] Over the years, as sound soil management has developed, we have continued to keep an appropriate focus on both the level and the dynamics of SOM.[3]

A major factor contributing to the level of SOM is annual input of plant residues. The productivity of native vegetation, tallgrass prairie in this region, once defined the annual carbon (C) input. Characteristic organic matter levels in soils were controlled by these vegetative inputs and by other interrelated factors such as climate and the nutrient release from the soil parent material. Under cultivated agriculture, crop residues serve as C inputs and thus influence both the level and the dynamics of organic matter in soil.

The level of SOM can be altered by changes in the ecosystem. The original native equilibrium levels were established as the soils developed over hundreds and thousands of years. In most cases, the current

0-8493-2802-0/97/$0.00+$.50

level of SOM in cultivated soils is significantly less than the level that existed under the native prairie conditions. When the prairies were first cultivated there was a relatively rapid loss of soil organic C. The decline slowed with time and a final equilibrium under a particular management system became established after 25 to 40 years.[4,5]

This slow establishment of a particular equilibrium level places extended time requirements on studies of changes in levels of SOM and its characterization. The usual experiment station project with a duration of 3 to 5 years is not a satisfactory framework for research of this nature, because even minor shifts in SOM may require 10 or more years of study to establish an altered level or different characteristics. During 100 years of Sanborn Field history, some shifts in SOM have been observed. The numerous plots of this field include different cropping and management systems which have been sampled periodically. These historical samples provided an opportunity for assessing long-term trends in the level of SOM.

Unfortunately, there were no direct data on crop residue inputs under the different cropping systems on Sanborn Field. The data that are available are grain and forage yields. Limited data are also available from the literature on above-ground residue yields of some crops grown in different ecological conditions. Below-ground residue production has often been ignored due to the difficulty in measuring it. Moreover, some of the data in the literature on this subject were collected in greenhouse pot experiments or water cultures and are not applicable to field condition.

In order to interpret SOM dynamics, one needs reliable data on crop residue yields, both above and below ground. Because corn, wheat, and soybean constitute major components of Sanborn Field management systems, we have concentrated on the biological productivity of these crops in small-scale field experiments. During the last decade, accurate measurements of postharvest crop residues have been a mandatory component of all our studies related to SOM dynamics.[6] Use of ^{14}C labeling in the field and subsequent measurements of CO_2 evolution from the soil have provided us with a comprehensive approach in studying C flow.[7]

Measurements using the ^{14}C technique, that we report in this paper, were carried out on small plots, usually 3.0×2.25 m, in three or four replicates. Regular farming practices have been simulated and no machinery was used due to the size of the plots. Above-ground biomass was collected from the entire plot areas, dried, weighed, and returned to the same plot. Root biomass was determined by washing soil cores, 10-cm diameter, taken to a depth of 50 cm. Cores were taken at several positions relative to the plants. A large body of data on annual C transfer from the crops to the soil has now been accumulated. We have used this information to estimate long-term biomass yields on Sanborn Field and to calculate the total amount of C input under different crops.

Another aspect of small-scale field studies that helped us to quantify C flow from crop residues into SOM was our work on the rate of decomposition of residues and their major components. Calculations of C budgets and direct measurements of ^{14}C in the soil, after the mineralization process of labeled residues had practically ceased, allowed us to approximate a humification rate of plant material. This index of humification is used in conjunction with data on that part of crop production that was added to the soil after harvest. To approximate the quantities of residues for the diverse management systems in long-term Sanborn Field experiments, we used grain and forage yields and data on above- and underground biomass obtained in small-scale field experiments.

II. PLANT PRODUCTIVITY AND POSTHARVEST RESIDUES

For many of the plots on Sanborn Field we have yield data for both grain and forage and this constitutes the above-ground production excluding stubble. The data were collected over a period of 100 years during which there were significant weather variations across the years. Several plots also represent different levels of fertility. For the first ~60 years of the experiment, the above-ground forages (stalks and straw) of the no-treatment and manured plots cropped continuously to corn and wheat have been removed at harvest. Therefore, the annual input of residues was only that of the stubble and the mature root system. For corn and wheat, this stubble residue was incorporated into the surface soil along with the roots by annual tillage.

The harvest of total above-ground biomass of the plots on Sanborn Field has yielded data that helps to estimate the amount of crop residues that are usually returned to the soil in contemporary farming. Of particular interest are the forage yields for timothy (Figure 1) and the yield of wheat straw (Figure 2) and corn stalks (Figure 3) computed as 15-year moving averages. Grain yields are reported in another chapter of this volume.[8]

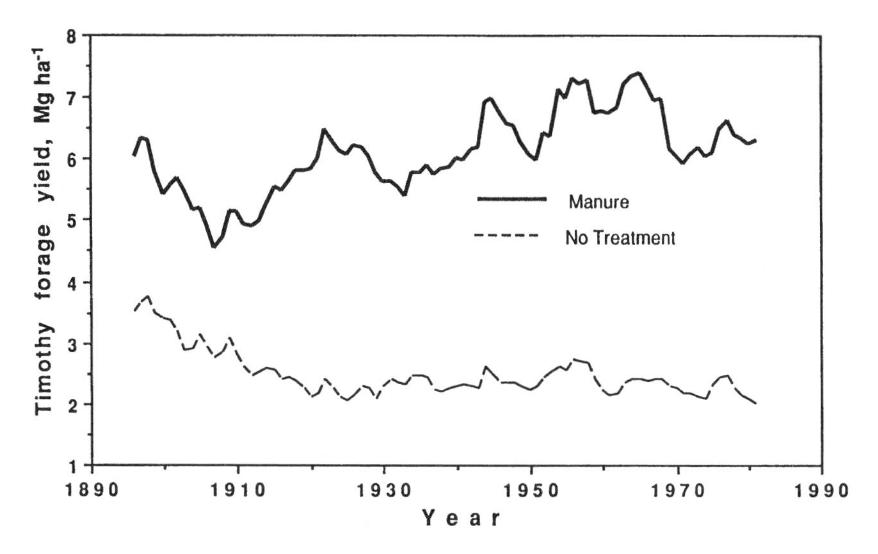

Figure 1 Timothy forage yield on Sanborn Field historical plots, as a 15-year moving average.

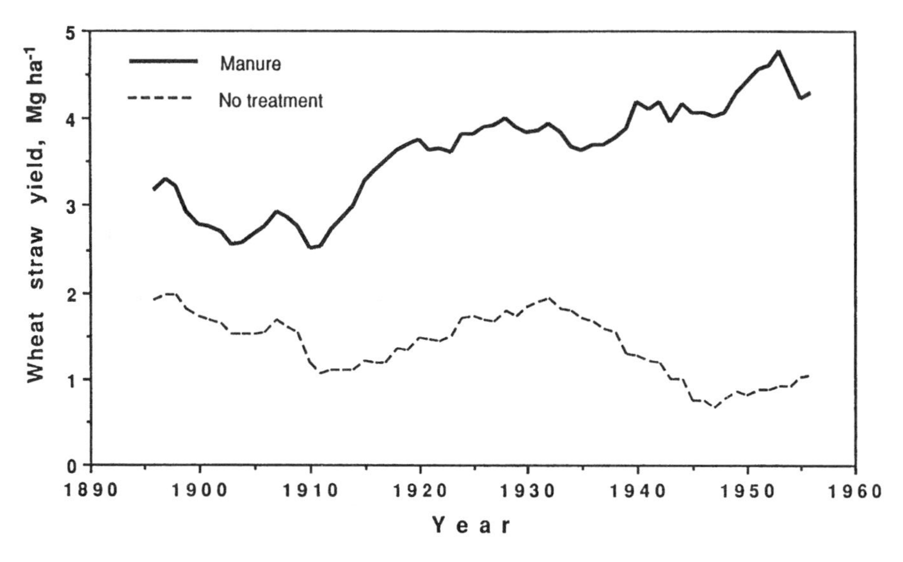

Figure 2 Wheat straw yield on Sanborn Field historical plots, as a 15-year moving average.

Total net annual production (TNAP) of fertilized winter wheat in our small-scale experiment varied from 13.4 to 14.0 Mg ha^{-1}, with grain yield ranging from 2.2 to 2.8 Mg ha^{-1} (Table 1). A best estimate of the ratio of net production to grain yield was 5:1. On this basis, the plant biomass, excluding grain, returned to the soil, was about 80% of total production. Root biomass of winter wheat was slightly less than the weight of straw. The vegetative shoot/root ratio which averaged 1.1:1 was relatively stable for the 3 years and was independent of weather conditions.

The estimate of C input obtained from ^{14}C deposition for straw (2.53 Mg ha^{-1})[7] corroborates that determined from direct measurements. The direct measurement method would exclude material lost by leaf drop but this was negligible for wheat during the growing period. For roots, however, significant amounts of exudates and sloughed-off material had to be accounted for when developing the total balance sheet of ^{14}C data. Total amounts of ^{14}C lost with soil respiration during 4 years from subplots with labeled roots indicated the initial quantity of below-ground carbon to be 2.9 Mg ha^{-1}. Root material at plant maturity, recovered from the washing procedure, contained only 55 to 60% of the labeled C estimated to have been deposited in the soil by the root system. The remaining below-ground production

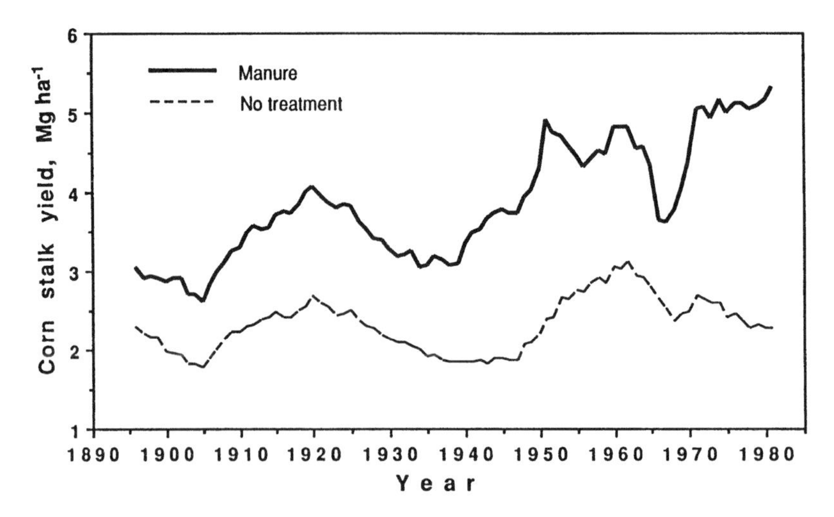

Figure 3 Corn stalk yield on Sanborn Field historical plots, as a 15-year moving average.

Table 1 Accumulation of Above- and Underground Biomass by Winter Wheat on Small-Scale Plots (Mg ha^{-1})

	1981	1982	1984	Mean ± SD
Grain	2.80	2.20	2.54	2.51 ± 0.30
Straw	5.90	6.16	5.63	5.90 ± 0.26
Roots (cm)				
0 – 10	3.66	4.36	3.82	3.95 ± 0.37
10 – 20	0.61	0.73	0.94	0.76 ± 0.17
20 – 30	0.28	0.34	0.32	0.31 ± 0.03
30 – 40	0.11	0.18	0.10	0.13 ± 0.04
40 – 50	0.11	0.18	0.05	0.11 ± 0.06
Total	4.77	5.62	5.27	5.21 ± 0.42
Total postharvest residue	10.67	11.78	10.87	11.11 ± 0.59
Shoots/roots ratio	1.23	1.09	1.07	1.13 ± 0.09
Total net annual production	13.47	13.98	13.41	13.62 ± 0.31
TNAP/grain ratio	4.8	6.3	5.3	5.5 ± 0.8

From Buyanovsky, G.A. and Wagner, G.M., *Plant Soil*, 93, 59, 1986. With permission.

was assumed to be in the form of exudates and sloughed-off material. Structural roots were concentrated primarily in the upper 20 cm of soil, where 90% of the total biomass was recovered.

Wheat yields on plots that had been fertilized since 1888 averaged 1 to 2 Mg ha^{-1} of grain for the period from 1910 to 1950. After 1950, grain yields were 2 to 3 Mg ha^{-1} as varieties changed, and these yields are similar to those obtained in the small-scale experiment (Table 1). Straw outputs for the same periods (pre- and post-1950) were about 3 Mg ha^{-1} and 4 to 4.5 Mg ha^{-1}, respectively. The values are similar to small plot observations (5 to 6 Mg ha^{-1} for straw cut just above the soil surface). Wheat straw harvested from the no-treatment plot was usually less than 2 Mg ha^{-1} (Figure 1) and that from manure-treated wheat was 3 Mg ha^{-1} during the first 25 to 30 years and then increased to 4 Mg ha^{-1} in more modern times.

Relationships between grain yield and total above-ground production for both long-term and small-scale experiments are also similar. For winter wheat on the main plots of Sanborn Field, grain yield averaged 37% of total grain plus forage production. The percent of yield as grain was higher under very low fertility and lowest for soil treated annually with 13.5 Mg ha^{-1} of manure (Figure 4 A to C). The average value from the main plots corroborates very well with our direct measurements (straw less stubble approximated at 5 Mg ha^{-1} and with grain yield 2.51 Mg ha^{-1} gave a factor of 33%). Assuming reliability of these relationships, the estimated carbon input from the root system in the Sanborn Field plots is presumed to be similar to our direct measurements (about 1.5 Mg ha^{-1}).

Net annual biomass production of corn averaged 35.4 Mg ha^{-1} and grain was 9.6 Mg ha^{-1} on the small-scale plots (Table 2). For the first 60-year period, manured corn on Sanborn Field yielded

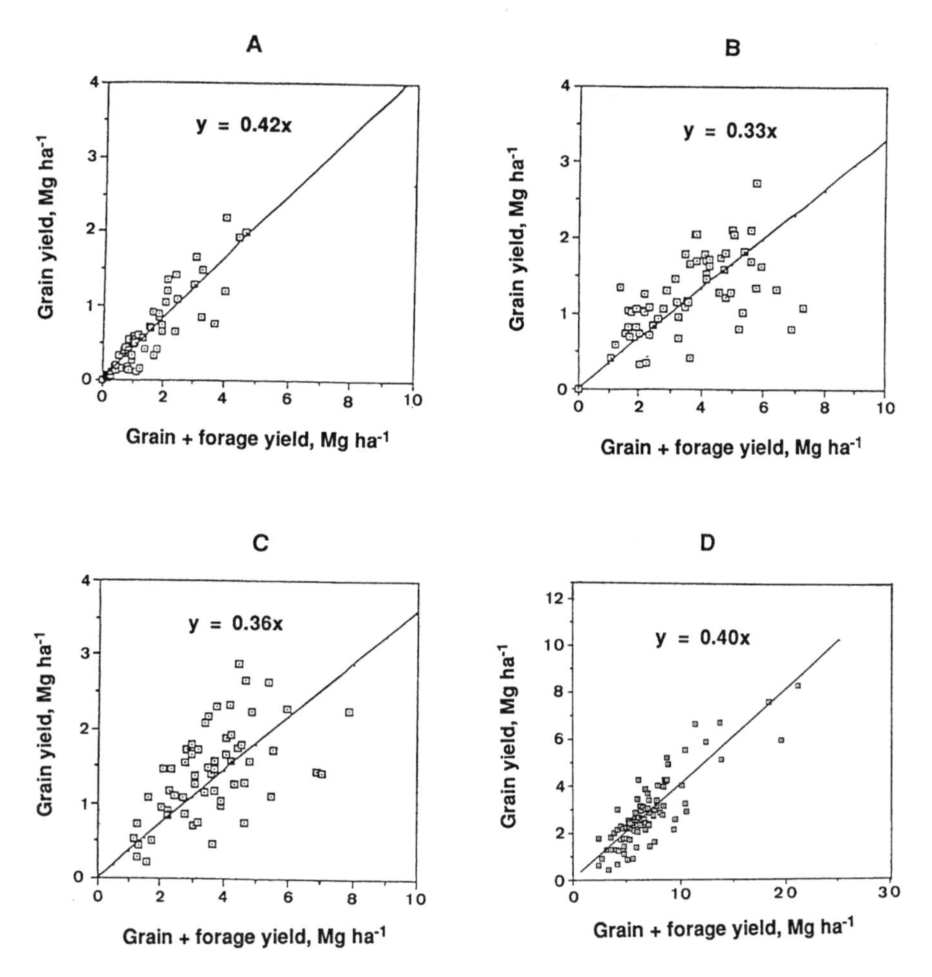

Figure 4 Relationships between grain yield and grain+forage on the Sanborn Field historical plots: (A) wheat — no treatment, (B) wheat plus 13.5 Mg ha^{-1} manure, (C) wheat plus mineral fertilizer (full treatment), (D) corn plus 13.5 Mg ha^{-1} manure. Manure additions values are for wet weight (up to 65% moisture).

approximately 2.5 Mg ha^{-1} of grain. Later, with improved varieties, it yielded between 2.5 and 5.5 Mg ha^{-1}, depending upon weather conditions. Higher yields and less variability in our small plot experiments were due to use of irrigation and mineral fertilizers.

Grain yield for the manure-treated corn was 40% of grain plus forage (Figure 4D). The same percentage was obtained from our direct measurements. The percentage calculated for no-treatment corn was 30%, but the individual data are not shown because the extremely low fertility level in this plot negates any value in interpreting contemporary data. The scatter of individual values for either culture was similar for corn and wheat. Although grain yields from chemically fertilized corn are available for 40+ years (since 1950), there have been no measurements of corn stalk residues which have been added back each year on this plot. The yields of corn stalk residue harvested from the no-treatment and manured plots are reported in Figure 3. Higher yields for hybrid varieties planted after 1950 are significant.

Corn developed a large and extensive root system, and accumulated 10 to 15 Mg ha^{-1} of underground biomass by the end of the growing season. The upper 10 cm of soil contained a sizable root crown and a considerable amount of root material with thickness of 2 to 3 mm in diameter that collectively comprised about 70% of the root mass. An additional 10% of the root biomass was fine roots in the upper 10 cm of soil.

Net annual biomass production of soybeans was 13.09 Mg ha^{-1} (Table 3). Considerable variation was demonstrated with annual values ranging from 10.63 Mg ha^{-1} in 1981 to 17.28 Mg ha^{-1} in 1983. Both above- and underground biomass varied in wide limits, although roots appeared to be more susceptible to variations in soil moisture and other abiotic influences. Soybean grain as a part of net annual production varied from year to year with an average of 77% of net production returned to the soil as residue.

Table 2 Accumulation of Above- and Underground Biomass by Corn on Small-Scale Plots (Mg ha^{-1})

	1981	1982	1992	Mean ± SD
Grain	9.54	9.47	9.82	9.61 ± 0.19
Cobs	1.62	1.53	1.97	1.71 ± 0.23
Stalks	13.78	11.46	10.94	12.06 ± 1.51
Above-ground residues	15.40	12.99	12.91	13.77 ± 1.42
Roots (cm)				
0–10 cm	7.94	12.59	7.97	9.50 ± 2.68
10–20 cm	1.81	1.52	1.27	1.53 ± 0.27
20–30 cm	0.45	0.57	0.38	0.47 ± 0.10
30–40 cm	0.31	0.34	0.24	0.30 ± 0.05
40–50 cm	0.34	0.21	0.19	0.25 ± 0.08
Total	10.85	15.23	10.05	12.04 ± 2.79
Total postharvest residue	26.24	28.22	22.96	25.81 ± 2.66
Shoots/roots ratio	1.42	0.85	1.28	1.18 ± 0.30
Total net annual production	35.78	37.69	32.78	35.42 ± 2.48
TNAP/grain ratio	3.75	3.98	3.34	3.69 ± 0.32

From Buyanovsky, G.A. and Wagner, G.H., *Plant Soil*, 93, 60, 1986. With permission.

Table 3 Accumulation of Above- and Underground Biomass by Soybeans on Small-Scale Plots (Mg ha^{-1})

	1981	1982	1983	1984	Mean ± SD
Seeds	2.80	3.89	2.84	2.33	2.97 ± 0.66
Pod hulls	1.25	1.65	1.32	1.20	1.36 ± 0.20
Leaves and stems	3.75	3.37	5.44	3.42	4.00 ± 0.98
Above-ground residues	5.00	5.02	6.76	4.62	5.35 ± 0.96
Roots (cm)					
0–10	1.96	3.09	4.00	3.10	3.04 ± 0.84
10–20	0.46	0.56	1.92	0.93	0.97 ± 0.67
20–30	0.24	0.32	1.36	0.22	0.54 ± 0.55
30–40	0.09	0.18	0.30	0.13	0.18 ± 0.09
40–50	0.08	0.05	0.10	0.05	0.07 ± 0.02
Total	2.83	4.20	7.68	4.43	4.78 ± 2.06
Total postharvest residue	7.82	9.22	14.44	8.99	10.12 ± 2.94
Shoots/roots ratio	1.76	1.20	0.88	1.04	1.22 ± 0.38
Total net annual production	10.63	13.11	17.28	11.32	13.09 ± 2.99
TNAP/grain ratio	3.79	3.37	6.08	4.85	4.52 ± 1.21

From Buyanovsky, G.A. and Wagner, G.H., *Plant Soil*, 93, 60, 1986. With permission.

Total annual inputs of C from plant residues into soil for the three different crops (Table 4) were calculated from the biomass data. For winter wheat this amounted to 3.72 Mg ha^{-1} of C. Lower C percentages of roots relative to that for tops in all crops presumably is due to the occurrence of clay particles that could not be washed free from the surfaces of small roots. Of the three crops, soybeans showed the smallest annual input of fresh carbon to the soil (3.43 Mg ha^{-1}) and, corn, the greatest (9.21 Mg ha^{-1}).

The grain yield factors allow an approximation of annual input of above-ground crop residues on the basis of the foregoing reported ratios of plant parts for the different crops. Shoot:root ratios obtained in our experiments (1.11 for wheat and 1.18 for corn) were used for root biomass computation. The ratios

Table 4 Carbon Content in the Plant Residues and Amounts of Postharvest Residues Added to Soil

	Wheat		Soybean		Corn	
	Tops	**Roots**	**Tops**	**Roots**	**Tops**	**Roots**
Carbon content (%)	37.0	29.6	40.5	26.2	40.9	26.1
Average annual input (Mg ha^{-1})	2.18	1.54	2.17	1.26	5.81	3.40
Tops plus roots (Mg ha^{-1})	3.72		3.43		9.21	

From Buyanovsky, G.A. and Wagner, G.H., *Plant Soil*, 93, 61, 1986. With permission.

of TNAP/grain (5.5 for wheat and 3.7 for corn) also assisted in establishing the biomass inputs. For soybeans this ratio was 4.5, a value intermediate between that for corn and wheat.

III. PLANT RESIDUE AS A SOURCE OF SOIL ORGANIC MATTER

The contribution of crop residues to SOM has been evaluated in a number of experiments which monitored the flow of C through the soil. The initial phase of the humification process is the decomposition of crop residues in the soil, and this has been evaluated for each crop under field conditions. In some studies, the proportion of residue C that enters the SOM pool was determined using ^{14}C measurements.

A. RESIDUE DECOMPOSITION

In one of our studies we monitored the decomposition of crop residues and enumerated the progression of the effecting microorganisms.[9] Residues of either corn, wheat, or soybean were placed in nylon mesh bags (1-mm mesh) and buried at 15-cm depth (to mimic tillage) in June 1982. At specified intervals, samples were exhumed, microbial counts were carried out, and a chemical analysis of the remains was made. Initial chemical composition of residues is reported in Table 5.

Table 5 Initial Chemical Composition of the Corn, Wheat, and Soybean Residue (g kg^{-1} dry wt) Buried in the Field

Component	Corn	Soybean	Wheat
Soluble components	293	557	288
Hemicellulose	268	90	184
Cellulose	284	222	361
Lignin	56	119	141
Ash	93	64	84
N	10	22	9

From Broder, M.W. and Wagner, G.H., *SSSA J.*, 52, 113, 1988. With permission.

The decay curves for the three residue types have been published elsewhere[10] and here we wish only to call attention to several differences in the rates of decay. Soybean residue had the most rapid rate of decomposition and wheat had the slowest. Thirty-two days after the initiation of the experiment, soybean retained 34% of its original ash-free dry weight compared with 58 and 53% for corn and wheat, respectively. Beyond this point, corn had a more rapid rate of decomposition than wheat. From November 1982 through July 1984, the rate of decay for all three residues decreased; however, the decrease was greater with soybean compared with wheat and corn, indicating that decomposition of soybean residue was nearly complete. At the July 1984 sampling, corn, soybean, and wheat residue retained 8, 5, and 16%, respectively, of their original weights.

The progressive loss of C for the 3-year period was analyzed using nonlinear regression to fit a two-component exponential model. The results, which define the percentage of the residue carbon remaining in the soil at any specified time, are as follows:

Corn $C = 48.6e^{-0.0464t} + 50.0e^{-0.0023t};\ r^2 = 0.994$

Soybean $C = 50.0e^{-0.1004t} + 50.0e^{-0.0100t};\ r^2 = 0.972$

Wheat $C = 47.9e^{-0.0392t} + 50.0e^{-0.0016t};\ r^2 = 0.988$

The first component represents the rate of mineralization of the readily decomposable fraction. This includes the simple sugars, soluble proteins, hemicellulose, and cellulose. The second component represents the mineralization of the resistant products of microorganisms which colonized the residues as well as a recalcitrant fraction of the plant residue. As indicated by the coefficients of determination, the equations account for nearly all the organic carbon remaining in the soil at various times. The half-lives for the components are reported in Table 6. The labile fraction demonstrated a half-life of 1 or 2 weeks with soybeans showing the briefest duration. The resistant fraction had a half-life for soybeans of nearly 7 weeks, whereas for corn this fraction showed a half-life of 10 months and for wheat it was 15 months.

Table 6 The Half-Life (d) of Organic Carbon of Various Residues Decomposing in Soil

	Labile	Resistant
Corn	15	301
Soybean	7	69
Wheat	18	433

Further information about decomposition as it related to chemical differences already mentioned (Table 5) was obtained by performing analyses of the residue remaining at several dates during decay. Hemicellulose and cellulose decayed rapidly, whereas lignin was observed to be considerably more durable. During the period 12 to 32 days, about half of the total hemicellulose and cellulose content was lost. During the next 49 days half of the remaining hemicellulose and 30% of the remaining cellulose was lost. In 6 months hemicellulose from the wheat and soybean residues was essentially exhausted but some small amounts persisted in corn residue beyond 1 year. The cellulose disappeared most rapidly during the initial 6 months and, in 2 years' time, about 5% of the original content remained in each of the three residues. Lignin had the slowest rate of decomposition and its progressive change with time is reported in Table 7. In 6 months, half of the original lignin had been decomposed. Wheat had the highest and soybean had the lowest total lignin content, and these differences influenced the relative order of the decay rates.

Table 7 Lignin Content Expressed as a Percentage of Corn, Soybean, and Wheat Residue, Initially and Following Decomposition in the Field

Sampling Date	Corn	Soybean	Wheat
Initial	5.6 ± 0.1	9.6 ± 0.3	14.0 ± 0.2
June 1982	12.0 ± 1.3	14.1 ± 1.3	14.5 ± 0.1
July 1982	13.9 ± 2.3	21.1 ± 3.2	18.9 ± 2.3
September 1982	16.8 ± 4.9	22.0 ± 3.0	21.0 ± 0.7
November 1982	14.4 ± 7.6	23.0 ± 0.2	19.6 ± 1.8
March 1983	17.2 ± 4.1	27.2 ± 1.9	21.7 ± 3.5
May 1983	22.4 ± 2.6	25.0 ± 0.5	18.8 ± 1.8
July 1983	20.4 ± 0.2	27.4 ± 6.8	23.6 ± 0.2
September 1983	20.2 ± 1.6	28.8 ± 3.0	18.2 ± 1.8
April 1984	26.2 ± 4.8	29.9 ± 2.7	20.1 ± 1.6
July 1984	22.4	31.9	22.6

B. INPUTS TO SOIL ORGANIC MATTER

The level of SOM under different long-term cropping systems of Sanborn Field is recognized to be related to the annual input of crop residues. However, the proportion of residue C flowing into the stable SOM pool is quite small. As the residues decay they undergo humification and this process converts residue C to stable SOM. At the same time, some of the existing SOM is mineralized each year and this rate varies around 2% per year. We have utilized our knowledge of SOM mineralization rates and observations on residue humification in order to construct balance sheets for C flowing into and out of the SOM pool.

The mineralization rates have been observed to vary somewhat with cropping and cultivation practices. Annual losses of soil organic C were estimated using measured soil C contents multiplied by the appropriate mineralization rates determined from earlier work on several of the historical plots.[11] The mean mineralization rate of 2% per year, as noted above, has been employed successfully for many years in soil testing procedures that evaluate available N for cultivated silt loam soils in this climatic region.[12] The rate is adjusted downward for sod crops and upward for crops such as corn where multiple tillage operations tend to more thoroughly aerate the soil. Several rates of mineralization have been used in constructing the balance sheet for carbon flow and this will be described subsequently after first presenting some discussion about the range of the rate used.

A SOM mineralization rate of 2%, representative for the fertilized wheat plots, is too rapid for soil with no fertilizer treatments during 100 years, where the SOM has dropped to a very low level and is, consequently, more recalcitrant. This recalcitrant SOM predominates in this plot and only a small pool of more active carbon is involved in the annual turnover so that, when mineralization is calculated relative to the total SOM, it is no more than 1% per year. Soils that have received annual additions of manure contain SOM that has been demonstrated to have a more rapid annual mineralization than soils without manure[11] and in each cropping system these increases approached one percentage point.

The C flowing each year into the soils of the historical plots of Sanborn Field was estimated from plant residue inputs multiplied by the coefficient of humification. This coefficient identifies the residue C that flows into the stable SOM pool where its residence time is 10 or more years. Information allowing the estimation of these coefficients has been obtained from several ^{14}C experiments.

In the case of wheat fertilized according to soil test, a humification coefficient has been obtained by the following approach. Mineralization and transfer of ^{14}C from wheat residues into humus was studied in specialized plots on Sanborn Field. The work showed that, by the end of the 1st year, 53% of the roots and 49% of the straw were mineralized (Table 8). Transfer to SOM corresponded to 100 to 110 g m^{-2} of plant C including 40 g from tops and 70 g from roots.[7] This flow constituted a mean of 20% of the initial C from the residue (humification coefficient of 0.2).

Table 8 Distribution of ^{14}C from Decomposing Wheat Residue by the End of the 1st Year after Labeling

		% of Starting ^{14}C		
Source	**Starting ^{14}C, ($kBq\ m^{-2}$)**	**CO_2**	**Residue**	**SOM**
Tops	1739	49	35	16
Roots	1990	53	23	24
Total	3729	52	28	20

From Buyanovsky, G.A. and Wagner, G.H., *Biol. Fertil. Soils*, 5, 76, 1987. With permission.

An annual C budget was constructed for wheat systems under steady-state conditions (Table 9). We have used data on the root and stubble C to determine the annual replacement of stable SOM C that was mineralized during the course of the year. For no-treatment wheat, the flow into humus is 144 kg ha^{-1} each year. With manure-treated wheat, the C humified each year from roots, stubble, and manure combined is 690 kg ha^{-1}.

Derived humification coefficients for fertilized wheat were 0.20 for above-ground residues and 0.24 for root systems. We are currently developing a value for corn root systems and assume that it is less than that for wheat roots. The humification coefficient for corn stalks also is probably lower than that for wheat straw. The balance data for corn without treatment and with annual addition of fertilizer (Table 9) is based on current assumptions. Current ^{14}C research for corn by our group is completed but the data are yet to be fully summarized and interpreted. The lowest index of humification, 0.1, is for barnyard manure, determined indirectly from ^{13}C studies.[13] These humification values and others utilized in drawing up the balance sheet are reported in Table 9. All of these balance data are based on annual input from roots and stubble only, because the above-ground residues were removed at harvest each year in these historical plots. Some balance data, derived using a somewhat different approach for fertilized corn and fertilized wheat (where above-ground residues were added back in recent years), are presented in Table 8 of another chapter of this volume.[8]

Table 9 Balance in Annual C Flow through SOM of Ap Horizon in the Historical Plots of Sanborn Field

Treatment	Total Soil C after 1940 (kg ha^{-1})	Annual Mineralization Rate (%)	Carbon in Roots and Stubble Added Annually[a] (kg ha^{-1})	Humification Coefficient	Carbon Humified (kg ha^{-1})
Wheat, no treatment	15,800	1	600	.24	144
Wheat, manure	34,800	2	1,500[b]	.24	360
			3,300	.10	330
Corn, no treatment	14,600	1.5	900	.24	216
Corn, manure	28,000	2.5	1,760	.22	380
			3,300	.10	330
Timothy	33,750	1.2	975	.40	390
Timothy, manure	49,750	2.0	1,800	.37	666
			3,300	.10	330

[a] Straw, stalks, and forage removed from all plots.
[b] Upper value is from crop, lower value from manure.

The apparent coefficients of humification for cultivated crops are significantly lower than that for the native prairie and do not appear appropriate for timothy grass. A value for root systems under perennial sod is more suitable and is taken from ^{14}C studies on native prairie grasses.[14,15] That work suggests we use a value of 0.4 for timothy sod that has had no treatment. The value is relevant in calculating humification of that portion of the root system produced annually. Total root biomass under a perennial crop may be nearly double the annual root production so that the actual value in relation to total living roots may be similar to that for wheat. For the preparation of an annual C balance, however, we must employ the apparent coefficient because we are using annual estimates of root production.

IV. CONCLUSIONS

Annual C input from major cultivated crops in Missouri is estimated at 3.7 Mg ha^{-1} for wheat, 3.4 Mg ha^{-1} for soybeans, and 9.2 Mg ha^{-1} for corn. The major part of the added C is mineralized during the first 2 to 3 years and is returned to the atmosphere as CO_2. The process is carried out by soil microorganisms and can be described by an exponential equation that suggests at least a two-component composition. A labile fraction demonstrated a half-life of 1 to 2 weeks and a more resistant fraction had a half-life extending several months. Residues rich in lignin experience the slowest decay.

Mean annual mineralization of SOM is about 2%, but the rate is more rapid with increased intensity of cultivation and is slower under sod crops. The loss of SOM C by mineralization is balanced by humification of a fraction of the residue C. This humification coefficient has been determined to be 0.2 for wheat. Other residues are assumed to have similar values whereas the value for barnyard manure is only 0.1.

REFERENCES

1. Upchurch, W.J., R.J. Kinder, J.R. Brown, and G.H. Wagner, Sanborn Field: Historical Perspective, Missouri Agricultural Station Research Bulletin, 1054, University of MO, Columbia, 1985.
2. Wagner, G.H. and G.E. Smith, Recovery of Fertilizer Nitrogen from Soils, Missouri Agricultural Experimental Station Bulletin, 738, 1960.
3. Wagner, G.H., Lessons in soil organic matter from Sanborn Field, in Proceedings of Sanborn Field Centennial, Missouri Agricultural Experimental Station Special Report, 415, 64, 1989.
4. Bauer, A. and A.L. Black, Soil carbon, nitrogen, and bulk density comparisons in two cropland tillage systems after 25 years and in virgin grassland, *SSSA J.*, 45, 1166, 1981.
5. Wagner, G.H., Humus under different long term cropping system, Colloque Humus-Azote, Reims, France, Assoc. Inter. Science du Sol Assoc. Fran. pour l'Etude du Sol., 1981, 23.
6. Buyanovsky, G.A. and Wagner, G.H., Post-harvest residue input into cropland, *Plant Soil*, 93, 57, 1986.
7. Buyanovsky, G.A. and Wagner, G.H., Carbon transfer in a winter wheat (*Triticum aestivum*) ecosystem, *Biol. Fertil. Soils*, 5, 76, 1987.

8. Buyanovsky, G.A., Brown, J.R., and Wagner, G.H., Sanborn Field: Effect of one hundred years of cropping on soil parameters influencing productivity, in *Soil Organic Matter in Temperate Agroecosystems: Long Term Experiments in North America*, Paul, E.A., Paustian, K., Elliott, E.T., and Cole, C.V., Eds., CRC Press, Boca Raton, FL, 1996, chap. 15.
9. Broder, M.W., Changes in the Chemical Composition of and Microbial Population on Corn, Wheat and Soybean Residue Decomposing in Sanborn Field, Ph.D. thesis, University of Missouri, Columbia, 1985.
10. Broder, M.W. and Wagner, G.H., Microbial colonization and decomposition of corn, wheat and soybean residue, *SSSA J.*, 53, 112, 1988.
11. Chahal, K.S. and Wagner, G.H., Decomposition of organic matter in Sanborn Field soils amended with ^{14}C glucose, *Soil Sci.*, 100, 96, 1968.
12. Graham, E.R., An Explanation of Theory and Methods of Soil Testing, University of Missouri, Agricultural Experimental Station Bulletin 734, 1959.
13. Wagner, G.H., Using natural abundance of ^{13}C and ^{15}N to examine soil organic matter accumulated during 100 years of cropping. Stable Isotopes in Plant Nutrition, Soil Fertility and Environmental Studies, International Atomic Energy Agency, Vienna (IAEA-SM-313), 1991, 261.
14. Dahlman, R.C. and Kucera, C.L., Root productivity and turnover in native prairie, *Ecology*, 46, 84, 1965.
15. Dahlman, R.C. and Kucera, C.L., Carbon-14 cycling in the root and soil components of a prairie ecosystem, in *Proceedings of the Second Symposium on Radioecology*, Publication Number 272, Oak Ridge National Laboratory, Oak Ridge, TN, 1969, 652.

Chapter 5

Biogeochemistry of Managed Grasslands in Central North America

Ingrid C. Burke, William K. Lauenroth, and Daniel G. Milchunas

CONTENTS

I. INTRODUCTION

Grasslands are among the most intensively managed ecosystem in the world, providing important land area for rangeland and for cultivation.[1] In the central grassland region of North America, a great deal of work has examined the influence of fire,[2] grazing,[3] and cultivation,[4,5] but little work has been done to draw the management responses of these grasslands into a conceptual framework. In this paper, we propose a conceptual framework for the biogeochemical responses of grasslands to management practices. In the context of this framework, we then review the state of knowledge regarding management impacts on ecosystems of the central grasslands.

0-8493-2802-0/97/$0.00+$.50

A. A CONCEPTUAL FRAMEWORK FOR THE EFFECTS OF MANAGEMENT ON GRASSLANDS

We have developed a conceptual framework to explain and predict the effects of human management practices on the biogeochemistry of grasslands in the central grassland region. The framework focuses on the potential effects of disturbance on altering the distribution and flux of biologically active elements in these grasslands. The fundamental theme is that disturbances are most likely to alter nutrient storage and cycling if they target large pools of nutrients. Small nutrient pools with short turnover times may be easily changed, but are likely to recover rapidly from disturbance. To support the framework, we have developed the relationship between carbon (C) pool sizes and turnover times for a shortgrass steppe site (Figure 1). Although other grassland sites in the central grassland region may vary significantly in the pool sizes, we suggest that the relative proportions of C in the pools and the turnover times are representative of grasslands throughout the region.

The components of the framework are as follows:

1. The pool of elements in plant biomass is relatively small (~10% of total) (Figure 1) and turns over relatively rapidly (2 to10 years). The above-ground biomass pool represents a small pool of elements that turns over rapidly (1 to 2 years), while the below-ground biomass pool is as much as ten times larger, and turns over more slowly (7 years).[13] Thus, perturbations of above-ground plant biomass alone are not likely to cause large changes in element storage, and recovery from such perturbations may be rapid (Figure 2).

This framework predicts that grazing or fire in any grassland system with elements distributed as described above will have relatively small impacts on biogeochemistry. However, there may be interesting interactions between plants and nutrient availability. For instance, grazing causes minimal changes in plant species composition in shortgrass steppe, but may force significant changes in the more mesic tallgrass prairie.[14] In systems such as the tallgrass prairie, species shifts may alter nutrient distribution or availability (see #4 below). Nonetheless, the influence of fire or grazing on element storage is likely to be small.

2. Most of the storage of biologically active elements (~90%, Figure 1) in grasslands is in soil organic matter (SOM), which has a slow rate of turnover, is well-protected from natural disturbance, and is thus generally resistant to change. Above-ground disturbances that do not directly influence this pool of organic matter have only minimal influences on total ecosystem carbon storage (Figure 2).

The soil organic matter may be conceptualized as consisting of several subpools that have been defined kinetically[9,15] (including[10] the Central Grasslands) as a very small active pool, an intermediate sized, partially stabilized but decomposable pool, and a large stable or recalcitrant pool. The recalcitrant is the largest of these pools, composing 60 to 80% of the SOM, with a very slow turnover rate, on the order of thousands of years. Plant inputs over relatively short time frames directly influence the smallest of these pools, the active SOM pool.

Most of the input to SOM from plants in grassland ecosystems is likely from below-ground organs. Further, most above-ground disturbances (e.g., grazing and fire) have indirect and unclear influences on root production.[3] Thus, any perturbation that directly influences only the above-ground plant compartment is unlikely to have a large effect on nutrient storage in grasslands. The exception occurs when removal of part or all of the above-ground biomass results in significant soil erosion.

Management practices such as cultivation that disrupt the soil and associated organic matter pools, however, have a very large impact on the storage of biologically active elements. This occurs through enhanced decomposition, losses of carbon, and thus reductions in the ability of the ecosystem to maintain biologically active elements such as N and P.[16,17] Conversely, because inputs to SOM are low relative to the pool size, element accumulation rates following disturbance are very slow.

Grasslands contrast with ecosystems in which a large amount of the organic carbon is stored above ground, such as forests.[18] A common characteristic of most ecosystems is a large pool of relatively inactive organic matter that provides resistance to disturbances.[19] A key element of our framework is the emphasis it places on the location of this large store of organic material relative to the effects of various kinds of disturbances. The C stored in above-ground biomass in forests can have turnover times comparable to the slowest component of C storage in grassland soils.[20] Despite the apparent vulnerability of the above-ground C stores to fire, Franklin and Waring[20] reported that wildfire consumes very little wood and therefore results in relatively small losses of C and nutrients. By contrast, harvesting and human-engineered fires can result in large losses of the carbon and nutrients stored in biomass. Such

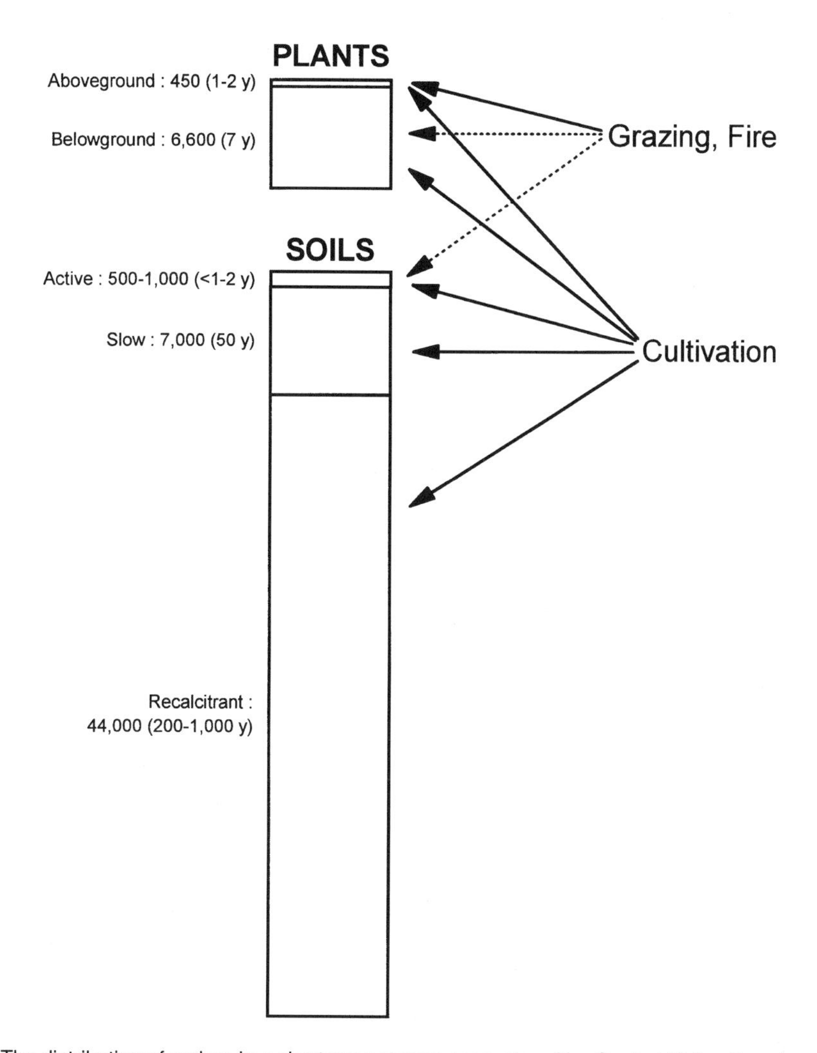

Figure 1 The distribution of carbon in a shortgrass steppe ecosystem (the Central Plains Experimental Range) to 1-m depth, showing the direct (solid lines) and indirect (dashed lines) influences of management on the component pools. Values are in kilograms per hectare; values in parentheses are turnover times of subpools. Estimates of above- and below-ground plant biomass are empirical harvest data;[6] turnover of plant biomass was determined from ^{14}C turnover.[7] We estimated active SOM C from potentially mineralizable C in 30-day incubations plus microbial biomass carbon determined from 10-cm samples[8] at 55 g m^{-2}; this may represent an underestimate of this pool for a 1-m depth. Turnover of the active pool is theoretical.[9,10] Estimates of "slow" or decomposable soil organic carbon and turnover are particulate organic matter estimates we recently measured using methods of Cambardella and Elliott[11] to a depth of 20 cm, likely representing the depth that contains most of the decomposable material.[6] Estimates of "recalcitrant" SOM were calculated by subtracting active and slow pools from the measured total C to a 1-m depth.[12] Turnover for this pool is theoretically based.[10]

losses may be maintained by continued human intervention, but, if the forest is allowed to regrow, a large proportion of the C and nutrients can be replenished over a relatively short period of time.[21]

3. Inorganic nutrients and labile SOM compose a small proportion of the biologically active elements in grasslands (Figure 1), and these pools turn over rapidly (days during the growing season). These pools are likely to be altered easily by management practices that manipulate plant inputs (Figure 2) or plant species, but will have only a minor influence on total ecosystem storage.

Current conceptual models of SOM include a component of the total that is termed "active",[15] comprising a small proportion (about 1%) of the biologically active elements, but conferring disproportionately important character with respect to soil fertility.[22] The inorganic pool of biologically active nutrients is smaller still, representing less than 0.1% of the total, and having a very high turnover rate.

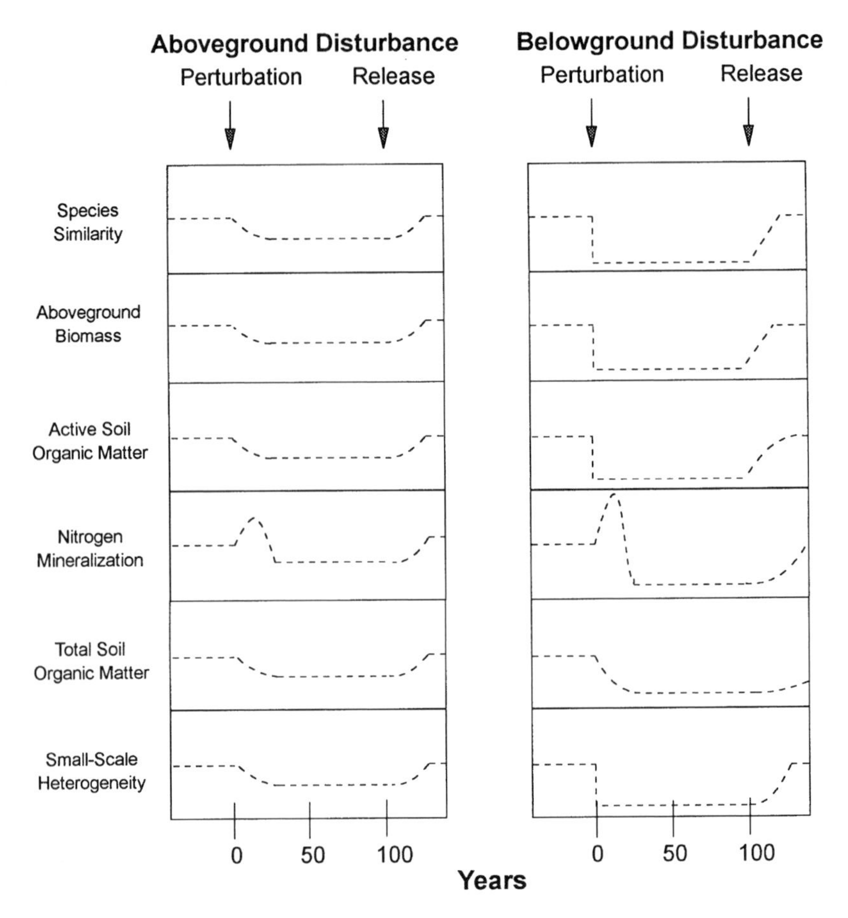

Figure 2 Conceptual framework for the response of grasslands to management practices. Management practices are viewed as a system level disturbance. Each plate in the figure represents the state of an ecosystem feature relative to a native grassland, its relative change in response to a disturbance, and release from that disturbance. We have generalized disturbances into two types to capture the biogeochemical impacts, disturbances that directly influence only the above-ground components such as grazing or fire and disturbances that directly influence the below-ground compartments, such as cultivation. See text for a description of each variable.

Alterations in litter inputs may significantly influence the character and size of the active SOM fraction and available nutrient pools. Although total element storage is unlikely to be influenced by management practices that only manipulate above-ground litter inputs, such manipulations may significantly influence the rate of nutrient mineralization (Figure 2). Grasslands differ from other systems in the ease in which above-ground litter inputs may be manipulated through protection from fire or grazing or through various rotations and intensities of fire and grazing. Such manipulations may alter the amount of litter being incorporated into the active SOM pool and the total energetic demand of microbes responsible for nutrient mineralization.[23,24] Our framework predicts that active SOM and inorganic nutrients may be easily decreased through fire or grazing and that these properties may recover rapidly following release from fire or grazing (Figure 2).

4. Small-scale heterogeneity of grasslands, at the scale of individual plants, may be reduced in response to management practices such as fire and cultivation that reduce plant cover or species diversity (Figure 2). Grazing may increase patch-scale heterogeneity by reinforcing patches with high fertility[25] or reduce it when a dominant species and total plant cover increase with grazing.[26]

Semi-arid grasslands are characterized by small-scale heterogeneity in soil nutrient pools and processes that correspond to the locations of individual perennial grasses.[8,27,28] More mesic grasslands are characterized by small-scale heterogeneity that corresponds more with differential species effects on litter quality and litter accumulation.[29,30] This heterogeneity is caused by both above- and below-ground litter inputs and may be decreased by any disturbance that reduces litter inputs or causes plant location to change over short time periods (e.g., shifts to annual species).

Below, we discuss the current state of knowledge with respect to the influence of management practices on perennial grasslands in the central grassland region. We focus on grazing, fire, and recovery from cultivation and first review the patterns of land use in the region. Effects of cultivation are covered in detail elsewhere in this volume. We wish to make clear that, although maps for the Canadian portion of the region were not available to us in digital form, we consider the entire North American central grassland region in our literature review.

B. OVERVIEW OF LAND USE IN THE CENTRAL GRASSLAND REGION OF NORTH AMERICA

The central grassland region of North America comprises a mosaic of native grasslands and croplands in the center of the continent between the Rocky Mountains and the Mississippi River. This region contains the major corn- and wheat-producing areas for the continent, in addition to important grazing lands for livestock.[31] Significant amounts of C are stored in the soils in this region; these reserves have been shown to be strongly affected by changes in land use.[4,5,32–34] The distribution of land use in the region is strongly influenced by climate; therefore, an understanding of land use pattern must begin with a description of climate.

The climate of the region shares many characteristics with other midcontinental areas in the temperate zone of the world.[35] Winters are dry, and summers provide most of the moisture. Seasonal and spatial climatic patterns are the result of interactions between the Rocky Mountains on the western border of the region and three air masses, each of which has different seasonal dynamics.[36] Ultimately, two important climatic gradients account for the distribution of ecosystems over the region: a west-east gradient of increasing precipitation and a north-south gradient of increasing temperature (Figure 3). Mean annual precipitation ranges from less than 300 mm y^{-1} to more than 1200 mm y^{-1} and mean annual temperature from less than 5°C to greater than 20°C. In addition to influencing ecosystem type, these gradients have large influences on net primary production and soil organic carbon. Above-ground primary production of native grasslands ranges from less than 100 g m^{-2} y^{-1} on the western edge of the region to more than 700 g m^{-2} y^{-1} on the eastern edge.[37] Soil organic carbon under native grasslands ranges from more than 8 kg m^{-2} in the northeast corner of the region to less than 3 kg m^{-2} in the southwest corner.[33]

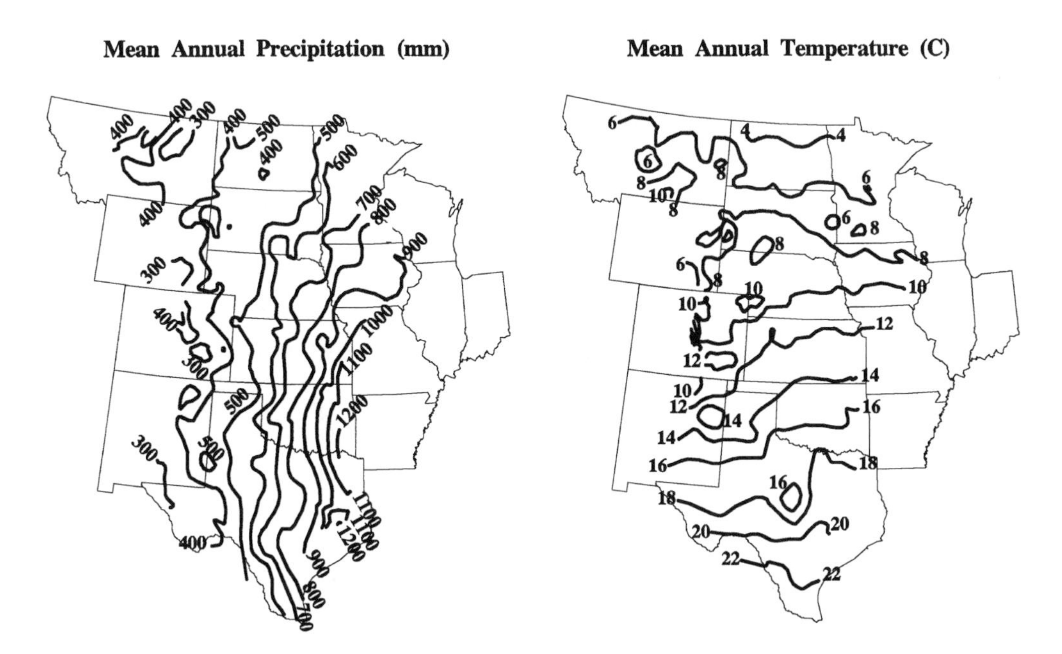

Figure 3 Climatic gradients of the central grasslands of the U.S.

Areas of native grassland can be identified with one of four major types found in the region (Figure 4). The types are defined based upon productivity and seasonality. Similarly, potential crop areas are defined based upon climate zones.[40] Seven crop/agricultural types fall within the central grassland region (Figure 3). The exact identity of a specific grassland or cropland depends upon local conditions and past management. For instance, central Kansas falls within the southern mixed prairie region yet heavily grazed pastures are often dominated by *Bouteloua gracilis* (H.B.K.) Lag. ex Griffiths, and *Buchloe dactyloides* (Nutt.) Engelm., the two most important species in the shortgrass steppe region.[42]

Prior to settlement by European-Americans in the 1800s, ecosystems in the central grassland region were heavily influenced by fires, wet and dry weather fluctuations, and by large numbers of native herbivores, including pronghorn, bison, elk, prairie dogs, and deer.[43] Settlement ultimately resulted in the replacement of many of the native herbivores by domestic cattle. In the late 1800s, farms began to replace ranches in many areas and by the early 1900s much of the cropland-grassland mosaic that we see today was established. The key factors controlling this mosaic of land uses are potential productivity,

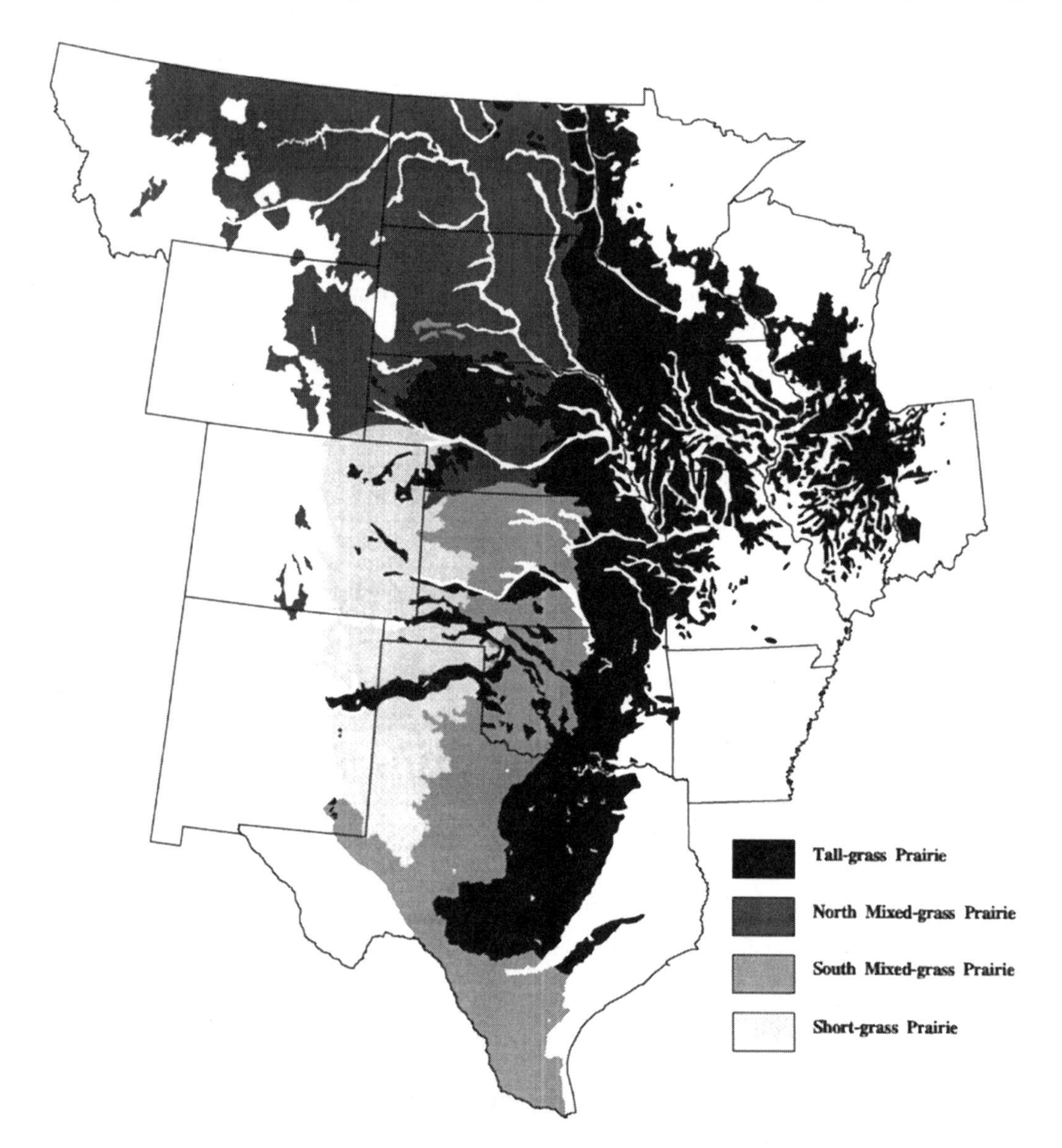

Figure 4 Geographic distribution of grassland types in the central grasslands region of the United States. (From Kuchler, A. W., Potential Natural Vegetation of the Coterminous United States, American Geographical Society Special Publication, New York, 1964.)

determined by annual precipitation and soil type, growing-season length, determined by annual temperature, availability of sources of irrigation water, and local social and economic conditions.

Currently, native grasslands are located primarily in the western part of the region (Figure 4) and cover an estimated 37%[44] of the U.S. portion. All but a very small proportion of these areas are grazed by domestic livestock. Areas that are managed for grazing tend to occur where the reliability of crop production is low, limited either by precipitation or the interaction of precipitation and soils.[45] In a recent analysis of the U.S. central Great Plains (Colorado, Nebraska, and Kansas), we found that precipitation, temperature, and soil texture could explain 67% of the variation in the spatial distribution of rangelands.[45] Proportion of land area in rangeland increases strongly as precipitation decreases, with this trend being greatest in regions with higher mean annual temperature, probably due to higher potential evapotranspiration (Figure 3). There are also interactions with soil texture; coarse-textured soils have the lowest likelihood of being cropped, at any precipitation level.[45] In dry areas, course-textured soils are not cultivated because of their vulnerability to soil erosion and in wet areas because they have low water and nutrient storage. Social and cultural factors probably account for the balance of the variation, not explained by environmental factors.

C. OVERVIEW OF GRASSLAND BIOGEOCHEMISTRY

To address the impacts of management on grasslands, we first need to review the basic structure of grasslands with respect to nutrient pools and fluxes. Several thorough reviews have been written regarding nutrient cycling in grasslands.[10,46–48] Below, we briefly review grassland biogeochemistry to provide a template for assessment of management effects.

Ecosystems dominated by perennial grasses are unique among ecosystems in their carbon and biologically active element storage in SOM.[48] Temperate grasslands have been estimated to contain 18% of global soil C reserves,[49] more than any other ecosystem except tundra. This large storage of soil C is attributed to low decomposition rates relative to net primary production.

Average above-ground net primary production in the central grassland region of the U.S. varies from 100 g m^{-2} y^{-1} in the western, low-precipitation extreme of the region to 700 g m^{-2} y^{-1} in the more mesic eastern areas.[37] Most estimates of below-ground net primary production have indicated that root production is higher than shoot production.[50,51] Dahlman and Kucera,[52] Warembourg and Paul,[53] and Milchunas and Lauenroth[13] used ^{14}C turnover analyses to evaluate below-ground net primary production and found that below-ground production was smaller than previously thought, ranging from 1 to 1.5 times that of above-ground production. These studies[13,52,53] spanned a large proportion of both the precipitation and temperature gradients in the central grassland region.

The early information on high below-ground net primary production was important in leading to the idea that most soil C in grasslands originates from below-ground detritus.[54] This is a difficult process to demonstrate, since it involves hundreds to thousands of years of soil development. However, two additional reasons suggest that roots are a more important source of detritus for soils than above-ground litter. First, the long-term presence of grazers in the central grassland region has selectively removed above-ground litter from the system, although some significant proportion of this is returned as feces. Second, root litter generally has higher C/N and lignin/N ratios than above-ground litter. We recently summarized data from the International Biological Program Shortgrass Steppe site to estimate above-ground C/N ratios of 46, lignin:N of 6, and below-ground C/N of 81, and lignin:N ratios of 27. High C/N or lignin:N ratios in forested systems have been found to lead to slower decomposition rates,[55] more effectively producing humus.

Total SOM in the central grassland region is determined by a combination of precipitation, temperature, soil texture, and land use management.[10,33] Regional patterns present an interesting paradox in the controls over SOM. Although temperature and precipitation are both positively correlated with decomposition and net primary production, organic matter levels increase with precipitation and decrease with temperature,[33] suggesting that decomposition is more constrained by temperature, and production by precipitation.

Nitrogen inputs to the central grassland region are dominated by wet deposition, rather than N fixation,[56] and are proportional to total precipitation.[10] Leaching is not an important loss vector for N except in the wettest areas, since precipitation is approximately balanced by actual evapotranspiration in all but the wettest grassland regions.[57] Losses of N from grasslands are dominated by N_2O loss,[47,58,59] accounting for 9 to 25% of wet deposition inputs. Ammonia volatilization from soils, plants, and herbivore urine[60,61] may balance the N cycle such that total losses are equal to inputs.

Nitrogen is cycled relatively slowly within the soil-plant system in grasslands, as a result of a large pool of inactive SOM.[10,47,62] In situ N mineralization may average 1 to 4% of the total soil organic N, depending upon the amounts of litter retained in the system.[63–66] Clark[62] demonstrated that, over the short term, a small proportion of the N in a shortgrass steppe ecosystem may cycle rapidly between plants and inorganic pools (Figure 1). Temporal dynamics of N mineralization are likely to be closely coupled to water availability,[60] showing large pulses following precipitation events.

The influence of soil texture on grassland biogeochemistry has been documented from both process work and regional patterns. A large body of work has demonstrated that fine-textured soils retain more SOM than coarse-textured soils,[33,63,67,68] and this pattern is evidenced both on landscape and regional scales. Microbial biomass and nutrient availability also tend to be highest on fine-textured soils, but are a higher proportion of total SOM on coarse-textured soils.[69] This trend is attributed to the effect of clay in physical and chemical protection of SOM.[70,71]

In addition, soil texture may influence net primary production through its influence on water availability.[37,72] Coarse-textured soils have more water for plant growth in dry regions and less in wet regions because of the effects of texture on evaporation and drainage beyond the root zone. Sala[37] reported higher above-ground net primary production on sandy soils compared to clay soils in dry portions of the central grassland region and the opposite effect in wet areas. The amount of annual precipitation at which the texture effect was neutral was 370 mm.

Recently, several investigators have demonstrated a significant influence of plant species composition on SOM accumulation and nutrient availability in grasslands.[8,29,73] Plant litter quality may have an important feedback to nutrient availability through its influence on decomposability and immobilization, and such effects have been demonstrated in both tallgrass prairie[29] and shortgrass steppe.[8] Such properties as biomass allocation and plant longevity may also influence organic matter accumulation.[73] Additionally, in semiarid regions, location of individual plants has an important impact on the accumulation and turnover of organic matter at small spatial scales.[8,27,74] Discontinuous plant cover that characterizes the semiarid grasslands such as the shortgrass steppe results in significant small-scale heterogeneity in total nutrient pools, available N, and microbial biomass.[8,27,28] Perennial plants accumulate C and N beneath their canopies as demonstrated in a large number of shrub-dominated arid and semi-arid systems,[75-79] at a scale that corresponds to individual grasses.

II. THE INFLUENCE OF GRAZING ON BIOGEOCHEMISTRY

Grazing by large domestic herbivores is the most common management strategy imposed on the remaining native grasslands of the central grassland region. We have organized our review of the effects of grazing into two sections, following Floate.[80] In the first section, we address the direct effects of large herbivores on biogeochemical pools and processes. These effects include the direct removal of above-ground biomass, the redistribution of nutrients via excrement deposition, volatile losses of N from this excrement, and erosion. In the second section, we discuss indirect effects of large herbivores, including changes in total SOM, root biomass, microbial dynamics, nutrient availability, plant allocation of nutrients, and fundamental ecosystem stoichiometry.

Studies addressing the long-term influence of grazing on ecosystems generally compare long-term grazing exclosures to adjacent grazed systems. Such studies are extremely useful; however, there is often a bias in the interpretation of results. Differences between grazed and exclosed areas are generally interpreted such that exclosed areas are assumed to represent native, "virgin", or pre-European influence conditions.[81,82] Milchunas et al.[83–85] found small responses in plant community dynamics to cattle grazing in shortgrass steppe; ungrazed plant communities were more similar to a variety of other disturbed communities than were grazed communities. They interpreted this relatively small response, and the direction of the response, as the result of a long evolutionary history of grazing. One might reasonably question whether exclosure from grazing is a "native" condition and rather, if ecosystems in exclosed areas might be aggrading, instead of the grazed areas degrading. Careful interpretation of differences between grazed and exclosed areas is necessary for understanding the long-term impacts of grazing on SOM pools and processes.

A. DIRECT EFFECTS

Milchunas and Lauenroth[3] recently reviewed the literature related to grazing effects on above-ground net primary production, root biomass, and SOM pools[3] across continents. Worldwide, there is not a simple response with respect to the influence of grazing on above-ground net primary production (ANPP), and there has been a great deal of debate regarding the response of ANPP to grazing intensity.[3,86–89] In general, systems with long evolutionary histories of grazing experience only slight reductions or even increases in ANPP under domestic grazing, while those that have short evolutionary histories have significant reductions in ANPP.[3]

Although grazing influences on ANPP are not consistent across grasslands, effective above-ground inputs into SOM are clearly reduced as a direct result of herbivory. Sims and Singh[51] estimated that 19 to 30% of ANPP is removed by cattle grazing grasslands of the western U.S.; Heitschmidt et al.[90] estimated that 29 to 40% of ANPP is removed by cattle in Texas shortgrass steppe. Lauenroth and Milchunas[91] calculated that cattle removed about 30% of above-ground net primary production at the Central Plains Experimental Range (CPER) in shortgrass steppe. Variation in above-ground biomass induced by grazing was roughly equivalent to that induced by landscape position.[92] Milchunas et al.[84] found that decreased litter cover, increased bare ground, and increased basal cover of the dominant species were the most significant influences of cattle on shortgrass steppe structure.

Of the biomass removed by grazers, a significant proportion (50 to 75%) is returned as feces and urine.[91,93,94] The return of this material in a new form may have at least three significant effects on ecosystems. First, the material is returned with a much different quality than the above-ground biomass that was removed, having undergone initial decomposition in the rumen of the cattle.[80] The urine and soluble fecal material are relatively labile, but much of the fecal mass has low content of carbohydrates, high cellulose concentration, and narrow C/N ratios due to the advanced decomposition stage of this material.[95] The return of soluble material may be responsible for many of the indirect effects that suggest more rapid cycling of nutrients as a result of grazing.[80,89] Second, feces and urine are distributed at the small scale such that high amounts are concentrated in small patches, and, at the landscape scale, the material is redistributed due to animal foraging behavior.[94,96] This may alter landscape patterns in total organic matter and nutrient capital after long periods of grazing. Areas that receive a great deal of animal activity, such as fencelines and water areas, have been found to accumulate N relative to grazed areas that receive less traffic, such as ridgetops.[94]

Third, urine is much more subject to volatile and leaching losses of N than the live biomass that was originally consumed.[14,56,60] Ammonia volatilization from urine in shortgrass steppe has been found to range from 1 to 15% of total N in urine.[14] Schimel et al.[60] found that volatile losses of N from urine were highly stratified by landscape position in shortgrass steppe. This pattern was likely due to high levels of plant uptake and microbial immobilization on lowland positions. Total volatile losses averaged across the landscape were estimated to be 0.07 g m^2 y^{-1}, about 25% of inputs via wet deposition. Thus, although earlier work suggested that cattle grazing caused significant volatile losses of N from grasslands,[56] it is currently thought that these losses are relatively insignificant relative to total system N capital. Parton and Risser[97] have estimated that the net removal of N in herbivore biomass is small in shortgrass steppe ecosystems, averaging only 0.06 g m^2 y^{-1}.

B. INDIRECT EFFECTS

Herbivory by large mammals results in the direct removal of above-ground plant biomass, which has significant effects on grassland biogeochemistry. However, if below-ground production is the source of most grassland soil organic matter, it may be that indirect effects of grazing on below-ground production and plant nutrient allocation will have at least as significant effects on the distribution and flux of nutrients within a grazed ecosystem as direct effects.

Milchunas and Lauenroth[3] and Lauenroth et al.[74] reviewed studies that addressed the responses of below-ground biomass and SOM to grazing across a range of continents, and in the central grassland region, respectively. At both the global scale and the regional scale, no consistent trends were found with respect to below-ground biomass SOM, or total nutrient pools. For example, at the same shortgrass steppe site, Milchunas and Lauenroth[92] found that long-term exclosures had significantly higher root biomass than long-term heavily grazed pastures, but Sims et al.[98] found that there was higher root biomass in grazed than ungrazed pastures, and Leetham and Milchunas[99] found no difference between long-term grazed and ungrazed treatments. In northern mixed prairie, Holland and Detling[100] reported significant decreases in root biomass with grazing in South Dakota, but Brand and Goetz[101] found significant increases in North Dakota. In areas where species shifts occur under grazing, such as in mixed prairie or tallgrass prairie, some root biomass changes may occur.

Few investigators have estimated the effects of grazing on below-ground net primary production, and none have used sophisticated methods such as ^{14}C turnover. Standing crop of root biomass may not indicate below-ground inputs, since it is not useful in assessing turnover of fine roots. There are numerous methodological problems with assessing root turnover;[13,83,102,103] however, this may be the key issue for understanding grazing effects on SOM. Jaramillo and Detling[104] found in a laboratory defoliation experiment that root/shoot ratios significantly decreased as a result of light and heavy defoliation, but that plants that had been subjected to grazing in the field over long time periods had higher root/shoot ratios in the laboratory than those that had not been grazed in the field. They suggested that grazing pressure selected for individual plants could offset shoot removal by having a large investment in roots.[105–107] Although root/shoot ratios were higher in long-term grazed sites than ungrazed, total root biomass was lower in grazed sites than ungrazed.[100,104] Thus, individual plants responded to grazing with an increase in below-ground allocation, but, on an ecosystem basis, there was less root biomass in the grazed sites. This result reiterates some of the complexity inherent in understanding the indirect effects of grazing on biogeochemistry of grassland ecosystems. Long-term responses may differ from short-term responses, and individual plant responses may differ from total ecosystem responses.

Similarly, there have been no consistent results with respect to the influence of grazing on total SOM or total soil nutrient pools.[3] Of the studies that have been conducted in the central grassland region, several have shown higher levels of SOM pools in grazed vs. ungrazed pastures,[108–111] some have shown lower organic matter in grazed pastures,[82,112–114] and others have shown no significant difference between grazed and ungrazed.[115] Interestingly, Bauer et al.[81] found that total soil C and P were significantly lower in grazed than ungrazed pastures, but that N was higher. Studies that found lower levels of SOM in grazed sites may be interpreted as organic matter decreasing with grazing, or as exclosed areas increasing in SOM due to increased levels of litter inputs, relative to grazed areas.

One of the most consistent observations regarding the influence of grazing in the central grassland region as well as elsewhere is that above-ground plant N concentrations increase.[104,116–119] Recent field and simulation studies[100,120] have attributed this increase in above-ground N concentration to microbial-plant interactions. Holland and Detling[100] found that N mineralization increased with grazing intensity, concomitant with decreases in root biomass. They attributed these trends to decreasing below-ground C allocation with grazing (see above), decreased microbial N immobilization, and increased net N mineralization. Simulation studies using the Century model[120] supported these interactions. This model accounts for several of the effects of grazing and their interactions at a particular site in South Dakota, including decreased root biomass, increased above-ground N concentrations, and increased N availability. However, Dormaar et al.[115] found decreased N mineralization with increasing grazing intensity in fescue grasslands of Canada, due to decreased active pools of organic matter.

Other indirect effects of grazing on grassland biogeochemistry are likely to occur. An interesting interaction may occur between plant species and nutrient cycling in systems in which shifts in community composition occur as a result of grazing. For example, tallgrass prairie sites under heavy grazing tend to shift in species composition toward shortgrass species.[121] Such shifts may change plant litter chemistry, decomposition dynamics, and nutrient cycling. In addition, the characteristic small-scale heterogeneity of shortgrass steppe[27] is apparently reduced to more homogeneous distributions of above- and below-ground biomass as a result of grazing.[92]

Why is there no general model for the influence of grazing on grassland biogeochemistry? It seems that there is an unusual amount of disarray in the observations of the effects of grazing on ecosystem structure and function, even for results obtained just within the central grassland region of North America. Each study must be interpreted based upon details of treatment, sampling, and particular vegetation, with no emerging generalizations for biogeochemical theory. Milchunas and Lauenroth[3] have been successful in developing a consistent model for the response of ANPP and community structure to grazing using arguments about the evolutionary history of grazing at individual locations. However, their analysis indicated no relationship among ANPP, community structure, and soil nutrients. Regional-scale studies, using similar consistent methods and focusing on sites with similar grazing histories, could provide significant insight into the effects of grazing on short- and long-term distribution and fluxes of elements in ecosystems. Certainly, at present the data do not suggest that there are dramatic and significant impacts of moderate levels of grazing on biogeochemistry in the central grassland region.

III. FIRE

Fire has been an important part of the evolutionary history of North American grasslands and, in particular, the tallgrass prairie.[122,123] Fires may have had both natural and human-induced origins, and carried well in this system due to a relatively high fuel load.[124,125] Such fires may have been important for the maintenance of tallgrass prairie which would otherwise be susceptible to woody plant invasion.[123,126]

A large number of studies have reviewed the influence of fire on ANPP in tallgrass prairie,[121] and relatively fewer have focused on fire effects on organic matter pools and nutrient availability.[25,66]

A. DIRECT EFFECTS

The most immediate, direct effect of burning is to remove above-ground biomass in the form of volatile gases. Ojima et al.,[66] reviewing data from the Konza Prairie in Kansas, found that biomass and C and N losses to combustion ranged from 63 to 89% per burn, or about 1 to 4 g m^{-2} N,[25,127] while P losses were relatively low. This level of loss is comparable to N inputs via precipitation plus fixation. Thus, frequency of burning is a large factor in determining net ecosystem losses of N. Combustion losses represent net losses from the ecosystem, compared with grazing, which returns some of the organic matter through excretion.[25] Some of the material that is combusted is litter. Knapp and Seastedt[64] and Hulbert[128] suggested that this removal of detritus has major consequences for tallgrass ecosystem structure and function through its indirect effects on light, N, and water availability.

B. INDIRECT EFFECTS

Removal of plant canopy and detritus by combustion decreases light interception, thereby increasing light and temperatures at the soil surface.[128,129] Increases in light reaching the soil surface have been demonstrated to increase soil temperatures,[129] particularly in spring, which result in increased evapotranspiration and reduced water availability.[125] These alterations in light and soil microclimate may have significant effects on plant phenology, plant community characteristics, net primary production, N availability, and plant nutrient allocation. For this review, we focus on the key processes responsible for the distribution and flux of elements, plant production and nutrient allocation, and soil microbial dynamics.

Increases in soil temperature and losses of detrital material both influence the rate of surface litter decomposition in burned prairie. Ojima et al.[127] found that short-term responses to annual burning (1 and 2 years after initiation of burning regime) include higher microbial C and N, higher standing pools of NH_4 and NO_3, and higher in situ net N mineralization rates relative to unburned prairie. Similar results were found for mountain grasslands.[130] Hobbs and Schimel[130] also suggested that short-term increases in N availability resulting from fire cause depressions in N-fixation rates.[131] Sites that were burned annually for long periods of time (18 years), however, showed significant reductions in microbial C and N and net N mineralization rates. Reductions in microbial activity were interpreted as being the result of long-term decreases in above-ground litter inputs.

Although a large number of studies have been conducted, there is not complete agreement on the impact of burning on ANPP. Studies in the tallgrass prairie of Kansas have consistently demonstrated increases in ANPP with burning.[25,65,66,127,132] However, several studies of mixed grass prairie have indicated that ANPP decreases as a result of fire.[64,121,133] The suggested reason for this trend is that fire increases ANPP in sites that gain from increased soil temperatures (growing season and N availability), but decreases ANPP where litter provides an added advantage of moisture retention. Few studies have reported the response of BNPP or below-ground biomass to fire. Ojima et al.[66] found a significant increase in below-ground live roots with annual burning on both long- and short-term burned sites in tallgrass prairie.

What are the long-term effects of burning on SOM and productivity? Simulation studies[134] suggest that the long-term (over 25 years) annual removal of above-ground biomass via combustion results in significant declines in total soil organic C, N, and P, with the least losses occurring with P, which is returned in ash. Simulated ANPP and net N mineralization, however, remained high in burned sites due to the reduced detritus and microbial immobilization dynamics. Field data supporting these results are not yet available.

The extent to which fire alters total C or nutrient capital may depend entirely on the frequency of burning and the extent to which pastures are grazed in combination with burning.

IV. FIRE–GRAZING INTERACTIONS

Recent studies at the Konza Prairie have indicated that fire and grazing may interactively influence grassland biogeochemistry.[25] Less above-ground biomass and litter accumulate in a grazed system, and this decreases fuel load and fire temperature. As a result, and somewhat counterintuitively, an ecosystem both burned and grazed lost less N via combustion than one that was burned but protected from grazing. In addition, fire increased the nutritional value of subsequent forage, probably through decreasing litter accumulation and microbial immobilization, and increasing nutrients available to plants. Further, frequent (annual) burning reduced patch heterogeneity, resulting in more uniform utilization of the landscape by herbivores. Such interactions may reflect the pre-settlement condition of the prairie, in which both fire and grazing occurred.

V. GRASSLAND RESPONSE TO AND RECOVERY FROM CULTIVATION

A very large proportion of the area that was native grasslands in North America is currently cultivated (Figure 5 — 63% of U.S. portion), but land use has been dynamic throughout the past 100 years. There have been several periods that resulted in widespread return of cropland to native grassland. A devastating drought during the mid-1930s[135] resulted in widespread abandonment of cropland throughout the western Great Plains. Other, shorter duration and less-intense droughts have resulted in smaller-scale cropland abandonment in marginal climatic or edaphic zones. In addition, cropland retirement policies such as the Soil Bank and the recent Conservation Reserve Program have resulted in the return of previously cultivated lands to perennial grasslands.[136]

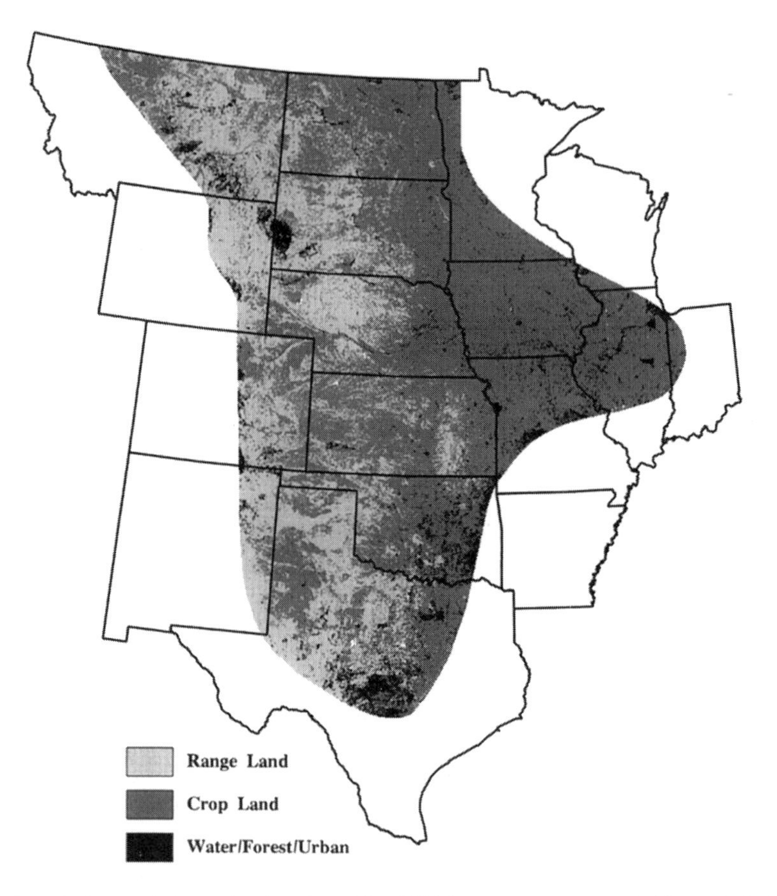

Figure 5 Land use in the central grassland region of the U.S. The map was generated from USDI U.S. Geologic Survey. (From USDI U.S. Geological Survey, Land Use and Land Cover Digital data from 1:250,000 and 1:100,000 Scale Maps, National Mapping Program Technical Instructions, Data Users Guide 4, Reston, VA, 1986.)

A. A BRIEF REVIEW OF CULTIVATION EFFECTS

Cultivation has the largest impact of any management system on native grassland ecosystem structure and biogeochemical dynamics.[5] It is fundamentally unlike anything in the evolutionary history of grasslands. Perhaps most striking is the notion that cultivation directly impacts two of the most important features of grasslands that we outlined above: the dominant below-ground organic matter pool and the presence of perennial, drought-resistant plants that apportion half or more of their net primary production below ground. Below, we briefly review the mechanisms by which cultivation changes grasslands. Our explanation is brief because agroecosystems are not strictly grassland ecosystems, but the context is important for a discussion of grasslands recovering from disturbance.

A large body of literature exists on the effects of cultivation management on ecosystem biogeochemistry, particularly focusing on SOM storage and nutrient availability.[4,5,32,33,137,138] The effects of cultivation are clear, resulting in widespread and significant losses of SOM. These losses are attributed to several mechanisms that both enhance outputs from SOM and diminish inputs. Cultivation increases decomposition rates by mixing, degrading soil aggregates, and enhancing contact of litter and inter-aggregate SOM with decomposing organisms.[139,140] Plant residue is sometimes removed from ecosystems, decreasing litter inputs.[141] In addition, perennial grass species are replaced with annual species that generally have lower root:shoot ratios. Finally, plowing greatly enhances erosive potential of soils.[137] The net effect of these processes is to reduce SOM contents between 20 and 50% over periods of several decades. Since these grasslands are at least several thousand years old, this represents a very rapid loss relative to the rate at which these soils developed. The total amounts of loss vary with climatic and edaphic factors, such that the largest absolute losses occur where the highest SOM pools are found.[33] Over large regions and long time periods, these losses in SOM have been demonstrated to represent significant reductions in terrestrial carbon storage.[34]

Losses in total SOM correspond with decreases in active SOM pools,[67] reducing microbial biomass and N mineralization. Reductions in nutrient availability correspond with declines in net primary production and grain yield in the absence of fertilizer applications.[142] Some investigators have found higher N turnover in cultivated systems as a result of decreased immobilization potential.[60]

B. RECOVERY FROM CULTIVATION

Long-term patterns of ecosystem recovery from cultivation in the central grassland region are not well known. Large-scale cropland abandonment has been an important historical property of the central grassland region, and understanding is important for predicting the effects of management strategies such as the Conservation Reserve Program.[136]

One of the fundamental components of our conceptual model is that plant species characteristics have a large impact on the structure and function of grassland ecosystems.[8] What are the important trends in plant succession following cultivation that may influence biogeochemical cycling? Although a number of studies have investigated plant successional dynamics on recovering fields in the central grassland region[143–147] few general trends have been advanced. One clear pattern is that during the early stages of succession (first several years), annual plants often dominate. The rate at which long-lived perennial grasses return to fields and begin to dominate is unclear, and may depend upon a number of factors, including past management (intensity of cropping and fertilization rates), current management (grazing and stocking rate), weather, and its influence on the seed pool present in adjacent areas.[147]

Recent work on recovering fields suggests that total SOM recovery may be extremely slow.[28,146,148,149] As plowing halts, decomposition rates likely decline, and inputs of litter to SOM increase as residue removal stops and perennial plants with high root inputs become established. However, the rate SOM increase is slow relative to the magnitude of past losses. Burke et al.[28] estimated that 25 g C m^{-2} (2.5% of surface soil) has recovered in fields abandoned for over 50 years, in soils that have lost 250 g m^{-2} (25% of surface soil) during approximately 20 years of cultivation. These results are consistent with recent conceptual models that suggest that pedogenetic processes are too slow to represent significant rates of global atmospheric carbon capture.[150]

The active SOM pool and nutrient availability, however, may recover much more rapidly than the total pools [28] (Figure 2). Microbial biomass responds over seasonal or yearly time scales to organic matter inputs, and, over several decades, the active pool is capable of supplying nutrients at rates similar to native fields. These results are consistent with the concept of the active pool having a turnover time of several years. In addition, the active pool and nutrient supply capacity regain spatial heterogeneity associated with perennial plants following 50 years of recovery (Figure 2).[28] This level of recovery is dependent upon the factors that govern plant community dynamics. Thus, an important feature regarding the spatial structure of ecosystems — the small-scale distribution of available nutrients — may recover relatively rapidly if native perennial plants become established.

VI. SUMMARY

Ecosystems of the central grassland region of North America are characterized by having a relatively small storage of C and associated nutrients in the above-ground plant biomass compartment, and a large storage of C and other elements stored in the SOM pool with a slow turnover rate. The plant communities of this region are relatively resistant to several of the most common human management strategies, including grazing and fire, because these management practices closely resemble natural disturbances. Grazing and fire have relatively minor influences on element distribution because there are small element pools associated with plants. Both grazing and fire tend to increase nutrient cycling rates and alter the active SOM pool, particularly in the high precipitation and productivity portions of the region. Recovery of ecosystems from grazing and fire is rapid (Figure 2) due to the rapid turnover time of active soil pools and because native vegetation generally persists through fire and grazing. In contrast, cultivation directly impacts the large pool of elements stored in SOM through physical and biological enhancements of losses. Recovery of ecosystems from cultivation is a very slow process because native perennial plant communities have been removed and, more importantly, because the total SOM pool has a very slow turnover time.

ACKNOWLEDGMENTS

The authors wish to express appreciation to the numerous reviewers of this manuscript, including M. A. Vinton, M. Aguiar, R. Kelly, J. Paruelo, M. Robles, E. Paul, and two anonymous reviewers.

REFERENCES

1. Coupland, R.T., Ed., Grassland ecosystems of the world: analysis of grasslands and their uses, in *International Biological Programme*, Vol. 18, Cambridge University Press, London, 1979.
2. Collins, S.L. and L.L. Wallace, Eds., *Proceedings of a Symposium: Fire in North American Grasslands*, University of Oklahoma Press, Norman, OK, 1990.
3. Milchunas, D.G. and W.K. Lauenroth, Quantitative effects of grazing on vegetation and soils over a global range of environments, *Ecol. Monogr.*, 63, 327, 1993.
4. Haas, H.J., C.E. Evans, and E.R. Miles, Nitrogen and Carbon Changes in Soils as Influenced by Cropping and Soil Treatments, USDA Technical Bulletin, 1164, U.S. Government Printing Office, Washington, DC, 1957.
5. Paustian, K., H.P. Collins, and E.A. Paul, Management controls on soil carbon, in *Soil Organic Matter in Temperate Agroecosystems: Long-Term Experiments in North America*, Paul, E.A., Paustian, K., Elliott, E.T., and Cole, C.V., Eds., CRC Press, Boca Raton, FL, 1996, chap. 5.
6. Liang, Y.M., D.L. Hazlett, and W.K. Lauenroth, Biomass dynamics and water use efficiencies of five plant communities in the shortgrass steppe, *Oecologia*, 80, 148, 1989.
7. Milchunas, D.G. and W.K. Lauenroth, Carbon dynamics and estimates of primary production by harvest, ^{14}C dilution, and ^{14}C turnover, *Ecology*, 73, 593, 1992.
8. Vinton, M.A. and I.C. Burke, Interactions between individual plant species and soil nutrient status in shortgrass steppe, *Ecology*, in press.
9. van Veen, J.A., J.N. Ladd, and M. Amato, Turnover of carbon and nitrogen through the microbial biomass in a sandy loam and a clay soil incubated with ^{14}C(U) glucose and [^{15}N]$NH_{4(2)}SO_4$ under different moisture regimes, *Soil Biol. Biochem.*, 17, 747, 1985.
10. Parton, W.J., D.S. Schimel, C.V. Cole, and D.S. Ojima, Analysis of factors controlling soil organic matter levels in Great Plains grasslands, *Soil Sci. Soc. Am. J.*, 51, 1173, 1987.
11. Cambardella, C.A. and E.T. Elliott, Particulate soil organic-matter changes across a grassland cultivation sequence, *Soil Sci. Soc. Am. J.*, 56, 777, 1992.
12. Yonker, C.M., D.S. Schimel, E. Paroussis, and R.D. Heil, Patterns of organic carbon accumulation in a semiarid shortgrass steppe, Colorado, *Soil Sci. Soc. Am. J.*, 52, 478, 1988.
13. Milchunas, D.G. and W.K. Lauenroth, Carbon dynamics and estimates of primary production by harvest, ^{14}C dilution, and ^{14}C turnover, *Ecology*, 73, 593, 1992.
14. Milchunas, D.G., O.E. Sala, and W.K. Lauenroth, A generalized model of the effects of grazing by large herbivores on grassland community structure, *Am. Nat.*, 132(1), 87, 1988.
15. van Veen, J.A. and E.A. Paul, Organic carbon dynamics in grassland soils. I. Background information and computer simulation, *Can. J. Soil Sci.*, 61, 185, 1981.
16. Redfield, A.C., The biological control of chemical factors in the environment, *Am. Sci.*, 46, 206, 1958.
17. Reiners, W.A., Complementary models for ecosystems, *Am. Nat.*, 127, 59, 1986.
18. Lieth, H., Primary productivity of the major vegetation units of the world, in *Primary Productivity of the Biosphere*, Lieth, H. and R.H. Whittaker, Eds., Springer-Verlag, New York, 1975, 203.
19. O'Neill, R.V. and D.E. Reichl, Dimensions of ecosystem theory, in *Proceedings of the 40th Annual Biology Colloquium*, Waring, R.H., Ed., Oregon State University Press, Corvallis, OR, 1979, 11.

20. Franklin, J.F. and R.H. Waring, Distinctive features of the northwestern coniferous forest: development, structure, and function, in *Proceedings of the 40th Annual Biology Colloquium*, Waring, R.H., Ed., Oregon State University Press, Corvallis, OR, 1979, 59.
21. Houghton, R.A., J.E. Hobbie, J.M. Melillo, B. Moore, B.J. Peterson, G.R. Shaver, and G.M. Woodwell, Changes in the carbon content of terrestrial biota and soils between 1860 and 1980: a net release of CO_2 to the atmosphere, *Ecol. Monogr.*, 53, 235, 1983.
22. Paustian, K. and T.A. Bonde, Interpreting incubation data on nitrogen mineralization from soil organic matter, *INTECOL Bull.*, 15, 101, 1987.
23. Paul, E.A. and R.P. Voroney, Nutrient and energy flows through soil microbial biomass, in *Current Perspectives in Microbial Ecology*, M.J. Klug, and C.A. Reddy, Eds., American Society of Microbiology, Washington, DC, 1984, 509.
24. van Veen, J.A., J.N. Ladd, and M.J. Frissel, Modelling C and N turnover through the mitobial biomass in soil, *Plant Soil*, 76, 257, 1984.
25. Hobbs, N.T., D.S. Schimel, C.E. Owensby, and D.S. Ojima, Fire and grazing in the tallgrass prairie: contingent effects on nitrogen budgets, *Ecology*, 72, 1374, 1991.
26. Milchunas, D.G. and W.K. Lauenroth, Three-dimensional distribution of plant biomass in relation to grazing and topography in the shortgrass steppe, *Oikos*, 55, 82, 1988.
27. Hook, P.B., I.C. Burke, and W.K. Lauenroth, Heterogeneity of soil and plant N and C associated with individual plants and openings in North American shortgrass steppe, *Plant Soil*, 138, 247, 1991.
28. Burke, I.C., W.K. Lauenroth, and D.P. Coffin, Recovery of soil organic matter and N mineralization in previously cultivated shortgrass steppe: implications for the conservation reserve program, *Ecol. Applic.*, in press.
29. Wedin, D.A. and D. Tilman, Species effects on nitrogen cycling: a test with perennial grasses, *Oecologia*, 84, 433, 1990.
30. Vinton, M.A. and I.C. Burke, Plant effects on soil nutrient dynamics along a precipitation gradient in Great Plains grasslands, submitted.
31. United States Department of Agriculture, Soil Conservation Service, Land Resource Regions and Major Land Resource Areas of the United States, Agriculture handbook 296, U. S. Government Printing Office, Washington, DC, 1981.
32. Tiessen, H., J.W.B. Stewart, and J.R. Bettany, Cultivation effects on the amounts and concentrations of carbon, nitrogen, and phosphorus in a grassland soils, *Agron. J.*, 74, 831, 1982.
33. Burke, I.C., C.M. Yonker, W.J. Parton, C.V. Cole, K. Flach, and D.S. Schimel, Texture, climate, and cultivation effects on soil organic matter content in U.S. grassland soils, *Soil Sci. Soc. Am. J.*, 53(3), 800, 1989.
34. Burke, I.C., T. G.F. Kittel, W.K. Lauenroth, P. Snook, and C.M. Yonker, Regional analysis of the central Great Plains: sensitivity to climate variability, *BioScience*, 41, 685, 1991.
35. Paruelo, J.M., W.K. Lauenroth, E.H. Epstein, I.C. Burke, M.R. Aguiar, and O.E. Sala. Regional climatic similarities in the temperate zones of North and South America, *Global Ecol. Biogeogr. Lett.*, submitted.
36. Lauenroth, W.K., and I.C. Burke, Climate variability in the Central Grasslands of the U.S., The Great Plains: Climate variability, *Encyclopedia of Environmental Biology*, Academic Press, in press.
37. Sala, O.E., W.J. Parton, L.A. Joyce, and W.K. Lauenroth, Primary production of the Central Grassland region of the United States, *Ecology*, 69, 40, 1988.
38. Kuchler, A.W., Potential Natural Vegetation of the Coterminous United States, American Geographical Society Special Publication, New York, 1964.
39. French, N.R., *Perspectives in Grassland Ecology*, Springer-Verlag, New York, 1979.
40. Dregne, H.F. and W.O. Willis, *Dryland Agriculture*, American Society of Agronomy, Madison, WI, 1983.
41. USDI U.S. Geological Survey, Land Use and Land Cover Digital Data from 1:250,000 and 1:100,000 scale maps, National Mapping Program Technical Instructions, Data Users Guide 4, Reston, VA, 1986.
42. Great Plains Flora Association, Flora of the Great Plains, University Press of Kansas, Lawrence, KS, 1986.
43. Stebbins, G.L., Coevolution of grasses and herbivores, *Ann. Mo. Bot. Gard.*, 68, 75, 1986.
44. Burke, I.C., unpublished data.
45. Burke, I.C., W.K. Lauenroth, W.J. Parton, and C.V. Cole, Regional assessment of land use, in *Integrated Regional Models*, Groffman, P.M., Likens, G.E., et al., Eds., Chapman and Hall, London, 1994, 79.
46. Clark, F.E., C.V. Cole, and R.A. Bowman, Nutrient cycling, in *International Biological Program*, Vol. 19, Breymeyer, A.I. and G.M. Van Dyne, Eds., Grasslands, systems analysis and man, Cambridge University Press, Cambridge, 1980, 659.
47. Parton, W.J., A.R. Mosier, and D.S. Schimel, Rates and pathways of nitrous oxide production in a shortgrass steppe, *Biogeochemistry*, 6, 45, 1988.
48. Clark, F.E. and R.G. Woodmansee, Nutrient cycling, Ecosystems of the World, 8A, Coupland R.T., Natural Grasslands: Introduction and Western Hemisphere, Elsevier, New York, 1992, 137.
49. Atjay, G.L., P. Ketner, and P. Duvigneaud, Terrestrial primary production and phytomass, SCOPE, 13, Bolin, B., E.T. Degens, S. Kempe, and P. Ketner, Eds., *The Global Carbon Cycles*, John Wiley & Sons, New York, 1979, 129.
50. Singh, J.S., W.K. Lauenroth, H.W. Hunt, and D.M. Swift, Bias and random errors in estimates of net root production: A simulation approach, *Ecology*, 65, 1760, 1984.
51. Sims, P.L. and J.S. Singh, The structure and function of ten western North American grasslands. III. Net primary production, turnover, and efficiencies of energy capture and water use, *J. Ecol.*, 66, 573, 1978.

52. Dahlman, R.C. and C.L. Kucera, Carbon cycling in the root and soil components of a prairie ecosystem, Proc. 2nd National Symposium on Radioecology, Ann Arbor, Mich., 1967, 652.
53. Warembourg, F.R. and E.A. Paul, Seasonal transfers of assimilated ^{14}C in grassland: plant production and turnover, soil and plant respiration, *Soil Biol. Biochem.*, 9, 295, 1977.
54. Dormaar, J.F., Decomposition as a process in natural grasslands, Ecosystems of the World 8A, Natural Grasslands Introduction & Western Hemisphere, R.T. Coupland, Ed., Elsevier, New York, 1992.
55. Melillo, J.M., J.D. Aber, and J.F. Muratore, Nitrogen and lignin control of hardwood leaf litter decomposition dynamics, *Ecology*, 63, 621, 1982.
56. Woodmansee, R.G., Additions and losses of nitrogen in grassland ecosystem, *BioScience*, 28(7), 448, 1978.
57. Rosenberg, N.J., *Microclimate: the Biological Environment*, John Wiley & Sons, New York, 1974.
58. Schimel, D.S., S. Simkins, R. Rosswall, A.R. Mosier, and W.J. Parton, Scale and the measurement of nitrogen-gas fluxes from terrestrial ecosystems, in *Scales and Global Change*, Rosswall, R., R.G. Woodmansee, and P.G. Risser, John Wiley & Sons, New York, 1988, 179.
59. Mosier, A., D. Schimel, D. Valentine, K. Bronson, and W. Parton, Methane and nitrous oxide fluxes in native, fertilized, and cultivated grasslands, *Nature*, 350, 330, 1991.
60. Schimel, D.S. and W.J. Parton, Microclimatic controls of nitrogen mineralization and nitrification in shortgrass steppe soils, *Plant Soil*, 93, 345, 1986.
61. Milchunas, D.G., W.J. Parton, D.S. Bigelow, and D.S. Schimel, Factors influencing ammonia volatilization from urea in soils of the shortgrass steppe, *J. Atmos. Chem.*, 6, 323, 1988.
62. Clark, F.E., Internal cycling of nitrogen in shortgrass prairie, *Ecology*, 58, 1322, 1977.
63. Schimel, D., M.A. Stillwell, and R.G. Woodmansee, Biogeochemistry of C, N, and P in a soil catena of the shortgrass steppe, *Ecology*, 66(1), 276, 1985.
64. Knapp, A.K. and T.R. Seastedt, Detritus accumulation limits productivity of tallgrass prairie, *BioScience*, 36, 662, 1986.
65. Seastedt, T.R., Mass, nitrogen, and phosphorus dynamics in foliage and root detritus in tallgrass prairie, *Ecology*, 69, 59, 1988.
66. Ojima, D.S., W.J. Parton, D.S. Schimel, and C.E. Owensby, Simulated impacts of annual burning on prairie ecosystem, in *Proceedings of a Symposium: Fire in North American Grasslands*, S.L. Collins and L.L. Wallace, Eds., Ecological Society of America and Botanical Society of America, University of Oklahoma Press, Norman, OK, 1990, 118.
67. Schimel, D.S., D.C. Coleman, and K.A. Horton, Soil organic matter dynamics in paired rangeland and cropland toposequences in North Dakota, *Geoderma*, 36, 201, 1985.
68. Aguilar, R. and R.D. Heil, Soil organic carbon, nitrogen, and phosphorus quantities in northern Great Plains rangelands, *Soil Sci. Soc. Am. J.*, 52, 1076, 1988.
69. Schimel, D.S., W.J. Parton, F.J. Adamsen, R.G. Woodmansee, R.L. Senft, and M.A. Stillwell, The role of cattle in the volatile loss of nitrogen from a shortgrass steppe, *Biogeochemistry*, 2, 39, 1986.
70. Sorenson, L.H., Carbon-nitrogen relationships during the humification of cellulose in soils containing different amounts of clay, *Soil Biol. Biochem.*, 13, 313, 1981.
71. Paul, E.A. and F.E. Clark, *Soil Microbiology and Biochemistry*, Academic Press, 1989.
72. Noy-Meir, I., Desert ecosystems: environment and producers, *Annu. Rev. Ecol. Syst.*, 4, 23, 1973.
73. Lauenroth, W.K., D.P. Coffin, and I.C. Burke, Effects of plant mortality on population dynamics and ecosystem structure: a case study, in *Plant Functional Types*, Smith, T.M., H.H. Shugart, and F.I. Woodward, Eds., Cambridge University Press, in press.
74. Lauenroth, W.K., D.G. Milchunas, J.L. Dodd, R.H. Hart, R.K. Heitschmidt, and L R. Rittenhouse, *Grazing in the Great Plains of the United States*, Society for Range Management, Vavra, M. and W.A. Laycock, Eds., Denver, 1994, 69.
75. Charley, J.L. and N. E. West, Plant-induced soil chemical changes in some shrub-dominated semi-desert ecosystems of Utah, *J. Ecol.*, 63, 94, 1975.
76. Charley, J. L. and N. E. West, Micro-patterns of nitrogen mineralization activity in soils of some shrub-dominated semi-desert ecosystems of Utah, *Soil Biol. Biochem.*, 9, 357, 1977.
77. Burke, I.C., Control of N mineralization in a sagebrush steppe landscape, *Ecology*, 70, 1115, 1989.
78. Bolton, H., J.L. Smith, and R.E. Wildung, Nitrogen mineralization potential of shrub-steppe soils with different disturbance histories, *Soil Sci. Soc. Am. J.*, 54, 887, 1990.
79. Schlesinger, W.H., J.F. Reynolds, G.L. Reynolds, G.L. Cunningham, L.F. Huenneke, W.M. Jarrell, R.A. Virginia, and W.G. Whitford, Biological feedbacks in global desertification, *Science*, 247, 1043, 1990.
80. Floate, M.J.S., Effects of grazing by large herbivores on nitrogen cycling in agricultural ecosystems, *Terres. Nitrogen Cycles. Ecol. Bull.*, 33, 585, 1981.
81. Bauer, A., C.V. Cole, and A.L. Black, Soil property comparisons in virgin grasslands between grazed and nongrazed management systems, *Soil Sci. Soc. Am. J.*, 51, 176, 1987.
82. Johnston, A., J.F. Dormaar, and S. Smoliak, Long-term grazing effects on fescue grassland soils, *J. Range Manage.*, 24, 185, 1971.
83. Milchunas, D.G., W.K. Lauenroth, J.S. Singh, C.V. Cole, and H.W. Hunt, Root turnover and production by ^{14}C dilution: implications of carbon partitioning in plants, *Plant Soil*, 88, 353, 1985.

84. Milchunas, D.G., W.K. Lauenroth, P.L. Chapman, and M.K. Kazempour, Effects of grazing, topography, and precipitation on the structure of a semiarid grassland, *Vegetation*, 80, 11, 1989.
85. Milchunas, D.G., W.K. Lauenroth, P.L. Chapman, and M.K. Kazempour, Community attributes along a perturbation gradient in a shortgrass steppe, *J. Veg. Sci.*, 1, 375, 1990.
86. McNaughton, S.J., On plants and herbivores, *Am. Nat.*, 128, 765, 1986.
87. Belsky, A.J., Does herbivory benefit plants? A review of the evidence, *Am. Nat.*, 127, 870, 1986.
88. Belsky, A.J., The effects of grazing: confounding of ecosystem, community, and organism scales, *Am. Nat.*, 129, 777, 1987.
89. Detling, J.K., Grasslands and savannas: regulation of energy flow and nutrient cycling by herbivores, *Concepts of Ecosystem Ecology — A Comparative View*, Ecological Studies 67, Pomeroy, L.R. and J.J. Alberts, Eds., Springer-Verlag, New York, 1988.
90. Heitschmidt, R.K., D.L. Price, R.A. Gordon, and J.R. Fasure, Short duration grazing at the Texas Experimental Ranch: effects on above-ground net primary productivity and seasonal growth dynamics, *J. Range Manage.*, 35, 367, 1982.
91. Lauenroth, W.K. and D.G. Milchunas, Short-grass steppe, in *Ecosystems of the World*, Vol. 8A, Coupland, RT., Ed., Elsevier, New York, 1992, 183.
92. Milchunas, D.G. and W.K. Lauenroth, Three-dimensional distribution of plant biomass in relation to grazing and topography and in the shortgrass steppe, *Oikos*, 55, 82, 1989.
93. Dean, R., J.E. Ellis, R.W. Rice, and R.E. Bement, Nutrient removal by cattle from a shortgrass prairie, *J. Appl. Ecol.*, 12, 25, 1975.
94. Senft, R.L., The Redistribution of Nitrogen by Cattle, Ph.D. thesis, Colorado State University, Fort Collins, 1983.
95. Pendleton, D.F., Degradation of Grassland Plants, M.S. thesis, Colorado State University, Fort Collins, 1972.
96. Stillwell, M.A. and R.G. Woodmansee, Chemical transformations of urea-nitrogen and movement of nitrogen in a shortgrass prairie soil, *Soil Sci. Soc. Am. J.*, 45, 893, 1981.
97. Parton, W.J., and P.G. Risser, Impact of management practices on tallgrass prairie, *Oecologia*, 46, 223, 1980.
98. Sims, P.L., J. S. Singh, and W.K. Lauenroth, The structure and function of ten western North American grasslands, *J. Ecol.*, 66, 251, 1978.
99. Leetham, J.W. and D.G. Milchunas, The composition and distribution of soil microarthropods in the shortgrass steppe in relation to soil water, root biomass, and grazing by cattle, *Pedobiologia*, 28, 311, 1985
100. Holland, E.A., and J.K. Detling, Plant response to herbivory and belowground nitrogen cycling, *Ecology*, 71, 1040, 1990.
101. Brand, M.D. and J. Goetz, Vegetation of exclosures in southwestern South Dakota, *J. Range Manage.*, 39, 434, 1986.
102. Lauenroth, W.K., H.W. Hunt, D.M. Swift, and J.S. Singh, Reply to Vogt et al., *Ecology*, 67, 580, 1986.
103. Vogt, K.A., C.C. Grier, S.T. Gower, D.G. Sprugel, and D.J. Vogt, Overestimation of net root production: a real or imaginary problem?, *Ecology*, 67, 577, 1986.
104. Jaramillo, V.J. and J.K. Detling, Grazing history, defoliation, and competition: effects on shortgrass production and nitrogen accumulation, *Ecology*, 69, 1599, 1988.
105. Detling, J.K., M.I. Dyer, and D.T. Winn, Net photosynthesis, root respiration, and regrowth of Bouteloua gracilis following simulated grazing, *Oecologia*, 41, 127, 1979.
106. Caldwell, M.M., J.H. Richards, D.A. Johnson, R.S. Nowak, and R.S. Dzurec, Coping with herbivory: photosynthetic capacity and resource allocation in two semiarid Agropyron bunchgrasses, *Oecologia*, 50, 14, 1981.
107. Coughenour, M.B., Graminoid responses to grazing by large herbivores: adaptations, expatiations, and interacting processes, *Ann. Mo. Bot. Gard.*, 72, 852, 1985.
108. Gardner, J.L., Effects of thirty years of protection from grazing in desert grassland, *Ecology*, 31, 44, 1950.
109. Ketling, R.W., Effects of moderate grazing on the composition and plant production of a native tall-grass prairie in central Oklahoma, *Ecology*, 35, 200, 1954.
110. Beebe, J.D. and G.R. Hoffman, Effects of grazing on vegetation and soils in southeastern South Dakota, *Am. Midl. Nat.*, 80, 96, 1968.
111. Smoliak, S., J.F. Dormaar, and A. Johnston, Long-term grazing effects on Stipa-Bouteloua prairie soils, *J. Range Manage.*, 25, 246, 1972.
112. Smith, C.C., The effect of overgrazing and erosion upon the biota of the mixed-grass prairie of Oklahoma, *Ecology*, 21, 381, 1940.
113. Brown, J.W. and J.L. Shuster, Effects of grazing on a hardlands site in the southern Great Plains, *J. Range Manage.*, 22, 418, 1969.
114. Wood, M.K. and W.H. Blackburn, Vegetation and soil responses to cattle grazing systems in the Texas Rolling Plains, *J. Range Manage.*, 37, 303, 1984.
115. Dormaar, J.F., S. Smoliak, and W.D. Wilms, Distribution of nitrogen fractions in grazed and ungrazed fescue grassland Ah horizons, *J. Range Manage.*, 43(1), 6, 1990.
116. Coppock, D.L., J.K. Detling, J.E. Ellis, and M.I. Dyer, Plant-herbivore interactions in a North American mixed-grass prairie. I. Effects of black-tailed prairie dogs on intraseasonal above-ground plant biomass and nutrient dynamics and plant species diversity, *Oecologia*, 56, 1, 1983.
117. Polley, H.W. and J.K. Detling, Herbivory tolerance of *Agropyron smithii* populations with different grazing histories, *Oecologia*, 77, 261, 1988.

118. Coughenour, M.B., Biomass and nitrogen responses to grazing of upland steppe on Yellowstone's northern winter range, *J. Appl. Ecol.*, 28, 71, 1991.
119. Milchunas, D.G., A.S. Varnamkhasti, W.K. Lauenroth, and H. Goetz, Forage quality in relation to long-term grazing history, current-year defoliation, and water resource, *Oecologia*, in press.
120. Holland, E.A., W.J. Parton, J.K. Detling, and D.L. Coppock, Physiological responses of plant populations to herbivory and their consequences for ecosystem nutrient flow, *Am. Nat.*, 140(4), 685, 1992.
121. Kucera, C.L., Tall-grass prairie, *Ecosystems of the World*, Vol. 8A, Coupland, R.T., Ed., Elsevier, New York, 1992, 227.
122. Weaver, J.E., *North American Prairie*, Johnson Publishing, Lincoln, NE, 1954.
123. Axelrod, D.I., Rise of the grassland biome, central North America, *Bot. Rev.*, 51, 163, 1985.
124. Bragg, T.B., Seasonal variation in fuel and fuel consumption by fires in a bluestem prairie, *Ecology*, 63, 7, 1982.
125. Knapp, A.K., Effect of fire and drought on the ecophysiology of *Andropogon gerardii* and *Panicum virgatum* in a tallgrass prairie, *Ecology*, 66(4), 1309, 1985.
126. Bragg, T.B. and L.C. Hulbert, Woody plant invasion of unburned Kansas bluestem prairie, *J. Range Manage.*, 29, 1976.
127. Ojima, D.S., D.S. Schimel, W.J. Parton, and C.E. Owensby, Long- and short-term effects of fire on nitrogen cycling in tallgrass prairie, *Biogeochemistry*, 24, 67, 1994.
128. Hulbert, L.C., Causes of fire effects in tallgrass prairie, *Ecology*, 69, 46, 1988.
129. Knapp, A.K., Post-burn differences in solar radiation, leaf temperature, and water stress influencing production in a lowland tallgrass prairie, *Am. J. Bot.*, 71, 220, 1984.
130. Hobbs, N.T. and D.S. Schimel, Fire effects on nitrogen mineralization and fixation in mountain shrub and grassland communities, *J. Range Manage.*, 37(5), 402, 1984.
131. Schimel, D.S., T.G.F. Kittel, A.K. Knapp, T.R. Seastedt, W.J. Parton, and V.B.B. Brown, Physiological interactions along resource gradients in a tallgrass prairie, *Ecology*, 72, 672, 1991.
132. Hulbert, L.C., Fire and litter effects in undisturbed bluestem prairie, *Ecology*, 50, 874, 1969.
133. Redmann, R.E., Plant and soil water potentials following fire in a northern mixed grassland, *J. Range Manage.*, 31, 443, 1978.
134. Ojima, D.S., W.J. Parton, D.S. Schimel, and C.E. Owensby, Simulating the long-term impact of burning on C, N, and P cycling in a tallgrass prairie, in *Current Perspectives in Environmental Biogeochemistry*, G. Giovannozzi-Sermanni and P. Nannipieri, Eds., C.N.R.-I.P.R.A. Via Nizza, Rome, 1988, 353.
135. Cannell, C.H. and H.E. Dregne, Regional setting, in *Dryland Agriculture*, (Agronomy 23), Dregne, H.E. and W.O. Willis, American Society of Agronomy, Madison, WI, 1983, 6.
136. Skold, M.D., Cropland retirement policies and their effects on land use in the Great Plains, *J. Prod. Agric.*, 2, 197, 1989.
137. Aguilar, R., E.F. Kelly, and R.D. Heil, Effects of cultivation on soils in northern Great Plains rangelands, *Soil Sci. Soc. Am. J.*, 52, 1081, 1988.
138. Davidson, E.T. and I.L. Ackerman, Changes in soil carbon inventories following cultivation of previously untilled soils, *Biogeochemistry*, 20, 161, 1993.
139. Doran, J.W. and M.R. Werner, Management and soil biology, in *Sustainable Agriculture in Temperate Zones*, Francis, C.A., C.B. Flora, and L.D. King, Eds., John Wiley & Sons, New York, 1990, 205.
140. Elliott, E.T., Aggregate structure and carbon, nitrogen, and phosphorus in native and cultivated soils, *Soil Sci. Soc. Am. J.*, 50, 627, 1986.
141. Doran, J.W., Soil microbial and biochemical changes associated with reduced tillage, *Soil Sci. Soc. Am. J.*, 44, 765, 1980.
142. Ridley, E.O. and R.A. Hedlin, Crop yields and soil management on the Canadian prairies, past and present, *Can. J. Soil Sci.*, 60, 393, 1980.
143. Judd, I.B. and M.L. Jackson, Natural succession of vegetation on abandoned farmland in the Rosebud soil area of western Nebraska, *Am. Soc. Agron. J.*, 39, 541, 1940.
144. Costello, D.F., Natural revegetation of abandoned plowed land in the mixed prairie association of northeastern Colorado, *Ecology*, 25, 312, 1944.
145. Judd, I.B., Plant succession of old fields in the Dust Bowl, *Southwest. Nat.*, 19, 227, 1974.
146. Dormaar, J.F. and S. Smoliak, Recovery of vegetative cover and soil organic matter during revegetation of abandoned farmland in a semiarid climate, *J. Range Manage.*, 38, 487, 1985.
147. Coffin, D.P., W.K. Lauenroth, and I.C. Burke, Recovery of vegetation in semiarid grassland 53 years after disturbance, *Ecol. Appl.*, submitted.
148. Ihori, T., I.C. Burke, W.K. Lauenroth, and D.P. Coffin, Effects of cultivation and abandonment on soil organic matter in northeastern Colorado, *Soil Sci. Soc. Am. J.*, in press.
149. Burke, I.C., E.T. Elliott, and C.V. Cole, Influence of macroclimate, landscape position, and management on active soil organic matter in agroecosystems, *Ecol. Appl.*, in press.
150. Schlesinger, W.H., Evidence from chronosequence studies for a low carbon-storage potential of soils, *Nature*, 348, 232, 1990.

Part II

Site Management Effects on Productivity and Soil Organic Matter in the Corn Belt

Chapter 6

Nitrogen Fertilizer and Legume-Cereal Rotation Effects on Soil Productivity and Organic Matter Dynamics in Wisconsin

M.B. Vanotti, L.G. Bundy, and A.E. Peterson

CONTENTS

0-8493-2802-0/97/$0.00+$.50

I. INTRODUCTION

Soil organic matter (SOM) content is the result of a steady-state equilibrium of formation and degradation, which can be affected by soil management practices. Where soils are subjected to similar practices over many years, the organic matter of the soil attains an equilibrium reflecting the cultural and cropping practices used.[1] The effects of management practices on SOM in agroecosystems can be better understood through their influence on soil fertility and crop productivity in long-term experiments.

The importance of the nitrogen (N) cycle in agricultural production systems has been recognized for centuries. Farmers have historically used legumes and crop rotations to enhance the productivity of the land, but only in relatively recent times have scientists discovered the value of legumes for fixing atmospheric N_2. Legumes in crop rotations were the most important supply of N in the Corn Belt region before commercial fertilizers became readily available.[3] In the 1940s and 1950s, farming systems changed toward monoculture and limited rotation, a technology requiring a continuous supply of N fertilizers.[4] There is a vast amount of information describing the immediate effect of agriculturally important legumes or inorganic N fertilizers on yield of cereal crops. However, little information is available on the long-term effects of continuous N fertilization or legume crops grown in rotation on soil N reserves or on SOM quantity and quality.

This paper describes the long-term effects of N fertilization and legume-cereal crop rotations on soil productivity and carbon (C) and N dynamics in two Wisconsin soils typical of the northern U.S. Corn Belt. We also contribute soil, crop production, and climatic data needed to integrate the numerous factors affecting SOM in agroecosystems on a regional scale. Special attention is given to the characterization of conceptually important soil organic fractions including the active pool, essential for evaluation of models designed to predict land use effects on SOM dynamics.

II. EXPERIMENTAL SITE CHARACTERISTICS

A. ARLINGTON SITE

The field study is located on the University of Wisconsin Arlington Agricultural Research Station in south central Wisconsin, approximately 25 km north of Madison (43°18′N; 89°21′W). The climate is humid continental with warm summers. Mean annual temperature is 7.6°C and monthly mean temperatures are –9.1 and 21.8°C for January and July, respectively. Mean annual precipitation totals 79.1 cm. The climate of this area normally provides 47.2 cm of precipitation and 2570 growing degree days (50°F base) from May 1 to September 30, which closely matches the normal frost-free period of 172 days (data for 1961 to 1990).[5] The site occurs in the Southeastern Upland Prairie region of Hole's Soils of Wisconsin.[6] The native prairie under which these soils formed is known as the Empire Prairie.[7] Soil at the site is a Plano silt loam (fine-silty, mixed, mesic, Typic Argiudolls) developed from loess deposits 90 cm or more in depth over calcareous loam glacial till. These dark, well-drained soils are some of the most fertile and productive in Wisconsin. The soil at the site was poorly managed for approximately 25 years before establishment of the experiment in 1958; corn stalks were even burned to facilitate plowing. The experimental site has been maintained in continuous corn with residues returned to the soil since 1958. Visual observations and penetrometer measurements (data not available) as well as workability of the soil (A. E. Peterson, personal communication) indicated the soil structure improved during the continuous corn experiment. These improvements in soil tilth were readily noticeable in the 1958 to 1963 period.[8]

1. Long-Term Continuous Corn × Nitrogen Fertilizer Study

Corn cultural trials evaluating plant density, fertility, and hybrid responses of continuous corn were conducted in the first 5 years of the study (1958 to 1963).[8] The initial average soil test values (1958 data) at the site were pH 6.75; available P, 20 mg kg^{-1}; exchangeable K, 87 mg kg^{-1}; and soil C, 18.8 g kg^{-1}. Fertility treatments during 1958 through 1962 involved three rates of fertilizer N (0, 56, and 112 kg N ha^{-1} as ammonium nitrate) and use of differential rates of starter fertilizer (6-24-24). Beginning in 1963, the annual N rates consisted of a control (0 kg N ha^{-1}), a medium rate (92 to 140 kg N ha^{-1}), and a high rate (184 to 280 kg N ha^{-1}) applied as anhydrous ammonia. The medium and high N fertilizer rates were selected to represent the current recommended N rate for corn production and twice this rate, respectively. Since 1963, all plots received 224 kg ha^{-1} y^{-1} of starter fertilizer (6-24-24) providing 13 kg N ha^{-1} y^{-1}. These three long-term N fertilizer treatments (LTN1, LTN2, and LTN3) were continued through the 1983 growing season without lime applications. Corn was harvested for grain each year, and the residues were returned to the soil. Grain yields were determined all years except for the 1963 to 1967 period. Average grain yields in treatments receiving recommended rates of fertilizer N were 5700 kg ha^{-1} at the beginning of the study (1958 to 1962) and gradually increased with time (8200 kg ha^{-1} for 1979 to 1983) due to improved corn hybrids, soil tilth, and cultural practices.

2. Soil Residual Nitrogen × Lime Study

In 1984, the long-term N treatments were discontinued and each of the original plots was subdivided to determine the residual effects of the LTN treatments on corn response to N fertilization.[9] Urea N was applied at rates of 0, 84, 168, and 252 kg N ha^{-1} and immediately incorporated by tillage in each of the LTN treatments from 1984 through 1991. These short-term N fertilizer treatments will be referred to as N0, N84, N168, and N252, respectively. As expected, the long history of N application generated soil acidity because of nitrification of ammonia fertilizer. The resulting pH values in 1984 were 6.1, 5.5, and 5.0 for LTN1, LTN2, and LTN3 plots, respectively. Lime treatments were imposed as an additional split treatment in spring 1985 to study the influence of liming on corn response to applied and residual soil N.[10] Variable lime rates were applied to raise soil pH to 6.9. Samples taken in 1990 indicated that all limed plots had a uniform pH of 6.6. Production measurements during 1985 through 1991 also included grain N removal and above-ground total dry matter and N yield. Details of LTN treatment effects on 1984 to 1990 crop yields and soil inorganic N in the limed plots have been described by Motavalli et al.[9]

B. LANCASTER SITE

The field study is located on the University of Wisconsin Lancaster Agricultural Research Station in southwest Wisconsin, 6 km southwest of Lancaster (42°51′N; 90°42′W). The climate is humid continental with warm summers. Mean annual precipitation is 83.3 cm and mean annual temperature is 7.8°C. Monthly temperature averages range from –9.3°C in January to 22.2°C in July. The climate of this area normally provides 50.5 cm of precipitation and 2429 growing degree days (50°F base) from May 1 to September 30, which closely approximates the normal frost-free period of 158 days (data for 1961 to 1990).[5] The site occurs in the driftless, hilly area identified by Hole[6] as the Southwestern Ridges and Valleys soil region. The soil and topography at Lancaster are representative of approximately 16 million hectares of the Upper Mississippi Valley, including parts of Iowa, Illinois, and Minnesota. Use of soil conservation practices in this area such as contour planting, conservation tillage and crop rotations are essential to control soil erosion. The experimental site is located on the top of an upland ridge. The soil is a Rosetta silt loam (fine-silty, mixed, mesic, Typic Hapludalfs) developed under forest cover in loess more than 150 cm thick over dolomite bedrock. An alfalfa-bromegrass pasture was grown on the site prior to establishment of the experiment in 1966.

1. Long-Term Crop Rotation × Nitrogen Fertilizer Study

The experiment began in 1967 to study yields of economically important crops in five crop rotations and to evaluate the N supplying capability of legumes to succeeding cereal crops.[11] The mean (and standard error) soil C and N contents at the start of the experiment were 16.1 (0.3) and 1.46 (0.03) g kg^{-1}, respectively (1966 data). The five rotations studied were continuous corn (c); corn-soybean-corn-oat-alfalfa (cscom); corn-corn-corn-oat-alfalfa (cccom); corn-corn-oat-alfalfa-alfalfa (ccomm); and corn-oat-alfalfa- alfalfa-alfalfa (commm). In any given year, there are 21 crop sequence-phase combination plots which constitute the experimental main plots. However, some crop sequences were modified over the years to address changes in farming systems. In 1977, the oat crop was removed from two of the rotations to study direct seeded alfalfa: cccom became corn-corn-corn-alfalfa-alfalfa (cccmm), and commm was modified into corn-corn-alfalfa-alfalfa (ccmm) and continuous alfalfa (m). The 4-year rotation was modified in 1987 to study 2-year legume-cereal crop rotations: ccmm became corn-alfalfa (cm) and corn-soybean (cs). Four levels of N fertilizer (LTN treatments) — 0, 84, 168, and 336 kg N ha^{-1} from 1967 to 1976, and 0, 56, 112, and 224 kg N ha^{-1} from 1977 to 1991 — were applied as ammonium nitrate to every phase of corn each year, but no fertilizer N was applied to any other crop. The annual experimental design is a split plot, where crop sequence-phases are assigned to main plots in a randomized, complete block design with the N treatments in subplots and two blocks (replicates). Crop residues were returned to the soil after grain harvest. Phosphorus and potassium fertilizers were applied frequently to maintain optimum fertility levels. Crop grain yields and alfalfa dry matter (DM) yields were determined annually. Continuous corn grain N removal was measured during the 1987 to 1991 period. A complete analysis of the legume × N fertilizer effects on crop yields for the first 10 years of the experiment is provided by Baldock et al.[12] A summary of crop yields for the second decade with economic comparisons is given by Higgs et al.[13]

III. SOIL ORGANIC MATTER MEASUREMENTS

A. SAMPLING

Soil samples were taken from both sites in early spring 1990 to investigate the effect of past management practices on SOM. Selected plots from the Arlington study included limed and unlimed plots receiving continuous N treatments from 1958 through 1989 (LTN1 N0, LTN2 N84, and LTN3 N168), and LTN plots that received no N fertilizer after 1983 (LTN2 N0 and LTN3 N0). Plots sampled in the Lancaster study included continuous corn at all LTN rates and ten crop sequence × phase plots with no history of N application. A composite sample of ten soil cores (20-cm depth) was collected from each replicate field plot, using a 3.5-cm diameter probe. Field-moist soil samples were passed through a 4-mm sieve and kept in plastic bags at 4°C (fresh soil) for incubation experiments. Subsamples were air-dried and ground for measurement of total soil organic C using a modified Mebius procedure[14] and total soil organic N.[15]

B. MICROBIAL BIOMASS CARBON AND NITROGEN

Carbon and nitrogen contents of the microbial biomass were determined by the chloroform fumigation-incubation method (CFIM).[16] Fifty grams (oven-dry weight) of fresh soil at approximately 60% WHC was exposed to ethanol-free $CHCl_3$ for 20 hours. Fumigated soils were inoculated with 0.2 g of unfumigated (control) soil after removal of $CHCl_3$. The inoculated samples were placed inside 1.9-L wide-mouth terephthalate (PET) jars (Cole-Parmer Instrument Company, Chicago) containing 20 mL of 0.5 *M* NaOH to absorb CO_2 and a vial of water to maintain soil moisture. Jars were held at 25°C for 10 days. The sorbed CO_2 was determined using an automated colorimetric method adapted from Chaussod et al.[17] The flush of CO_2-C from decomposition of microbial cells (F_C) was calculated as the difference between the amount of CO_2-C evolved from fumigated and control soil samples in the 10 days after fumigation. The quantity of microbial biomass C was calculated by dividing F_C by a k_C factor of 0.45.[18] Inorganic N was determined by automated analysis of 2 *M* KCl soil extracts (1:10 w/v).[19] The flush of N mineralized (F_N) was calculated as the difference between the net amount of NH_4-N mineralized by fumigated and control soil over the 10-day period. Biomass N content was calculated by dividing F_N by a k_N of 0.57.[20]

C. CARBON AND NITROGEN MINERALIZATION

Nitrogen mineralization was measured using a procedure based on the Stanford and Smith[21] leaching-incubation technique. The apparatus was similar to the one used by Bonde and Rosswall,[22] modified to include simultaneous measurements of net C and N mineralization. Fresh samples (30 g oven-dry weight) were uniformly mixed with equal amounts of acid-washed silica sand and transferred to filter units (Falcon 7111, Becton Dickson Co., Lincoln Park, NJ) fitted with cellulose acetate filter membranes having a 0.22-μm pore size and fiberglass prefilters with a 0.5-μm pore size. This prefilter prevents soil particles from clogging the membrane pores and avoids microbial growth on the surface of the membrane. The soil was covered with two layers of 1-mm mesh fiberglass screen and a 1.3-mm thick fiberglass prefilter, which were permanently positioned in the filter unit to minimize soil dispersion during leaching. Filter units were placed on a 250-mL Erlenmeyer filtering flask connected to a vacuum manifold. The soil was leached with 100 mL of 0.01 *M* $CaCl_2$ followed by 25 mL of minus-N nutrient solution,[21] using a drip irrigation system.[22] The leaching solution corrected original soil pH in Arlington samples to optimum values. The flow rate was 150 mL h^{-1} and the applied suction (80 kPa) extended for 0.5 h after the end of leaching. The soil samples were leached at 0, 7, 14, 21, 28, 42, 56 or 63, 77, 98, 126, 154, 182, 210, 238, and 280 days. Between leachings the filter units were held at 35°C inside PET jars containing alkali traps and water as was described for the biomass procedure. Inorganic N (NO_3^- and NH_4^+) in the leachates and CO_2 sorbed in the traps were determined using automated colorimetric methods.[17,19]

IV. RESULTS AND DISCUSSION

A. PAST MANAGEMENT EFFECT ON SITE PRODUCTIVITY AND SOIL ORGANIC MATTER

1. Long-Term Continuous Corn × N Fertilizer Treatments

a. Soil Organic Carbon

Corn grain yields were markedly affected by N application rate at both sites (Table 1). Average corn stover additions to soil were estimated from grain yields, based on a close relationship found between above-ground dry matter and grain yields using Arlington data collected in 1985 through 1991. These results suggest that a higher proportion of above-ground dry matter production was channeled into the grain (HI) and removed from the system with increasing N application rates. Final soil C concentrations increased with the higher LTN treatments at Arlington, but were not significantly different between LTN treatments at Lancaster (Table 1).

Table 1 Carbon Balance for Continuous Corn at Arlington and Lancaster, Wisconsin, after Long-Term N Fertilization

Treatment	Grain Yield[a]	Above-Ground Biomass[b]	Stover Biomass	HI[c]	Total Residue C Returned to Soil[d]		Soil Organic C[e] (g kg^{-1})	
	(kg ha^{-1} y^{-1})				t ha^{-1}	g kg^{-1}	Initial	Final
			Arlington, 1958–1983					
LTN1	3,760	8,924	5,747	0.36	67	23	18.8	19.6
LTN2	6,632	13,353	7,749	0.42	91	31	18.8	22.0
LTN3	7,047	13,993	8,038	0.43	94	32	18.8	22.2
$LSD_{0.05}$								1.7
			Lancaster, 1967–1989					
LTN1	3,261	8,154	5,400	0.34	56	19	16.1	14.0
LTN2	6,106	12,542	7,382	0.41	76	26	16.1	12.2
LTN3	6,920	13,797	7,949	0.42	82	28	16.1	13.6
LTN4	7,447	14,609	8,316	0.43	86	30	16.1	13.3
$LSD_{0.05}$								NS

[a] 15.5% H_20.
[b] 0% H_20, above-ground biomass (AGB) = 3126 + 1.542 grain yield. (Arlington 1985 to 1991 data, R^2 = 0.80).
[c] Harvest index, fraction of AGB into grain.
[d] Stover biomass = 0.45% C; bulk density (0 to 20 cm) = 1.47 g cm^{-3} (Arlington), 1.45 g cm^{-3} (Lancaster).
[e] Arlington: 1958 (initial), 1984 (final). Lancaster: 1966 (initial), 1990 (final).

The relative losses or gains of soil C at both sites were related to the soil C level at the start of the experiment and the quantity and quality of residues returned to the soil. Initial soil C at the Arlington site was very low at 18.8 g kg^{-1} compared with an uncultivated site which had 22.6 g kg^{-1} soil C, the result of previous management practices (Section II.A). Under continuous corn, removal of corn residues can severely deplete organic matter content of Corn Belt prairie-derived soils.[1,23,24] After 26 years of corn monoculture with residues returned to soil, the soil C content increased in all N treatments at Arlington (Table 1). Changes in soil C were proportional to the amount of residue returned, increasing at an annual rate of 0.24% (0.045 g kg^{-1}) for every 1000 kg ha^{-1} y^{-1} of corn stover above 5100 kg ha^{-1} y^{-1}. This is in agreement with data reported by Larson et al.[23] after 11 years of continuous corn production on an Iowa Mollisol with an initial soil C content similar to that of the Arlington site. In the Iowa experiment, soil C increased at an annual rate of 0.24% (0.043 g kg^{-1}) per 1000 kg ha^{-1} y^{-1} of corn stover returned in excess of 5700 kg ha^{-1} y^{-1}. It was estimated that net soil C losses will occur for quantities of corn residues returned below 5700 kg ha^{-1} y^{-1}.[23]

The decline in soil C at Lancaster after 23 years of corn monoculture contrasts with the results obtained at Arlington because equivalent amounts of residue C were returned to the soil in both experiments (Table 1). The alfalfa-bromegrass meadow grown at the Lancaster site for many years prior to establishment of the experiment may have had a major role in the relative losses of C observed during the subsequent continuous corn culture. Previous cropping practices at Lancaster undoubtedly benefited initial soil C levels, which in 1966 were 7% higher than that of a nearby virgin Rosetta soil sampled under native oak trees. When soil C dynamics at both sites are considered, these results indicate that the capacity of soils to incorporate corn residues into stabilized SOM products is governed primarily by the initial soil C status relative to indigenous equilibrium soil C levels (i.e., degraded at Arlington by residue burning, improved at Lancaster by high-quality residues from continuous meadow) and secondarily by the quantity of corn residue added to the soil as affected by N fertilization.

b. Soil Organic N

Total soil N was significantly affected by LTN rates at both Arlington and Lancaster (Table 2). The range of total net N input to soil across LTN rates was similar at both sites. The high LTN treatment at Arlington resulted in 735 kg N ha^{-1} (0 to 20 cm) more soil N content than LTN1. At Lancaster, the difference in organic N between the highest and lowest LTN treatments was 320 kg N ha^{-1} (0 to 20 cm). The higher efficiency of N incorporation into the humus fraction in the Arlington soil is likely due to a higher total amount of N returned in residues and to a synchrony between N and C stabilization processes as reflected by final soil C/N ratios (Table 2).

Table 2 Nitrogen Balance for Continuous Corn at Arlington and Lancaster, Wisconsin, after Long-Term N Fertilization

	kg ha^{-1}									
	Applied N		Grain N[a]		Stover N[b]		Net N[c]		SOM[d] (g kg^{-1})	
Treatment	Total	Avg.[e]	Total	Avg.	Total	Avg.	Total	Avg.	N	C/N
				Arlington, 1958–1983						
LTN1	315	12	1,053	40	676	26	–738	–28	1.56	12.6
LTN2	3,088	119	2,038	78	1,404	54	1,050	40	1.74	12.7
LTN3	5,825	224	2,180	84	1,534	59	3,645	140	1.81	12.3
$LSD_{0.05}$									0.14	NS
				Lancaster, 1967–1989						
LTN1	0	0	659	29	621	27	–664	–29	1.10	12.7
LTN2	1,568	68	1,773	77	966	42	–205	–9	1.02	12.0
LTN3	3,136	136	2,092	91	1,127	49	1,044	45	1.18	11.5
LTN4	6,272	273	2,298	100	1,265	55	3,974	173	1.21	10.9
$LSD_{0.05}$									0.07	1.4

[a] Arlington: grain N = –9 + 0.0132 grain yield (1985 to 1991 data, $R^2 = 0.86$). Lancaster: grain N = –27 + 0.0170 grain yield (1987 to 1991 data, $R^2 = 0.90$).
[b] Stover N = above-ground N yield (AGN) – grain N. AGN calculated from: grain yield = 200 + 59.2 AGN –0.080 AGN^2 (1985 to 1991 data, $R^2 = 0.70$).
[c] Net N = N applied – grain N harvest.
[d] Measured at the end of the period (0 to 20 cm).
[e] Average per year.

Low LTN rate additions at Lancaster resulted in significantly lower soil N than the check plot. Fertilizer N has been reported in some cases to increase mineralization of soil N.[25–27] Westerman and Tucker[26] suggested that this is the result of the stimulatory effect of added N on microbial decomposition of SOM or a "priming effect". The pattern of decrease in soil C with the LTN2 rate at Lancaster (Table 1) and the lack of expected soil C response to increased stover biomass support evidence that a priming effect caused the observed differences in soil N at low LTN rates.

2. Short-Term N × Lime Treatments

a. Soil Productivity

Corn yields during 1984 through 1991 when no fertilizer N was applied were significantly higher in plots with a history of high fertilizer additions. In 1984, this response was likely a result of residual inorganic N carryover from previous fertilizer addition (86, 176, and 306 kg nitrate-N ha^{-1}, 0 to 90 cm, in LTN1, LTN2, and LTN3 treatments, respectively). However, differences in profile inorganic-N content in subsequent years were not large enough to account for the marked response to LTN treatments observed.[9] These responses were likely due to changes in the soil's mineralizable N pool due to long-term N fertilizer additions. Treatments with a history of high long-term N fertilization but with a 0 N rate during 1984 through 1991 (LTN3 N0) produced average grain yields 2110 kg ha^{-1} y^{-1} higher than plots receiving no N since 1958 (LTN1 N0) (Table 3). Above-ground dry matter production and grain N after 1985 were also affected by the residual effects of long-term N fertilization, with an average increase for the same treatments of 2630 kg DM ha^{-1} y^{-1} and 26 kg N ha^{-1} y^{-1}.

Grain yields, above-ground dry matter, and grain N removal were also enhanced by the 1985 lime application. Corn response to liming was greatest at the high LTN treatment and low 1985 to 1991 N rate (Table 3). Addition of N fertilizer offset most of the lime effect, suggesting that yield response to liming was in part due to differences in soil N availability. Since the effect of liming on corn productivity was more pronounced in plots where corn N nutrition depended on soil N reserves, it is possible that liming provided a better environment for microbial activity and N cycling, resulting in higher N mineralization rates. Cornfield[28] compared the mineralization rates of numerous acid soils in Great Britain with a range of pH values and found that the positive response to liming decreased as original soil pH increased. Results from field experiments on newly cultivated acid soils in Canada indicate that, although liming increases mineralization of N, it is generally a temporary effect lasting no more than 3 years.[29] In the Arlington plots, however, the liming effect on N cycling, as measured by corn total N uptake and grain N removal, was still noticeable 6 years after lime application.

b. Soil Organic N

Split treatments imposed in 1984 (N rates) and 1985 (lime) provided information on the stability of SOM pools formed after long-term N fertilization. Total soil N was much more sensitive to changes in management practices than total soil C. Soil N content was significantly reduced 6 years after discontinuing N fertilization (Figure 1), consistent with the residual effects on corn production observed during the same period (Table 3). In the N treatments, soil N changes between 1984 and 1990 were similar in magnitude to those produced after 32 years of differential N fertilization. For example, 1990 soil N level in LTN3-N plots decreased to that found in 1984 LTN2 plots, while soil N level in LTN2-N plots approached the 1984 LTN1 level (Figure 1). This is contrary to the general assumption that fertilizer N immobilized in humus is very stable, i.e., increases a passive soil nitrogen pool where remineralization will proceed for several decades or even centuries.[30]

Table 3 Effect of Long-Term N Treatments and Lime Applications on Corn Grain Yield, Grain N, and Above-Ground Biomass Responses to Applied N, Arlington, 1985–1991

	Long-Term N Treatments (1958–1983)					
	LTN1		LTN2		LTN3	
1985–1991 N Rates	Without Lime	With Lime[a]	Without Lime	With Lime	Without Lime	With Lime
	Grain Yield (kg ha^{-1} y^{-1})					
0	4,411	4,999	5,816	5,944	6,188	7,448
84	8,742	8,780	9,125	9,059	9,148	9,536
168	9,749	9,765	9,587	9,673	9,276	9,682
252	9,575	9,812	9,309	9,549	9,830	9,936
	Grain N Harvest (kg ha^{-1} y^{-1})					
0	45	51	59	60	65	82
84	103	103	105	112	109	119
168	122	123	120	121	120	123
252	123	126	116	119	127	126
	Above-Ground Biomass (kg ha^{-1} y^{-1})					
0	10,100	10,760	11,810	12,230	12,250	13,860
84	17,100	17,090	16,230	18,000	16,540	18,120
168	18,180	18,410	18,120	17,400	17,220	17,850
252	19,080	17,780	17,380	18,850	18,480	18,100

Note: Grain yield reported at 15.5% H_2O and above-ground biomass at 0% H_2O. The standard errors of these LTN × N × lime treatment means are 151, 3, and 406 for grain yield, grain N harvest, and above-ground biomass, respectively. Analysis of variance summaries are shown in the appendix (Section VI.C., Arlington site).

[a] Split lime treatment applied in spring 1985.

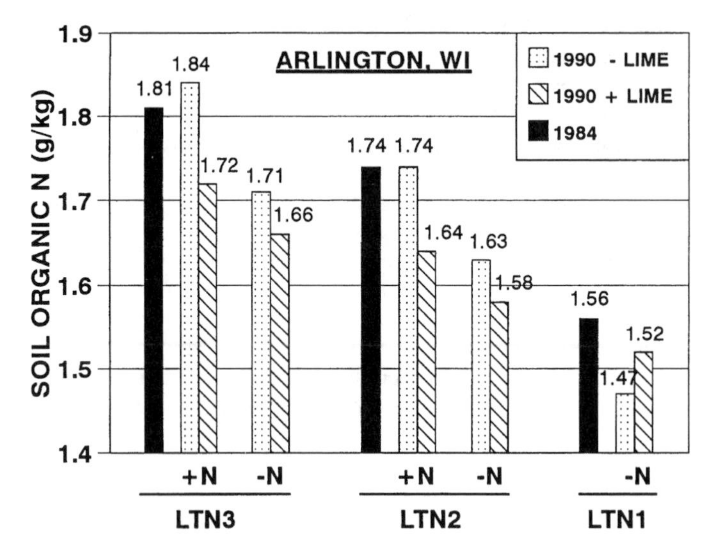

Figure 1 Short-term N and lime treatments effect on total soil organic N. LTN denotes long-term N treatments (1958 to 1983).

In plots receiving continuous N treatments from 1958 through 1989 (LTN2 N84 and LTN3 N168), the soil N content between 1984 and 1990 was maintained in unlimed plots, but reduced by 6% where lime was added (Figure 1). Fu et al.[31] found a twofold reduction in N turnover from corn residues mixed and incubated with three Iowa soils that were adjusted to pH 4 relative to the same soils adjusted to pH 6. The rapid decline in soil N levels observed in the limed plots at Arlington suggest that a significant fraction of the soil N formed after a long history of NH_3-N fertilization is made up of highly labile materials, and its N turnover is limited by a fertilizer induced acid environment.

3. Crop Rotation × N Fertilizer Treatments

a. Soil Productivity

The sequential order in which crops are grown often affects the yield of succeeding crops, a phenomenon known as the rotation effect. Usually the largest and best-documented rotation effect of legumes is their contribution of symbiotically fixed N for the next crop. After the legume plant residues are left in the soil to decompose, the N stored in those residues is gradually released and is available to succeeding cereal crops. Yield data for the Lancaster study summarized in Table 4 indicate that the amount of N supplied by a legume crop to first-year corn is equivalent to 168 to 224 kg N fertilizer ha^{-1} in continuous corn. For subsequent years of corn following the legume, the direct effect of legume residues on corn yields dissipated, while yield responses to N fertilizer increased (Table 4).

Where no N was applied, yields of corn grown continuously were halved in less than 10 years. In contrast, yields of corn grown in rotation with legumes but without N additions were maintained or improved through the length of the study (Table 4). This illustrates the beneficial effects of legume-cereal crop rotations on soil productivity in addition to the direct N contribution. These results are by no means unusual, for many workers [32–37] have found that the main value of the legume in crop sequences is long term, in that it maintains soil N concentrations to ensure adequate delivery of N to future cereal crops.

A third positive rotation effect of legumes that is not directly associated with N is also evident at the Lancaster site, where sufficient N fertilizer was applied to both rotational and continuous corn plots (Table 4). This non-N rotation effect, called the AL effect by Baldock et al.,[12] may include reduced disease incidence,[38] addition of growth promoting substances,[39] and improvement of SOM content,[13,36,37] with its many indirect effects on soil fertility.[40]

b. Soil Organic Matter

Legume-cereal crop rotations maintained soil C content at initial levels, while soil C decreased in all plots under corn monoculture at Lancaster (Tables 1 and 5). Soil N content in rotational plots without external N additions were even higher than in continuous corn plots that were heavily fertilized with N (Table 5). Although legume-cereal rotations without N treatment maintained soils in a high state of

Table 4 Nitrogen Fertilizer and Rotation Effect on Corn Grain Yields at Lancaster, Wisconsin

Crop Sequence	Year	N Rate Applied to Corn (kg ha^{-1})[a] 0 / 0	84 / 56	168 / 112	336 / 224
		Grain Yield (kg ha^{-1})[b]			
Continuous corn	1967–1976	4415	6773	7370	7085
	1977–1991	2400	5580	6890	7935
Corn following alfalfa[c]					
First year	1967–1976	7613	8168	8289	8183
	1977–1991	8009	8306	8435	8372
Second year	1967–1976	6136	7664	7683	7741
	1977–1991	6591	7702	8382	8323
Third year	1967–1976	4981	6981	7533	7562
	1977–1991	4904	7281	7962	8097
Corn following soybean[c]					
First year	1967–1976	7437	8065	8222	7844
	1977–1991	7778	8313	8539	8274

[a] Nitrogen fertilizer applied only to corn: 0, 84, 168, and 336 N rates from 1967 to 1976, and 0, 56, 112, and 224 N rates from 1977 to 1991.
[b] The standard errors of these N treatment × phase means are 182 kg ha^{-1} for 1967 to 1976 data and 191 kg ha^{-1} for 1977 to 1991 data.
[c] Average of all rotations in the study.

Table 5 Nitrogen Fertilizer and Crop Rotation Effect on C and N Content Measured in Various Soil Fractions

Crop Rotation N Treatment	Total SOM (g kg^{-1}) C	Total SOM (g kg^{-1}) N	Microbial Biomass (mg kg^{-1}) C	Microbial Biomass (mg kg^{-1}) N	Total Net Mineralization[a] (mg kg^{-1}) C	Total Net Mineralization[a] (mg kg^{-1}) N	Inorganic[b] (mg kg^{-1}) N
			Arlington Study, 1990[c]				
Continuous corn							
LTN1-N0	19.2	1.47	164	23	2234	83	12
LTN2-N84	22.8	1.74	168	23	3429	132	25
LTN3-N168	21.5	1.84	140	26	3721	140	35
			Lancaster Study, 1990				
Continuous corn							
LTN1	14.0	1.10	143	15	2611	95	8
LTN2	12.2	1.02	144	14	2986	86	11
LTN3	13.6	1.18	178	20	3223	100	14
LTN4	13.3	1.21	203	23	3611	125	21
Legume-cereal sequences[d]							
CCOMMx	16.0	1.27	276	30	2887	177	25
CCxOMM	16.3	1.33	218	24	3214	159	14
CSxCOM	16.2	1.28	203	21	2860	153	18
MxC	15.3	1.33	213	27	3211	164	31

[a] 280 days, 35°C (Figure 2).
[b] Before incubation.
[c] Unlimed plots.
[d] Plots receiving 0 N rate; "x" indicates position in the crop sequence where sample was taken (March 1990). C = corn, O = oat, S = soybean, M = alfalfa.

productivity in the Lancaster study (1967 to 1990), soil N in these plots declined to about 91% of initial level (1.46 g N kg^{-1}, 1966). The C/N ratios of SOM in rotational plots, therefore, increased from initial values. This change in SOM composition probably reflects the changes in average plant residue quality before and after 1966.

B. PAST MANAGEMENT EFFECT ON LABILE FRACTIONS OF SOIL ORGANIC MATTER

The labile fractions of the SOM are expected to have the greatest sensitivity to N management and cropping practices.[41–44] In this section, we describe the effect of previous N treatments and legume-cereal rotations on several SOM pools involved in rapid C and N turnover in soil. Experimental data consisted of microbial biomass (CFIM) and N and C net mineralization obtained by aerobic incubations (Section III). To provide biologically meaningful active SOM fractions from C and N mineralization measurements, data were analyzed and interpreted with a multicompartment mechanistic model that simultaneously takes into account N and C transformations in soil, including microbial growth from decomposing plant residues.

1. Microbial Biomass Carbon and Nitrogen

Microbial biomass comprises a small but comparatively labile part of the organic fraction in soil that relates to soil productivity.[44] Its role as a source and sink of plant nutrients in addition to its traditional role in transformation of nutrients has been recently recognized.[45] Biomass-N at both sites ranged from 14 to 30 mg N kg^{-1} (41 to 87 kg N ha^{-1}, 0 to 20 cm). The lowest concentrations were in the continuous corn plots that received low LTN rates (Table 5). The C/N ratios of the biomass were related to soil type, but insensitive to cropping practices or N management: Arlington (5.4 to 7.2), Lancaster (8.0 to 10.0). The biomass-N comprised 1.7 to 2.4% of soil N in the rotational plots and 1.0 to 1.8% of soil N in continuous corn plots at both sites.

2. Carbon and Nitrogen Mineralization

The effect of long-term N fertilization on N and C mineralization was more pronounced than its effect on total soil N and C (Table 5). For example, the difference in soil N between the highest and lowest LTN treatments was 25% at Arlington and 10% at Lancaster, while for N mineralization it was 69% at Arlington and 32% at Lancaster, or about three times as much. Although the differences in the amounts of corn residue returned to soil had no significant effect on total soil C at Lancaster (Table 1), its effect was clearly shown on C mineralization (Table 5).

The N and C dynamics in soil are closely linked. This interdependence of C and N transformations is reflected in the similarity of the net C and N mineralization data shown in Figure 2. The effect of N management on the mineralized C fraction exceeds the effects due to soil characteristics. This is illustrated in Figure 2 by the similar amounts and pattern of C mineralized from continuous corn plots at Lancaster and Arlington, even though total soil C levels in the prairie-derived Arlington soil were about 50% higher than those in the forest-derived Lancaster soil.

3. Mineralization Model: NCSOIL

Net N and C mineralization was simulated with the process-oriented model NCSOIL[46] to characterize pools comprising the active SOM or SOM involved in rapid N and C turnover,[47] following general procedures described by Houot et al.[48] The version of NCSOIL used for this study included three organic pools (Figure 3): (1) plant residues, (2) microbial biomass (pool I), and (3) the humads (pool II). The humads pool has also been defined as the heterogeneous chemicals that are stabilized by humification and adsorption, but still amenable to biodegradation (McGill et al.[49]).

Highly humified materials making up the resistant (or passive) SOM phase (C decay rate constant = 1.0E-5 day^{-1}) were assumed to provide no significant interaction with the dynamics of the mineralization systems studied. The size of the active SOM N fraction was quantified as the initial N content of the model's pool I plus pool II.[48] The term "initial" in this section refers to the status at the start of incubation experiments using soil samples taken from the field in early spring of 1990. Table 6 shows the kinetic constants used in NCSOIL to perform the simulations.[46,48,50]

Initial C levels and C/N ratios of pool I were set equal to the microbial biomass values obtained by CFIM (Table 5). The C/N ratios of pool II were determined from the slope coefficients of linear regression equations relating observed net C and N mineralized after 11 weeks (Figure 2 and Table 7). Optimum values for the initial concentrations of plant residue C and N and pool II C were selected using the Marquardt algorithm[51] to achieve minimum divergence between NCSOIL simulations and experimental data, consisting of mineralized C and N (Figure 2 and Table 7). The use of a process-oriented model within an optimization program to estimate the active fraction of soil N from net C and N mineralization data presents a real advantage over the use of a single equation (N_0)[52] on N data alone when finely divided plant residues in various stages of decomposition remain in the soil at the time of sampling. Under such conditions, it is to be expected that inclusion of C flow data as an additional dependent variable will improve final parameter values.[51,52]

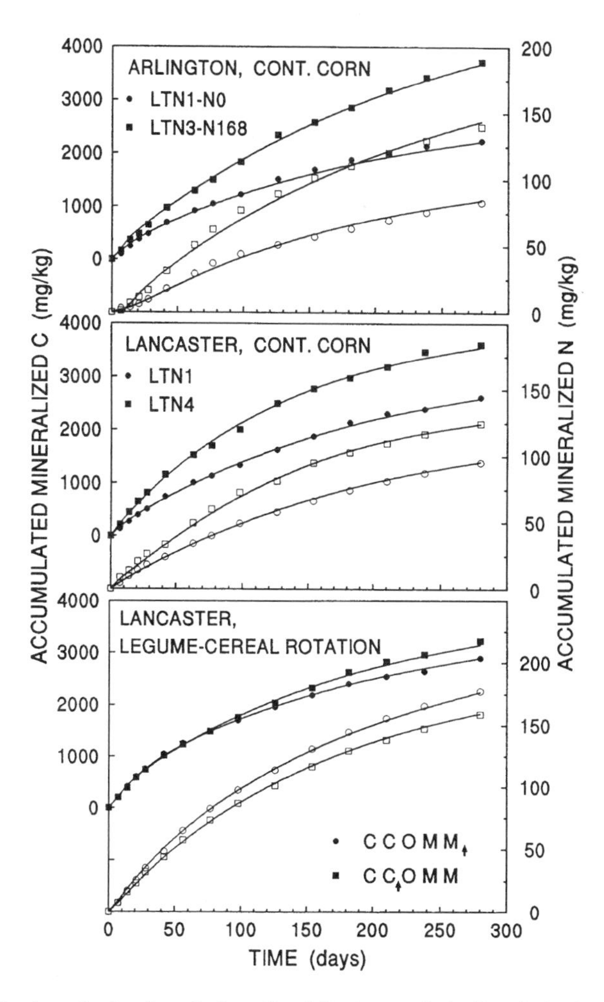

Figure 2 Net C and N mineralization in soils from the Arlington study (unlimed plots) and the Lancaster study; measured: CO_2-C (filled symbols), inorganic N (empty symbols); simulated (line). Simulated kinetics obtained with constants and initial values shown in Tables 5 to 7.

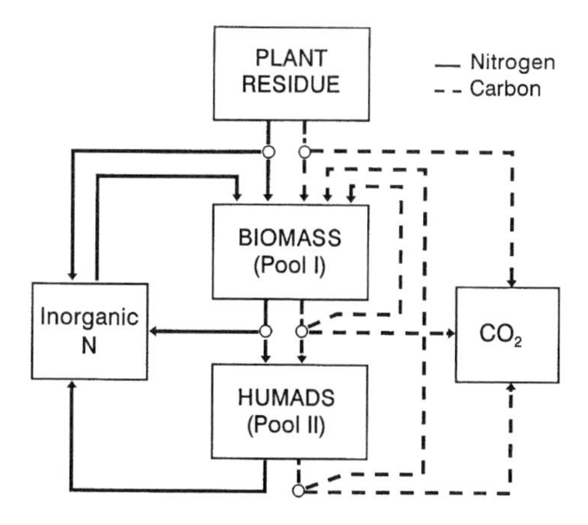

Figure 3 Carbon and nitrogen flow in NCSOIL.

Table 6 NCSOIL Kinetic Constants Used to Simulate N and C Mineralization in Three SOM Pools

	Residue I	Pool I	Pool II
C decay rate constants (day^{-1})			
Labile component	0.10[a]	0.33	—
Resistant component	—	0.04	0.0065
Fraction of labile component in pool I	—	0.56	—
Fraction of decomposed C recycled in pool I	0.60	0.60	0.38
Fraction of decomposed C from pool I incorporated in pool II	—	0.17	

[a] 0.30 for continuous corn at Lancaster, 0.80 for LTN3 at Arlington.

Table 7 Nitrogen Fertilizer and Crop Rotation Effect on the Active Fraction of Soil Organic N

Crop Rotation N Treatment	Plant Residue[a] (mg kg^{-1})		Humads (Pool II) (mg kg^{-1})			χ^2	Active Soil Organic N
	C	N	C[a]	N	C/N[b]	10^{-2}	Pool/Soil N[c]
Arlington Study, 1990							
Continuous corn							
LTN1-N0	317	5	2433	91	27	0.83	0.08
LTN2-N84	151	3	4038	155	26	0.23	0.10
LTN3-N168	486	8	4744	162	29	0.43	0.10
Lancaster Study, 1990							
Continuous corn							
LTN1	127	4	2980	104	29	0.13	0.11
LTN2	50	1	3085	89	35	0.99	0.10
LTN3	297	10	3036	105	29	0.87	0.11
LTN4	162	6	3738	125	30	0.37	0.12
Legume-cereal sequences[d]							
CCOMMx	910	46	2524	162	16	0.07	0.15
CCxOMM	412	15	3249	165	20	0.09	0.14
CSxCOM	470	26	2975	161	18	0.08	0.15
MxC	614	31	3049	158	19	0.20	0.14

[a] Initial concentration of plant residue and humads pools were determined by calibration of NCSOIL against the observed kinetics of soil net N and C mineralization (Figure 2), whereby a low figure-of-merit (χ^2) indicates a good fit. Initial pool I: Experimental data obtained by CFIM (Table 5).

[b] Calculated from slope of linear regression function relating cumulative amounts of net C and N mineralized after 77 days of incubation.

[c] Active soil organic N = humads plus biomass N pools.

[d] Plots receiving 0 N rate; "x" indicates position in the crop sequence where sample was taken (March 1990). C = corn, O = oat, S = soybean, M = alfalfa.

a. Plant Residue and Inorganic Nitrogen

Plant residue and inorganic N pools had the greatest impact on the N nutrition of cereal crops planted in 1990. When samples were taken in early 1990, the soil inorganic N pool consisted of N fertilizer carryover from previous year fertilization and N released from plant residues and humus (Table 5). The inorganic N pool in the rotational plots shown in Table 5 originated only from organic N sources since no N was ever applied to those plots. Finely divided plant residues were still in the soil at sampling time and their presence affected the kinetics of C and N mineralization (Figure 2). In plots where legumes were incorporated the previous autumn, the values obtained by calibration showed that 26 to 46 mg N kg^{-1} (75 to 133 kg N ha^{-1}, 0 to 20 cm) were present in early spring in the form of plant residues (Table 7), which in turn will mineralize and add to the 18 to 31 mg N kg^{-1} (52 to 90 kg N ha^{-1}, 0 to 20 cm) already present in the inorganic N pool. The total of both pools provides an estimate of legume N potentially available to the first cereal crop: alfalfa = 206 kg N ha^{-1} (CCOMM$_x$) and 180 kg N ha^{-1} (M$_x$C); soybean = 128 kg N ha^{-1}. Although some of this N will be incorporated into more stable organic fractions, a significant portion will be absorbed by the first cereal crop consistent with the direct N contribution effect on yields shown in Table 4.

b. Active Soil Organic Matter

The rotational plots had a pool II C/N ratio lower (16 to 20) than the continuous corn plots (26 to 35), indicating the importance of the chemical composition of plant residues on the quality of the SOM formed (Table 7). Pool II C and N were markedly affected by N treatments at both sites. The larger N contribution of legumes grown in rotation is reflected in the non-biomass active SOM fraction (pool II). Comparison between plots that received 0 N but were cropped under continuous corn or legume-cereal rotations showed that the rotational plots had 65% more N in pool II, or a difference of 285 kg N ha^{-1} (0 to 20 cm). This difference corresponds with the observed long-term N effect of legumes on succeeding corn yields (Table 4), and it is in general agreement with the results of several investigators[32–35] reporting that legume crops can affect the yield of succeeding cereal crops through enrichment of the active SOM N pool. Soils from continuous corn plots had a similar proportion of organic N in the active SOM pool including the biomass (8 to 12% of total organic N); their active SOM pool was a much lower proportion of total organic N than in soils from legume-cereal rotations (14 to 15% of total organic N) (Table 7). The range of these values compares to estimates of the active SOM N pool in arable mineral soils obtained by Jansson[53] (10 to 15%), Jenkinson and Parry[54] (14%), and Paul and Juma[55] (8%).

V. SUMMARY

We analyzed the interactions between cropping sequences, N fertilization, and crop productivity as they affect SOM dynamics and soil N reserves in two continuing long-term experiments in southern Wisconsin. Sequences included continuous corn and legume-cereal rotations. Prior cropping and soil management practices had a major influence on the changes in SOM content. At the Arlington site, a shift in corn stover residue management from burning to soil incorporation resulted in a net gain in SOM. Changes in soil C were linear with respect to annual C input rates, which were associated with N fertility management. In contrast, corn monoculture at the Lancaster site following a history of alfalfa meadow resulted in a net decrease in SOM, irrespective of N fertilizer addition. However, legume-cereal crop sequences without external N inputs maintained SOM at initial levels.

Long-term NH_3 fertilizer additions to continuous corn at Arlington significantly increased soil organic N reserves. This practice also increased soil acidity. On plots where long-term N additions were halted, the supply of mineralized N was an important factor affecting corn yields in subsequent harvests. Lime amendments resulted in higher N mineralization rates, and consequently in a more rapid decline in soil N levels. The magnitude of soil N changes after 6 years without N fertilization at optimum soil pH were equivalent to those realized after 32 years of differential N fertilization. This indicated that most of the fertilizer N was stored in relatively labile soil organic fractions. At Lancaster, the patterns of soil organic N and C content across N treatments suggested that N additions caused a priming effect on mineralization of indigenous SOM.

As expected, the effect of N fertilization and cropping sequences were more pronounced on labile SOM pools than on total soil organic C and N contents. Mineralized C in long-term aerobic incubations reflected the effect of previous soil management on newly formed SOM, but it was insensitive to changes in the size of a much larger passive SOM pool at equilibrium with the native vegetation. Mineralization and microbial biomass data (C and N) were interpreted with the NCSOIL model to estimate an active SOM N pool size. Results of simulations showed that the size of the active N pool was about 10% of soil organic N in all continuous corn plots irrespective of N treatment and 15% in legume-cereal plots without external N additions.

Legumes grown in rotation with cereals affected the yield of succeeding crops through three distinct mechanisms or rotation effects: (1) a direct, immediate N effect; (2) an indirect, long-term N effect; and (3) a non-N effect. Each of these rotation effects was evident in the yield data presented and was also shown in the context of changes in SOM pools. The largest impact of legumes in agroecosystems was to maintain soils at a high state of productivity. This was achieved by N-enrichment of the active SOM pool and by the legume contribution to soil fertility through maintaining higher SOM content.

REFERENCES

1. Barber, S.A., Corn residue management and soil organic matter, *Agron. J.*, 71, 625, 1979.
2. King, F.H., *The Soil*, MacMillan, New York, 1896.
3. Peterson, T.A. and Russelle, M.P., Alfalfa and the nitrogen cycle in the Corn Belt, *J. Soil Water Conserv.*, 46, 229, 1991.
4. Power, J.F. and Follett, R.F., Monoculture, *Sci. Amer.*, March, 79, 1987.
5. U.S. Department of Commerce, Monthly Station Normals of Temperature, Precipitation, and Heating and Cooling Degree Days 1961–1990: Wisconsin, National Climatic Data Center, Asheville, NC, 1992.
6. Hole, F.D., *Soils of Wisconsin*, The University of Wisconsin Press, Madison, WI, 1976.
7. Engel, M.S. and Hopkins, A.W., The Prairie and its People, Wis. Agric. Exp. Stn. Bull. 520., University of Wisconsin-Madison, 1956.
8. Andrew, R.H., Arawinko, Z.M., Love, J.R., and Peterson, A.E., Population, Fertility and Varietal Response for Continuous Corn with Minimum Tillage, Wis. Agric. Exp. Stn. Bull. 244, University of Wisconsin-Madison, 1963.
9. Motavalli, P.P., Bundy, L.G., Andraski, T.W., and Peterson, A.E., Residual effects of long-term nitrogen fertilization on nitrogen availability to corn, *J. Prod. Agric.*, 5, 363, 1992.
10. Bundy, L.G., Andraski, T.W., and Peterson, A.E., Effect of long-term nitrogen application on soil pH, *Proc. Wis. Fert. Aglime and Pest Mgmt. Conf.*, 27, 48, 1988.
11. Higgs, R.L., Paulson, W.H., Pendleton, J.W., Peterson, A.E., Jackobs, J.A., and Shrader, W.D., Crop Rotations and Nitrogen: Crop Sequence Comparisons on Soils of the Driftless Area of Southwestern Wisconsin, Wis. Agric. Exp. Sta. Bull. R2761, University of Wisconsin-Madison, 1976.
12. Baldock, J.O., Higgs, R.L., Paulson, W.H., Jackobs, J.A., and Shrader, W.D., Legume and mineral N effects on crop yields in several crop sequences in the Upper Mississippi Valley, *Agron. J.*, 73, 885, 1981.
13. Higgs, R.L., Peterson, A.E., and Paulson, W.H., Crop rotations: sustainable and profitable, *J. Soil Water Conserv.*, 45, 68, 1990.
14. Yeomans, J.C. and Bremner, J.M., A rapid and precise method for routine determination of organic carbon in soil, *Commun. Soil Sci. Plant Anal.*, 19, 1467, 1988.
15. Nelson, D.W. and Sommers, L.E., A simple digestion procedure for estimation of total nitrogen in soils and sediments, *J. Environ. Qual.*, 1, 423, 1972.
16. Jenkinson, D.S. and Powlson, D.S., The effects of biocidal treatments on metabolism in soil. V. A method for measuring soil biomass, *Soil Biol. Biochem.*, 8, 209, 1976.
17. Chaussod, R., Nicolardot, B., and Catroux, C., Mesure en routine de la biomasse microbienne des sols par la methode de fumigation au chloroforme, *Sci. Sol*, 24, 201, 1986.
18. Jenkinson, D.S. and Ladd, J.S., Microbial biomass in soil: measurement and turnover, in *Soil Biochemistry*, Vol. 5, Paul, E.A. and Ladd, J.N., Eds., Marcel Dekker, New York, 1981, 415.
19. Bundy, L.G. and Meisinger, J.J., Nitrogen availability indices, in *Methods of Soil Analysis*, Part 2, 3rd ed., Weaver, et al., Eds., ASA and SSSA, Madison, WI, 1993.
20. Jenkinson, D.S., Determination of microbial biomass carbon and nitrogen in soil, in *Advances in Nitrogen Cycling in Agricultural Ecosystems*, Wilson, J.R. Ed., C.A.B. International, Wallingford, 1988, 368.
21. Stanford, G. and Smith, S.J., Nitrogen mineralization potentials of soils, *Soil Sci. Soc. Am. Proc.*, 36, 465, 1972.
22. Bonde, T.A. and Rosswall, T., Seasonal variation of potentially mineralizable nitrogen in four cropping systems, *Soil Sci. Soc. Am. J.*, 51, 1508, 1987.
23. Larson, W.E., Clapp, C.E., Pierre, W.H., and Morachan, Y.B., Effects of increasing amounts of organic residues on continuous corn. II. Organic carbon, nitrogen, phosphorus, and sulphur, *Agron. J.*, 64, 204, 1972.
24. Bloom, P.R., Schuh, W.M., Malzer, G.L., Nelson, W.W., and Evans, S.D., Effect of N fertilizer and corn residue management on organic matter in Minnesota mollisols, *Agron. J.*, 74, 161, 1982.
25. Broadbent, F.E., Effect of fertilizer nitrogen on the release of soil nitrogen, *Soil Sci. Soc. Am. Proc.*, 29, 692, 1965.
26. Westerman, R.L. and Tucker, T.C., Effects of salts and salts plus nitrogen-15-labeled ammonium chloride on mineralization of soil nitrogen, nitrification, and immobilization, *Soil Sci. Soc. Am. Proc.*, 38, 602, 1974.
27. Sapozhnikov, N.A., Nesterova, E.I., Rusinova, I.P., Sirota, L.B., and Livanova, T.K., The effect of fertilizer nitrogen on plant uptake of nitrogen from different podzolic soils, Trans. 9th Int. Congr. Soil Sci., Vol. 2, Adelaide, 467, 1968.
28. Cornfield, A.H., The mineralization of the nitrogen of soils during incubation: influence of pH, total nitrogen, and organic carbon contents, *J. Sci. Food Agric.*, 3, 343, 1952.
29. Nyborg, M. and Hoyt, P.B., Effects of soil acidity and liming on mineralization of soil nitrogen, *Can. J. Soil Sci.*, 58, 331, 1978.
30. Hauck, R.D., Nitrogen fertilizer effects on nitrogen cycle processes, in *Terrestrial Nitrogen Cycles. Processes, Ecosystem Strategies and Management Inputs*, Clark, F.E. and Rosswall, T., Eds., Ecol. Bull. 33, Stockholm, 1981, 551.
31. Fu, M.H., Xu, X.C., and Tabatabai, M.A., Effect of pH on nitrogen mineralization in crop-residue-treated soils, *Biol. Fertil. Soils*, 5, 115, 1987.
32. Pierce, F.J. and Rice, C.W., Crop rotation and its impact on efficiency of water and nitrogen use, in *Cropping Strategies for Efficient Use of Water and Nitrogen*, ASA Spec. Publ. 51, Hargrove, W.H., Ed., ASA, CSSA, and SSSA, Madison, WI, 1988, 21.

33. Ladd, J.N., Oades, J.M., and Amato, M., Distribution and recovery of nitrogen from legume residues decomposing in soils sown to wheat in the field, *Soil Biol. Biochem.*, 13, 251, 1981.
34. Ladd, J.N., Butler, J.H.A., and Amato, M., Nitrogen fixation by legumes and their role as sources of nitrogen for soil and crop, *Biol. Agric. Hort.*, 3, 269, 1986.
35. Harris, G.H. and Hesterman, O.B., Quantifying the nitrogen contribution from alfalfa to soil and two succeeding crops using nitrogen-15, *Agron. J.*, 82, 129, 1990.
36. Odell, R.T., Melsted, S.W., and Walker, W.M., Changes in organic carbon and nitrogen of Morrow plot soils under different treatments, 1904–1973, *Soil Sci.*, 137, 160, 1984.
37. Moldenhauer, W.C., Wischmeier, W.H., and Parker, D.T., The influence of crop management on runoff, erosion, and soil properties of a Marshall silty clay loam, *Soil Sci. Soc. Am. Proc.*, 31, 541, 1967.
38. Reeves, T.G., Ellington, A., and Brooke, H.D., Effects of lupin-wheat rotations on soil fertility, crop disease and crop yields, *Aust. J. Exp. Agric. Anim. Husb.*, 24, 595, 1984.
39. Sanford, J.O. and Hairston, J.E., Effects of N fertilization on yield, growth, and extraction of water by wheat following soybeans and grain sorghum, *Agron. J.*, 76, 623, 1984.
40. Vaughan, D. and Ord, B.G., Soil organic matter — a perspective on its nature, extraction, turnover and role in soil fertility, in *Soil Organic Matter and Biological Activity*, Vaughan, D. and Malcolm, R.E., Eds., Martinus Nijhoff/Dr. W. Junk, Dordrecht, 1985, 1.
41. McGill, W.B., Cannon, K.R., Robertson, J.A., and Cook, F.D., Dynamics of soil microbial biomass and water-soluble organic C in Breton L after 50 years of cropping to two rotations, *Can. J. Soil Sci.*, 66, 1, 1986.
42. Powlson, D.S., Brookes, P.C., and Christensen, B.T., Measurement of soil microbial biomass provides an early indication of changes in total soil organic matter due to straw incorporation, *Soil Biol. Biochem.*, 19, 159, 1987.
43. Schnurer, J., Clarholm, M., and Rosswall, T., Microbial biomass and activity in an agricultural soil with different organic matter contents, *Soil Biol. Biochem.*, 17, 611, 1985.
44. Carter, M.R. and Rennie, D.A., Changes in soil quality under zero tillage farming systems: distribution of microbial biomass and mineralizable C and N potentials, *Can. J. Soil Sci.*, 62, 587, 1982.
45. Paul, E.A. and Voroney, R.P., Nutrient and energy flows through soil microbial biomass, in *Contemporary Microbial Ecology*, Ellwood, D.C., Hedger, J.N., Lathan, M.J., Lynch, J.M., and Slater, J.H., Eds., Academic Press, London, 1980, 216.
46. Molina, J.A.E., Clapp, C.E., Shaffer, M.J., Chichester, F.W., and Larson, W.E., NCSOIL, a model of nitrogen and carbon transformations in soil: description, calibration and behavior, *Soil Sci. Soc. Am. J.*, 47, 85. 1983.
47. Jansson, S.L. and Persson, J., Mineralization and immobilization of soil nitrogen, in *Nitrogen in Agricultural Soils*, Stevenson, F.J., Ed., ASA, Madison, WI, 1982, 229.
48. Houot, S., Molina, J.A.E., Chaussod, R., and Clapp, C.E., Simulation by NCSOIL of net mineralization in soils from the Deherain and 36 Parcelles plots at Grignon, *Soil Sci. Soc. Am. J.*, 45, 451, 1989.
49. McGill, W.B., Hunt, H.W., Woodmansee, R.G., and Reuss, J.O., Phoenix, a model of dynamics of carbon and nitrogen in grassland soils, in *Terrestrial Nitrogen Cycles. Processes, Ecosystem Strategies and Management Inputs*, Clark, F.E. and Rosswall, T., Eds., Ecol. Bull. 33, Stockholm, 1981, 49.
50. Vanotti, M.B., Bundy, L.G., and Kurakov, A.V., Variation of potentially mineralizable nitrogen in legume-cereal crop sequences, in *Agronomy Abstracts*, ASA, Madison, WI, 1991, 303.
51. Barak, P., Molina, J.A.E., Hadas, A., and Clapp, C.E., Optimization of an ecological model with the Marquardt algorithm, *Ecol. Model.*, 51, 251, 1990.
52. Ellert, B.C. and Bettany, J.R., Comparison of kinetic models for describing net sulfur and nitrogen mineralization, *Soil Sci. Soc. Am. J.*, 52, 1692, 1988.
53. Jansson, S.L., Tracer studies on nitrogen transformations in soil with special attention to mineralization-immobilization relationships, *Ann. R. Agric. Coll. Sweden*, 24, 101, 1958.
54. Jenkinson, D.S. and Parry, L.C., The nitrogen cycle in the Broadbalk Wheat Experiment: a model for the turnover of nitrogen through the soil microbial biomass, *Soil Biol. Biochem.*, 21, 535, 1989.
55. Paul, E.A. and Juma, N.G., Mineralization and immobilization of soil nitrogen by microorganisms, in *Terrestrial Nitrogen Cycles. Processes, Ecosystem Strategies and Management Inputs*, Clark, F.E. and Rosswall, T., Eds., Ecol. Bull. 33, Stockholm, 179, 1981.
56. Olsen, R.J., Effect of Various Factors on Movement of Nitrate Nitrogen in Soil Profiles and on Transformations of Soil Nitrogen, Ph.D. thesis, University of Wisconsin-Madison, 1969.

Chapter 7

Long-Term N Management Effects on Corn Yield and Soil C of an Aquic Haplustoll in Minnesota

D.R. Huggins and D.J. Fuchs

CONTENTS

I. INTRODUCTION

Levels of soil organic matter (SOM) are positively correlated to carbon (C) additions from above-ground residue,[1–3] root biomass, and rhizodeposition.[4] Nitrogen (N) management can elicit changes in SOM levels by affecting crop productivity and the return of unharvested residues to the soil. The magnitude, direction, and rate of management induced changes in SOM, however, are controlled by climate, relief, soil parent material, and organisms that function as driving variables for soil processes and the expression of soil properties.[5,6] The influence of N treatments on SOM levels is best evaluated in long-term studies where the interaction between N management and driving variables is strongly expressed.

Field research on urea and ammonium nitrate management was initiated in 1960 at the Southwest Experiment Station (SWES), Lamberton Minnesota, to evaluate the effects of N form, placement, rate, and time of application on corn (*Zea mays* L.) yields.[7] The study continues as of 1993 after 34 years of corn production.[8] Additional studies have included the effects of N fertilizer rate on corn root growth

0-8493-2802-0/97/$0.00+$.50

(unpublished) and on SOM.[9] This paper (1) summarizes the long-term effects of N management on corn yield and soil carbon levels; (2) estimates annual returns of crop residues based on corn grain yield data; and (3) provides climatic and soil information that can be used to model SOM changes.

II. EXPERIMENTAL SITE CHARACTERISTICS

The field study is located in southwest Minnesota on the University of Minnesota, Southwest Experiment Station (95° 18′ E and 44° 14′ N) near Lamberton. The weather station at the SWES was established in 1961 and the following statistics are for the period 1961 to 1992. The climate is continental and monthly averages of air temperature range from −11.3°C in January to 21.2°C in July. The average annual air temperature is 6.2°C. Growing degree days for corn from 1 May to 1 October average 1372 (base temperature 10°C, maximum temperature 50°C). Monthly averages of precipitation range from 14.0 mm in February to 89.4 mm in July and have a similar distribution pattern as temperature (Figure 1). The average annual precipitation is 632 mm with 63% occurring from 1 May to 1 October.

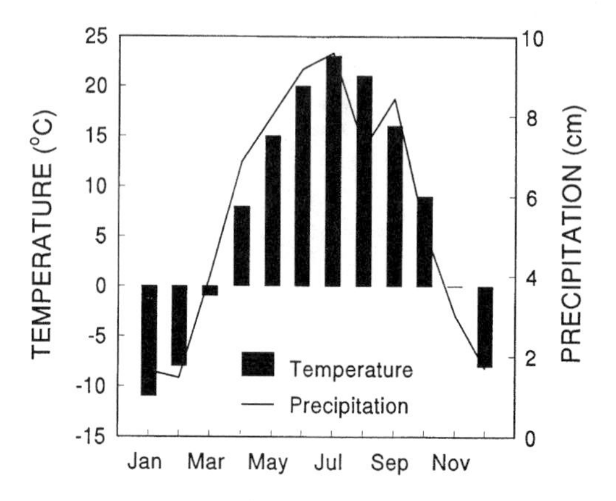

Figure 1 Average monthly temperature and precipitation from 1961 to 1992 for the University of Minnesota Southwest Experiment Station, Lamberton.

Native vegetation was tall-grass prairie and the soil is currently classified as Normania loam (fine-loamy, mixed, mesic Aquic Haplustolls) with minor inclusions of Webster clay loam (fine-loamy, mixed, mesic Typic Haplaquolls). The soil formed in Late Wisconsin glacial drift derived from the Des Moines lobe ice advance and occurs in the Blue Earth Till Plain region.[10] The site is nearly level with little history of soil erosion or deposition.

III. DESCRIPTION AND MANAGEMENT HISTORY OF SELECTED STUDIES

A. LONG-TERM (1960 TO 1993) N MANAGEMENT OF CORN

Nitrogen management treatments consist of N form (urea, 46-0-0, and ammonium nitrate, 34-0-0), annual N rate (0, 45, 90, and 180 kg N ha^{-1}), application timing (fall, spring, and sidedress), and N placement (fall surface broadcast, fall broadcast, and incorporated). Details of the continuous corn study have been described by Nelson and MacGregor.[7] Briefly, the experimental design consists of 18 treatments in a randomized block with 4 replications. The experiment is an incomplete factorial design as each N rate does not occur for every N form, placement, or timing. All treatments have annually received banded starter fertilizer (5 cm to the side and 5 cm below the seed) during planting that has averaged 16 kg N (ammonium form) ha^{-1}, 31 kg P_2O_5 ha^{-1}, and 15 kg K_2O ha^{-1}. Spring and sidedressed applications of N were soil incorporated. All of the corn stover has been returned to the soil following harvest by moldboard plowing to a depth of 25 to 30 cm. Stover biomass was collected in 1960, 1961, 1962, 1978, 1979, and 1992.

B. ORGANIC MATTER STUDY

A 1978 study examined N management effects on soil organic C after 19 seasons of continuous corn.[9] Soil samples were collected from 0 to 20 cm in each plot (four samples per plot, six cores per sample).

The samples were air-dried, ground, and sieved to pass 2 mm. Organic carbon was determined by the Walkley-Black method.[11]

C. ROOT BIOMASS STUDY

Corn root systems were excavated using a soil monolith method[12] on August 12, 1970, and August 2, 1971. One monolith (35 × 43 × 152 cm) was obtained from each of three plots that had received 16 kg N ha^{-1} at planting (starter fertilizer) and 0, 90, and 180 kg N ha^{-1}, respectively, of sidedressed ammonium nitrate. The monoliths were divided into 7.6-cm depth increments and root biomass was collected from each increment to a depth of 152 cm. Root lengths and biomass data for each increment were obtained for the three root samples. These data were not evaluated statistically due to lack of replication.

IV. RESULTS AND DISCUSSION

A. CORN GRAIN YIELD AND ABOVE-GROUND BIOMASS

Nitrogen application rate has had the greatest influence on corn yield (1960 to 1993, Table 1). Yield averages have ranged from 3.98 Mg ha^{-1} for control plots (16 kg N ha^{-1}) to 7.29 Mg ha^{-1} for the greatest N rate (195 kg N ha^{-1}, sidedress treatment). Grain yields have also been increased by 3 to 17% with spring and sidedressed applications of N as compared to fall N applications (Table 1). There has been little difference in grain yield due to N form or fall placement (surface broadcast before plowing vs. surface broadcast after plowing).

Table 1 Average Corn Yields (1960 through 1993) as Influenced by N Rate, Form, Timing, and Placement

N Form and Rate[a] (kg ha^{-1})	Nitrogen Timing and Placement (Yield SE[b], Mg ha^{-1})							
	Fall				Incorporated			
	Surface		Plowdown		Spring		Sidedressed	
Urea								
61	5.51	0.18	5.63	0.18	5.75	0.18	6.05	0.18
106	nd[c]	nd	6.42	0.20	6.77	0.19	7.00	0.18
195	nd	nd	7.05	0.19	nd	nd	nd	nd
Ammonium nitrate								
61	5.43	0.18	5.19	0.18	5.79	0.18	6.06	0.17
106	nd	nd	6.38	0.18	6.63	0.19	6.49	0.19
195	nd	nd	6.94	0.18	nd	nd	7.29	0.19

Note: $LSD_{0.005}$ = 0.31 Mg ha^{-1}. Check plots (16 kg N ha^{-1}) averaged 3.98 Mg ha^{-1} with an SE of 0.16 Mg ha^{-1}.

[a] N rates include 16 kg N ha^{-1} applied as starter fertilizer.

[b] SE = standard error.

[c] nd = no data.

Although grain yield responded significantly to applied N, changes in SOM due to N management are dependent on the N response of unharvested above-ground and root biomass. Cropping system studies often lack above-ground biomass data and estimates based on yield or an assumed harvest index (HI) are required to assess the amount of residue annually returned to the soil. The greater grain yield associated with increased N rates suggests that greater stover biomass was also annually returned to the soil. Stover biomass has been calculated from grain yield by assuming a constant HI (grain mass/total above-ground biomass).[9] HI, however, may not remain constant over a range of grain yields. Moreover, grain yield may be a better predictor of stover biomass than HI.

Above-ground biomass data for this study are incomplete; however, data from 1960 to 1962, 1978, 1979, and 1992 revealed that HI increased linearly with increased above-ground biomass (Figure 2). In addition, as above-ground biomass increased, grain yield increased at a greater rate than stover biomass. Increases in both above-ground biomass and harvest index with applied N have been reported for grain sorghum (*Sorghum bicolor* L. Moench).[14] Simultaneous increase in above-ground biomass and HI contrasts, however, with studies of small grains where increased above-ground biomass were associated with decreased HI.[13] The range in HIs estimated in this study were similar to those reported for corn in a variety of Minnesota environments.[15]

Table 2 Corn Grain Yield, Stover, and HI as Influenced by Applied N (Mg ha^{-1})

	Applied N[a]											
	16 kg ha^{-1}			61 kg ha^{-1}			106 kg ha^{-1}			195 kg ha^{-1}		
Year	Grain Yield	Stover Biomass	HI	Grain Yield	Stover Biomass	HI	Grain Yield	Stover Biomass	HI	Grain Yield	Stover Biomass	HI
1960	3.28	4.08	0.45	3.66	4.19	0.46	4.12	4.23	0.49	4.19	4.52	0.47
1961	5.84	6.79	0.46	6.09	8.15	0.43	6.06	7.95	0.43	6.79	9.30	0.42
1962	1.46	2.53	0.37	1.96	3.56	0.35	2.36	3.72	0.38	3.50	4.67	0.43
1978	4.57	7.58	0.38	6.82	8.29	0.46	8.10	7.63	0.55	8.38	8.67	0.49
1979	2.84	4.79	0.37	5.23	7.20	0.43	6.56	6.28	0.53	8.49	7.73	0.52
1992	4.26	4.67	0.48	7.62	8.06	0.49	9.07	9.35	0.49	9.88	9.75	0.50

[a]Total N applied (N rate + 16 kg N ha^{-1} starter fertilizer).

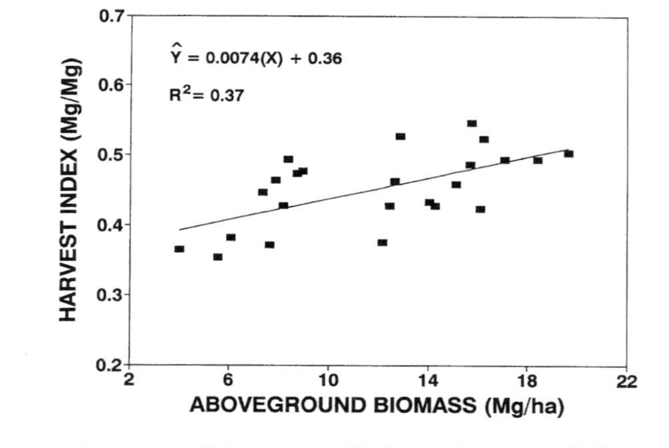

Figure 2 Relationship between above-ground biomass and HI (data include years 1961 and 1962, 1978, 1979, and 1992).

Stover biomass and grain yield were linearly related and had a greater R^2 than the relationship between above-ground biomass and HI (Figure 3). The relationship between grain and stover biomass was extrapolated over the entire experiment to provide estimates of stover biomass for each N rate (Table 3). Stover biomass estimated from grain yield increased by 26% as applied N increased from 16 kg N ha^{-1} to 61 kg N ha^{-1}. Nitrogen rates greater than 61 kg N ha^{-1} showed diminishing response of stover biomass estimates to applied N. The lower HI with 16 kg N ha^{-1} as compared to greater N rates showed that greater stover biomass relative to grain was produced at low N rates. This indicates that low levels of N can result in greater proportions of above-ground biomass returned to the soil as compared to greater N rates.

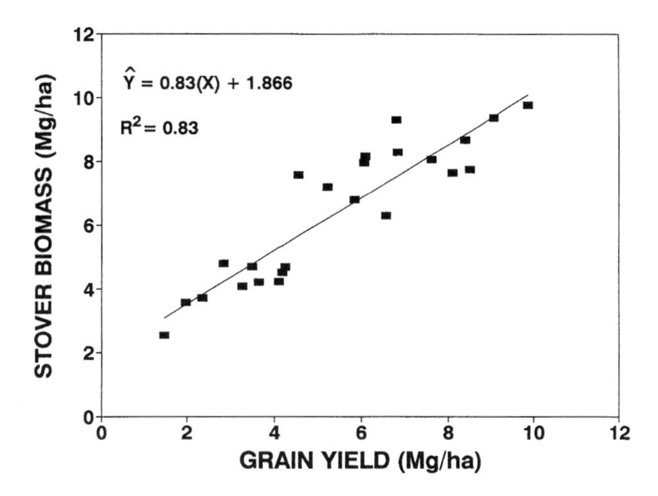

Figure 3 Relationship between grain yield and stover biomass (data include years 1961 and 1962, 1978, 1979, and 1992).

Table 3 Effects of Applied N on Yield, Stover Biomass Estimates, and Soil Organic Carbon after 19 Years of Continuous Corn

Applied N[a] (kg ha^{-1})	Grain Yield[b] (Mg ha^{-1})	Estimated Stover Biomass[c] (Mg ha^{-1})	HI[d]	Estimated Stover Biomass[e] (Mg ha^{-1})	Constant HI	Soil Organic Carbon (g kg^{-1})
16	4.47	5.57	0.44	5.04	0.47	26.3
61	6.22	7.03	0.47	7.02	0.47	27.4
106	7.13	7.79	0.48	8.04	0.47	27.0
195	7.42	8.02	0.48	8.37	0.47	27.5

[a] Total N applied (N rate + 16 kg N ha^{-1} starter fertilizer).
[b] Grain yield averaged from 1960 through 1978.
[c] Stover biomass estimated from linear relationship with grain yield (Figure 2).
[d] HI based on grain yield and stover estimated from grain yield or held at a constant value (0.47).
[e] Stover estimated from constant HI (0.47).

The range of stover biomass estimated from grain yield was less than the range estimated from HIs that were constant across N rate (Table 3). A constant HI of 0.47 estimated stover biomass at the 15 kg N ha^{-1} applied N rate to be 10% less and at 195 kg N ha^{-1} 5% more than estimates based on grain yield and a variable HI. The sensitivity of stover biomass estimated from HI becomes critical in long-term studies when calculating annual returns of crop residue to the soil. Not assessed were changes in HI that may have occurred over time with corn hybrid improvements.[16]

B. ROOT BIOMASS

The effects of N rate on root biomass were greatest in the upper 20 cm of soil (Figure 4). Differences in total root biomass among the N rates were less apparent (Table 4). Although these results indicate that differences in root biomass due to N nutrition can occur in the upper soil profile, the magnitude is much smaller than that associated with differences in grain and stover biomass. Rhizodeposition could contribute to the maintenance of soil carbon levels,[4] but was not assessed in this study.

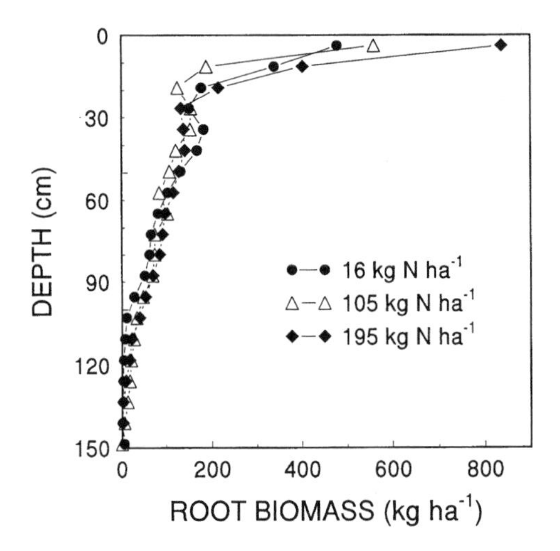

Figure 4 Accumulated root biomass of corn on 12 August, 1970, as affected by applied N.

Table 4 Effect of Applied N on Corn Root Biomass and Grain Yield in 1970 and 1971

Applied N[a] (kg ha^{-1})	Root Biomass[b] (Mg ha^{-1}) 1970	Root Biomass[b] (Mg ha^{-1}) 1971	Grain Yield (Mg ha^{-1}) 1970	Grain Yield (Mg ha^{-1}) 1971	Root Biomass-to-Grain Yield Ratio 1970	Root Biomass-to-Grain Yield Ratio 1971
16	2.05	2.60	5.14	2.51	0.40	1.03
106	1.99	2.86	8.16	6.34	0.24	0.45
195	2.60	2.72	8.42	6.27	0.31	0.43

[a] Total N applied (N rate + 16 kg N ha^{-1} starter fertilizer).
[b] Root biomass in a 35 × 43 × 152 cm soil volume (above-ground brace roots not included).
Unpublished data courtesy of R.R. Allmaras and W.W. Nelson.

C. SOIL CARBON

The levels of soil carbon were not significantly different between N rates after 19 years of continuous corn (Table 3).[9] There was, however, a trend of 4.5% less soil carbon for the N treatment with 16 kg N ha^{-1} as compared to applications of 195 kg N ha^{-1}. The trend in reduced soil carbon corresponded to 2.45 Mg ha^{-1} less stover biomass estimated to be annually returned to the soil in the low N treatment as compared to the high N treatment.

Several studies have indicated a linear relationship between changes in soil C and the amount of unharvested C inputs.[1–3] Converting the data of Larson et al.[1] from Iowa, yields the equation $Y = 0.280X - 611$, where Y is the change in soil organic C (kg C ha^{-1} y^{-1}) and X is organic C additions (kg ha^{-1} y^{-1}. Assuming a C concentration of 4.0 g kg^{-1} for corn stover, decreases in stover biomass additions from 8.02 to 5.57 kg ha^{-1} y^{-1} would be predicted to reduce soil C by 274 kg C ha^{-1} y^{-1}. Rasmussen et al.[2] presented a similar linear equation for a long-term study in eastern Oregon where $Y = 0.186X - 393$. This equation would predict a decrease of 182 kg C ha^{-1} y^{-1} for the study. The trend in organic C decline in the 195 vs. the 16 kg N ha^{-1} treatments is 0.12 g kg^{-1}. Assuming a soil bulk density of 1.3 g cm^{-3} for the 0 to 20 cm depth, the trend in soil organic C reduction is 164 kg C ha^{-1} y^{-1}. These calculations suggest that the effect of residue inputs on the rate of change in organic C for a mollisol in southwest Minnesota is less than that found for a mollisol under Iowa's or eastern Oregon's climatic and cultural systems. This is particularly apparent considering that no significant difference in soil organic C has occurred despite substantial differences in estimated stover biomass returned to the soil.

Carbon levels for the same soil series under native prairie are 56.3 g kg^{-1}.[17] These data indicate that slightly over 50% of the soil C has been lost since the original prairie was converted to agricultural cropping systems in the late 1800s. The remaining soil C levels are still relatively high (Table 3) and are likely composed of a high proportion of stable organic fractions that are resistant to microbial decomposition. The studies of soil C in Iowa[1] and eastern Oregon[2] had lower initial soil C levels of 18.1 and 12.5 g kg^{-1}, respectively, as compared to soil C levels averaging 27.1 g kg^{-1} after 19 years of continuous corn in southwest Minnesota. Changes in overall SOM percentage would require greater additions or losses of C given higher initial soil C levels.

Larson et al.[1] estimated that 5000 kg ha^{-1} y^{-1} of crop residues were needed to maintain soil C levels at 18 g kg^{-1} under a conventional tillage regime. Because initial soil C levels are not known for this study, it is not apparent if soil C levels are increasing, decreasing, or at equilibrium with the imposed N treatments. If soil C levels are declining in the treatment receiving 16 kg N ha^{-1} y^{-1}, the annual stover inputs of greater than 5500 kg ha^{-1} y^{-1} would likely slow the decline in soil C levels and help maintain SOM. The relatively high production of corn stover biomass at low applied N rates could be due to N mineralization from organic matter which is estimated to be 90 kg N ha^{-1} y^{-1}.[18] The high rate of N mineralization is associated with the high SOM levels and favorable within-season temperatures and moisture (Table 3, Figure 1). Due to the nature of stover biomass response to N, high levels of soil N mineralization would reduce differences in above-ground biomass production among N treatments and may buffer against changes in soil C.

V. SUMMARY AND CONCLUSIONS

Corn yields were influenced by N rate and timing but not form (urea vs. ammonium nitrate). Grain yield and stover biomass were highly correlated; therefore, estimates of stover biomass based on grain yield were used to estimate stover biomass returned to the soil. Corn stover biomass was less responsive than grain yield to applied N; however, low levels of applied N (16 kg N ha^{-1}) reduced estimated stover dry matter by 31% as compared to the greatest N rate (195 kg N ha^{-1}). Despite the difference (2.45 Mg ha^{-1}) in annual returns of stover dry matter to the soil, levels of soil C were not significantly different after 19 years of continuous corn. Studies in Iowa[1] and eastern Oregon[2] observed greater rates of soil C change than those that had occurred in this study. The relatively high percentage of SOM coupled with favorable seasonal climatic conditions that stimulate N mineralization could buffer against changes in soil C due to N fertilization. To determine changes in organic matter levels induced by N management will require additional measurements at longer time intervals.

ACKNOWLEDGMENTS

The authors express their gratitude to Dr. P. R. Bloom, J. A. Lamb, and two anonymous reviewers for their helpful suggestions in the preparation of this manuscript. The authors also wish to recognize Dr. W. W. Nelson for his foresight and perseverance in continuing this long-term experiment.

REFERENCES

1. Larson, W.E., Clapp, C.E., Pierre, W.H., and Morachan, Y.B., Effects of increasing amounts of organic residues on continuous corn. II. Organic carbon, nitrogen, phosphorus, and sulfur, *Agron. J.*, 64, 204, 1972.
2. Rasmussen, P.E., Allmaras, R.R., Rohde, C.R., and Roager, N.C., Jr., Crop residue influences on soil carbon and nitrogen in a wheat-fallow system, *Soil Sci. Soc. Am. J.*, 44, 596, 1980.
3. Paustian, K., Parton, W.J., and Persson, J., Modeling soil organic matter in organic-amended and nitrogen-fertilized long-term plots, *Soil Sci. Soc. Am. J.*, 56, 476, 1992.
4. Barber, S.A., Corn residue management and soil organic matter, *Agron. J.*, 71, 625, 1979.
5. Jenny, H., *Factors of Soil Formation*, McGraw-Hill, New York, 1941.
6. Jenny, H., *The Soil Resource: Origin and Behavior*, Springer-Verlag, New York, 1980.
7. Nelson, W.W. and MacGregor, J.M., Twelve years of continuous corn fertilization with ammonium nitrate or urea nitrogen, *Soil Sci. Soc. Am. J.*, 37, 583, 1973.
8. Fuchs, D.J., Huggins, D.R., and Nelson, W.W., Thirty-two years of continuous corn with variable nitrogen source, placement, and time of application at the southwest experiment station, in Field Research in Soil Science 1992, Soil series #134, Miscellaneous publication 75-1992, Minnesota Agricultural Experiment Station University of Minnesota, St. Paul, 1992, 56.
9. Bloom, P.R., Schuh, W.M., Malzer, G.L., Nelson, W.W., and Evans, S.D., Effect of N fertilizer and corn residue management on organic matter in Minnesota Mollisols, *Agron. J.*, 74, 161, 1982.

10. Wright, H.E., Jr., Physiography of Minnesota, in *Geology of Minnesota, A Centennial Volume*, Simms, P. K. and Morey, G.B., Eds., Minnesota Geological Survey, University of Minnesota, St. Paul, 1972, 561.
11. Allison, L.E., Organic Carbon, in *Methods of Soil Analysis: Chemical and Microbiological Properties*, Black, C.A., Ed., American Society of Agronomists, Madison, WI, 1965.
12. Nelson, W.W. and Allmaras, R. R., An improved monolith method for excavating and describing roots, *Agron. J.*, 61, 751, 1969.
13. Donald, C.M. and Hamblin, J., The biological yield and harvest index of cereals as agronomic and plant breeding criteria, in *Advances in Agronomy*, Vol. 31, Academic Press, New York, 1976, 361.
14. Roy, R.N. and Wright, B.C., Sorghum growth and nutrient uptake in relation to soil fertility. I. Dry matter accumulation patterns, yield, and N content of grain, *Agron. J.*, 65, 709, 1973.
15. DeLoughery, R.L. and Crookston, R.K., Harvest index of corn affected by population density, maturity rating, and environment, *Agron. J.*, 71, 577, 1979.
16. Russell, W.A., Genetic improvement of maize yields, in *Advances in Agronomy*, Vol. 46, Academic Press, New York, 1991, 245.
17. Huggins, D.R., unpublished data, 1993.
18. Rehm, G.W. and Schmitt, M.A., Fertilizing Corn in Minnesota, AG-FO-3790, Minnesota Extension Service, University of Minnesota, St. Paul, 1989.

Chapter 8

Long-Term Effects of Fertilizer and Manure on Corn Yield, Soil Carbon, and Other Soil Chemical Properties in Michigan

M.L. Vitosh, R.E. Lucas, and G.H. Silva

CONTENTS

I. INTRODUCTION

Historically, manure has been used as a major source of nutrients for crop production.[1–3] Manure is usually considered as a source of nitrogen (N) and, to a lesser extent, a source of phosphorus (P), potassium (K), and micro nutrients. It is also known as a soil conditioner because it improves soil structure, water infiltration, and cation exchange capacity. The long-term benefits of manure can be appreciable when large quantities of manure are applied.[3,4] The effects of manure on the storage of carbon (C) relative to global C cycling are also now being investigated.

Changes in the agricultural industry over the past 40 years have had a significant impact on manure management practices: (1) fertilizer N has become readily available and relatively inexpensive compared to manure; (2) crop yields have increased greatly requiring an increasing amount of N to achieve optimum

0-8493-2802-0/97/$0.00+$.50

yields; (3) because of the concentration of livestock into larger units, many livestock operations have excess manure creating a disposal problem rather than providing a practical source of nutrients.

Livestock operations often purchase feed from distant sources. This renders nutrient recycling ineffective and impractical. Particularly disturbing is the potential for nitrate contamination of groundwater in livestock intensive areas. The ultimate goal of manure management should be to recycle the nutrients so contamination of groundwater and surface water, odor, and spread of disease are minimized. In this context, manure research has centered on disposal in an economical and environmentally sound manner rather than on evaluating manure as a nutrient source.[5]

One hundred years of continuous chemical fertilizer use on wheat, on a clay loam soil at the Rothamsted Experimental Station in England, has proven to be as effective as farm yard manure, producing equivalent grain yields.[4,6] Nonetheless, there is concern that continuous use of fertilizers may diminish soil quality in terms of long-term sustainability. Many farmers have recently adopted farming systems which include a combination of manure and fertilizers. These farmers are interested in knowing the amount of nutrients that manures can furnish to crops and the amount of fertilizer that manure can economically replace.

Changes in soil C due to manure and other management practices can only be determined with long-term experiments.[6,7] Data from such experiments can then be extrapolated to larger agricultural areas where manure is applied in large quantities. This would greatly assist in the evaluation of manure practices in relation to soil C and global CO_2 changes. The specific objectives of this long-term study were (1) to determine the optimum rate of annual cattle manure application to a loamy sand soil in a continuous corn production system where both grain and silage are removed; and (2) to compare the long term effects of fertilizer and manure on corn yield, soil C, and other chemical properties.

II. EXPERIMENTAL SITE AND PROCEDURE

The experiment was located at the Michigan State University Agronomy Farm at East Lansing (Latitude 42.5°, Longitude 85.5°). The climate is a semi-humid temperate with a mean annual precipitation of 782 mm, of which 396 mm occurs during the active growing period from May to September. The mean annual temperature is 8.6°C. The soil type is a Metea loamy sand (Arenic Hapudalfs loamy, mixed, mesic) with an initial pH of 6.5.

The study was established in 1963. From 1963 until 1982, five manure and fertilizer treatments, A to E (Table 1), were evaluated in a randomized split block design with three replications. Each block was divided into two grain and two silage plots (Figure 1). The individual plots were 12.8 m long and 9.8 m wide. Starting in 1974, one half of each of block was sprinkler irrigated. From 1963 to 1982, each year prior to spring plowing, treatment A received 168 kg ha^{-1} N broadcast as ammonium sulfate. Treatment B received 168-74-140 kg ha^{-1} N-P-K as ammonium sulfate and 0-8.8-16.6 N-P-K fertilizer. Treatment A was considered to be a low P and K treatment whereas treatment B was considered to be a high P and K treatment. Manure obtained from the loose housing beef barns at MSU was applied in the fall of each year to treatments C, D, and E at rates of 22, 45, and 67 Mg ha^{-1}, respectively. All plots including the manured plots received 224 kg ha^{-1} of 5-8.8-16.6 N-P-K starter fertilizer banded 5 cm to the side and 5 cm below the seed at planting. In the fall of 1982, the manure and fertilizer treatments were terminated. The plots have since been maintained in a corn-soybean rotation where only grain was removed.

Manure used from 1963 to 1981 had a considerable amount of straw bedding. In 1982, manure from the same source without straw was applied. The manure used from 1971 to 1982 was analyzed for chemical constituents and dry matter content using total kjeldahl and spectrophotometric methods. The mean nutrient concentrations in the manure on a dry weight basis expressed as kilograms per megagrams were 18.4 N, 4.6 P, 18.9 K, 7.7 Ca, and 3.8 Mg. One megagram of manure contained an average of 260 kg of dry matter.

The field was moldboard plowed each spring and corn was planted after minimum secondary tillage. Initially, the field was plowed to a depth of 20 to 23 cm. In 1976, the field was plowed to a depth of 25 to 30 cm due to larger and more powerful equipment. Weeds in the corn were controlled with atrazine and alachlor herbicides. Corn varieties used in the experiment changed periodically, depending on the availability of adapted hybrids. The same corn variety was used for grain and silage purposes except in 1970 and 1971. A row spacing of 105 cm was used at the start of the experiment in 1963 but reduced to 70 cm in 1967. The plant population was gradually increased from 33,500 in 1963 to 69,000 plants ha^{-1} in 1982. Both grain and silage yields were measured in irrigated and nonirrigated plots between

Table 1 Annual Fertilizer and Manure Treatments Evaluated at East Lansing, 1963 to 1982

Treatment	N-P-K (kg ha^{-1})	Manure (Mg ha^{-1})
A	179-20-37	0
B	179-94-177	0
C	11-20-37	22
D	11-20-37	45
E	11-20-37	67

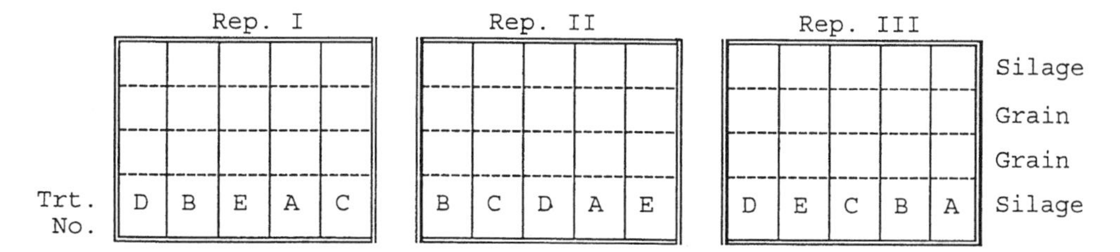

Figure 1 Field layout of experimental treatments.

1974 and 1982. Corn earleaf samples were taken at early silking from irrigated grain and silage plots from 1974 to 1979 for total N determinations. Soil samples from the plow layer were taken in June of each year and analyzed for pH, Bray P_1 P, and ammonium acetate exchangeable K. Soil C and N were determined by dry combustion on duplicate samples using a Carlo Erba CHN analyzer Model 1104 (Carlo Erba Instruments, Milano, Italy). Soil C and N analyses were performed on the irrigated half of the experiment in 1972, 1976, 1982, and 1991.

III. RESULTS AND DISCUSSION

A. CORN YIELDS

A summary of the corn grain and silage yields in response to fertilizer and manure treatments is presented in Table 2 for the 11- and 9-year nonirrigated periods (1963 to 1973 and 1974 to 1982) and 9-year irrigated period (1974 to 1982). Nonirrigated yields obtained during the last 9 years were higher than the first 11 years. This was primarily due to improved corn hybrids and increased plant densities from 1972 to 1982.

Treatment C which received 22 Mg ha^{-1} manure generally produced the lowest grain and silage yields. These plots often showed visual N deficiency symptoms. Nonirrigated grain yields were not significantly

Table 2 Effects of Long-Term Fertilizer and Manure Treatments on Corn Grain and Silage Yield

	Grain Yield[a] (Mg ha^{-1})			Silage Yield[b] (Mg ha^{-1})		
	Nonirrigated		Irrigated	Nonirrigated		Irrigated
Treatment	1963–1973	1974–1982	1974–1982	1963–1973	1974–1982	1974–1982
A	5.3_a[c]	7.5_{ab}	10.1_{ab}	30.2_d	45.9_b	54.4_{bc}
B	4.8_b	7.3_{bc}	10.5_a	33.5_b	46.3_b	58.0_{ab}
C	5.2_b	6.9_c	9.6_b	31.8_c	42.2_c	52.9_c
D	5.2_a	8.0_a	10.5_a	33.1_{bc}	47.6_b	58.5_a
E	5.1_{ab}	7.8_{ab}	10.6_a	35.0_a	50.3_a	59.6_a
Mean	5.1	7.5	10.3	32.7	46.5	56.7

[a] Adjusted to 15.5% moisture.

[b] Adjusted to 70% moisture.

[c] Any two means within a column followed by different letters are significantly different, as measured by the LSD method (p = 0.05).

increased by the addition of P and K beyond the starter application (treatment B vs. A). In fact, the 11-year average nonirrigated grain yield was significantly lower when additional P and K were added. The additional K salt and dry soil moisture conditions on this sandy soil may have created the potential for salt injury and reduced grain yields. The salt accumulation assumption was supported by the K soil test data (Table 8). When the crop was irrigated, grain yields were not reduced by treatment B. In contrast, the yield of the 11-year nonirrigated silage was significantly increased by the addition of P and K beyond the starter application rate (treatment B vs. A). Greater annual nutrient removal and lower soluble salt accumulation in the soil are possible reasons for the differences in grain and silage response to treatment B. In 1982, soil samples from the grain plots of treatment B contained nearly 77 mg kg^{-1} more exchangeable K than comparable silage plots (Table 8).

Irrigated and nonirrigated grain yields for treatments A, D and E were not significantly different (Table 2). We conclude that 45 Mg ha^{-1} of manure (treatment D) was optimum for grain production at this site. The data also indicated that commercial fertilizers when applied at appropriate rates can be just as effective as manure for grain production. Yields of nonirrigated silage increased with each rate of manure applied. The 67-Mg manure (treatment E) yielded more silage than fertilizer treatments A or B. When irrigated, however, treatment B produced silage yields equivalent to treatments D and E. The optimum manure rate for silage production appeared to be slightly higher than 45 Mg ha^{-1}. Again, when nutrients are properly balanced, comparable corn silage yields can be obtained with either commercial fertilizer or manure.

B. SOIL CARBON

The soil C level of the manured plots was directly related to the rate of manure applied (Table 3). In 1982, after 20 annual applications of manure, soil C in the manured plots was significantly higher than nonmanured plots. The average soil C content in 1982 in the three manured grain and silage plots was 12.4 g kg^{-1}, while in the two nonmanured plots, it was only 7.8 g kg^{-1}. The total quantity of manure added to the three manure plots over 20 years averaged 900 Mg ha^{-1} (fresh weight) or 234 Mg ha^{-1} (dry weight). Manured plots had an overall increase of 4.6 g kg^{-1} (12.4-7.8) in soil C compared to nonmanured plots.

Table 3 Effects of Long-Term Fertilizer and Manure Treatments on Total Soil Carbon

	Total Soil Carbon (g kg^{-1})							
	Grain Plots				Silage Plots			
Treatment	**1972**	**1976**	**1982**	**1991**	**1972**	**1976**	**1982**	**1991**
A	9.1_{d}	9.2_{ef}	7.9_{d}	7.5_{def}	9.7_{d}	8.4_{f}	7.2_{d}	6.2_{f}
B	9.5_{d}	10.1_{def}	8.2_{cd}	7.8_{de}	9.7_{d}	8.3_{f}	7.8_{d}	6.4_{ef}
C	12.4_{bc}	11.5_{cd}	10.7_{b}	9.8_{bc}	12.0_{c}	10.8_{de}	10.0_{bc}	7.3_{def}
D	12.7_{abc}	13.0_{bc}	14.0_{a}	10.8_{b}	12.1_{c}	11.3_{cd}	11.1_{b}	8.5_{cd}
E	14.9_{a}	15.6_{a}	15.0_{a}	12.3_{a}	14.5_{ab}	13.6_{ab}	13.8_{a}	10.6_{b}
Mean	11.7	11.9	11.2	9.7	11.6	10.5	10.0	7.8

Note: Any two means within a year followed by different letters are significantly different, as measured by the LSD method ($p = 0.05$).

It was not possible to calculate the exact amount of C lost due to decomposition, leaching, and erosion because the manure was not analyzed for C. The average N content of manure used between 1971 and 1982 was 18.4 kg Mg^{-1}. If we assume a C:N ratio of 10:1 for the manure, then the C content of manure could be estimated to be 184 kg Mg^{-1}. Based on the addition of 234 Mg ha^{-1} dry matter, we estimate that the addition of C is equivalent to 43,056 kg ha^{-1} (234×184). Using an average soil weight of 3.36 million kg ha^{-1} 25 cm deep, we estimate that the manured plots gained 15,456 kg C ha^{-1} ($0.0046 \times 3.36 \times 10^6$) compared to the fertilized plots. Thus, during the period of manure additions, we estimate that 64% of the C added was lost through various means such as decomposition to CO_2, leaching below the plow layer, or losses by wind and water erosion.

In 1991, after 9 years without manure and a less-intensive rotation of corn and soybeans, and changing from plowing to no-till, the soil C level had declined significantly in all plots but especially in the manured plots. The soil C attained during the manuring phase of the study reached a steady state in 1982 but, due to its transitory nature, the soil C declined when no manure was added in subsequent years. Lucas and Vitosh[8] discussed the development of the steady-state condition in a previous paper using data from this experiment and other long-term crop rotation studies.

Although similar trends for soil C were evident both in grain and silage plots, the overall accumulation of soil C in silage plots was less than in grain plots. The average soil C in 1982 was 11.6 g kg^{-1} for the three manured silage plots and 7.5 g kg^{-1} for the two nonmanured plots. This gives an overall increase of 4.1 g kg^{-1} soil C for manured silage plots compared to 4.6 g kg^{-1} for the manured grain plots. The difference of 0.5 g kg^{-1} represents a small loss of 25 mg C kg^{-1} of soil per year due to the lack of corn residue incorporation into the soil.

C. TOTAL SOIL N

The total soil N changes were very similar to soil C changes (Table 4). The addition of manure significantly increased the soil N content. The rate of manure application tended to increase soil N, although the differences were not statistically significant. The soil N content of fertilized plots (treatment A and B) remained relatively constant during the first 20 years of the study but declined rapidly once the less-intensively managed corn-soybean rotation without manure was introduced. The removal of all above-ground residues in the silage treatment tended to reduce the level of soil N in all plots, but many of the differences among treatments were not significant. When the data were combined and statistically analyzed over years, the silage area had significantly less total soil N than the grain area (0.84 vs. 0.97 g kg^{-1}). The yearly mean for soil N, combined over treatments and areas, was also significantly lower in 1991 than in previous years (0.75 for 1991 vs. 0.93, 0.98, and 0.94 g kg^{-1} for 1972, 1976, and 1982, respectively). This would indicate that either soybeans in the rotation consumed some soil N or inadequate N fertilizer was added to the corn crop.

Table 4 Effects of Long-Term Fertilizer and Manure Treatments on Total Soil Nitrogen

	Total Soil Nitrogen (g kg^{-1})							
	Grain Plots				Silage Plots			
Treatment	1972	1976	1982	1991	1972	1976	1982	1991
A	0.75_b	0.84_{cd}	0.72_{bc}	0.66_{bcd}	0.70_b	0.69_d	0.65_c	0.51_d
B	0.78_b	0.95_{bcd}	0.73_{bc}	0.67_{bcd}	0.78_b	0.67_d	0.68_{bc}	0.52_{cd}
C	1.00_{ab}	1.05_{abc}	0.95_{abc}	0.86_{abcd}	0.93_{ab}	0.91_{bcd}	1.00_{abc}	0.63_{bcd}
D	1.04_{ab}	1.20_{ab}	1.18_a	0.96_{ab}	0.94_{ab}	0.98_{bcd}	1.02_{ab}	0.73_{abcd}
E	1.23_a	1.39_a	1.27_a	1.08_a	1.13_a	1.14_{abc}	1.25_a	0.86_{abc}
Mean	0.96	1.09	0.97	0.84	0.90	0.88	0.92	0.65

Note: Any two means within a year followed by different letters are significantly different, as measured by the LSD method (p = 0.05).

D. SOIL CARBON:NITROGEN RATIO

The soil C to N ratio was not significantly affected by any of the treatments (Table 5). When the data were combined over treatments and areas, however, the C:N ratio was significantly different by year having the highest C:N ratio in 1972 and lowest C:N ratio in 1982 (12.6 vs. 11.2, respectively). A small but significant difference between grain and silage areas (11.5 vs. 12.0, in favor of the silage area) was also observed from the combined statistical analysis.

E. SOIL pH

The long-term effect of fertilizer and manure treatments on soil pH, Bray P_1 P, and exchangeable K levels for the grain and silage plots is presented in Tables 6, 7, and 8, respectively. The soil pH in 1963 was 6.5. Due to the addition of ammonium sulfate, the acidity in the fertilized plots increased more rapidly than in the manured plots. To overcome this increasing acidity caused by ammonium sulfate, lime (6.7 Mg ha^{-1}) was applied to treatments A and B in 1971 and 1972. Apparently, the manure contained sufficient bases to neutralize the acidity produced by the oxidation of organic N and, to a lesser extent, organic S during manure decomposition. Olsen et al.[9] reported a similar basic reaction of cattle manure in Wisconsin soils. In 1982, the average pH in treatment B of the grain and silage plots was significantly lower than treatment A. Only a slight difference was observed in previous years. The lower soil pH was probably due to K salts applied to this plot each year and because the pH was measured in water where salt effects are not minimized. Soil samples were taken in June of each year, 1 to 2 months after potash was applied.

Table 5 Effects of Long-Term Fertilizer and Manure Treatments on Total Soil Carbon to Nitrogen Ratio

	Total Soil Carbon:Nitrogen Ratio							
	Grain Plots				Silage Plots			
Treatment	**1972**	**1976**	**1982**	**1991**	**1972**	**1976**	**1982**	**1991**
A	12.1	11.0	11.0	11.4	13.8	12.2	11.0	12.1
B	12.2	10.8	11.2	11.7	12.7	12.4	11.3	12.3
C	12.4	10.9	11.3	11.6	12.9	11.8	10.3	11.6
D	12.2	10.9	11.8	11.3	13.1	11.6	10.9	11.7
E	12.1	11.2	11.8	11.4	12.9	11.9	11.3	12.2
Mean	12.2	10.9	11.4	11.5	13.1	12.0	11.0	12.0

Note: None of the means within a year were significantly different as measured by the LSD method ($p = 0.05$).

Table 6 Effects of Long-Term Fertilizer and Manure Treatments on Soil pH

	Soil pH							
	Grain Plots				Silage Plots			
Treatment	**1972**	**1976**	**1982**	**1991**	**1972**	**1976**	**1982**	**1991**
A	5.7_b	6.0_b	6.1_b	6.2_b	6.0_b	6.5_b	6.3_b	6.5_b
B	5.5_b	5.8_b	5.8_c	6.2_b	5.6_b	6.3_b	5.9_c	6.4_b
C	6.8_a	6.8_a	6.6_a	6.5_a	6.8_a	7.1_a	6.8_a	6.7_a
D	6.7_a	6.9_a	6.7_a	6.6_a	6.7_a	7.1_a	6.8_a	6.7_a
E	6.5_a	6.7_a	6.6_a	6.5_a	6.6_a	7.0_a	6.7_a	6.7_a
Mean	6.2	6.5	6.4	6.4	6.3	6.8	6.5	6.6

Note: Any two means within a column followed by different letters are significantly different, as measured by the LSD method ($p = 0.05$).

Table 7 Effects of Long-Term Fertilizer and Manure Treatments on Soil Phosphorus

	Bray P_1 Phosphorus (mg kg^{-1})							
	Grain Plots				Silage Plots			
Treatment	**1972**	**1976**	**1982**	**1991**	**1972**	**1976**	**1982**	**1991**
A	70_b	102_c	119_e	124_e	82_b	87_d	109_e	115_e
B	196_a	251_a	336_a	279_a	193_a	228_a	302_a	257_a
C	105_b	131_c	159_d	154_d	105_b	125_c	168_d	157_d
D	104_b	150_{bc}	204_c	204_c	116_b	144_{bc}	186_c	176_c
E	161_a	192_b	260_b	220_b	160_a	180_b	249_b	207_b
Mean	127	165	215	196	131	152	203	184

Note: Any two means within a column followed by different letters are significantly different, as measured by the LSD method ($p = 0.05$).

F. SOIL TEST PHOSPHORUS

Soil test P was considerably higher when broadcast P and high rates of manure were applied (treatments B to E), compared to treatment A which received only banded P. In both grain and silage plots, there was a rapid buildup of P from 1972 to 1982. Removing silage rather than grain appeared to have only a minimal effect on the soil test P level. In the high P plots (treatments B and E) a larger decline of P (40 to 50 mg kg^{-1}) occurred from 1982 to 1991 compared to other treatments. It is hypothesized that some of this excess P measured in 1982 has since moved into the subsoil.

G. SOIL EXCHANGEABLE POTASSIUM

Soil exchangeable K buildup for treatment B and the manured plots was much greater than for treatment A. The K buildup in grain plots was also greater compared to silage plots. In 1972, the mean exchangeable

Table 8 Effects of Long-Term Fertilizer and Manure Treatments on Soil Potassium

	Exchangeable Potassium (mg kg^{-1})							
	Grain Plots				Silage Plots			
Treatment	**1972**	**1976**	**1982**	**1991**	**1972**	**1976**	**1982**	**1991**
A	92_c	78_d	90_d	65_d	46_c	42_d	38_e	43_d
B	208_b	168_{bc}	207_c	96_{bc}	174_b	112_{bc}	130_c	71_{bc}
C	249_b	165_c	195_c	91_c	184_b	103_c	107_d	65_c
D	256_b	207_b	276_b	102_b	188_b	148_{ab}	173_b	74_b
E	446_a	262_a	383_a	121_a	294_a	181_a	221_a	91_a
Mean	250	176	230	95	177	117	133	69

Note: Any two means within a column followed by different letters are significantly different, as measured by the LSD method ($p = 0.05$).

K was 73 mg kg^{-1} higher where grain was removed compared to silage. By 1982, this difference was 97 mg kg^{-1}. After 20 years of applying 67 Mg ha^{-1} y^{-1} of manure (treatment E), the exchangeable K level exceeded 380 mg kg^{-1} in the grain plots, and 220 mg kg^{-1} in silage plots. In 1991, the K level had declined to 121 mg kg^{-1} in grain plots and 91 mg kg^{-1} in silage plots. This decline in exchangeable K over the last 9 years represents losses by crop removal and possibly some K leaching into the subsoil. We have observed similar movements of P and K into the subsoil in other Michigan studies when these nutrients were applied in excess.

H. N-P-K NUTRIENT BALANCE

Estimates for N, P, and K recovery from the fertilized and manured grain and silage plots are shown in Table 9. Nutrient recovery was calculated from total additions of fertilizer and manure nutrients using average nutrient composition values for manure, grain, and silage determined periodically between 1971 and 1982 (data not shown). Silage recovered 16% more N, 21% more P, and 91% more K than grain. The highest recovery of N by both grain and silage came from treatment C (22 Mg ha^{-1} manure), while treatment A gave the highest P and K recovery. Treatment C had the lowest annual N addition (108 kg N ha^{-1}) while treatment A had the lowest annual application of P and K (20 and 37 kg P and K ha^{-1}, respectively).

The greater N removal by silage plots resulted in slightly lower soil N values for the silage area (0.84 vs. 0.97 g kg^{-1} when data were combined over years, Table 4). Silage plots showed the greatest soil N loss from 1982 to 1991 when compared to grain plots. The removal of silage resulted in more than 50% N recovery from all treatments except treatment E.

Estimates of P recovery from the fertilized grain plots ranged from a low of 22% for treatment B to a high of 90% for treatment A. For manured plots, grain P recovery ranged from 42% for treatment C to a low of 22% for treatment E. Phosphorus recovery by silage exceeded 100% in treatment A. These plots also exhibited the lowest soil test P values at the end of the study (Table 7).

Estimates of K recovery from the fertilized grain plots ranged from a low of 15% for treatment B to a high of 65% for treatment A. For manured plots, grain K recovery ranged from 16% for treatment C

Table 9 Recovery of N-P-K Nutrients by Grain and Silage from Long-Term Fertilizer and Manure Plots

	Percent Nutrients Removed by the Crop					
	Grain Plots			Silage Plots		
Treatment	**N**	**P**	**K**	**N**	**P**	**K**
A	57	90	65	74	126	241
B	59	22	15	82	35	104
C	77	42	16	94	67	102
D	46	28	10	58	46	68
E	34	22	7	45	37	54
Mean	55	41	23	71	62	114

Note: Calculated from total additions of fertilizer and manure nutrients and the average nutrient composition of manure, grain, and silage determined periodically between 1971 and 1982.

to a low of 7% for treatment E. Potassium recovery by silage exceeded 100% in treatment A, B, and C. Only treatment A, however, was severely inadequate in K additions when compared to removal. The soil test K in these plots declined with time; however, yields were only slightly affected (Table 2). More than twice as much K was removed as was added to these plots. This soil appears to have a large capacity to supply K from nonexchangeable forms. The extra K of treatment B increased the N recovery of silage by 8% compared to treatment A.

I. SOIL ORGANIC CARBON DYNAMICS

The soil organic matter dynamics for the loamy sand site reported in this paper differ from those of the silty clay site reported in Chapter 10. The two sites, although 130 km distant from each other, have a similar climate. This allows a direct comparison of the overall decay rates as affected by soil texture.

Our comparison is based on the plant and soil decay rates used in the model of Jenkinson et al.[11] This model shows two major decay groupings: decomposable plant material (DPM) and resistant plant material (RPM). The DPM amounts to about 77% of the plant material. The decay rate of soil incorporated DPM material is affected by temperature and moisture but not by soil texture. The RPM is about 23% and the decay rate is modified by soil aggregates and by clay. The Jenkinson model breaks the RPM into two components, which we do not show in our calculations.

In our calculations we assume plant material has about 40% carbon. The residues from an individual year include roots, turnover, exudates, etc., which will produce about 33% soil C. The carryover amounts to about 22, 16, 13, and 11%, respectively, for the next 4 years. This 5-year input is described as "active" soil C. If the input or rotation are averaged then the makeup of the active soil C is 95% of the average yearly plant C residue. The soil C over 5 years in age is described as "old" C. This description differs from that of many scientists, who place "old" at 100 years or more in age. In our estimations, 11% of the annual plant C input is added to the old soil C pool. Table 10 illustrates the soil C changes for continuous corn growing on the loamy sand and Table 12 for a crop rotation on a silty clay.

Using the above estimates, it was found that the old C decay rate of 2.1%, annually, came the closest to matching the inorganic fertilized corn growing on a loamy sand. Calculations for the six rotation

Table 10 Soil Carbon Content of a Metea Loamy Sand Growing Continuous Corn from 1963 to 1982 and a Soybean, Corn Rotation 1983 to 1993

Year	Active Soil C ($kg\ ha^{-1}$)	Old C ($Mg\ ha^{-1}$)	Total Soil C ($Mg\ ha^{-1}$)	Amount Added to Old ($kg\ ha^{-1}$)	Soil C ($g\ kg^{-1}$)
	1958–1972 PCR = 3500 kg ha^{-1}, Yield 5.3 Mg ha^{-1}				
63	3500 × 0.95	28.60	31.92	3500 × 0.11	9.5
64	3500 × 0.95	28.38	31.70	385	9.4
67	3500 × 0.95	27.73	31.06	385	9.2
	1973–1977 PCR = 4000 kg ha^{-1}, Yield 6.1 Mg ha^{-1}				
73	3500 × 0.95	26.56	29.88	385	8.9
75	3500 × 0.40 + 4000 × 0.55	26.20	29.80	385	8.8
77	3500 × 0.11 + 4000 × 0.84	25.86	29.60	385	8.8
	1978–1982 PCR = 4800 kg ha^{-1}, Yield 8.3 Mg ha^{-1}				
78	4000 × 0.95	25.69	29.49	4000 × 0.11	8.8
80	4000 × 0.40 + 4800 × 0.55	25.47	29.71	440	8.8
82	4000 × 0.11 + 4800 × 0.84	25.27	29.74	440	8.9
	1983–1993 PCR = 3200 kg ha^{-1}, (corn-soybeans)				
83	4800 × 0.95	25.17	29.70	4800 × 0.11	8.8
85	4800 × 0.40 + 3200 × 0.55	25.15	28.83	528	8.6
87	4800 × 0.11 + 3200 × 0.84	25.12	28.34	528	8.4
89	3040	24.93	27.97	3200 × 0.11	8.3
91	3040	24.58	27.62	352	8.2
93	3040	24.24	27.28	352	8.1

Note: PCR, plant carbon residue annual input from various sources. Soil in 1963 estimated to test 0.95% C for 3.36 million kg ha^{-1}. Annual decay rate of old C: 2.1% — Simulates changes for inorganic fertilizer A and B treatments.

studies for the silty clay soil show a decay rate of 1.2% annually. Two factors probably account for the difference. One is that more clay or clay aggregates means greater bound C mineral particles. The other is the total old C of 49 Mg ha^{-1} for the silty clay is much greater than the 32 Mg ha^{-1} of old C for the loamy sand.

The plant equivalent for the manure treatment can be calculated if the percentages of completely decomposable and resistant material can be calculated for the animal waste and straw. Because the manure plots on the loamy sand nearly reached a soil C steady-state level in 10 years and with some rough estimates of plant residue equivalent inputs, the decay rate appears to be about 2.5% annually for old C where manure is applied at the rate of 45 Mg ha^{-1} annually. In the calculations each megagram of manure contained 260 kg of dry matter. Because of the loss of much of the DPM, each megagram of the manure was given an estimated value equivalent to 360 kg of plant material. Table 11 illustrates the calculation. It shows the delay of 5 years before the manure changes the decay rate of the old soil C. The 2.5% decay rate for old C appears high. Increasing the decay rate of the active 1% each year and using a 2.2% decay rate for old soil C, similar soil C values are obtained as reported in Table 11. This adjustment may better describe the changes when long-time applications of manure are made to loamy sands.

The results of the Michigan trials show that the soil C decay rates are high for coarse-textured soils. They differ greatly from the Broadbalk winter wheat trials at Rothamsted, England,[12] where farm manure has been applied annually since 1843 at the rate of 35 Mg ha^{-1}. In 1843, soil contained about 32 Mg ha^{-1} of organic C (about 60 Mg soil organic matter) in the surface 23 cm. It now exceeds 80 Mg ha^{-1} of C and continues to slowly increase.

IV. SUMMARY

The application of manure to soil can have a profound affect on soil C and other nutrient accumulations. A great deal of C is lost through microbial decomposition soon after manure is added to soil. The removal

Table 11 Soil Carbon Content of a Metea Loamy Sand Receiving Annually 45 Mg ha^{-1} of Farm Manure and Growing Continuous Corn from 1963 to 1982 and Soybeans-Corn without Manure from 1983 to 1993

Year	Active Soil C (kg ha^{-1})	Old C (Mg ha^{-1})	Total Soil C (Mg ha^{-1})	Amount Added to Old C (kg ha^{-1})	Soil C (g ha^{-1})
1963–1972 PCR = 3500 kg ha^{-1}, (corn) + 6500 kg (manure), Yield 5.3 Mg ha^{-1}					
63	3500 × 0.95 (3,325)	28.60	31.90	3500 × 0.11	9.5
65	3325 + 6500 × 0.55	28.16	35.06	385	10.4
67	3325 + 6500 × 0.84	27.73	36.52	385	10.9
1973–1977 PCR = 4000 kg ha^{-1} + 6500 kg (manure), Yield 6.1 Mg ha^{-1}					
73	6500 × 0.95 + 3500 × 0.95	29.36	38.86	10,000 × 0.11	11.6
74	6175 + 3500 × 0.62 + 4000 × 0.33	29.69	39.36	1,100	11.7
76	6175 + 3500 × 0.24 + 4000 × 0.71	30.35	40.21	1,100	12.0
1978–1982 PCR = 4800 kg ha^{-1} + 6500 kg ha^{-1}, Yield 8.3 Mg ha^{-1}					
78	6175 + 4000 × 0.95	30.97	40.95	10,500 × 0.11	12.2
79	6175 + 4000 × 0.62 + 4800 × 0.33	31.03	41.56	1,155	12.4
82	6175 + 4000 × 0.11 + 4800 × 0.84	32.32	42.91	1,155	12.8
1983–1993 PCR = 3200 kg ha^{-1} corn-soybean rotation					
83	6500 × 0.95 + 4800 × 0.95	32.64	43.32	11,300 × 0.11	12.9
84	6500 × 0.62 + 4800 × 0.62 + 3200 × 0.33	33.04	41.04	1,243	12.2
87	6500 × 0.11 + 4800 × 0.11 + 3200 × 0.84	34.17	38.05	1,243	11.3
90	3200 × 0.95	33.50	36.49	3200 × 0.11	10.9
93	3040	32.05	35.09	352	10.4

Note: Manure dry matter × 1.4 = estimated plant residue equivalent. Soil in 1963 estimated to test 0.95% C for 3.36 million kilograms per hectare. Simulates changes for D treatment. Decay rate 2.1% 1963 to 1968; 2.5% 1969 to 1993. PCR, plant carbon residue annual input from various sources.

Table 12 Soil Carbon Content of a Misteguay Silty Clay Growing in Rotation Oats (PCR = 1800 kg), Alfalfa (PCR = 3200), Navy Beans (PCR = 1280), Sugar Beets (PCR = 1780) which Averages a Yearly Input of 2015 kg ha^{-1}

Year	Active Soil C (kg ha^{-1})	Old C (Mg ha^{-1})	Total Soil C (Mg ha^{-1})	Amount Added to Old C (kg ha^{-1})	Soil C (g kg^{-1})
72	2360[a] × 0.95	52.68	54.92	2360 × 0.11	16.9
73	2360 × 0.62[b] + 2015 × 0.33[b]	52.31	54.44	260	16.8
74	2360 × 0.40 + 2015 × 0.55	51.94	53.99	260	16.6
75	2360 × 0.24 + 2015 × 0.71	51.57	53.57	260	16.4
76	2360 × 0.11 + 2015 × 0.84	51.21	53.12	260	16.3
77	2015 × 0.95	50.85	52.76	2015 × 0.11	16.2
78	1914	50.46	52.37	222	16.1
80	1914	49.69	51.60	222	15.8
85	1914	47.85	49.76	222	15.3
90	1914	46.12	48.03	222	14.8

[a] The estimated PCR amount for the 1968–71 period.

[b] Note these two values total 0.95.

Note: Soil C based on 3.25 million kg ha^{-1} of soil, testing 16.9 g kg^{-1} C in 1972, and had a decay rate of 1.2 g kg^{-1} for old C. PCR, plant carbon residue annual input from various sources.

of silage compared to grain had a small but noticeable negative affect on the accumulation of soil C (25 mg kg^{-1} of soil per year). After 20 years of manure additions, soil C and N appeared to reach a steady-state condition. The termination of manure applications and the introduction of a less-intensive corn-soybean rotation resulted in a rapid decline in both soil C and N. A similar decline in soil test P and K also occurred when manure applications ceased. Most of the P and K losses from manured grain plots may be attributed to a lack of the soil's ability to hold these nutrients in excess and subsequent leaching to the subsoil.

A nutrient balance for this study indicated that treatment A was inadequately fertilized with P and K, particularly when silage was removed. Recovery of P and K over the 20-year period showed more than 100% removal from these plots. The ability of this soil to supply the necessary K for 20 years in treatment A is very significant in light of the good yields obtained in these plots. The extra K supplied in treatment B improved the N recovery in silage by 8% over treatment A. The removal of silage resulted in more than 50% N recovery from all treatments except treatment E. The excellent recovery observed in these plots is confirmation of the importance of applying optimum rates of nutrients to prevent nutrient losses to the environment.

This research demonstrates that manure is a good source of crop nutrients. The data indicate that, when both inorganic fertilizer and manure are managed properly, equivalent corn grain and silage yields could be obtained. The management of a continuous silage system compared to a continuous grain system should be adjusted for the greater nutrient removal from the silage system. Soil accumulations of nutrients and salts from fertilizer or manure are greater under the grain system and must be managed accordingly. Routine soil testing and the matching of nutrient inputs to crop removal are highly recommended best management practices. Excessive applications of nutrients, whether they come from fertilizer or manure, can have detrimental effects on the soil environment as well as the growing crop.

REFERENCES

1. Salter, S. and K. Schollenberger, Farm Manure. Ohio Agric. Exp. Stn. Bull. 605, 1939.
2. Turk, L.M. and A.G. Weidemann, Farm Manure, Michigan Agric. Ext. Bull. 300, 1949.
3. Vitosh, M.L., J.F. Davis, and B.D. Knezek, Long Term Effects of Fertilizer, Manure and Plowing Depth on Corn, Michigan Agric. Exp. Stn. Res. Rept. 198, 1972.
4. Garner, H.V. and G.V. Dyke, The Broadbalk Wheat Yields: Comparison of Farmyard Manure with Complete Fertilizers. 1852–1967, Rothamsted Report, part 2, 30, 1968.
5. Gillham, R.W. and L.R. Webber, Nitrogen contaminations of groundwater by barnyard leachates, *J. Water Pollut. Contr. Fed.*, 41, 1752, 1969.

6. Robertson, L.S., R.L. Cook, P.J. Rood, and L.M. Turk, Ten year results from the Ferden rotation and crop sequence experiment, *Proc. Am. Soc. Sugarbeet Tech.*, 7, 172, 1952.
7. Jenkinson, D.S., Studies on the decomposition of plant material, *J. Soil Sci.*, 16, 104, 1965.
8. Lucas, R.E. and M.L. Vitosh, Soil Organic Matter Dynamics, Research report 358, Michigan State University, East Lansing, 1978.
9. Olsen, R.J., R.F. Hensler, and O.J. Attoe, Effect of manure applications, aeration, and soil pH on soil nitrogen transformations and on certain soil test value, *Soil Sci. Soc. Am. Proc.*, 34, 222, 1970.
10. Schulte, E.E., Recommended soil organic matter tests, in *Recommended Chemical Soil Test Procedures for the North Central Region*, North Central Regional Publication No. 221, North Dakota Agricultural Experiment Station, North Dakota State University, Fargo, 1988, chap. 12.
11. Jenkinson, D.S., P.B.S. Hart, J.H. Rayner, and L.C. Parry, Modeling the turnover of organic matter in long-term experiments at Rothamsted, *INTECOL Bull.*, 15, 1 1987.
12. Jenkinson, D.S., The Rothamsted long-term experiments: are they still of use?, *Agron. J.*, 83, 2, 1991.

Chapter 9

Long-Term Tillage and Periodic Plowing of a No-Tilled Soil in Michigan: Impacts, Yield, and Soil Organic Matter

F.J. Pierce and M.-C. Fortin

CONTENTS

I. INTRODUCTION

In the United States, the 1970s and 1980s were characterized by growing interest in conservation tillage systems, both from their adoption and adaptation by farmers and their development and evaluation by the agricultural research community. Of particular interest has been the no-tillage (NT) system. NT offers the greatest potential for erosion control while at the same time posing an interesting set of management problems related to the lack of tillage and the management of crop residues on the soil surface. While important to soil conservation, the management of crop residues and the lack of tillage have significant impacts on important properties and processes in soil that regulate productivity and the material and energy flows within agroecosystems.

Many of the soil changes associated with NT evolve over the long term. This is particularly true of organic C and N.[1] Hence, interest has been in the maintenance of long-term tillage comparisons. The problem is that, while long-term changes in continuous NT can be substantial,[2–5] in practice, long-term continuous NT in agriculture is limited. Farmers, for numerous reasons, periodically rotate tillage with their NT system. Often, periodic tillage is planned as part of a crop rotation or used to correct a pest or

0-8493-2802-0/97/$0.00+$.50

soil management problem. Therefore, it is also interesting and important to examine the effects of periodic tillage on NT in such long-term comparisons.

Under periodic tillage, the question arises as to the fate of short-term changes in soil properties when a NT soil is periodically plowed. Typically, NT soils are characterized by stratified nutrients, organic C, and pH and can have higher bulk densities in the surface soil layer than tilled soils.[1–7] Cycling of nutrients and C are quite different under long-term NT when compared to plowed systems.[1,8] The stratified layer should "break up" when plowed periodically and begin to reform when returned to NT. There are limited data on the effects of periodic tillage on soil properties created by NT, although a few studies have focused on the use of in-row tillage methods in NT.[9–11]

Our objective is to summarize a long-term tillage experiment that compares conventional tillage (CT) with continuous NT over a 13-year period during which NT soils were periodically plowed. Details of this study are reported in Pierce et al.[12] and Staton.[13] This paper focuses on long-term crop yields and the impacts of CT systems and periodic plowing of NT soil organic matter and nitrogen dynamics.

II. GEOGRAPHY

East Lansing is situated in the south-central part of the lower peninsula of Michigan, approximately equidistant from three of the Great Lakes — Michigan, Huron, and Erie. The soils and landscapes in Michigan owe their origin to the Wisconsin glaciation. While a majority of Michigan soils were formed in glacial outwash and are coarse textured, the more productive agricultural soils were formed in loamy glacial till, on till plains and moraines, and in lake bed soils associated with glacial lakes. The productivity of these soils is frequently limited by the need for artificial drainage. Consequently, many of these soils are tile drained.

The first permanent white settlements began in south-central Michigan in the decade 1830 to 1839. The area was heavily forested; thus, logging was an initial industry. Agriculture grew as a major part of the economy as land under cultivation increased rapidly until 1880 when the rate of increase declined.[14] The climate and soils of the region have favored the production of cereal grains, large seeded legumes, hay, and root crops.

The experimental site is located on the campus of Michigan State University (42° 40′N, 84° 28′W). The site elevation is approximately 268 m and the landscape is gently rolling. The climate is continental, moderated by the presence of the Great Lakes. Historically, the weather record extends to the earliest published records for January, 1864.[15] The weather record for the 30-year period 1951 to 1980 for the East Lansing 4 S station was summarized by Nurnberger[15] as follows (Table 1). Average annual precipitation at East Lansing 4 S is 728 mm. Pan evaporation for the April to November period averages 1318 mm. Fall and winter precipitation is sufficient for complete soil water recharge. About 62% of the total annual precipitation is received during the April to September period, with June being the wettest month, receiving 89.9 mm on the average. This area of southern Michigan receives the lowest average annual growing season precipitation east of the Mississippi River. The average season snowfall is 988 mm, with 61 days per season averaging 25 mm or more snow on the ground. The mean annual air temperature at East Lansing 4 S is 8.8°C. The average date of the first and last freeze is May 8 and October 4, respectively, giving an average frost-free growing season of 148 days annually. The average cumulative growing degree days for the site is 1340°C, calculated on a base temperature of 10°C for the May through September period.

III. MATERIALS AND METHODS

A tillage experiment was initiated in 1980 by Dr. A. E. Erickson at the Michigan State University Research Farm, East Lansing, comparing CT with NT managed with clover and rye cover crops. The soil is a Capac loam, (fine-loamy, mixed, mesic, Aeric Ochraqualf) consisting of very deep, somewhat poorly drained soil formed in loamy glacial till on the low parts of moraines and till plains. These soils have moderate to moderately slow permeability. The experimental site is flat and is well tilled.

The initial experimental design consisted of CT and three NT/cover crop management treatments: NT with a red clover (*Trifolium pratense*) cover crop except for a killed strip in the corn row (NTc), NT with a rye (*Secale cereale*) cover crop strip killed in the corn row (NTr), and NT with a rye cover crop completely killed at planting corn (NT). The experimental design was a randomized complete block design with four replications and a plot size of 6.1 × 15.2 m. CT consisted of fall moldboard plowing to a depth of 0.2 m with secondary tillage in the spring consisting of one pass of a disk followed by

Table 1 Climatological Summary and Statistics for the Michigan State University Research Farm, East Lansing, (East Lansing 4 S Weather Station, 42°40′N, 84°28′W) for the Period 1951 to 1980

		Pan Evaporation[a] (mm)		Air Temperature[b] (°C)			
Month	**Precipitation (mm)**	**Mean**	**Std**	**Mean**	**Min**	**Max**	**GDD**
January	36	—	—	–6	–10	–2	—
February	31	—	—	–5	–9	0	—
March	53	—	—	1	–4	6	6
April	71	105	19	8	2	14	46
May	69	151	27	14	8	21	151
June	90	176	24	19	13	26	288
July	77	186	21	22	15	28	360
August	79	156	16	21	14	27	332
September	64	110	19	17	10	23	208
October	56	68	14	11	5	17	78
November	56	35	3	4	–1	8	12
December	46	—	—	–3	–7	1	1
Annual	728	988	—	9	3	14	1481

[a] Pan evaporation is for the period May 1957 to October 1986 as summarized by the National Climatic Data Center (NCDC) for the East Lansing 4 S station.

[b] Mean refers to mean daily temperature; Min and Max refer to daily minimum and maximum temperatures, respectively.

From Nurnberger, F. B., 1951–1980 Climatological Summary and Statistics, Michigan Department of Agriculture Climatology Program, 1989.

one pass of a spring-toothed harrow. The NT treatments were planted with a no-till slot planter; cover crops were killed with a burndown herbicide. The clover cover crop was changed to rye in 1985 and killed with a burndown herbicide at planting. The cover crop treatments were terminated after the 1985 growing season but the tillage systems were retained.

For the 1986 growing season, the NTr treatment was conventionally tilled and subsequently returned to NT (renamed NTP86). The NTc and NT treatments were continued in NT without cover crops. The NTc treatment was plowed in the fall of 1986 along with CT and was subsequently returned to NT in 1987 (renamed NTP87). After 1987, the NT, NTP86, and NTP87 plots remained in NT until the NTP87 was plowed again in the fall of 1991 and subsequently returned to NT. Corn (*Zea mays* L.) was grown continuously from 1980 to 1988. A corn-soybean rotation [*Glycine max* (L.) Merr.] was introduced in 1989.

Soil physical properties were measured on intact soil cores (76-mm diameter) from the 0- to 76- and the 76- to 152-mm depths and analyzed for bulk density, porosity, and pore size distribution.[12] Composite soil samples were taken at the 0- to 50-, 50- to 100-, 100- to 150-, and 150- to 200-mm depths on 10 April 1987 and 22 April 1991 and analyzed for pH and extractable P and K.[12] Organic carbon (OC) for the 1987 samples was determined on duplicate soil samples using the method of Snyder and Trofmow.[16] Total OC and N were determined on the 1991 samples by dry combustion of duplicate subsamples on a Carlo Erba CHN analyzer Model 1104 (Carlo Erba Instruments, Milan).

IV. RESULTS AND DISCUSSION

The long-term yield record for this tillage study is plotted in Figure 1 (Pierce and Erickson, unpublished data). During the period 1980 to 1985, cover crops were part of the NT systems. In the living mulch treatments (NTc and NTr), corn yields were considerably lower than either CT or NT in 1980 and 1981 and again in 1984. The NTr treatment was the only living mulch treatment in 1985 and had a lower yield than CT. Continuous NT was lower than CT during this period of the study only in 1981.

Plowing NTr in 1986 had little effect on corn yield. Grain yield was both highest for the NT with a clover cover crop history (NTc) and lowest for continuous NT, both in NT for 6 years. This can possibly be attributed to residual benefits of 5 years of red clover managed as a living mulch, although the NTc treatment was planted to a rye cover and killed at planting in the 1985 cropping year. In 1987, corn yields for both periodically plowed treatments were higher than NT but not different from continuous plowing. Stover yield, harvest index, and N content of plant tissue were relatively unaffected by tillage system (Table 2). A severe drought in 1988 limited corn yields across Michigan, and data for this year

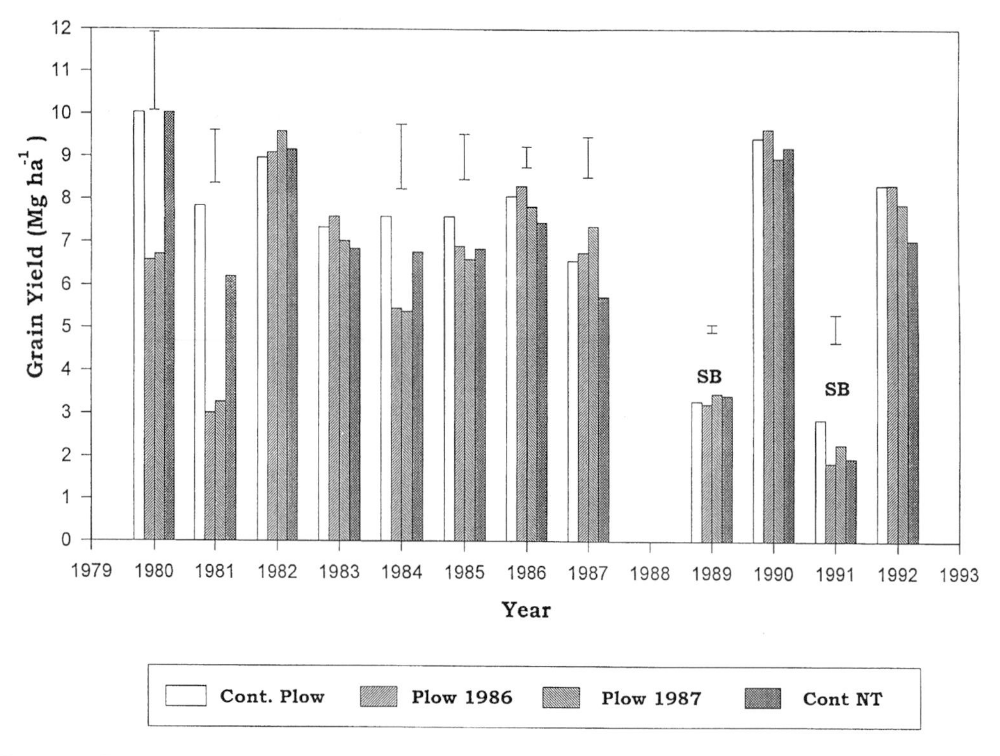

Figure 1 Tillage effects on grain yield of corn and soybeans over the period 1980 to 1992 on a Capac soil in East Lansing, MI. From 1980 to 1985, the NTP86 tillage treatment was NT with a rye cover crop (NT_r) and NTP87 was NT with a clover cover crop (NT_c). Data for 1988 were misplaced and unavailable. (From Pierce, F.J. and Erickson, unpublished data.)

Table 2 Immediate Effects of Plowing No-Till Soil in 1986 (NTP86) Compared to Continuous NT and CT on Stover Yield, Harvest Index, and Tissue N Content of Corn in 1986 and 1987

					N Content (g kg^{-1})		
Year	Tillage System	Stover (Mg ha^{-1})	Total Biomass (Mg ha^{-1})	Harvest Index	Ear Leaf	Grain	Stover
1986	CT	4.89	11.71	0.58	26.6	13.1	7.4
	NTP87	4.95	11.97	0.59	27.6	12.6	7.0
	NTP86	5.37	11.99	0.55	24.7	13.3	7.6
	NT	4.91	11.21	0.56	24.7	12.3	7.2
	LSD (0.05)	1.13			3.3	1.1	1.0
1987	CT	3.04	8.59	0.65			
	NTP87	3.45	9.15	0.62			
	NTP86	4.34	10.57	0.59			
	NT	3.22	8.07	0.60			
	LSD (0.05)	0.75					

From Staton, M.J., Conservation Tillage Effects on Soil Properties and Corn Yields in Central Michigan, M.Sc. thesis, Michigan State University, East Lansing, 1988. With permission.

were not available. The only remaining effect of tillage treatment on crop yield was a reduction in soybean yield in 1991 in the NTP87 and NT treatments. This was attributed to a heavy weed infestation in this study with higher weed populations in the NT treatments than CT.

A. SOIL PHYSICAL PROPERTIES

Pierce et al.[12] reported on the changes in soil properties resulting from periodic plowing of NT soil. Continuous NT soil had higher bulk densities than plowed treatments in all years in the nontrafficked rows. The immediate effects of plowing a NT soil were to reduce bulk density by 0.02 to 0.06 Mg m^{-3} below that obtained with continuous plowing in CT in both the 0- to 76-mm depth and in 1986 in the 76- to 152-mm depth. Plowing NT altered the soil moisture release curve and hence the airfilled porosity of the Capac soil in the year of tillage; this effect remained significant 1 year after tillage (Figures 2 and 3). The NT soil retained more water at all but the lowest suctions and therefore had lower airfilled porosities at higher suctions. This indicates that aeration may be limited in the NT soil at high moisture contents but NT soils have a higher available water capacity than plowed systems. The plowed Capac soil has higher total porosity, primarily due to higher macroporosity but the NT soil has higher microporosity (Figure 4). These differences are retained 1 year after tillage but are diminished 4 years after tillage.[12]

B. SOIL CHEMICAL PROPERTIES

A summary of the effects of plowing soil managed in NT was reported by Pierce et al.[12] Soil pH was 0.7 units lower in NT than CT in the surface 50 mm. Periodic plowing of NT increased pH to levels in CT and maintained the pH at the 1987 levels. In 1991, CT had a higher pH than NT in the surface 50 mm but the periodically plowed soils were intermediate between CT and NT. The soil test P levels were high and variable in all treatments and no differences were measured in 1987. In 1991, P levels had declined from 1987 levels at all depths, reflecting no P fertilization since 1987. By 1991, the redistribution of P by plowing the soil managed in NT was evident 4 to 5 years after plowing, even though it was not apparent in 1987. Soil P was higher in the surface 50 mm of all NT treatments than CT and higher in the 50- to 100- and 100- to 150-mm depths in the NTP86 and NTP87 soils than CT and NT. Thus, the residual effects of plowing NT on P distribution in soils was to redistribute P below the soil surface and stratify P upon return to NT.

Soil test K levels were in the medium range for Michigan soils. In 1987, soil test K in NT was highly stratified in the surface 200 mm. Plowing soils managed in NT reduced K in the surface 50 mm to levels intermediate between CT and NT and redistributed K in the 50- to 200-mm depths. Soil test K in NTP87 was higher than NT throughout the 50- to 200-mm depth and higher than CT in the 50- to 100-mm depth. The distribution of soil test K in NTP86 was similar to CT but lower than NT in the 50- to 100-mm depth and higher than NT in the 150- to 200-mm depth. By 1991, soil test K had declined from levels measured in 1987, reflecting the lack of K fertilization. Stratification of soil test K in the 0- to 50-mm depth was present in all NT treatments but there were no differences between NT and NTP86

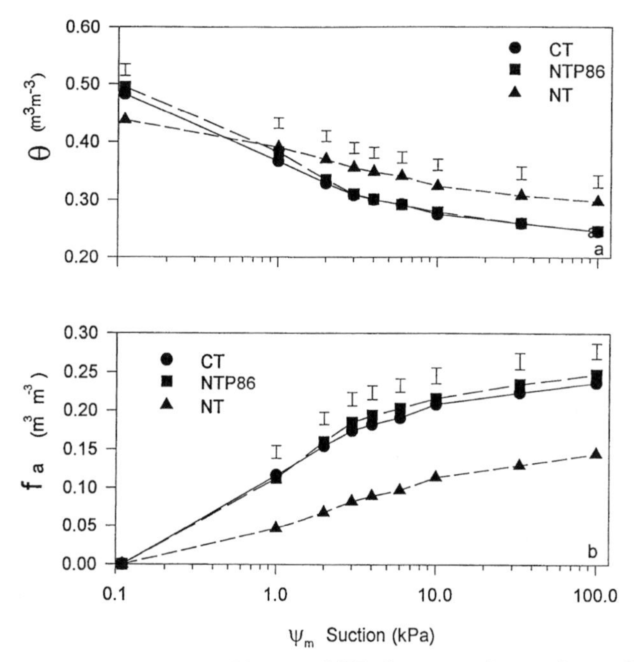

Figure 2 Immediate effects of plowing NT soil in 1986 (NTP86) compared to continuous NT and CT on the soil moisture (θ) characteristic curve and corresponding airfilled porosity (f_a) of the 0- to 76-mm depth of a Capac loam soil sampled on 14 July 1987. (From Staton, M.J., Conservation Tillage Effects on Soil Properties and Corn Yields in Central Michigan, M.Sc. thesis, Michigan State University, East Lansing, 1988. With permission.)

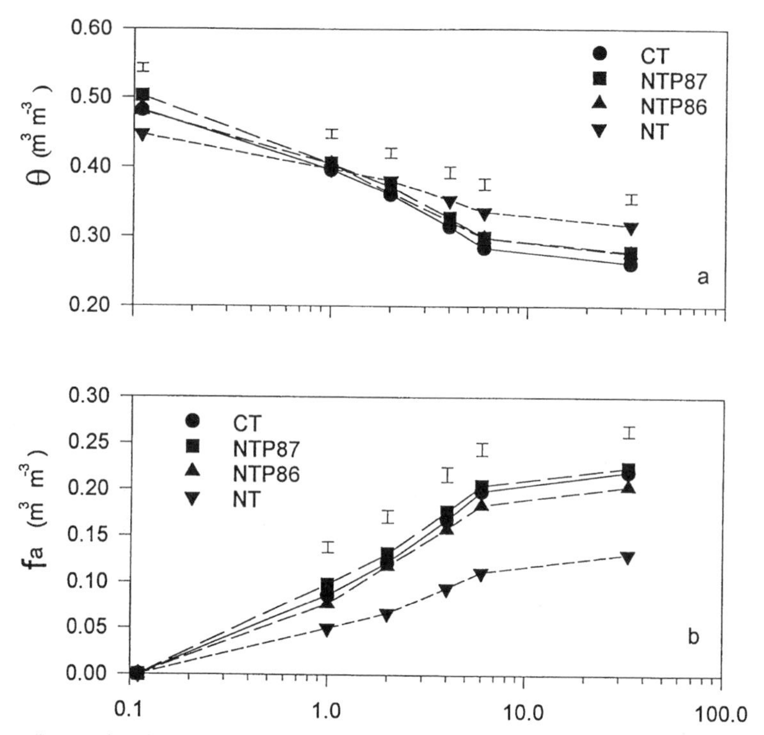

Figure 3 Immediate and residual effects of plowing NT soil in 1986 and 1987 (NTP86 and NTP87) compared to continuous NT and CT on the soil moisture (θ) characteristic curve and corresponding airfilled porosity (f_a) 0- to 76-mm depth of a Capac loam soil sampled on 14 July 1987. (From Staton, M.J., Conservation Tillage Effects on Soil Properties and Corn Yields in Central Michigan, M.Sc. thesis, Michigan State University, East Lansing, 1988. With permission.)

and NTP87. Contrary to soil test P, the redistribution of K in the 50- to 200-mm depth due to plowing soil managed in NT was dissipated by 1991. Only the difference between CT and NT in the 150- to 200-mm depth remained in 1991. The difference between P and K relative to residual effects of plowing and returning to NT may relate in part to the fact that soil test K levels were in the medium range and soil test P levels remained in the high range over the period of study. Regardless, the effects of plowing on the soil fertility of soil managed in NT were still evident 4 to 5 years after plowing.

C. ORGANIC C AND N

In 1987, OC was 8.7 and 3.5 kg m^{-3} higher in the 0- to 50- and 50- to 100-mm depths, respectively, in the NT treatment than in CT (Table 3).[12] NTP86 and NTP87 redistributed some of the OC to the lower depths while maintaining higher OC in the surface 50 mm of the NTP86 and NTP87 treatments than in CT. The OC content of the NTP87 was 5.8 and 4.1 kg C m^{-3} higher in 50- to 150-mm depth increments than CT. One year after plowing, OC in the 50 to 100 mm of the NTP86 treatment had 3.0 kg C m^{-3} more than CT. There were no differences on OC in the 150- to 200-mm depth.

The OC contents measured in 1991 were considerably lower than those measured in 1987 (Table 3). Pierce et al.[12] concluded that there was no plausible explanation for the magnitude of the differences in OC between 1987 and 1991. The differences were not due to differences in analytical method between years, to sampling time (same for both years), or to the cover crop history of the treatments. By 1991, the differences in OC between treatments in the 50- to 200-mm depth had dissipated. However, NT had 7.1 kg OC m^{-3} more than CT in the surface 50 mm. Both the NTP86 and NTP87 had higher OC contents than CT. The NT had 4.2 kg OC m^{-3} more than NTP86 but similar OC content to NTP87. Total N ranged from 1.57 kg N m^{-3} in the surface 50 mm of CT to 2.12 kg N m^{-3} in the surface 50 mm of NT. The treatment differences followed a similar pattern to OC, with treatment differences only in the 0- to 50-mm depth. However, differences in total N at the lower depths were significant at $p<0.10$, which would indicate that plowing soil managed in NT increased total N 4 to 5 years after tillage.

Thus, the immediate effects of plowing soil managed in NT were to redistribute OC throughout the 0- to 150-mm soil depth. Residual effects of periodic plowing were evident after 1 year for OC in NTP86

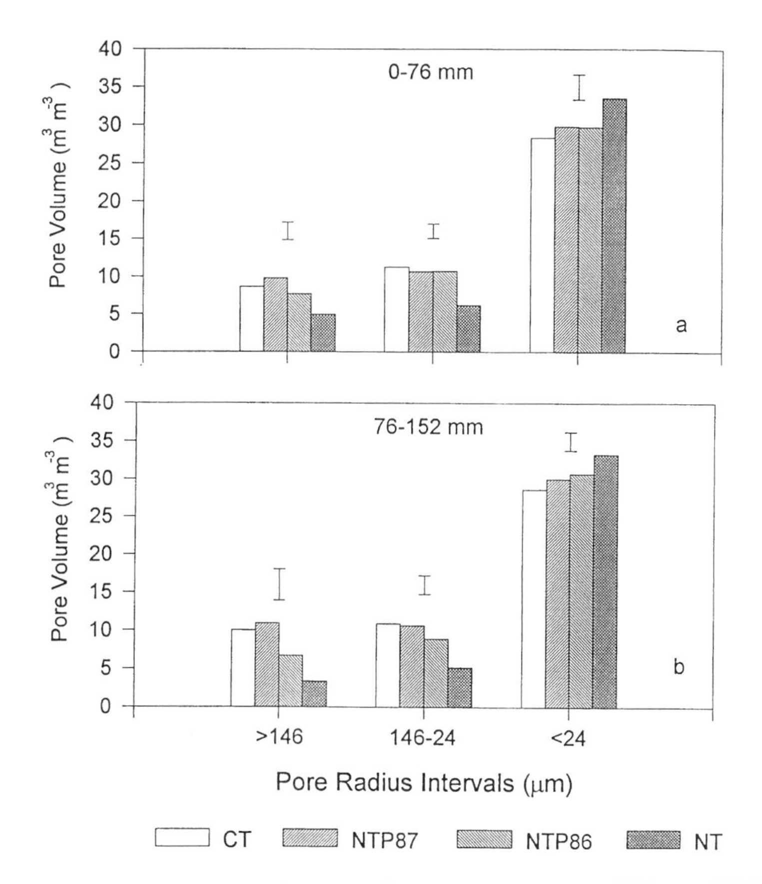

Figure 4 Immediate and residual effects of plowing NT soil in 1986 and 1987 (NTP86 and NTP87) compared to continuous NT and CT on the on macro- and microporosity of the 0- to 152-mm depth of a Capac loam soil sampled on 14 July 1987. (From Staton, M.J., Conservation Tillage Effects on Soil Properties and Corn Yields in Central Michigan, M.Sc. thesis, Michigan State University, East Lansing, 1988. With permission.)

(measured in 1987) and 4 to 5 years after plowing for OC and total N in the surface 50 mm and potentially in the 50- to 200-mm depth.

V. SUMMARY AND CONCLUSIONS

Living mulches reduced corn yields 3 out of 5 years while continuous NT reduced yields relative to CT 3 of 13 years. Period plowing in 1986 and 1987 increased corn yields relative to NT but not CT. Plowing soils managed in NT created soil physical conditions similar to continuous plowing systems in the year of tillage. Upon return to NT, the soil consolidated and the physical properties were intermediate between CT and NT for at least an additional year. Within 4 to 5 years, the effects of plowing were dissipated and the soil was physically similar to NT, as residual effects of plowing were not found. Wheel traffic quickly returned soil physical properties to those characteristic of NT.

Plowing a soil in NT management redistributed soil nutrients within the plow layer. The surface stratification of NT was not entirely lost when plowed and was increased over time, relative to CT. While the redistribution of P and K was still evident 4 to 5 years after plowing, differences in OC and pH had dissipated. It appears that the long-term benefit from plowing soils in long-term NT management is found in soil fertility, particularly in the redistribution of P and K and increase in surface pH. Accumulation of C and N in NT was lost some time after the plowing disturbance, although they appeared to re-establish rather quickly near the soil surface. Continuous NT had lower pH and stratified K, OC, and total N in the surface 50 mm after 12 years. Thus, periodic plowing had the immediate effect of improving soil physical properties, increasing surface pH, increasing mineralization of NO_3-N, and redistributing nutrients and organic matter within the plow layer. Residual effects of periodic plowing were minimal for soil physical properties but significant for soil chemical properties including OC.

Table 3 Organic C or Total N Content of a Capac Loam after Plowing following 6 (NTP86) and 7 (NTP87) Years of NT Management and Subsequent Return to NT Compared to Continuous CT and Long-Term NT

Year	Tillage	Depth (mm)			
		0–50	50–100	100–150	150–200
			kg C m^{-3}		
1987	CT	20.7	22.7	24.1	24.7
	NTP87	26.3	28.5	28.2	27.3
	NTP86	24.8	25.7	26.1	26.8
	NT	29.4	26.2	25.3	24.6
	LSD	2.8	2.0	2.4	NS
			kg C m^{-3}		
1991	CT	17.6	19.9	19.5	19.4
	NTP87	22.1	21.9	22.4	22.6
	NTP86	20.5	20.4	21.7	21.7
	NT	24.7	19.9	20.6	20.6
	LSD	2.7	NS	NS	NS
			kg N m^{-3}		
1991	CT	1.57	1.75	1.66	1.68
	NTP87	1.99	1.95	1.99	1.99
	NTP86	1.83	1.82	1.97	1.78
	NT	2.12	1.82	1.89	1.76
	LSD	0.26	NS	NS	NS

From Pierce, F.J., M.-C. Fortin, and M.J. Staton, *Soil Sci. Soc. Am. J.*, 58, 1785, 1994.

ACKNOWLEDGMENTS

The authors would like to thank Dr. A. Earl Erickson for initiating the tillage study in 1980, for Michael Staton's efforts, and for Brian Long for his technical support in maintaining the experiment since 1985. This research was supported in part by the Michigan Agricultural Experiment Station at Michigan State University.

REFERENCES

1. Dick, W.A., Organic carbon, nitrogen, and phosphorus concentrations and pH in soil profiles as affected by tillage intensity, *Soil Sci. Soc. Am. J.*, 47, 102, 1983.
2. Edwards, W.M., M.J. Shipitato, L.B. Owens, and L.D. Norton, Water and nitrate movement in earthworm burrows within long-term no-till cornfields, *J. Soil Water Conserv.*, 44, 240, 1989.
3. Karlen, D.L., D.C. Erbach, T.C. Kaspar, T.S. Colvin, E.C. Berry, and D.R. Timmons, Soil tilth: a review of past perceptions and future needs, *Soil Sci. Soc. Am. J.*, 54, 153, 1990.
4. Radcliffe, D.E., E.W. Tollner, W.L. Hargrove, R.L. Clark, and M.H. Golabi, Effect of tillage practices on infiltration and soil strength of a Typic Hapludult soil after ten years, *Soil Sci. Soc. Am. J.*, 52, 798, 1988.
5. Voorhees, W.B. and M.J. Lindstrom, Long-term effects of tillage method on soil tilth independent of wheel traffic compaction, *Soil Sci. Soc. Am. J.*, 48, 152, 1984.
6. Gantzer, C.J. and G.R. Blake, Physical characteristics of LeSueur clay loam soil following no-till and conventional tillage, *Agron. J.*, 70, 853, 1978.
7. Hill, R.L., Long-term conventional and no-tillage effects on selected soil physical properties, *Soil Sci. Soc. Am. J.*, 54, 161, 1990.
8. Rice, C.W., M.S. Smith, and R.L. Blevins, Soil nitrogen availability after long-term continuous NT and conventional tillage corn production, *Soil Sci. Soc. Am. J.*, 50, 1206, 1986.
9. Pierce, F.J., M.-C. Fortin, and M.J. Staton, Immediate and residual effects of zone tillage in rotation with no tillage on soil physical properties and corn performance, *Soil Tillage Res.*, 24, 149, 1992.

10. Campbell, R.B., D.L. Karlen, and R.E. Sojka, Conservation tillage for maize production in the U.S. Southeastern Coastal Plain, *Soil Tillage Res.*, 3, 135, 1984.
11. Sene, M., M.J. Vepraskas, G.C. Naderman, and H.P. Denton, Relationships of soil texture and structure to corn yield response to subsoiling, *Soil Sci. Soc. Am. J.*, 49, 422, 1985.
12. Pierce, F.J., M.-C. Fortin, and M.J. Staton, Periodic plowing of a no-till soil: immediate and residual effects on soil properties, *Soil Sci. Soc. Am. J.*, 58, 1782, 1994.
13. Staton, M.J., Conservation Tillage Effects on Soil Properties and Corn Yields in Central Michigan, M.Sc. thesis, Michigan State University, East Lansing, 1988.
14. Veatch, J.O., H.G. Adams, E.H. Hubbard, C. Dorman, L.R. Jones, J.W. Moon, and C.H. Wonser, Soil Survey Ingham County Michigan. USDA, Bureau of Plant Industry and Michigan Agricultural Experimental Station, Washington, DC, 1941.
15. Nurnberger, F.B., 1951–1980 Climatological Summary and Statistics, Michigan Department of Agriculture, Climatology Program, 1989.
16. Snyder, J.D. and J.A. Trofmow, A rapid accurate wet oxidation diffusion procedure for determining organic and inorganic carbon in plant and soil samples, *Commun. Soil Sci. Plant Anal.*, 15, 587, 1984.

Chapter

Soil Organic Matter in Sugar Beet and Dry Bean Cropping Systems in Michigan

D.R. Christenson

CONTENTS

I. INTRODUCTION

Long-term cropping systems and crop rotation research is important in helping understand how these systems affect crop yields, yield stability, and soil physical and biological properties. Since the beneficial effects of forage legumes on soil properties are recognized, most long-term rotation studies include forage legumes. Studies commonly include evaluation of soil N and C, but P and S are also included as part of various investigations.[1] Cropping systems that conserve soil C and N often increase crop yields because they promote the formation of stable aggregates and in turn increase capillary pore space.[2]

There have been three rotation or cropping systems studies involving the production of sugar beets and dry edible beans (navy and other classes) conducted in Michigan since 1940. The first two studies[2–6] were designed to evaluate the effects of forage legumes, farm yard manure, and green manure crops on yield of these crops in rotations of 5 years in length. The third study,[7] which is the focus of this paper, was designed after farming systems had changed in the navy bean- and sugar beet-producing area of Michigan. The change in farming systems was accompanied by a significant reduction in the amount of forage legumes grown and farm yard manure used on the land.

A crop rotation study involving the production of sugar beets and dry beans was initiated in Michigan in 1940.[3] The study was located in Saginaw County on a Sims sandy clay loam (Mollic Haplequept) containing 25 to 28% clay, 50 to 58% sand, and a pH of 6.8. The rotations were 5 years in length. Quantity of crop residue returned to the soil was varied by changing the amount of forage legumes and green manure crops grown in the rotation.

The above study was terminated in 1970 and data presented in Table 1[4,5] show the effect of various rotations on crop yields averaged for the period 1961 to 1970. All forage legumes and green manure

0-8493-2802-0/97/$0.00+$.50

Table 1 Sugar Beet and Navy Bean Yield as Affected by Crop Rotation on a Sims Sandy Clay Loam

		Sugar Beet		Navy Bean	
Rotation Number	Crop Rotation[a]	Yield (Mg ha^{-1})	Increase (%)[b]	Yield (kg ha^{-1})	Increase (%)
1	Ba-A-A-NB-SB	46.7	18	2510	38
2	C(gm)-NB-W-O(sw)-SB	47.7	21	2320	28
3	C(gm)-Sb-W(gm)-NB-SB	44.9	13	2250	24
4	NB-W-A-C-SB	45.1	14	2350	30
5	Sb-W-O(sw)-NB-SB	48.6	23	2280	26
6	Ba-NB-W-C-SB	39.4	0	1810	0
7	C(gm)-NB-W(gm)-Sb-SB	45.7	16	1960	8

Note: The experiment was initiated in 1940 and yields are the average from 1961 to 1970.[4,5]

[a] Ba = barley; A = alfalfa; NB = navy beans; SB = sugar beets; C = Corn; W = wheat; O = oats; Sb = soybeans; sw = sweet clover interseeded with the cash grain; gm = green manure — mixture of alsike and June and sweet clover interseeded with the cash grain. Both were plowed down before the next year's crop was planted.

[b] Percent increase from Rotation Number 6.

crops were plowed before the next crop was planted. A systematic comparison of the effects of soil-building crops is difficult because of the choice of treatments and the lack of statistical analysis. However, some general comments can be made about the effect of green manure and forage legumes on yield of sugar beets and navy beans.

Yields of sugar beets and navy beans were increased up to 23 and 38%, respectively, when forage legumes or green manure crops were included in the rotation compared to a cash crop rotation (#6). However, there was a difference in yield response of navy beans and sugar beets between sweet clover as a green manure crop and alfalfa. Sugar beet yields were increased 22% (average of rotations #2 and #5) when sweet clover was grown and 16% (average of #1 and #4) where alfalfa was included in the rotation. Navy bean yields were increased 27% when sweet clover was included and 34% where alfalfa was included. The reason for this difference in response is not clear. Alfalfa grown for 2 years in a rotation increased sugar beet yield 4% and navy bean yields 8% compared to 1 year of alfalfa (rotations #1 and #4).

Robertson[2] reported that aggregates were more stable and pore space was increased when alfalfa was included in the rotation. Furthermore, crusting was more of a problem on the cash crop sequence (#6). Higher yields where forage legumes were included in the sequence were partially attributed to improvement of physical conditions of the soil.

A separate study on the Sims sandy clay loam soil was initiated in 1941 to measure the effects of farm yard manure and alfalfa on yield of navy beans and sugar beets.[6] Rotation #1 included corn, sugar beets, barley, and 2 years of alfalfa-bromegrass. Farm yard manure at a rate of 22.4 Mg ha^{-1} was applied to the corn in each of the four cycles of the rotation for a total of 89.6 Mg of manure. Rotation #2 consisted of corn, sugar beets, barley, navy beans, and wheat. Soil organic N changes were compared in 1958 for the two rotations. The soil organic N was 4600 and 4100 kg ha^{-1} for the rotations #1 and #2, respectively. Lucas and Vitosh[8] estimated the increase to be 435 kg ha^{-1} of organic matter per year from the alfalfa and manure. No other evaluation of changes in organic matter was made over the course of either of the two experiments on the Sims soil.

In the 1950s, the number of dairy and beef operations in the dry bean and sugar beet production area of Michigan declined. Farms were converted to cash crop systems with corn (grain), dry beans, soybeans, sugar beets, and small grains as the major crops. A new experiment was established in 1972 on a Misteguay silty clay soil (Aeric Haplaquept) at the Saginaw Valley Bean and Beet Research Farm. Included in the rationale for the new study was the need for such an experiment to include cropping systems which reflected the change from livestock to cash crop farming systems. The initial goals of the study were to (1) evaluate how often dry bean and sugar beets could be grown in a rotation and (2) evaluate how much organic matter needed to be returned to the soil to maintain soil productivity. This paper (1) evaluates changes in soil C and N concentration in the plow layer for 6 of the 12 cropping systems after 9 and 19 years of cropping and (2) evaluates these changes in view of yield and economic considerations.

II. METHODS

The experiment consists of four types of cropping systems each placed in a 2-, 3-, or 4-year rotation for a total of 12 systems (Table 2). System 1, a corn-sugar beet system, returns relatively large amounts of crop residue to the soil. Systems 2 and 3 are corn, dry bean, and sugar beet systems with smaller amounts of crop residues returned. System 4 is a cash grain system with oats representing a cash grain crop. In the 4-year cash grain system, alfalfa is clear seeded in the spring and plowed down in the fall. This allows for the comparison of a forage legume grown for a single season with corn as a contributor of crop residue.

Each crop in each system has an annual entry point and the experiment is replicated 4 times giving a total of 144 experimental units. Each experimental unit is 6 × 21 m. There has been a concerted effort to keep the management consistent over the course of the study. However, it has been necessary to change varieties, herbicides, and fertilizer as conditions changed (Part IV, Tables 2 and 3).[7] Rates of P and micronutrients were reduced as soil test concentrations increased as a result of fertilization practices. In addition, nutrient needs changed because new cultivars were not responsive to some of the applied nutrients. For example, the sugar beet variety was changed in 1985 and the new variety does not need boron for maximum yields.[9] Nitrogen rates given in Table 2 were adjusted to give the same annual N rate within each of the four types of cropping systems. Rainfall during the growing season ranged from 7.3 to 11.2 cm from April to September (Table 3). Average maximum temperature ranged from 13°C in April to 28°C in July, while the average minimum temperature ranged from 7 to 16°C.

Soil samples (20-mm diameter cores) to a depth of 20 cm were taken from the same experimental units in the spring of 1972, 1981, and 1991. The samples were dried, ground, and stored in sealed containers. Oxidizable carbon was determined by the Walkley-Black (WBC) method.[10] These values were converted to total soil organic C using a recovery factor of 77%.[11] Total soil N was determined using the Kjeldahl method.[12] To eliminate errors from different technicians and procedures, all analyses were done on stored samples at the same time.

The experiment was fall plowed and tilled once in the spring with a field cultivator type of tillage tool. A complete description of cropping methods and analysis of yields are reported elsewhere.[7] Data were analyzed as a randomized complete block with four replications. Mean differences were tested using the Duncan's New Multiple Range test.

Table 2 Cropping Systems and N Rates for the Cropping Systems Study Conducted on a Misteguay Silty Clay Soil from 1972 to 1991

	2 Year		3 Year		4 Year	
Type of System[a]	**Crop**	**N Rate (kg N ha^{-1})**	**Crop**	**N Rate (kg N ha^{-1})**	**Crop**	**N Rate (kg N ha^{-1})**
1	Corn[a]	224	Corn	196	Corn	168
(140)[b]	Beets	56	Corn	168	Corn	168
			Beets	56	Corn	168
					Beets	56
2	Corn	185	Corn	235	Corn	168
(105)	Beans	28	Beans	28	Corn	168
			Beets	56	Beans	28
					Beets	56
3	Beans	28	Beans	56	Corn	168
(65)	Beets	84	Beans	56	Beans	28
			Beets	84	Beans	28
					Beets	56
4	Oats	56	Oats	56	Oats	56
(45)	Beans	28	Beans	28	Alfalfa	0
			Beets	56	Beans	28
					Beets	56

[a] Corn (*Zea mays* L.), beets (*Beta vulgaris* L.), beans (*Phaseolus vulgaris* L.), oats (*Avena sativa* L.), alfalfa (*Medicago sativa* L.).

[b] Average annual nitrogen applied (kg N ha^{-1}).

Table 3 Average Monthly Precipitation and Maximum and Minimum Temperatures for a Cropping Systems Study Conducted on a Misteguay Silty Clay Soil

Month	1972–1991 Precipitation (cm)	Temperature (1974–1991, °C)	
		Maximum	Minimum
January	3.1	–3	–10
February	2.7	–1	–9
March	5.7	5	–4
April	7.2	13	7
May	7.3	20	16
June	8.4	25	13
July	6.6	28	16
August	8.7	26	15
September	11.2	21	10
October	6.8	14	4
November	6.9	7	–1
December	4.2	0	–7

III. RESULTS

The amount of crop residue returned to the soil was estimated based on considerations formulated by Lucas et al.[13] These estimates for the periods 1972 to 1980 and 1972 to 1990 along with the fertilizer N applied are given in Table 4. There are wide differences in the amount of residue returned among the various cropping systems. The total amount returned between 1972 and 1990 is estimated to be between 70 Mg ha^{-1} for the navy bean-sugar beet system and 190 Mg ha^{-1} for the corn-corn-corn-sugar beet system. Nitrogen applied ranged from 0.62 Mg ha^{-1} for the oats-alfalfa-navy bean-sugar beet system to 2.28 Mg ha^{-1} for the corn-corn-corn-sugar beet system.

A. CARBON

At the initiation of the experiment in 1972 soil C was not significantly different among cropping systems (Table 5). After 9 years of cropping under the different systems, there were significant differences in soil C due to cropping system. The O-A-NB-SB, NB-SB, and O-NB-SB systems lost a significantly larger amount of soil C than the C-C-C-SB system. None of the six systems showed a significant increase in soil C. However, cropping systems containing corn resulted in up to 10% more soil C than those systems without corn.

The above trends were still apparent after 19 years of cropping (Table 5). Soil C losses were greatest for the NB-SB, O-NB-SB, and O-A-NB-SB and least for the systems which contained corn. These results are in contrast to those reported by Larson et al.[1] They found no difference in soil C where corn and alfalfa residues were applied at the same rate. In the current study, it is estimated that greater amounts of crop residues were returned by corn (10 Mg ha^{-1}) compared to 8 Mg ha^{-1} for alfalfa. Only 25% of the rotation had alfalfa grown compared to 50 to 67% for corn. It is apparent that alfalfa grown in this manner does not maintain soil C as corn does when grown more frequently.

The smallest rate of soil C loss for the first 9 years was measured for the C-C-NB-SB and C-C-C-SB systems with rates less than 0.50% y^{-1} (Table 6). The largest average rate of C loss was measured for the O-A-NB-SB, NB-SB, and O-NB-SB systems. The rate of loss for the second 10 years and for the 19 years of the study was less than the first 9 years. However, the relative ranking of the various cropping systems was similar for the two periods of time. In all three comparisons, the 4-year system with alfalfa was intermediate to those systems with and those without corn. The relative rate of soil C disappearance was related to the estimated amount of residues returned to the soil ($r^2 = 0.82$).

The rate of disappearance of soil C was faster in the first 9 years than when averaged over the course of the 19 years of this study. This fits with the general understanding of the effect of cropping systems on soil organic C. The data suggest that these cropping systems may be reaching a new equilibrium with respect to C. This new equilibrium concentration may be unique for each cropping system or type of system. However, the duration of the study is insufficient to substantiate that a new equilibrium exists.

Table 4 Estimated Crop Residues Returned and Fertilizer N Applied to a Cropping Systems Study Conducted on a Misteguay Silty Clay Soil

Cropping System[a]	Residues[b] ($Mg\ ha^{-1}$)		Nitrogen ($Mg\ ha^{-1}$)	
	1972–1980	1972–1990	1972–1980	1972–1990
C-SB	68	140	1.19	2.44
NB-SB	33	70	0.42	0.92
O-NB-SB	38	80	0.36	0.79
C-C-C-SB	90	190	1.08	2.28
C-C-NB-SB	58	128	0.80	1.74
O-A-NB-SB	44	95	0.27	0.62[c]

[a] C = corn, SB = sugar beet, NB = navy beans, O = oats, A = alfalfa.
[b] Estimated residue returned:[13] corn 10 Mg ha^{-1}; sugar beets, 4.5 Mg ha^{-1}; navy beans 3.5 Mg ha^{-1}; oats, 5.0 Mg ha^{-1}; alfalfa, 8.0 Mg ha^{-1}.
[c] Does not include nitrogen contribution from alfalfa.

Table 5 Effect of Cropping System on Carbon Concentration in a Misteguay Silty Clay Soil after 9 and 19 Years of Cropping

Cropping System[a]	Carbon ($g\ kg^{-1}$)			Δ Carbon ($g\ kg^{-1}$)		
	1972	1981	1991	1972–1981	1981–1991	1972–1991
C-SB	16.8a[b]	15.4ab	15.6a	1.4abc	–0.2a	1.2ab
NB-SB	16.3a	14.1c	14.1c	2.2a	0.0a	2.2a
O-NB-SB	16.3a	14.2bc	14.2c	2.1a	0.0a	2.1a
C-C-C-SB	16.4a	15.7a	16.1a	0.7bc	–0.4a	0.3b
C-C-NB-SB	15.8a	15.3ab	15.3ab	0.5c	0.0a	0.5b
O-A-NB-SB	16.5a	14.7abc	14.4bc	1.7ab	0.3a	2.0a

[a] C = corn, SB = sugar beet, NB = navy beans, O = oats, A = alfalfa.
[b] Means followed by the same letter within a column are not significantly different, alpha = 0.05 (Duncan's new multiple range test).

Table 6 Effect of Cropping System on the Rate of Carbon Loss from a Misteguay Silty Clay Soil after 9 and 19 Years

Cropping System	k[a]		
	1972–1981	1981–1991	1972–1991
C-SB	0.90abc[b]	–0.13a	0.37ab
NB-SB	1.53a	–0.05a	0.71a
O-NB-SB	1.42a	0.01a	0.68a
C-C-C-SB	0.45bc	–0.28a	0.08b
C-C-NB-SB	0.38c	–0.01a	0.18b
O-A-NB-SB	1.16ab	0.21a	0.66a

[a] $k = 100(C_o - C_t)/(t * C_o)$, where t = time in years.
[b] Means followed by the same letter within a column are not significantly different, alpha = 0.05 (Duncan's new multiple range test).

Zielke and Christenson,[14] reporting on the first 9 years of the study, suggested the following relationship of these systems for soil C conservation and productivity (yield):

$$\text{C-C-C-SB} = \text{C-C-NB-SB} \geq \text{O-NB-A-SB} = \text{C-SB} > \text{O-NB-SB} = \text{NB-SB}$$

There was a similar ranking for these systems in 1991. However, discussion later in the chapter suggests that practices that conserve soil C do not give the highest economic return.

B. NITROGEN

After 19 years of cropping, soil N concentration was highest in the C-SB system which had received large amounts of fertilizer N from commercial sources (Table 7). Conversely, soil N was lowest in the NB-SB and O-NB-SB systems which had received less applied N over the course of the study. The O-A-NB-SB system was intermediate, receiving less fertilizer N than systems containing corn.

Table 7 Soil N Concentration and the C:N Ratio at the Initiation of a Cropping Systems Study and after 19 Years of Cropping on a Misteguay Silty Clay Soil

Cropping System	Total N Applied	Soil Nitrogen ($g\ kg^{-1}$)		C:N Ratio		
		1972	1991	1972	1991	Δ C:N[b]
C-SB	2.44	1.75a[a]	1.69ab	9.58a	9.24bc	0.34abc
NB-SB	0.92	1.73a	1.60cd	9.50a	10.1a	–0.59c
O-NB-SB	0.79	1.64a	1.56d	9.66a	9.78ab	–0.12bc
C-C-C-SB	2.28	1.79a	1.73a	9.13a	8.16d	0.97a
C-C-NB-SB	1.74	1.76a	1.65bc	9.34a	8.60cd	0.74ab
O-A-NB-SB	0.62	1.72a	1.66abc	9.60a	8.70cd	0.90a

[a] Means followed by the same letter within a column are not significantly different, alpha = 0.05 (Duncan's new multiple range test).

[b] 1972 C:N ratio minus 1991 C:N ratio.

Collectively, the changes in soil C and N between 1972 and 1991 were small and were not correlated ($r^2 = 0.01$). Similarly, changes in C:N ratios for the six cropping systems are small and reflect the amount of fertilizer nitrogen applied (Table 7). The increase in C:N ratio for the NB-SB and O-NB-SB systems reflect the smaller amount of fertilizer N applied. Conversely, C-SB, C-C-C-SB, and C-C-NB-SB, which showed a decrease in C:N ratio, had lower soil C loss and greater amount of fertilizer N applied. The C:N ratio for the O-A-NB-SB system declined, reflecting both the loss of soil C and N supplied to the system by grain and forage legumes grown.

Nitrogen balance was calculated by subtracting the N added from the N removed. N added was the total from fertilizer plus that added from navy beans and alfalfa. Nitrogen fixed by legumes was conservatively estimated at 40 and 50 kg ha^{-1} for navy beans[15] and alfalfa,[16] respectively. Nitrogen removed was the total of the loss of soil N (Table 7) plus crop removal. Nitrogen removed by crops was estimated based on a number of sources summarized by Christenson et al.[17] Values used are given in the footnote of Table 8. Measured yields (Part IV, Table 6) were used to calculate crop removal.

The majority of the N lost from the system was from crop removal. Corn removed 50 to 75% of the total in those cropping systems containing corn, reflecting the proportion of the system devoted to this crop. N removed by the NB-SB and O-NB-SB systems was less than that in systems containing corn.

Table 8 Nitrogen Balance for Six Cropping Systems after 19 Years of Cropping on a Misteguay Silty Clay Soil

Cropping System	N Removed ($kg\ ha^{-1}$)			N Added ($kg\ ha^{-1}$)			Difference[d] ($kg\ ha^{-1}$)
	Soil N[a]	CR[b]	Total	Legume[c]	Fertilizer	Total	
C-SB	146b[e]	2255	2401	—	2440	2440	–39
NB-SB	291a	1723	2014	400	920	1320	694
O-NB-SB	174b	1516	1690	240	790	1030	660
C-C-C-SB	146b	2070	2216	—	2280	2280	–64
C-C-NB-SB	235a	2134	2369	200	1740	1940	429
O-A-NB-SB	286a	2048	2334	400	620	1020	1314

[a] Nitrogen lost from decline in soil N (Table 7). Bulk density (1.30 Mg m^{-3}).[18]

[b] Crop removal:[17] corn, 16 kg Mg^{-1}; sugar beets, 2 kg Mg^{-1}; navy beans, 36 kg Mg^{-1}; oats, 19 kg Mg^{-1}; alfalfa, 22 kg Mg^{-1}.

[c] Nitrogen fixed: navy beans[15] = 40 kg ha^{-1}; alfalfa[16] = 50 kg ha^{-1}.

[d] Nitrogen removed minus nitrogen added.

[e] Means followed by the same letter within a column are not significantly different, alpha = 0.10 (Duncan's new multiple range test).

Nitrogen added from fertilizer ranges from 690 for the O-A-NB-SB system to 2440 kg ha^{-1} for the C-SB system. The amount added from legumes was small because dry beans do not add large quantities and alfalfa was grown for a single summer in each cycle of the rotation. Where no legumes were grown, there was a loss of 2 to 3 kg soil N per hectare per year. Where legumes were grown, the soil contributed between 22 and 80 kg ha^{-1} y^{-1} more than was added.

C. YIELD RELATIONSHIPS AND ECONOMIC CONSIDERATIONS

Examination of these results in light of crop yield and the economic return for the various systems is important. Crop yields are summarized in a previous publication[7] and Part IV, Table 6.

Yield of navy beans was increased 18% by the 4-year C-C-NB-SB and O-A-NB-SB systems over the 2-year NB-SB system (Table 9). Yield of navy beans for the O-NB-SB system was intermediate to the other systems. This is in direct agreement with the results of Robertson et al.,[5] where navy beans yielded best in systems where crop residues were added in largest amounts.

Yield of sugar beets was 6.8% higher for the O-A-NB-SB cropping system than for the other five systems (Table 9). The reduced yield for the NB-SB and O-NB-SB systems is probably related to the lower amount of crop residue returned to the soil and the resulting loss of organic matter. Based on previous rotation studies by Robertson et al.,[2,4] sugar beet yields were expected to be high for the cropping systems that contained corn because of the amount of crop residues returned. However, there was a yield reduction compared to the O-A-NB-SB cropping system. The reduced yield was associated with a reduced leaf area index plus a partitioning of greater quantities of carbon to fibrous roots than those systems without corn.[18] This was reported to be related to an interaction of residual quantities of cyanazine applied to corn with the herbicide pyrazon applied to sugar beets.[7] While corn residues helped maintain soil C, the apparent herbicide interaction depressed yield of sugar beets.

From the grower's perspective, the economic return is the driving force behind the selection of a cropping system. In another report,[19] systems that contained a greater proportion of sugar beets and navy beans were shown to have greater gross margins over fixed and variable costs. Systems that contained forage legumes did not compete because of the large investment in machinery, the low price of hay, and the additional labor requirement involved in harvest.

Christenson et al.,[20] conducted an analysis of costs and returns for these cropping systems. Gross margins over fixed and variable costs were calculated using 15 years of collected yield data and current commodity prices and input costs. Comparison of these data with carbon loss shows a poor relationship between conservation of carbon and economic return (Table 9). The highest gross margin was associated with the largest carbon loss.

Since farming is a very competitive business, growers have adopted cropping systems which are economically more advantageous. The results presented here show that shorter rotations in these conventionally tilled cropping systems are more competitive than systems which are longer in length and may have better soil building characteristics.

The final point of this work is that adopting strategies in conventional tillage that conserve soil carbon may not be in the farmers' best interest for economic survival. Development of alternative strategies that conserve soil C and allow the grower to compete is worthy of additional research. Conservation

Table 9 Yield of Navy Beans and Sugar Beets (Average of 1975 to 1990) and the Associated Carbon Loss and Gross Margin over Fixed and Variable Costs for a Cropping Systems Study Conducted on a Misteguay Silty Clay Soil

Cropping System	Navy Bean Yield (kg ha^{-1})	Sugar Beet Yield (Mg ha^{-1})	Carbon Loss (g kg^{-1})	Gross Margin (\$ ha^{-1})
C-SB	—	55.6b[a]	1.2ab	842b
NB-SB	1830b	56.7b	2.2a	1000a
O-NB-SB	2070ab	56.4b	2.1a	719c
C-C-C-SB	—	55.3b	0.3b	452f
C-C-NB-SB	2160a	56.9b	0.5b	624d
O-A-NB-SB	2180a	60.0a	2.0a	548e

[a] Means followed by the same letter within a column are not significantly different, alpha = 0.05 (Duncan's new multiple range test).

tillage systems that reduce soil erosion and oxidation due to tillage hold some promise. Incorporation of forage legumes into cropping systems without losing a cash crop would also be advantageous. Increasing N from legumes and the addition of organic matter would be useful. However, work here at Michigan State University[21] with no-till on these fine-textured soils has not demonstrated adequate yield to merit adoption for sugar beet production. Even though emergence was not a problem, early season growth rates were slower than conventional tillage, resulting in smaller roots at harvest.

IV. SUMMARY

Nitrogen and C concentrations in this soil declined for all systems even though forage legumes and corn were included in the rotation. Where corn was grown over 50% of the time, C concentration declined less than where a 1-year stand of alfalfa was included. The rate of decline in soil C was more rapid in the first 9 than in the second 10 years. This suggests that a new equilibrium is being established. The data further suggest that the new equilibrium level may be unique for each cropping system or type of system.

Nitrogen balance reflected the amount of legume in the system and the amount of fertilizer nitrogen applied.

Economically, growers cannot compete with systems that do not provide the maximum return. Gross margins over fixed and variable costs calculated here do not take into account managerial salary, taxes, or interest on the investment. Land values exceed \$5000 ha^{-1} and, therefore, an annual gross margin of less than \$600 ha^{-1} makes it difficult for the grower to stay in business. Even at a modest interest on the investment in land of 7%, and taxes of \$200 ha^{-1}, the \$600 ha^{-1} gross margin for a 400-ha farm would return less than \$20,000 for a manager's salary.

Adopting traditional cropping strategies in conventional tillage that conserve soil C may not be in the farmer's economic interest. However, development of alternative strategies in conservation tillage systems that conserve soil C and allow the grower to compete is worthy of additional research. Early season growth may be enhanced with strategically managed fertilizer application and soil manipulation to aid in warming the soil faster in the spring. Ridge tillage offers one possibility.

In addition to conserving soil C by reducing tillage-induced oxidation, reduced tillage systems conserve soil C by reducing erosion. The research reported here is applicable to fine-textured soils in the Great Lakes Drainage Basin. Reduction of soil moving into the lakes will reduce nutrient loading in this natural resource. It has the potential of conserving soil C as well.

REFERENCES

1. Larson, W.E., Clapp, C.E., Pierre, W.H., and Horachan, Y.B., Effects of increasing amounts of organic residues on continuous corn. II. Organic carbon, nitrogen, phosphorus and sulfur, *Agron. J.*, 64, 204, 1972.
2. Robertson, L.S., A study of the effects of seven systems of cropping upon yields and soil structure, *Am. Soc. Sugar Beet Tech.*, 7, 255, 1952.
3. Cook, R.L., Millar, C.E., and Robertson, L.S., A crop rotation field layout with an illustration of the statistics involved in combining several years data, *Soil Sci. Soc. Am. Proc.*, 10, 213, 1945.
4. Robertson, L S., Cook, R.L., and Davis, J.F., The Ferden Farm Report: Part II. Soil Management for Sugar Beets, Mich. State Univ. Agric. Exp. Stn. Res. Rep., 324, 1976.
5. Robertson, L.S., Cook, R L., and Davis, J.F., The Ferden Farm Report: Part IV. Soil Management for Navy Beans, Mich. State Univ. Agric. Exp. Stn. Res. Rep., 350, 1978.
6. Singh, B.N., Mineral and Organic Forms of Nitrogen in Some Michigan Soils, Ph.D. dissertation, Michigan State University, East Lansing, 1960.
7. Christenson, D.R., Bricker, C.E., and Gallagher, R.S., Crop Yield as Affected by Cropping System and Rotation, Mich. State Univ. Agric. Exp. Stn. Res. Rep. 516, 1991.
8. Lucas, R.E. and M.L. Vitosh, Soil Organic Matter Dynamics, Mich. State Univ. Agric. Exp. Stn. Res. Rep. 358, 1978.
9. Christenson, D.R., Bricker, C.E., and Hubbell, L., Yield and Quality of Sugar Beets as Affected by Applied Boron, Mich. State Univ. Agric. Exp. Stn. Res. Rep. 518, 1991.
10. Schulte, E.E., Recommended Soil Organic Matter Tests, Recommended chemical soil test procedures for the North Central Region, N. Dak. State Univ. Agric. Exp. Stn. Bull., 49, 29, 1988.
11. Walkley, A. and Black, I.A., An examination of Degtjareff method for determining soil organic matter and a proposed modification of the chromic acid titration method, *Soil Sci.*, 37, 29, 1934.
12. Bremner, J.M., Total nitrogen. in *Methods of Soil Analysis*, American Society of Agronomists, Madison, WI, 1965, 1149.

13. Lucas, R.E., Holtman, J.B., and Connor, L.J., Soil carbon dynamics and cropping practices, in *Agriculture and Energy*, Academic Press, 1977, 333.
14. Zielke, R.C. and Christenson, D.R., Organic carbon and nitrogen changes in soil under selected cropping systems, *Soil Sci. Soc. Am. J.*, 50, 353, 1986.
15. Piha, M.I. and Munns, D.N., Nitrogen fixation capacity of field bean compared to other grain legumes, *Agron. J.*, 79, 690, 1987.
16. Hesterman, O.B., Griffin, T.S., Williams, P.T., Harris, G.H., and Christenson, D.R., Forage legume-small grain intercrops: nitrogen production and response of subsequent corn, *J. Prod. Agric.*, 5, 340, 1992.
17. Christenson, D.R., Warncke, D.D., Vitosh, M.L., Jacobs, L.W., and Dahl, J.G., Fertilizer Recommendations for Field Crops in Michigan, Mich. State Univ. Coop. Ext. Bull., E-550A, 1992.
18. Momen, N.M., Effect of Cropping System on Soil Structure and Growth and Development of Sugar Beets, Ph.D. thesis, Michigan State University, East Lansing, 1985.
19. Christenson, D.R., Helsel, Z., Meints, V., Black, R., Hoskins, R., Wolak F., and Burkhardt, T., Agronomics and economics of some cropping systems for fine textured soils, *Dry Bean Digest*, 4(2), 6, 1980.
20. Christenson, D.R., Harrigan, T.M., Black J.R., and Gallagher, R.S., A comparison of the economic productivity of twelve navy bean and sugar beet cropping systems, *J. Prod. Agric.*, 8, 276, 1995.
21. Xu, C., Tillage and row spacing effects on the development and growth of dry bean (*Phaseolus vulgaris* L.) and sugar beet (*Beta vulgaris* L.) on a Parkhill loam soil, M.S. thesis, Michigan State University, East Lansing, 1991.

Chapter 11

Soil Organic Carbon Changes through Time at the University of Illinois Morrow Plots

R.G. Darmody and T.R. Peck

CONTENTS

I. INTRODUCTION

The Morrow Plots are the oldest continuous field study in the United States. They were designed to answer questions concerning the effects of crop and soil management on crop yields and soil properties. The effects of crop rotation as a management technique is evaluated on the three major plots. Plot 3 has been cropped to continuous corn (*Zea mays* L.) from 1876 to the present. Plot 4 was cropped to a corn-oats rotation from 1876 to 1966. In 1967, soybeans were substituted for oats in the rotation. Plot 5 has been cropped to a 3-year rotation of corn-oats-clover hay from 1876 to the present.

Soil fertility as a management tool is investigated on subplots of the three major plots. Soil fertility treatments have changed over the years. To accommodate the changes, the plots were subdivided into eight subplots. During the first 28 years, 1876 through 1903, no soil fertility studies were done on the plots. In 1904, treatments of manure, limestone, and phosphorus (MLP) were applied to the subplots designated south. Rock phosphate was the carrier of P applied to subplots South A and B, and bone phosphate was the carrier of P to subplots South C and D.

The second change in soil fertility treatment was the spring 1955 addition of lime, nitrogen, phosphorus, and potassium (LNPK) fertilizers to the previously untreated area of North B and the treated area of South B. The last soil fertility change occurred in the spring of 1967. The soil fertility treatments for subplots North B and South B were based on recommendations from soil test levels. Subplot South

0-8493-2802-0/97/$0.00+$.50

A had a higher soil fertility treatment to increase soil test levels above recommended levels. Currently, subplots North A, C, and D remain unfertilized, awaiting a time when a new treatment is needed.

II. HISTORY

The Morrow Plots on the campus of the University of Illinois are reputed to be the oldest continuous agronomic research study in the United States.[1] Some evidence indicates that the plots were established in 1876 and this is the accepted date. However, the first yield records start in 1888.[2] Included with the records are notes stating that the plots were cropped during the previous 14 years in continuous corn and in a corn and oats rotation and a corn, oats, and clover hay rotation.[2] Also, the year 1888 was the start of the Illinois Agricultural Experiment Station, which may account for the data record starting that year.[3] Additionally, in early field research studies it was common to not record data until the study came into equilibrium with the treatments.

Thus, the exact origin of the Morrow Plots is obscure, but the historical record of the intent to start a long-term field cropping rotation involving corn is clear. Corn yields in the late 1800s were declining. This was attributed by some to the exploitation of virgin soil fertility,[4] by others to "toxics" with claims this could be corrected by crop rotation.[5] In addition, there was an awareness of the field studies by Lawes and Gilbert in Rothamsted, England, which began in 1856.[6]

The plots were established to settle a controversy about depletion of the dark prairie soils. That question was settled 28 years later. By 1904, corn yields of the continuous corn plots were averaging only 2007 kg ha^{-1} (32 bu/acre) while corn yields of the 3-year rotation plots (corn, oats, clover hay) were 3135 kg ha^{-1} (50 bu/acre). Interpretations arrived at were (1) soils can be exploited and depleted and (2) soils can be cropped and preserved.

The next soil management milestone occurred in 1955 with additions of LNPK fertilizers to previously unfertilized subplots (North B). Questions then asked were (1) Is the loss of soil productive capacity due to exploitative cropping methods temporary or permanent? (2) Can chemical fertilizer restore the soil productivity? (3) If so, how long will it take?

Newspaper headlines of the day proclaimed "Fertilizers Will Revive Worn-out Soils".[7] The Morrow Plots show that low yields caused by continuous cropping of noneroded soils were due to loss of plant food nutrients and not to irreversible changes in soil fertility. While the increases in crop yields with fertilization were immediate and dramatic, fertilization has not completely offset the effects of past treatment. The MLP + LNPK has consistently produced higher yields than LNPK alone.[1] Likewise, fertilization did not completely offset the effects of crop rotation. Under each of the different fertilization systems, corn yields were higher with the 2- and 3-year crop rotations than with continuous corn. The highest crop yields were obtained with crop rotation and suitable fertilization.

III. CLIMATE

For some time during and after the retreat of the last Wisconsinan glacial ice from Illinois, some 12,000 years ago, the climate in Illinois was cooler and wetter than at present.[8] A rather warm, dry period 4000 to 6000 years ago led to an expansion of grassland in the area. Since that time, the climate seems to have remained similar to that of today.

Weather record keeping in Urbana, IL, at the Morrow Plots did not begin until 1888.[9] The records indicate significant fluctuations in temperature and rainfall over the years. Droughts occurred in 1933, 1934, 1936, and 1988. The present climate in Urbana is temperate continental, with hot summers and cold winters (Table 1). The average annual temperature range is about 19°C (67°F). January is normally the coldest month: the mean temperature is –4°C (26°F). The mean temperature in July (usually the hottest month) is about 24°C (75°F). The average annual precipitation is 940 mm (37 in.) with wide annual variations. About 85% falls from April through September. The number of frost-free days in a year ranges from 178 to 205. The elevation at the plots is about 225 m (742 ft) above sea level.

IV. SOIL

The soil of the plots is Flanagan silt loam, a fine, montmorillonitic, mesic Aquic Argiudoll.[8] Flanagan silt loam is a nearly level, dark colored, somewhat poorly drained soil. The A horizon is about 46 cm thick. It developed under tall grass prairie vegetation in 102 to 152 cm (40 to 60 in.) of Peoria loess over Wisconsinan age, calcareous loam glacial till. Before the 1904 season, underground tile lines were

Table 1 Climatic Summary of Urbana, IL

	Temperature						Precipitation Average	
	Maximum		Minimum		Average			
	°F	°C	°F	°C	°F	°C	in.	mm
January	33.2	0.7	18.0	–7.8	25.6	–3.6	1.56	39.6
February	37.4	3.0	21.0	–6.1	29.3	–1.5	1.87	47.5
March	47.9	8.8	30.5	–0.8	39.2	4.0	3.24	82.3
April	62.2	16.8	41.7	5.4	52.0	11.1	3.79	96.3
May	73.0	22.8	51.6	10.9	62.3	16.8	3.57	90.7
June	82.7	28.2	61.0	16.1	71.8	22.1	4.40	111.8
July	85.0	29.4	64.8	18.2	74.9	23.8	4.81	122.2
August	83.8	28.8	62.7	17.1	73.2	22.9	3.06	77.7
September	78.7	25.9	55.7	13.2	67.2	19.6	2.99	75.9
October	66.4	19.1	44.6	7.0	55.6	13.1	2.79	70.9
November	50.1	10.1	33.3	0.7	41.7	5.4	2.45	62.2
December	38.4	3.6	24.3	–4.3	31.4	–0.3	2.40	61.0
Year	61.6	16.4	42.4	5.8	52.0	11.1	36.93	938.0

From Bryan, A.A. and Wendland, W.M., Local Climatological Data Summary Champaign-Urbana, Illinois 1888–1986, Illinois Water Survey Miscellaneous Publication 98, Illinois State Water Survey, Champaign, 1987.

installed in the plots to provide supplemental drainage. Available moisture holding capacity is about 0.2 cm water per centimeter of soil (0.2 in. of water per inch of soil). The cation exchange capacity of Flanagan soils is about 18 cmol kg^{-1} of soil of cation exchange capacity in the surface, increasing to 25 cmol kg^{-1} in the subsoil. The natural soil reaction is slightly acidic with pH about 5.8 to 6.0.

V. PLOT LAYOUT

The Morrow Plots are a registered national historic landmark. They are located on the Urbana campus of the University of Illinois at 40° 06′ 15″ north latitude and 88° 13′ 32″ west longitude. The existing plots are arranged in a north to south orientation with Plot 3 to the North, Plot 4 in the middle, and Plot 5 to the South (Figure 1). Each Plot is subdivided into subplots North A (northwest corner), South A, North B, South B, North C, South C, North D (northeast corner), and South D, respectively. Each subplot is 101 m^2 (1/40 acre). When the field study was established in 1876, it consisted of ten 0.2-ha (1/2-acre) plots and was known as Rotation Experiment No. 23.[2] Of these ten plots, only parts of three remain.

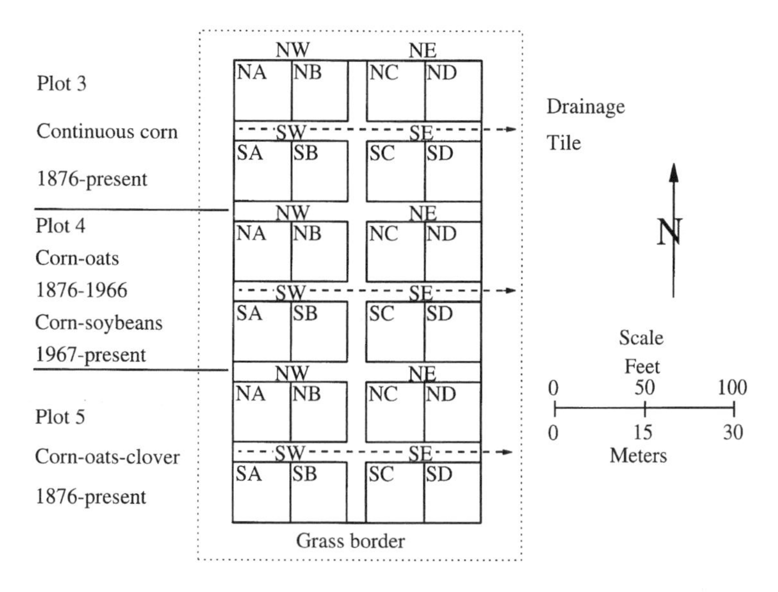

Figure 1 Layout of the Morrow Plots at the University of Illinois.

In 1895 an observatory was built upon the area that had been Plots 1 and 2.[10] In 1903 Plots 6 through 10 were taken out of the experiment and seeded to lawn grasses. The present grass border around the plots apparently originated then.[11] Gregory Drive and Mumford Hall were constructed in 1923 on plot areas 6 through 10. Also after the 1903 crop, Plots 3, 4, and 5 were divided into four quarters of 202 m^2 (1/20 acre) each. A 2.4-m (half-rod) division strip or alley was added between the plots. Drain tile was installed in a west to east direction between each north-south subplot pair. No further changes have been made in the overall configuration of the plots since then. In 1955, however, each subplot was divided in a north to south direction to give the present 24 101-m^2 (1/40 acre) plots.

Figure 2 shows the topography of the plots. The southwest corner is the highest. The northeast corner is the lowest. There is a surface drain located in that corner. The drain establishes the local base level and all runoff exits the plots through it. Eroded material moves in a north and east direction toward the drain. The plots themselves lie in a shallow basin surrounded by a grass border (Figure 3). Loss of soil from the plots due to wind and water erosion and to soil sampling and cultivation has lowered the plots about 15 cm (6 in.) relative to the grass border. In addition, the grass border has had soil added to it by dust additions and as a consequence of soil falling off of tillage equipment and personnel.

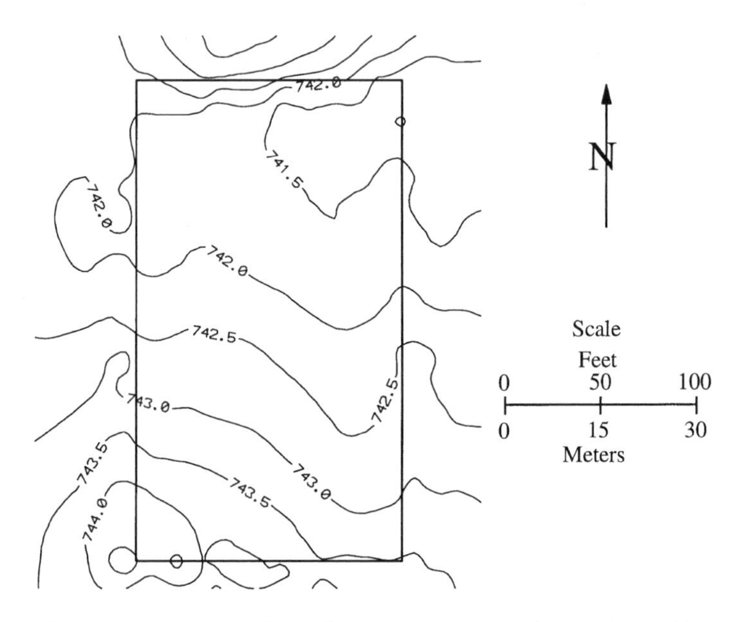

Figure 2 Topographic map of the Morrow Plots. Elevations are in feet. Contour interval is 15 cm (0.5 ft).

VI. CORN YIELD

Every 6 years the entire Morrow Plot is cropped to corn. These corn yields are shown in Figure 4.[1] The three lower lines show corn yields on plots (subplots North A, C, and D) that have never been fertilized or treated with a liming material. The lowest of the three is the continuous corn yield. The middle line is the corn yield in the present corn-soybean rotation (corn-oats before 1967). The higher of the three lower lines is the corn yield in the 3-year rotation of corn-oats-clover hay. Year to year variation is due to weather effects, but there is a gradual increase in corn yield in the long term. This trend is attributed to genetic improvement of corn such as the introduction of hybrid varieties in 1937. The upper three partial lines beginning in 1955 show the corn yields with good fertilizer and liming practices (North B subplots). These subplots were split off to determine if yields on "worn-out" soils could be restored by amendments.

VII. SOIL ORGANIC CARBON

In the years 1904, 1913, 1923, 1933, 1944, 1953, 1955, 1961, 1967 to 1969, 1973, 1974, and 1980 to 1990, soil samples (0 to 15 cm) were collected from the Morrow Plots for soil organic carbon (SOC) analysis. The analytical method is a modification of the method of Schollenberger,[13] which involves an oxidizing solution of concentrated sulfuric acid/potassium dichromate at 175°C for 90 seconds.

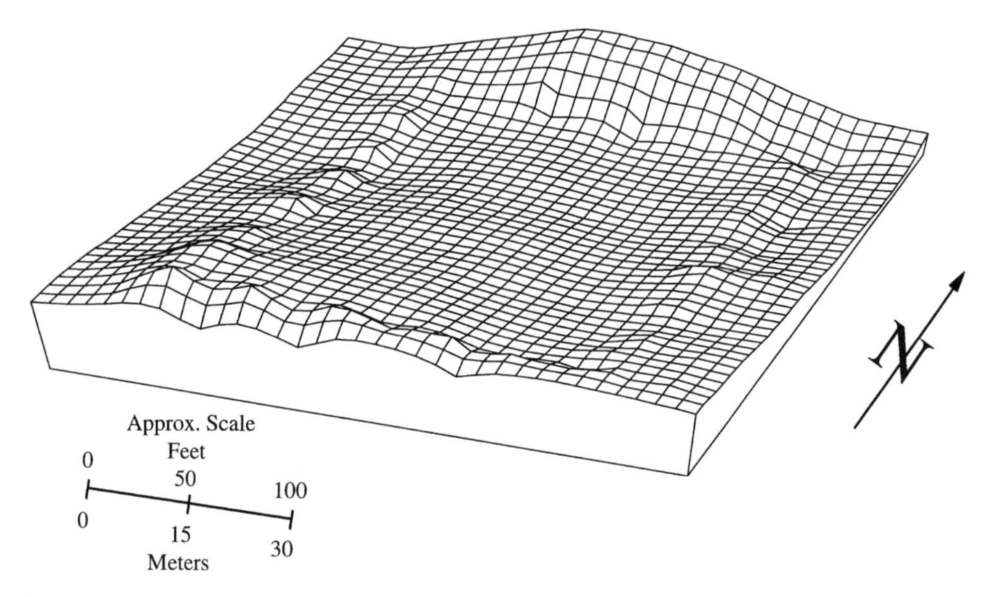

Figure 3 Topography of the area around the Morrow Plots. The actual plots lie in the central depression. Vertical exaggeration is 30×.

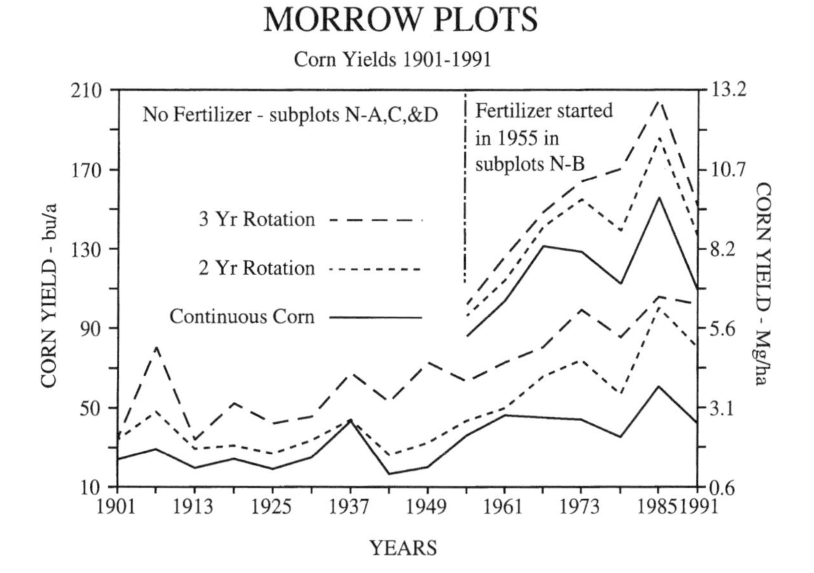

Figure 4 Corn (*Zea mays*) yields on the North subplots of the Morrow Plots. Three-year rotation is corn-oats-hay; two-year rotation is corn-soybean.

Figure 5 shows the surface (0 to 15 cm) SOC levels of the North C subplots that have never been fertilized. The lower line is from Plot 3 that is cropped to continuous corn. The middle line is from Plot 4 that is cropped to corn-soybean rotation (corn-oat rotation before 1967). The upper line is from plot 5 that is cropped to the 3-year rotation of corn-oats-clover hay. Figure 6 shows the surface (0 to 15 cm) SOC trends of the fertilized subplots (South B). The same relative rankings hold for the crop treatments; continuous corn has the least SOC, with the most in the 3-year rotation. The initial SOC levels for all crop treatments, however, were higher. Unfortunately, the first SOC data available for the Morrow Plots come from the 1904 sampling, which is 29 years into the study. There are no background or baseline initial SOC data available. If the assumption is made that SOC distribution was uniform in 1876, by 1904 SOC levels appear to have already responded to the cropping systems. Since that time, SOC levels have not changed much and have declined slightly at about the same rate in the three cropping systems.

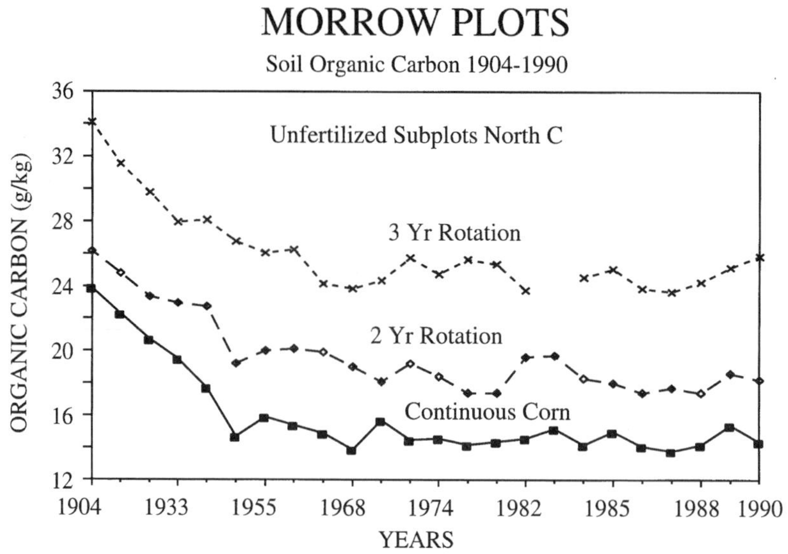

MORROW PLOTS
Soil Organic Carbon 1904-1990
Unfertilized Subplots North C
3 Yr Rotation
2 Yr Rotation
Continuous Corn
ORGANIC CARBON (g/kg)
36
32
28
24
20
16
12
1904
1933
1955
1968
1974
1982
1985
1988
1990
YEARS

Figure 5 Surface (0 to 15 cm) SOC of unfertilized subplots. Three-year rotation is corn-oats-hay; two-year rotation is corn-soybean.

MORROW PLOTS
Soil Organic Carbon 1904-1990
Fertilized Subplots South B
3 Yr Rotation
2 Yr Rotation
Continuous Corn
ORGANIC CARBON (g/kg)
36
32
28
24
20
16
12
1904
1933
1955
1968
1974
1982
1985
1988
1990
YEARS

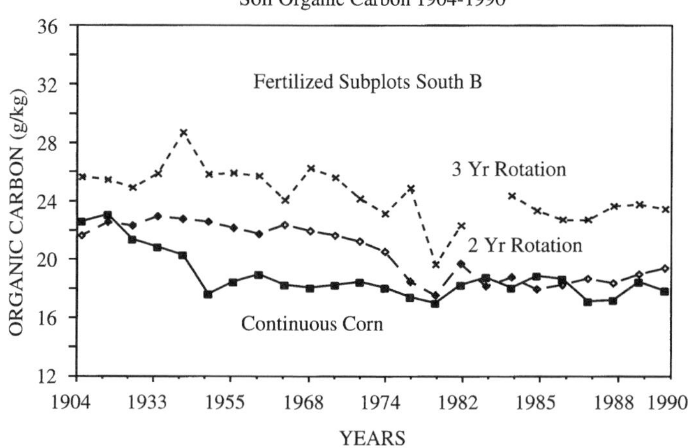

Figure 6 Surface (0 to 15 cm) SOC of fertilized subplots. Three-year rotation is corn-oats-hay; two-year rotation is corn-soybean.

Assumptions about the homogeneity of the plot area may not be valid. Table 2 and Figure 7 show (0 to 15 cm) SOC levels measured in a systematic sampling of areas in the grass borders, in the alleys between subplots, and within the plots. The SOC level of the grass border is higher on the east than on the west. Also, the sampling between the plots in the alleys shows an increase in SOC west to east as does the trend of the average SOC for the plot soil sampling.

The trend in a west-east gradient in SOC has been observed in previous analyses of the data.[1] The east and west grass borders were sampled in 1944. The results were similar to those of 1990. The west border SOC was 31.9 g kg^{-1} both in 1944 and 1990. The east border SOC was 37.7 g kg^{-1} in 1944 and 37.1 g kg^{-1} in 1990.

An inspection of Figure 7 graphically shows the trend for SOC levels to increase toward the southeast corner. This area of the plots also has the most favorable soil treatment and crop rotation for maintaining SOC levels. This anomaly confounds interpretations of the data. The grass border gives the best estimate of the SOC level in the soil at the start of the experiment. However, the original SOC levels probably

Table 2 SOC (g kg^{-1}) at the Morrow Plots

	A	B	Subplots	C	D	
30.6	30.4	27.6	Grass Border	29.9	31.1	35.5
	13.6	14.9	Plot 3N	14.0	16.2	
31.5	16.0	15.1	Alley	16.1	18.5	38.2
	18.0	16.8	Plot 3S	18.0	19.2	
32.6	17.5	17.4	Alley	17.9	18.8	32.3
	16.8	18.0	Plot 4N	17.5	19.7	
32.0	18.4	19.3	Alley	20.2	22.8	35.2
	18.8	19.5	Plot 4S	22.0	23.8	
31.7	17.9	19.8	Alley	22.6	24.7	39.6
	18.4	20.8	Plot 5N	24.0	25.5	
35.2	21.7	24.6	Alley	28.0	28.3	38.9
	23.3	24.3	Plot 5S	28.4	28.2	
31.2	26.7	29.1	Grass Border	33.9	32.8	38.8

Note: The grass borders and alleys between Plots were sampled October 16, 1990. Plot SOC values are averages of samples from 1986, 1987, and 1988.

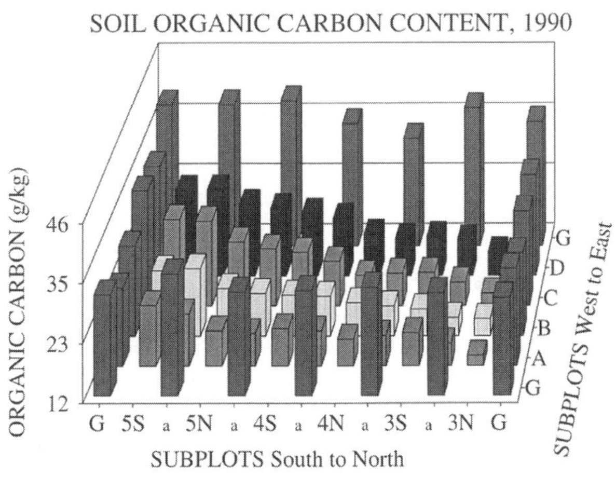

Figure 7 SOC (0 to 15 cm) content of the Morrow Plots. Note the decline in SOC toward the north and west in the plots, alleys (a), and the grass border (G).

were higher than those now found in the border. The border has been in bluegrass sod since 1904.[11] Before that it was part of the larger original plot area. The border has presumably never been fertilized because soil treatments were not started until 1904 and there is no record of fertilizing the sod. The bluegrass is a poor analog of the original tall grass prairie. It is mowed periodically and the water table has been artificially lowered for 87 years. In addition, there could have been a loss in SOC in the 40 or so years the area may have been cultivated before 1904 and a dilution of surficial SOC due to additions of lower SOC material originating as a consequence of cultivation of the plots.

Values for SOC from the 1990 sampling of the border range from 26.7 to 39.4 g kg^{-1}. The difference between these numbers and the plot SOC averages for the last 3 years gives an estimate of the range in minimum loss of SOC. Calculating SOC loss in this manner, the loss of SOC (0- to 15-cm depth) due to cultivation on the Morrow Plots ranges from about 16.2 to 38.3 g kg^{-1}. The continuous corn unfertilized plots (3NA, 3NC, and 3ND) have the greatest loss (Table 3). These plots show a decrease of between 12.8 and 24.9 g kg^{-1} in measured SOC. The 3-year rotation fertilized plots (5SB, 5SC, and 5SD) show the least decrease in measured SOC of between 1.2 and 12.8 g kg^{-1}.

Cropping system and fertilizer management also had some effect on soil nitrogen.[12] The C:N ratio is 12.6 averaged for all the plots and 13.3 for the sod border.

Table 3 SOC Loss Estimates on the Morrow Plots

Crop Treatment	Plot	Soil Treatment (C Loss, g kg^{-1})	
		None	Since 1904
Continuous corn	3	12.8–24.9	9.9–22.0
2-y rotation	4	9.9–22.0	5.8–18.0
3-y rotation	5	5.2–16.8	1.2–12.8

Note: The estimates are numerical differences between 3-y averages of three treatment subplots and grass border low and high C contents. Included were subplots N-A, C, and D as untreated and S-B, C, and D as treated.

Table 4 Soil (0 to 15 cm) Nitrogen (g kg^{-1}) Content of Selected Morrow Plots

Crop Treatment	Plot	Soil Treatment			
		None		Since 1904	
		N	C:N	N	C:N
Continuous corn	3	1.39	12.3	1.65	12.5
2-y rotation	4	1.66	12.6	2.03	12.8
3-y rotation	5	1.91	12.7	2.33	12.6

Note: Organic C and N were analyzed for the years 1904, 1913, 1923, 1933, 1944, 1953, 1955, 1961, 1967, and 1973. Included were subplots N-A, C, and D as untreated and S-B, C, and D as treated.

Data adapted from Odell, R. T., Melsted, S. W., and Walker, W. M., *Soil Sci.*, 137(3), 1984.

VIII. CONCLUSIONS

The Morrow Plots at the University of Illinois are the oldest agronomic research fields in America and include the oldest continuous corn plots in the world. They were designed about 1876 to evaluate the effects of different cropping systems and soil treatments on yield. These treatments include continuous corn with no soil amendments. This paper presents data on SOC content in the surface 15 cm of the plots that reflect soil treatments and cropping systems over the last 115 years.

The take-home lesson of the Morrow Plots is that these fertile Mollisols can provide food and fiber continuously if they are well treated. While SOC levels initially decline under continuous cultivation, proper management can overcome some of the deleterious effects of cultivation. Including legumes in a rotation is beneficial to long-term productivity and addition of fertilizer can restore yields to continuously cropped soils.

Care must be applied to interpretations of data from the Morrow Plots given the unreplicated treatments, the lack of good early records and baseline data, and the natural soil variability of the site. However, the general lessons are clear: soils that are poorly managed do not maintain their productivity; soils that are well managed can sustain high productivity in perpetuity.

REFERENCES

1. Odell, R.T., Walker, W.M., Boone, L.V., and Oldham, M.G., The Morrow Plots — A Century of Learning, Agric. Exp. Stn. Bull. 775, College of Agriculture, University of Illinois, Urbana, 1984.
2. Hunt, T.F. and Morrow, G.E., Field Experiments with Corn, 1889, Agric. Exp. Stn. Bull. No. 8, University of Illinois, Champaign, 1890.
3. Peabody, S.H., First Annual Report, Ag. Exp. Sta. Bull. No. 1, Trustees of the University, Champaign, IL, 1888.
4. Miles, M., Commercial fertilizers and Indian corn, *Prairie Farmer*, 47(24), 1876.
5. Hopkins, C.G., The Duty of Chemistry to Agriculture, Agric. Exp. Stn. Circ. No. 105, University of Illinois, Urbana, 1906.
6. Silver, C.W., Abstract of the results of the field experiments by Lawes and Gilbert, Rothamsted, England, *Illini*, IV(5), 129, 1875.

7. Russell, M.B., All the way back in one year?, *Plant Food Rev.*, 12(1), 18, 1956.
8. Fehrenbacher, J.B., Alexander, J.D., Jansen, I.J., Darmody, R.G., Pope, R.A., Flock, M.A., Scott, J.W., Andres, W.F., and Bushue, L.J., Soils of Illinois, Agric. Exp. Stn. Bull. 778, University of Illinois, Urbana, 1984.
9. Bryan, A.A. and Wendland, W.M., Local Climatological Data Summary Champaign-Urbana, Illinois 1888–1986, Illinois Water Survey Miscellaneous Publication 98, Illinois State Water Survey, Champaign, 1987.
10. DeTurk, E.E., Bauer, F.C., and Smith, L.H., Lessons from the Morrow Plots, Agric. Exp. Stn. Bull. No. 300, University of Illinois, Urbana, 1927.
11. Stauffer, R.S., Muckenhirn, R.J., and Odell, R.T., Organic carbon, pH and aggregation of the soil of the Morrow Plots as affected by type of cropping and manurial addition, *J. Am. Soc. Agron.*, 32(11), 819, 1940.
12. Odell R.T., Melsted, S.W., and Walker, W.M., Changes in organic carbon and nitrogen of Morrow Plot soils under different treatments, 1904–1973, *Soil Soc.*, 137(3), 160, 1984.
13. Schollenberger, C.J., Determination of soil organic matter, *Soil Sci.*, 59, 53, 1945.

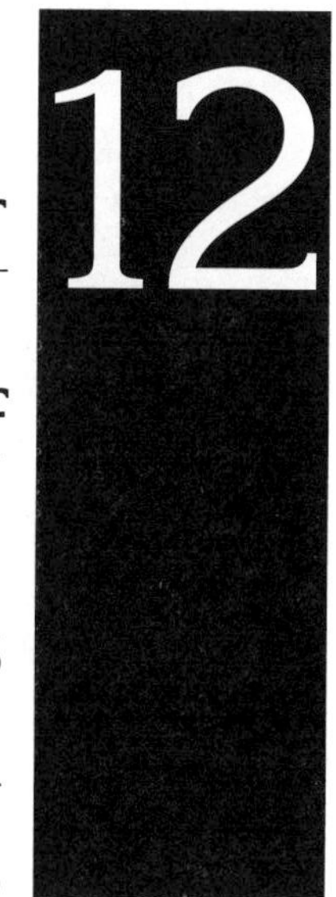

Chapter

Continuous Application of No-Tillage to Ohio Soils: Changes in Crop Yields and Organic Matter-Related Soil Properties

W.A. Dick, W.M. Edwards, and E.L. McCoy

CONTENTS

I. INTRODUCTION

Agriculture presumably began as a no-tillage (NT) system where a pointed stick was used to place seed directly into untilled soil. In many parts of the tropics, NT is still a part of slash and burn agriculture. After clearing an area of forest by controlled burning, seed is placed into the soil without benefit of tillage. However, as humans developed a more systematic agriculture, cultivation of the soil became an accepted practice for preparation of a more suitable environment for plant growth. Pictures in ancient Egyptian tombs portray a farmer tilling his fields using a plow and oxen prior to planting of the seed. Indeed, tillage, as symbolized by the moldboard plow, became almost synonymous with agriculture.

Since the 1950s research has indicated that, in many cases, less tillage than traditionally performed may be possible and may actually result in greater crop yields. The most extreme form of the "less-tillage"

0-8493-2802-0/97/$0.00+$.50

systems developed is NT. NT is defined as a crop production system where the soil is left undisturbed prior to planting and weed control is accomplished by means other than mechanical tillage, primarily by using herbicides. A historical record of NT development in the United States has been written by Triplett and Van Doren.[1]

Adoption of NT production methods has been steadily increasing and, in 1992, NT was practiced on approximately 11.4 million hectares of cropland in the United States.[2] Of this total, 51% (i.e., 5.8 million hectares) was located in the corn belt states of Iowa, Missouri, Illinois, Indiana, and Ohio. NT was practiced on 18.6% of all the cropland in the corn belt states, with Ohio being the third largest NT state in the region with approximately 1 million hectares under NT management in 1992. Although the actual reasons for adopting a NT crop production system vary among farmers, the trend toward greater use of NT seems to be linked to time and labor savings, excellent erosion control, and recent investment by farmers in NT equipment.

Some of the earliest work on NT was conducted in Ohio.[1] Between 1962 and 1964, four different experimental sites were developed to study NT crop production. The original questions to be answered included

1. How much tillage is required to produce crops with satisfactory yields?
2. To what extent are crop rotations required for corn (*Zea mays* L.) yields to remain optimum?
3. Do tillage and rotation interact to influence crop yields?

Maintaining these plots over the past 30 years has permitted us to study other tillage-related problems not originally considered when the experiments were first established. Questions now being asked, in addition to those related to crop yields, include

- How do continuous long-term applications of NT practices change the amount of C that can be sequestered in soil while still maintaining high productivity levels?
- How do the changes in soil profile properties associated with continuous application of NT, especially soil organic matter concentrations, affect other properties such as biological activity, nutrient stratification within the profile, and surface and ground water hydrology and quality?

This paper presents results obtained during the past 30 years of maintaining NT on four soil types at four locations in Ohio. It describes some of the changes observed in soil properties that are affected by long-term maintenance of NT. In addition, we have proposed how these changes in soil properties may influence both the long-term sequestering of C (an environmental question related to the issue of global warming) and crop productivity (an economic and agronomic question).

II. SITES DESCRIPTION

The four experimental sites in Ohio are widely separated (see Introduction to this volume for site locations). The soil at the Wooster site (Wooster silt loam, Typic Fragiudalf) developed under forest and is characterized as a deep, well-drained, moderately permeable soil on uplands. The soil was formed in glacial till and contains a fragipan at a depth of 50 to 90 cm. The Hoytville site contains Hoytville silty clay loam soil (Mollic Ochraqualf) which is very poorly drained. It developed in fine-textured, calcareous glacial till under forest conditions. The South Charleston site contains Crosby silt loam soil (Aeric Ochraqualf). Crosby soils are somewhat poorly drained and developed under forest on nearly level topography in loamy calcareous till of Wisconsin age. The soil at the Coshocton site is a Rayne silt loam (Typic Hapludalf). This soil is a highly weathered, well-drained forest soil. Organic matter content, clay content, internal drainage, and other properties of these four soils are typical of those used for crop production in Ohio.

Descriptions of the location, climate, management, and data obtained for these four sites have been previously published.[3] However, for clarity and completeness, a summary of soil and site characteristics is also provided in Table 1.

A. CLIMATE

Ohio's climate is continental with a wide range of air temperatures.[4] Meterological information for the study sites is summarized in Table 2. Generally, higher precipitation amounts fall in the spring and summer months and lower amounts in the autumn and winter months. Storm systems often form along the boundary between the major cold air masses from northern Canada and warm, tropical, southern air

Table 1 Soil and Site Characteristics Describing the Long-Term Tillage and Rotation Experiment in Ohio

	Description of Parameter at Ohio Site Specified			
	Wooster	**Hoytville**	**South Charleston**	**Coshocton**
Common soil name	Wooster silt loam; fine loamy, mixed, mesic	Hoytville silty clay loam; fine, illitic, mesic Mollic	Crosby silt loam; fine, mixed, mesic Aeric	Westermorelan silt loam; fine-loamy, mixed, mesic
Taxonomic soil name	Typic Fragiudalf	Ochraqualf	Ochraqualf	Ultic Hapludult
Slope, %	2.45–4.5	<1	1	2–6 (upper 1/3 watershed), 6–12 (lower 2/3 watershed)
Soil properties, Ap horizon				
Organic C, g kg^{-1}	14	23	10	30
Sand, g kg^{-1}	250	210	150	175
Clay, g kg^{-1}	150	400	200	140
Available water-holding capacity in rooting zone, cm	24	20	20	20
Minimum saturated hydraulic conductivity, cm h^{-1}	0.6 (120–150)	0.1 (55–85)	0.2 (30–60)	1.5
Tile spacing, m	None	17	16	None
Prior cropping	6-y grass meadow	6-y corn-oats-meadow	6-y corn-soybean	25-y corn-wheat-meadow-meadow
Prior tillage	None for 6 y	Plow + disk for 4 of 6 y	6-y plow + disk	Plow + disk for corn and wheat years
Experimental design	Randomized block (site A) Factorial, randomized block (site B)	Factorial randomized block	Randomized block	—
Variables	Tillage[a] (site A) Tillage[a] and rotation[b] (site B)	Tillage[a] and rotation[b]	Tillage[a]	—
Plot size	6.4 × 37 m (site A) 4.3 × 22 m (site B)	6.4 × 31 m	5.0 by 61 m	0.5 ha watershed

[a] NT, minimum tillage, and plow tillage.

[b] Continuous corn; corn and soybean in a 2-y rotation; and corn, oat, and meadow in a 3-y rotation.

Table 2 Location of Sites and Meteorological Information

Site	Geographical Location (Longitude/Latitude)	Mean Annual Precip.[a] (mm)	Mean Annual Temp. (°C)	Freeze-Free Days
Wooster	82.0 (W) 40.8 (N)	905(563)	9.1	173
Hoytville	84.0 (W) 41.0 (N)	845(522)	9.5	173
South Charleston	83.5 (W) 39.8 (N)	952(600)	11.9	189
Coshocton	81.8 (W) 40.4 (N)	999(560)	10.5	179

[a] Value in parenthesis is the amount of precipitation that occurs during the growing season (April through September).

masses from the Atlantic Ocean and Caribbean Sea. These storm systems often produce brief intense storms that provide much of the moisture during the growing season in Ohio.

Most soils are saturated during March and early April. Although the average annual rainfall ranges from 750 to more than 1100 mm, the average precipitation during the growing season is usually only 450 to 650 mm. This amount of precipitation is generally inadequate for maximum crop yields and drought stress often occurs during the months of June, July, and August. Average length of the freeze-free period (number of growing season days) ranges from 200 days along the Lake Erie shore to 140 days in east-central Ohio.

B. TREATMENTS

Continuous monoculture NT corn is grown at the Coshocton site. At the Wooster Site B and Hoytville site, all combinations of three tillage and three rotation treatments are applied each year (Table 1). Sufficient numbers of plots (54) were established so that each crop in rotation appears each year. At the Wooster Site A and the South Charleston site, continuous (monoculture) corn is grown using three different tillage treatments.

The tillage treatments at all sites, except Coshocton, consist of (1) NT other than that accomplished by the planter or (at the Hoytville site only) fertilizer injection system, (2) minimum tillage that originally included the use of a moldboard plow 20 to 25 cm deep with no other tillage prior to planting — this treatment has changed several times over the years to include paraplow (1983 and 1984), NT (1985 and 1986), and chisel plow to 20-cm depth (since 1987), and (3) plow tillage which involves spring moldboard plow (fall plow at Hoytville) 20 to 25 cm deep plus one or two other 10-cm deep secondary tillage treatments. The rotation treatments at the Wooster Site A and Hoytville sites include (1) continuous corn, (2) corn and soybeans (*Glycine max* L.) in a 2-year rotation, and (3) corn, oats (*Avena sativa* L.), and meadow in a 3-year rotation. The meadow crop at the Wooster Site B and Hoytville site has varied over the years and has included mixed grasses, alfalfa (*Medicago sativa* L.) or clover (*Trifolium pratense* L.). Most recently the meadow crop has been alfalfa at Wooster Site B and clover at the Hoytville site. The meadow crop is cut two to four times per year, depending on productivity, and removed from the plots.

C. MANAGEMENT

Detailed records of management practices have been kept at all of the sites since the experiments were started.[5–7] Fertility is maintained by broadcasting N, primarily as ammonium nitrate (NH_4NO_3) in the spring prior to tillage. Since 1986, anhydrous ammonia has been the primary N source at the Hoytville site. During the first few years P and K (at all sites except Coshocton) were applied with the planter. Beginning in 1967 most of the P and K has been broadcast in the late fall or spring. Generally, a small amount of starter fertilizer was also applied with the planter for corn. Manganese was foliar applied to soybeans as $MnSO_4$ at the Hoytville site. Lime was broadcast in the late fall or winter to maintain pH values of 6 or higher in the continuous corn plots and then all other plots generally received the same lime treatment. At the Coshocton site strawy manure was applied in the winter months during most years.

Insecticides were used primarily to control rootworm, (*Diabrotica* spp.) and cutworm (*Agrotis ipsilon* say) in corn and were applied at planting. Herbicides were applied shortly after planting with follow-up control measures often required, especially during the initial years of continuous NT. The corn and soybean crops received the greatest amount of herbicide with atrazine (6-chloro-*N*-ethyl-*N*′-[1-methylethyl]-1,3,5-triazine-2,4-diamine) being the most often used; lesser amounts of dicamba (3,6-dichloro-2-methoxybenzoic acid), alachlor (2-chloro-*N*-[2,6-diethylphenyl]-*N*-[methoxymethyl] acetamide), glyphosphate (*N*-[phosphonomomethyl]glycine), and 2,4-D ([2,4-dichlorophenoxy] acetic acid) were also used. Weed control was generally good for the tilled plots and not as good for the NT plots during the early years of the experiment, but good weed control for all treatments is now achieved.

Cultivars of crops planted changed with time as better yielding cultivars became available. All plots at a site were planted on the same day with the same planter. Row width for corn was originally 100 cm and changed to 75 cm in 1968. Row width for soybeans has been primarily 75 cm at the Hoytville site and, since 1977, 38 cm at the Wooster site. Oats were seeded using 18-cm row widths. Meadow crops were generally started using the oat crop as a nursery and then allowing the meadow crop to grow up through the stubble after oat harvest. When stand establishment was poor, the meadow crop was replanted the following spring.

D. DATA COLLECTED

Emergent corn populations were recorded each year at each site. Plots were thinned to achieve a common plant density across all treatments. Plots where plants were below the threshold level, which varied from year to year, were ignored in the thinning process. Plots with adequate weed control and the required plant densities were harvested for grain yield measurements, with corn yields being reported at 15.5% moisture and soybean and oat yields at 13.5% moisture. Hay yields were determined by weighing the total amount harvested from a swath of known area cut from the center of the plot. A subsample was dried to determine moisture content so that yields could be expressed on a dry-weight basis.

Crop yields were analyzed by the least-squares method.[8,9] All other data were analyzed using analysis of variance (ANOVA) and the appropriate statistical model corresponding to the experimental design established at a site.

Soils were periodically sampled and various chemical, biological, and physical parameters measured. In addition, several on-site measurements, e.g., residue cover, were made. The procedures used to make specific measurements have been previously described.[5–6,10–13]

III. RESULTS AND DISCUSSION

A. CROP YIELDS

Corn grain yields have averaged between 6 and 10 Mg ha^{-1} between 1963 to 1991. A previously reported study at the Coshocton site showed greater grain and biomass production under NT than for a plowed treatment.[14] For this report, the comparative impact of tillage on continuous (monoculture) corn yields was determined at the Wooster Sites A and B and the Hoytville site. The tillage effect was determined by subtracting the 5-year average value of the NT yields from the 5-year average values of the plowed grain yields. A difference value near zero indicates that tillage had no effect on yield while larger positive and negative values indicate a positive or negative effect of NT, respectively, on corn grain yields as compared to plow tillage.

At the Wooster Sites A and B, the 5-year means of the NT yields were always higher than the yields for the plowed treatments (Figure 1). There were five negative values and eight positive values associated with NT at the South Charleston site with the recent trend (since 1980) being corn grain yields being higher under NT than where plowing was practiced. At the Hoytville site, the combination of continous corn and continous NT caused a severe yield depression during the first 20 years of the experiment. However, more recently the trend has been to obtain equal yields under both the NT and plowed treatments. The data show similar yield trends for soybeans at the Wooster Site B and the Hoytville site (Figure 2).

Several working hypotheses are presented to explain these trends, especially the trend of equal or higher yields associated with continuous application of NT since 1980: (1) The climate during the past 10 years (during the 1980s) was hotter and drier than normal. These are conditions that favor NT crop production even in heavy soils that generally respond negatively to NT. (2) The soil profile has drastically changed with continuous application of NT, with respect to nutrient stratification and other plant growth-related parameters so that the NT soil now supports increased crop production above that

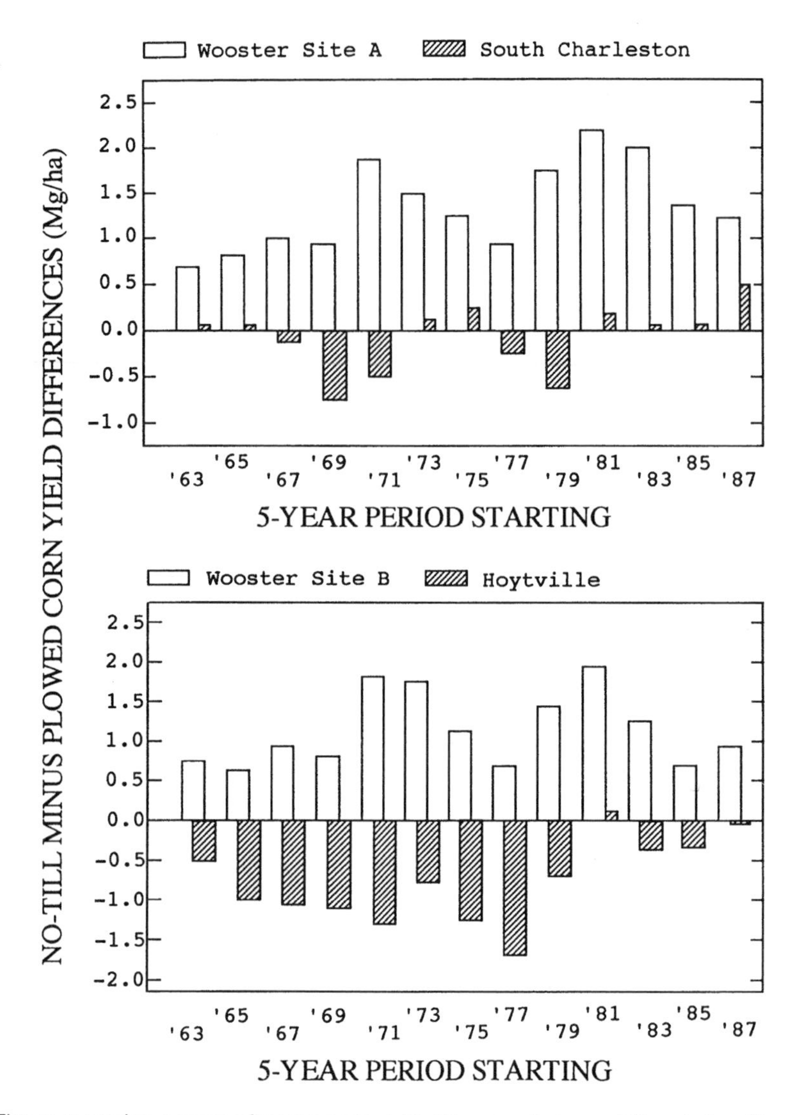

Figure 1 Five-year running averages for corn grain yield differences between NT and plow tillage. Differences are calculated by subtracting the plow tillage from the NT grain yields. Standard errors are 0.18, 0.21, 0.24, and 0.21 Mg ha^{-1} for the Hoytville, Wooster Site A, Wooster Site B, and South Charleston sites, respectively.

found in the plowed soil. (3) We have observed that removing the disease variable by growing phytophthera-resistant/tolerant soybean cultivars immediately improved soybean yields where NT is continuously practiced to levels equal to or greater than that in the plowed soil. It is possible that corn hybrids have also been inadvertantly developed that have a greater resistance or tolerance to plant diseases, especially root diseases, that once severely curtailed corn yields under continuous NT. (4) The soil biological activity has changed in the long-term NT plots due to chemical and soil organic matter changes in the soil profile so that a more diverse and larger biomass is now present in the surface soil where most disease organisms reside. This larger and more biologically active microbial biomass acts as a naturally suppressive means of controlling plant diseases as well as providing a more sustained level of nutrient availability from organic matter mineralization throughout the growing season.

B. RESIDUES

The most visible effect of applying continuous NT cropping practices is the rapid development of a residue cover on the soil surface. It takes only a few years for a stable equilibrium level of residue cover to develop, and the amount of cover measured in midsummer seems to stabilize between 50 to 85% of the soil surface (Table 3). Crops that produce less residues and more easily degraded residues, such as

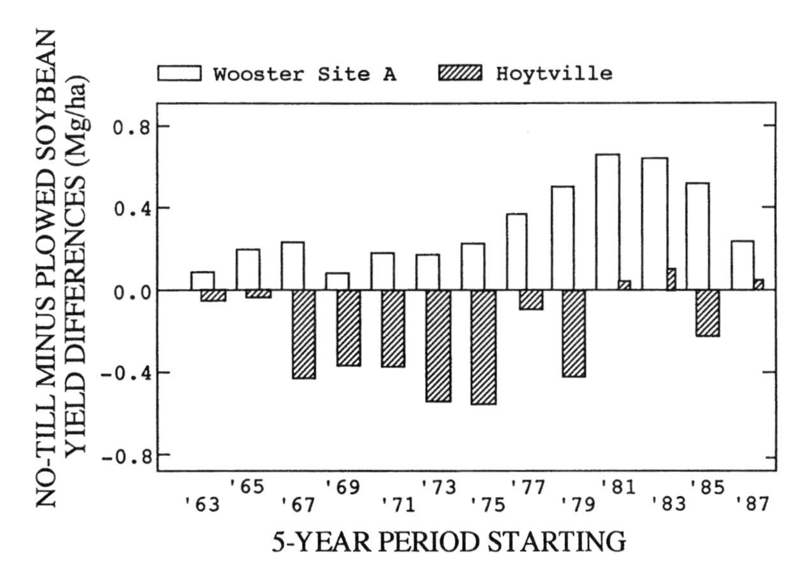

Figure 2 Five-year running averages for soybean grain yield differences between NT and plow tillage. Differences are calculated by subtracting the plow tillage from the NT grain yields. Standard error is 0.10 Mg ha^{-1} for both sites.

soybeans, when grown in rotation with corn will result in a lower equilibrium residue cover level than when corn is grown in monoculture. The amount of C sequestered in the residue cover of a continuous NT field may be considered as being at steady state as long as soil disturbance is avoided. Based on field observations, as soon as the soil becomes even slightly disturbed, though the disturbance may still be defined as an appropriate conservation tillage practice (i.e., a 30% residue cover at time of planting is maintained), much of the residue C will be oxidized to CO_2.

C. SOIL ORGANIC MATTER

The residues deposited on the soil surface during the first year of NT seem to have little permanent impact on organic matter-related soil properties. With the passage of time, the continual addition and decay of residue enriches the surface soil layer with humified organic matter. The C in this organic matter is more resistant to decomposition than the residues from which it is derived. The distribution of organic C within the soil profile also changes from that observed in a plowed soil, i.e., a stratification occurs (Figure 3). Carbon is lost from the subsurface soil layers, even though root biomass is added on an annual basis, while at the same time C is gained in the soil surface layer due to stabilization of C deposited as plant residues.

Carbon sequestration seems to be stimulated by NT, at least during the early years of NT, due to increased immobilization of N and humification of C in the surface soil layer. At the same time, however, the subsurface soil layers continue to mineralize C, lowering the C/N ratio in these soil layers.[10] The net result is a redistribution of C with a slight increase in total amount of C in the soil profile. Once a redistribution of C is established in the NT soil, there seems to be a slow and steady increase in overall C sequestration in the NT soil, provided disturbance of this C-enriched surface layer is minimized.

Studies at the Wooster Site B and Hoytville site indicate that the total amount of organic C in the soil profile can be maintained at a higher level when continuous NT, as compared to annual plowing, is practiced (Table 4). The concentrations of organic C in the NT soil profile at the Hoytville site remained unchanged after 18 years (from 1962 to 1980) and, compared to concentrations measured in the plowed soil, were 16% greater. For the Wooster soil (Wooster Site B), continuous application of NT over the same 1962 to 1980 period resulted in a 10% decrease in organic C concentrations in the soil profile as compared to a 25% decrease for plow tillage. The soils were sampled in November and December when the differences in bulk density between tillage treatments were not statistically different.

Higher rates of fertilizer N generally are required to achieve the same level of yield for NT as compared to plow tillage.[15,16] This is attributed to greater immobilization of N fertilizer by soil microorganisms in the surface soil layer during the decomposition of freshly deposited plant residues. In the long-term tillage experiments described, equal amounts of fertilizer N were applied to each tillage

Table 3 Percentage of Soil Surface Covered with Residues in the Wooster, Hoytville and South Charleston, OH, NT Corn Plots

	Wooster[a]				Hoytville (Residue Cover, %)				South Charleston (Residue Cover, %)	
	Site A	Site B								
Date	CC	CC	CS	COM	Date	CC	CS	COM	Date	CC
3 July 1964	61	57	79	60	29 July 1964	36	12	28	17 June 1964	88
10 September 1968	96	92	76	92	19 August 1965	66	81	94	15 August 1973	92
16 June 1975	45	66	45	66	21 June 1968	77	47	87	8 August 1978	90
10 June 1987	78	81	74	84	9 June 1987	80	80	31	11 June 1987	78
Means	72	75	62	80		70	52	58		84

[a] CC, continuous corn; CS, corn and soybean in a 2-y rotation; and COM, corn, oat, and meadow in a 3-y rotation.

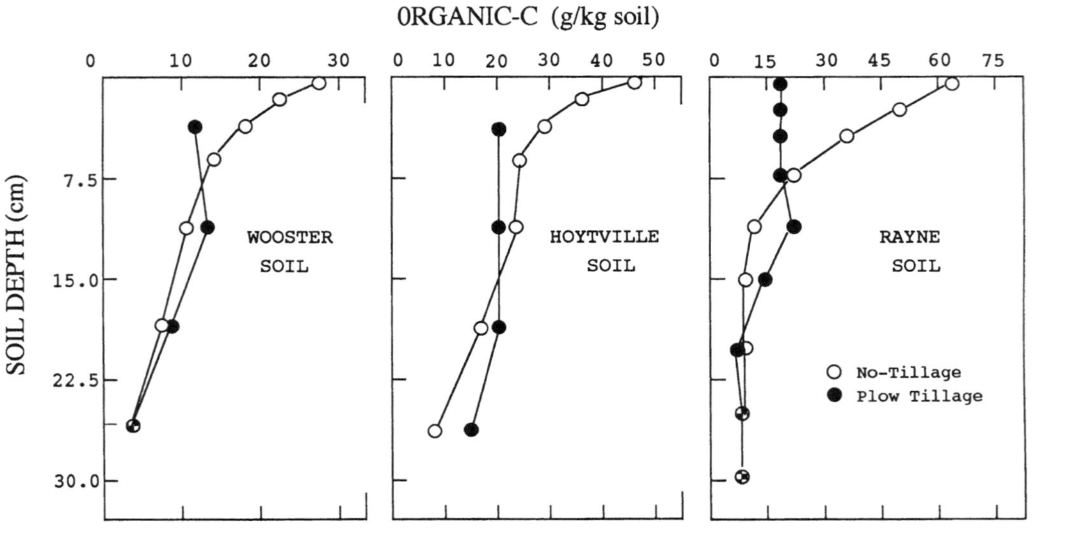

Figure 3 Organic C concentrations in Wooster and Hoytville soil profiles after 18 years of continuously applying an NT or plow tillage treatment.

Table 4 Effect of Tillage Intensity on Organic C Concentrations in the Plow Layer (0 to 22.5 cm) of Hoytville and Wooster Site B Soils

Soil	Initial Organic C Concentration[a] (g kg^{-1})	Organic C Concentrations after 18 Years under Tillage Treatment[b] (g kg^{-1})		
		NT	MT	CT
Wooster	14	12.5(11)[c]	10.8(23)	10.5(25)
Hoytville	23	23.1(0)	20.2(12)	20.0(14)

[a] Concentrations measured when tillage treatments were imposed.
[b] NT, no-tillage; MT, minimum tillage; and CT, plow tillage.
[c] Values in parentheses are the percentage decreases in organic C calculated from the equation (1-B/A)100 where A equals the initial and B the present organic C concentrations.

treatment. The total amount of N in the 0- to 30-cm soil profile, however, was significantly ($P<0.05$) greater under NT than where the soil had been plowed.[10]

Mean C/N ratios for the 0- to 7.5-cm soil layer were 10.7 and 9.7 for the Hoytville and Wooster NT soils, respectively, and 10.5 and 9.1 for the plowed soils.[10] The highest C/N ratios were found at the top (0- to 1.25-cm layer) of the NT profile and were 11.4 and 11.0 for the Hoytville and Wooster soils, respectively. Below 7.5-cm depth, the plowed plots had a higher C/N ratio, especially in the Wooster soil.

Concentrations of organic C and nutrients in the surface layer of the NT soil can be two to seven times higher than for the surface layer of a plowed soil.[5,6,10] Loss of C and nutrients from this enriched surface layer, caused by erosion, might be expected. However, losses are minimized because the NT practice is an excellent technology for erosion control.[17,18] In Ohio, erosion occurs primarily as a result of a few intense storms. Studies conducted at Coshocton at the North Appalachian Experimental Watershed have shown that, even for a storm with an expected return period of >200 years, infiltration was sufficient to delay runoff until 25 minutes after the storm began and only 12% of the total storm rainfall was lost as runoff.[3,18]

Soil surface elevation measurements at Wooster Site A, which has a very moderate slope of 2.5 to 4.5%, have been used to estimate relative differences in soil erosion as affected by long-term NT (Table 5). After 18 years, cumulative soil erosion had resulted in statistically lower elevations for the plowed plots compared to the NT plots.[19] This elevation difference was evident even though plowed soils are generally considered to be less compact and dense than NT soils.

Table 5 Soil Erosion Estimates from Reduction in Surface Elevations

Years	Rotation[a]	Reduction in Surface Elevation as Affected by Tillage[b] (cm)		
		NT	MT	CT
1971–1976	CC	0.3	–1.2[c]	1.5
	CS	–0.1	0.2	2.6
	COM	0.2	0.2	2.4
	Mean	0	-0.05	2.2
LSD (0.05)	Tillage = 1.3	Rotation = 1.3	Til × Rot = 2.2	
1971–1980	CC	0.2	1.5	3.6
	CS	0	2.5	4.3
	COM	–0.3	2.0	3.2
	Mean	0	2.0	3.7
LSD (0.05)	Tillage = 1.4	Rotation = 1.4	Til × Rot = 2.4	

Note: The mean of the NT treatment was adjusted to zero and the other tillage treatments were compared relative to the NT treatment.

[a] CC, continuous corn; CS, corn and soybean in a 2-y rotation; and COM, corn, oat, and meadow in a 3-y rotation.
[b] NT, no-tillage; MT, minimum tillage; and CT, plow tillage.
[c] A negative value indicates a higher surface elevation compared to the mean NT treatment.

D. BIOLOGICAL ACTIVITY

The number of earthworms in soil greatly increases when using NT practices.[20–22] Work conducted at the Coshocton site has quantified the number of earthworm channels within a specified area of a NT soil and the amount of water and chemicals that percolate through these channels.[20,21,23] We have observed rapid transport of dyes and pesticides through the earthworm macropores when these compounds are applied immediately prior to a precipitation event.[23] However, the extent of the loss of dissolved organic C compounds of natural origin via vertical transport through the soil, both as a result of matrix flow and preferential or macropore flow, is not known at this time. While one may postulate the amount of dissolved organic C is relatively small compared to the total amount of C in the soil, this information needs to be measured to calculate an accurate C budget for soil systems. Comparison of loss of soluble organic C from NT and plowed soils, however, is frought with difficulty because of the many hydrologic and chemical variables associated with its transport.

The microbial biomass C pool size is also affected by the continued application of NT. Data from the Coshocton site indicate that, as the number of years of continuous NT increases, biomass C concentrations in the 7.5- to 15-cm and 15- to 30-cm layers are significantly reduced.[7] There was a gradual decrease in biomass C concentrations under continuous NT from the maximum level observed after the first year of NT (786 kg/ha furrow-slice) in the 0- to 7.5-cm soil layer, which reached an apparent equilibrium level after about 10 years that was 30% higher than in the plowed soils. In contrast, significant increases in biomass C concentrations were observed with increasing years of continuous NT in the 30- to 45-cm soil layer. In this soil layer, biomass C levels increased from approximately 150 to 280 kg/ha furrow slice after 20 years. These data suggest that biomass C may be a useful parameter to assess long-term trends associated with the continous application of NT.

A study of total organic C and biomass C concentrations at a number of long-term tilled sites (including the Hoytville site and the Wooster Site B) in North America, presumably at equilibrium levels, yielded a significant relationship between microbial biomass C per unit of organic C and the quotient between mean annual precipitation and annual pan evaporation (Figure 4).[24] This equation was considered representative of a baseline level of microbial biomass C in soil and confirmed that microbial biomass C was best described by an equilibrium function rather than an equilibrium constant. Any deviation of the microbial biomass C to organic C ratio from the equilibrium function for a given soil would indicate that its organic C content is increasing or decreasing. A similar study in which only long-term NT soils are used has not been conducted.

The ratio of microbial biomass C to total organic C was higher at the Wooster site than at the Hoytville site, even though both sites had similar histories.[24] However, the textural class differs between the two sites with the soil at the Hoytville site having more than twice as much clay than the soil at the Wooster site. Clay is known to impart a protective effect on soil organic C. The difference in the microbial biomass C/total organic C ratios was caused by a higher total organic C content in the soil at the Hoytville site (21 g/kg) than at the Wooster Site B (12 g/kg), rather than by differences in microbial biomass C content.

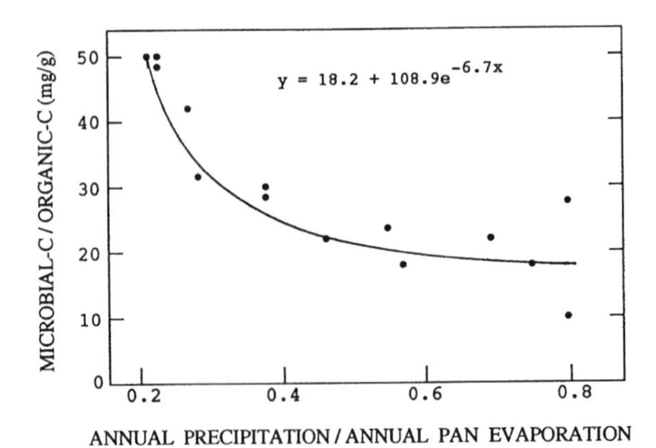

Figure 4 Amount of microbial biomass carbon per unit of organic C plotted against the quotient between annual precipitation (mm) and annual pan evaporation (mm) for tilled study sites in North America. (From Shipitalo, M. J., Edwards, W. M., Dick, W. A., and Owens, L. B., *Soil Sci. Soc. Am. J.*, 54, 1530, 1990. With permission.)

Some soil enzymes reflect biological activity in soil[25] and the continuous application of NT enriches the surface layer in soil enzyme activity.[11] This increased enzyme activity can influence the dynamics of nutrient cycling in soil. For example, N loss from urea-based fertilizers will be much more susceptible to loss under NT than plow tillage (Figure 5). The higher enzyme activities are directly related to higher organic C levels in NT soils. The higher enzyme activities coupled with the accumulation of organic matter in the surface layer of the NT soil suggest the ability of the soil enzymes (and the microbial biomass as well) to effectively degrade the organic matter and catalyze nutrient cycling reactions seems to be lower under NT than CT.

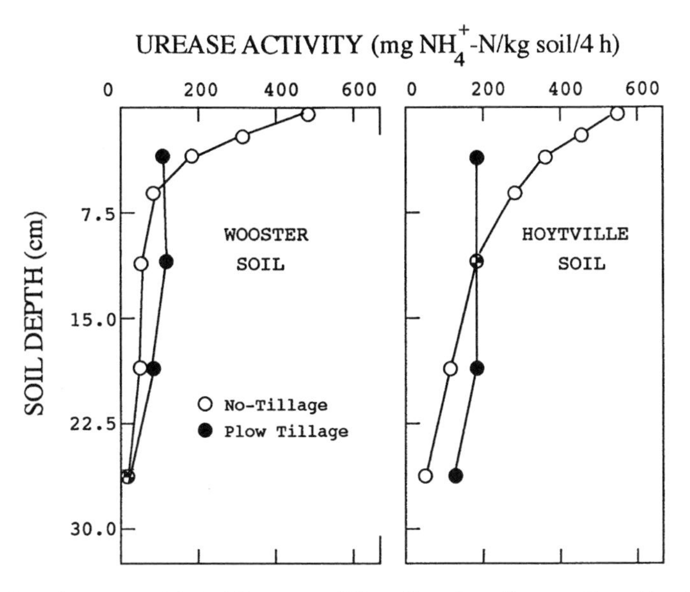

Figure 5 Distribution of urease activity in Wooster and Hoytville soil profiles as affected by plow tillage and NT.

IV. CONCLUSIONS

NT, when continuously practiced, promises to be an effective technology for removing C from the atmosphere and sequestering it in soil. Our data indicate that long-term application of NT results in soil C concentrations that are greater than where plowing is practiced. However, based on long-term tillage and rotation records such as ours, models need to be developed to estimate the impact of converting the majority of crop production land to NT on C levels in the soil and atmosphere. Continuous and long-term application of NT over a 30-year period representing a wide variation of climatic conditions, to both well-drained light soils or a very poorly drained heavy soil, also resulted in crop yields that were consistently equal to or higher than where plowing was practiced. This suggests that NT is a promising technology for maintaining high crop yields if climatic changes, such as hotter and drier growing seasons, occur.

Results presented in this paper also indicate that biomass C may be a useful parameter to assess long-term trends associated with the continuous application of NT. When studying biomass C levels in tilled sites, an equilibrium function was obtained that was considered representative of a baseline level of microbial biomass C in soil. Any deviation of the microbial biomass C to organic C ratio from the equilibrium function for a given soil would indicate that its organic C content is increasing or decreasing. A similar study in which only long-term NT soils are used has not been conducted. Additional research is needed to improve our understanding of C dynamics in NT systems and to better predict how long-term NT will affect soil chemical, biological, and physical parameters and, ultimately, crop productivity. Managment decisions that involve adoption of NT farming practices, however, must be assessed with the idea that other environmental impacts may arise when converting large areas of cropland to continuous NT practices.

REFERENCES

1. Triplett, G.B., Jr. and D.V. Van Doren, Jr., Agriculture without tillage, *Sci. Am.*, 236, 28, 1977.
2. Conservation Impact, Conservation Technology Information Center, West Lafayette, IN, 10(8), 1, 1992.
3. Dick, W.A., McCoy, E.L., Edwards, W.M., and Lal, R., Continuous application of no-tillage to Ohio soils, *Agron. J.*, 83, 65, 1991.
4. Ohio Agronomy Guide (12th edition), The Ohio Cooperative Extension Service and The Ohio State University, Columbus, Bulletin 472, 1988.
5. Dick, W.A., Van Doren, D.M., Jr., Triplett, G.B., Jr., and Henry, J.E., Influence of Long-Term Tillage and Rotation Combinations on Crop Yields and Selected Soil Parameters. I. Results Obtained for a Mollic Ochraqualf Soil, Research Bulletin 1180, The Ohio State University and The Ohio Agricultural Research and Development Center, Wooster, 1986.
6. Dick, W.A., Van Doren, D.M., Jr., Triplett, G.B., Jr., and Henry, J.E., Influence of Long-Term Tillage and Rotation Combinations on Crop Yields and Selected Soil Parameters. II. Results Obtained for a Typic Fragiudalf Soil, Research Bulletin 1181, The Ohio State University and The Ohio Agricultural Research and Development Center, Wooster, 1986.
7. Staley, T.E., Edwards, W.M., Scott, C.L., and Owens, L.B., Soil microbial biomass and organic component alterations in a no-till chronosequence, *Soil Sci. Soc. Am. J.*, 52, 998, 1988.
8. Harvey, W.R., Least Squares Analysis of Data with Unequal Numbers, USDA-ARS 20-8, U.S. Government Printing Office, Washington, DC, 1960.
9. Dick, W.A. and Van Doren, D.M., Jr., Continuous tillage and rotation combinations effects on corn, soybean and oats yields, *Agron. J.*, 77, 459, 1985.
10. Dick, W.A., Organic carbon, nitrogen and phosphorus concentrations and pH in soil profiles as affected by tillage intensity, *Soil Sci. Soc. Am. J.*, 47, 102, 1983.
11. Dick, W.A., Influence of long-term tillage and rotation combinations on soil enzyme activities, *Soil Sci. Soc. Am. J.*, 48, 569, 1984.
12. Page, A.L., Miller R.H., and Keeney, D.R., Eds., *Methods of Soil Analysis, Part 2 — Chemical and Microbiological Properties*, 2nd ed., American Society of Agronomy, Madison, WI, 1982.
13. Klute, A., Ed., *Methods of Soil Analysis, Part 1 — Physical and Mineralogical Methods*, 2nd ed., American Society of Agronomy, Madison, WI, 1986.
14. Dick, W.A., Stehouwer, R.C., Edwards, W.M., and Eckert, D.J., Maize yield and nitrogen uptake after established no-tillage fields are plowed, *Soil Till. Res.*, 24, 1, 1992.
15. Fox, R.H. and Bandel, V.A., Nitrogen utilization with no-tillage, in *No-Tillage and Surface-Tillage Agriculture*, Sprague, M.A. and Triplett, G.B., Eds., John Wiley & Sons, New York, 1986, 117.
16. Rice, C.W., Smith, M.S., and Blevins, R.L., Soil nitrogen availability after long-term continuous no-tillage and conventional tillage corn production, *Soil Sci. Soc. Am. J.*, 50, 1206, 1986.
17. Van Doren, D.M., Moldenhauer, W.C., and Triplett, G.B., Influence of long-term tillage and crop rotations on water erosion, *Soil Sci. Soc. Am. J.*, 40, 636, 1984.
18. Dick, W.A., Roseberg, R.J., McCoy, E.L., Edwards, W.M., and Haghiri, F., Surface hydrologic response of soils to no-tillage, *Soil Sci. Soc. Am. J.*, 53, 1520, 1989.
19. Roseberg, R.J. and McCoy, E.L., Time series analysis for statistical inferences in tillage experiments, *Soil Sci. Soc. Am. J.*, 52, 1771, 1988.
20. Edwards, W.M., Norton, L.D., and Redmond, C.E., Characterizing macropores that affect infiltration into nontilled soil, *Soil Sci. Soc. Am. J.*, 52, 483, 1988.
21. Edwards, W.M., Shipitalo, M.J., and Norton, L.D., Contribution of macroporosity to infiltration into a continuous corn no-tilled watershed: implications for contaminant movement, *J. Contam. Hydrol.*, 3, 193, 1988.
22. Stinner, B.R., Edwards, W.M., and Shipitalo, M.J., Soil invertebrates in conservation tillage and lower-input systems, in *Agronomy Abstracts*, American Society of Agronomy, Madison, WI, 1989, 295.
23. Shipitalo, M.J., Edwards, W.M., Dick, W.A., and Owens, L.B., Initial storm effects on macropore transport of surface-applied chemicals in no-till soil, *Soil Sci. Soc. Am. J.*, 54, 1530, 1990.
24. Insam, H., Parkinson, D., and Domsch, K.H., Influence of macroclimate on soil microbial biomass, *Soil Biol. Biochem.*, 21, 211, 1989.
25. Frankenberger, W.T., Jr. and Dick, W.A., Relationships between enzyme activities and microbial growth and activity indices in soil, *Soil Sci. Soc. Am. J.*, 47, 945, 1983.

Chapter

Management Impacts on SOM and Related Soil Properties in a Long-Term Farming Systems Trial in Pennsylvania: 1981–1991

S.E. Peters, M.M. Wander, L.S. Saporito, G.H. Harris, and D.B. Friedman

CONTENTS

I. INTRODUCTION

Advocates of sustainable agriculture strive to identify and promote alternative management practices that maintain soil quality.[1,2] There is increasing focus on the relationship between soil management practices and soil organic matter (SOM) characteristics because organic matter is closely associated with soil productivity and because soil C storage impacts the global C cycle and mediates the fate and transport

0-8493-2802-0/97/$0.00+$.50

of pesticides and synthetic organic compounds applied to soils.[3] SOM must be maintained to sustain soil quality.

While the amount of organic carbon in soils is generally controlled by climate (moisture and temperature regimes) and soil texture,[4] farming practices in most agricultural systems have led to significant soil carbon losses.[5] It must be determined whether management options exist that can improve SOM characteristics so that adequate soil fertility is supplied while desirable soil physical and biological properties are maintained. Important factors dictated by management that affect SOM include the quantity and quality of plant above-ground and root residues, manures and other organic amendments returned to the soil, and the timing and method of incorporation of these materials. The primary management variables influencing these factors include cropping sequence, plant species diversity, tillage, weed control, fertilization practices, and crop/livestock interactions.

A long-term experiment, known as the Farming Systems Trial (FST), was initiated in 1981 at the Rodale Institute Research Center to examine the process of converting from a conventionally managed cropping system (i.e., synthetic fertilizers, pesticides, simple rotation, no cover crops) to an organically managed system.[6] The three distinct cropping systems represented in the study have led to differences in SOM contents and in several organic matter-dependent soil properties. Several management variables unique to each system (Table 1) could have affected SOM. These management practices (crop rotation, weed control, fertilization, and tillage practices) all influence soil physical, biological, and chemical properties which in turn impact SOM characteristics. In this paper the relationship between observed changes in SOM and associated soil characteristics and various aspects of the three cropping systems' management practices is discussed.

Table 1 Comparison of Three Cropping Systems (1981–1991)

Cultural Practices	Low Input Animal	Low Input Cash Grain	Conventional Cash Grain
Crops grown	Corn (grain), corn silage, soybean, wheat, legume hay	Corn (grain), soybean, wheat, oats, barley, legume green manure/hay	Corn (grain), soybean
Ground cover[a]	Live 73%, dead –20%, bare –7%	Live –70%, dead –22%, bare –8%	Live –42%, dead –50%, bare –8%
Primary tillage	Moldboard plow (4 times every 5 years)	Moldboard plow (once per year)	Moldboard plow (once per year)
Weed control	Rotary hoe, cultivate (corn and soybean only)	Rotary hoe, cultivate (corn always, soybean) 1981–1985, 1991	Herbicides (corn and soybean)
Nitrogen fertility[b]	Beef manure (applied before corn only), 3rd-year legume hay	Green manure (red clover/alfalfa, hairy vetch)	Starter fertilizer, urea, ammonium nitrate (corn only)
Potassium fertility	139 kg ha^{-1} K as potassium sulfate (1989 only), 69 kg ha^{-1} y^{-1} as beef manure (corn only)	139 kg ha^{-1} K as potassium sulfate (1989 only)	9–28 kg ha^{-1} y^{-1} K as starter fertilizer (corn only), 139 kg ha^{-1} K as potassium sulfate (1989 only)
Phosphorus fertility	30 kg ha^{-1} y^{-1} P as beef manure (corn only)	None	10–15 kg ha^{-1} y^{-1} P as starter fertilizer (corn only)
Lime	3363 kg ha^{-1} Ca limestone (1989 only)	3363 kg ha^{-1}Ca limestone (1989 only)	8968 kg ha^{-1} Mg limestone (1989 only)
Additional features	Frost seed legumes into small grains	Frost seed legumes into small grains, relay crop soybean into small grains	None

[a] Expressed as percentage of time that the ground was covered by living plants, plant residues, or was bare. If green manure crop or hay was established before the onset of winter, this period was considered to be under live cover even though plants were dormant. Soil surface was assumed to be bare for 1 month after plowing and planting a new crop.

[b] See Table 3 for N input levels.

II. SUMMARY OF SITE, MANAGEMENT PRACTICES, AND AGRONOMIC RESULTS

A. SITE

The FST experiment was conducted on a 6.1-ha field in the Great Valley section of the Ridge and Valley province in southeastern Pennsylvania. A summary of the geographic, soil, topographic, and climatic characteristics are as follows:

Location: 40° 33′ N (latitude), 75° 43′ W (longitude).
Soil texture: Silty-clay loam (17% sand, 53% silt, 30% clay); plow layer (0 to 20 cm) includes 10 to 20% shale fragments (2- to 10-mm size).
Soil classification: 90% typic fragiudalf or hapludalf, 10% typic dystrochrept.
Clay type: Illite (2:1) with significant amount of fixed ammonium.
Parent material: Unglaciated colluvium and residuum from gray shale and small amounts (<10%) of limestone, sandstone, and quartzite.
Topography: Gently rolling, facing southwest, 1 to 5% slope; erosion likely impacts soil texture over 10-year period.
Precipitation: 1045 mm annually; 50% occurring from May 1 to September 30.
Temperature: 12.4°C annual average; hot, humid summers and relatively short, mild winters.
Growing season: 180 frost-free days; 1600° growing degree days (based on 10°C).

The site had been used continuously since the original forest was cleared, approximately 250 years ago, to provide pasture and produce hay, corn, and small grains.

B. MANAGEMENT PRACTICES

Beginning in 1981, three cropping systems, based on either 3- or 5-year rotations, were established. The three systems included

1. The Low Input/Animal system (LIP-A), which simulated the cropping system in a beef or dairy operation and produced red clover (*Trifolium pratense* L.)/alfalfa (*Medicago sativa* L.)/orchardgrass (*Dactylis glomerata* L.) hay, oats (*Avena sativa* L.), winter wheat (*Triticum aestivum* L.), corn (*Zea mays* L.) grain and silage, and soybeans [*Glycine max* (L) Merr.]. Nitrogen was provided by cattle manure and 3rd-year hay crops plowed down prior to the planting of corn.
2. The Low Input/Cash Grain system (LIP-CG), which did not contain an animal component and produced a cash grain every year. It included corn, soybeans, oats, winter wheat, and spring barley (*Hordeum distichum* L.). Plowdown legumes were the primary N source. Both low-input systems avoided herbicide use and instead employed rotary hoeing, row-cultivation, crop rotation, green manuring, and relay cropping for their weed control.
3. The Conventional Cash Grain system (CONV) produced corn and soybeans using Pennsylvania State University recommendations for fertilizers and pesticides application.

The three rotation sequences are summarized in Figure 1. Conventional tillage (moldboard plow, disk, field cultivator, cultipacker) was used in all three cropping systems. The primary differences among the farming systems are summarized in Table 1. The experiment included eight replicates (blocks) divided into 18.3 × 91.5 m farming system main plots. These main plots were further divided into three 6.1 × 91.5 m crop-entry point subplots. Grass buffer strips (1.5 m wide) were maintained to minimize the exchange of soil, fertilizers, and pesticides between main plots.

C. AGRONOMIC RESULTS

During the first 4 years of the study, a "transition effect" or adjustment in the equilibrium between soil processes and plant growth occurred in the low-input systems. Both low-input systems had lower corn yields relative to the conventional system until 1984, but since then corn yields have been similar in all systems.[6,7] The effect was primarily the result of a lack of plant-available N in the low-input systems during the early years of the study. Although soil N may have been accumulating during the transition period, it was unavailable to the crops. After 4 or 5 years, the corn plants, in particular, appeared to begin utilizing these N reserves as reflected in improved growth and higher yields. Soybeans had nearly equal yields in all three systems throughout the 11-year period, and thus seemed to be relatively unaffected by the transition.

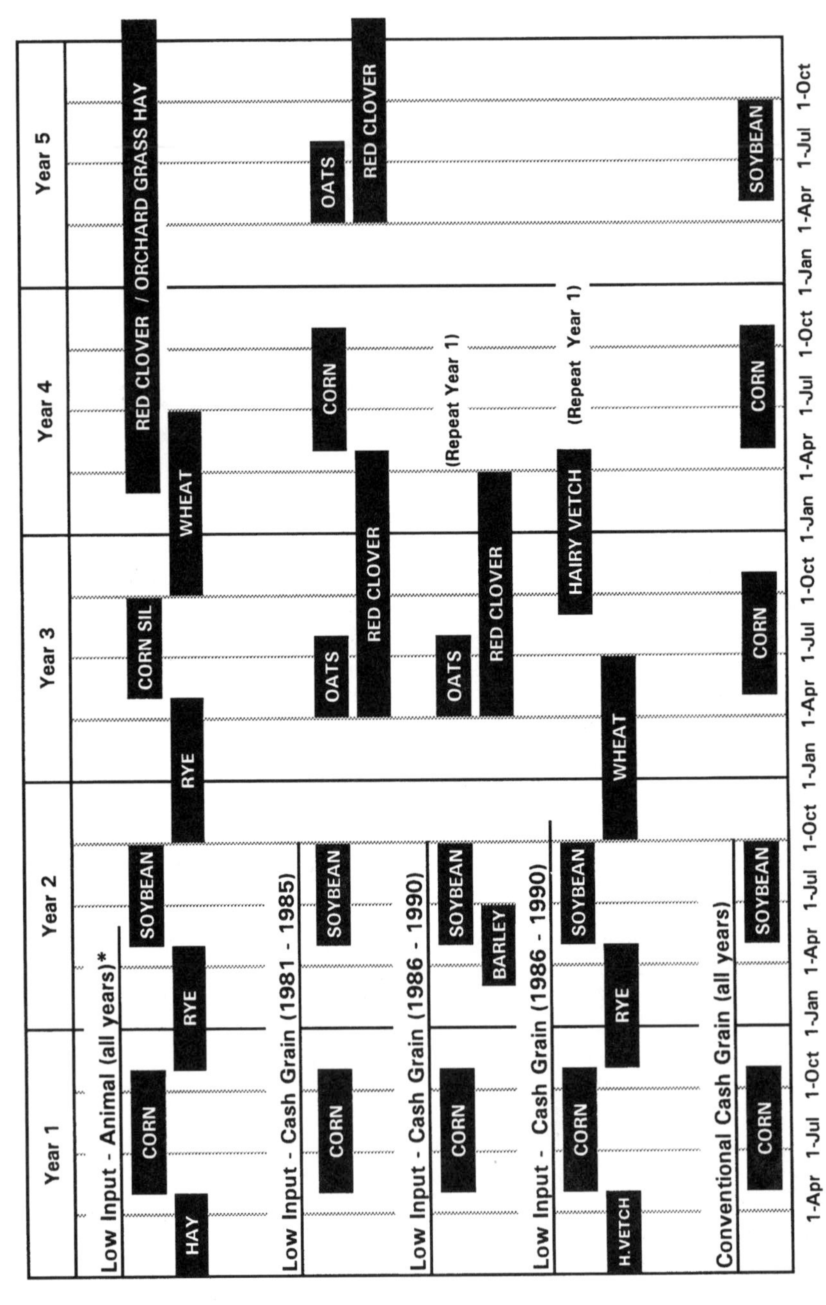

Figure 1 Farming systems trial rotations.

Total above-ground biomass production over the duration of the experiment was about 10 to 15% less in the LIP-CG system than in the other systems (Figure 2). This result, however, was not necessarily due to an inherently lower productivity level in the LIP-CG, but rather the particular sequence of crops grown in this system.

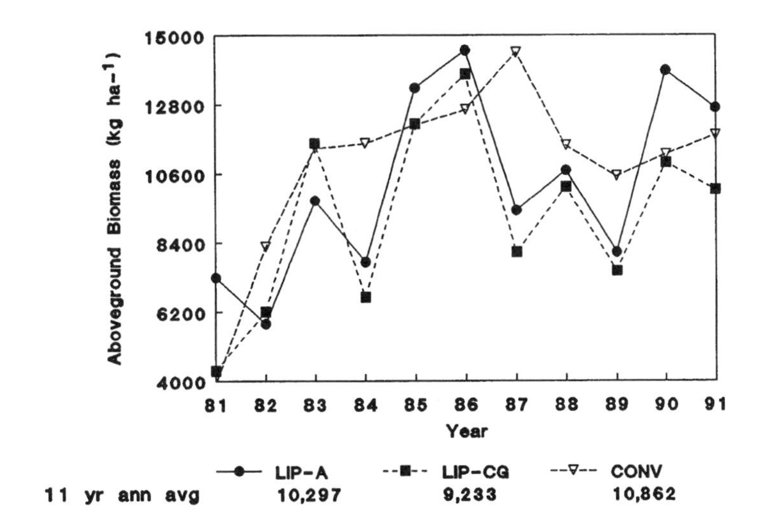

Figure 2 Annual above-ground biomass production.

III. CHANGES IN SOIL PROPERTIES

A. TOTAL C AND N

During the course of the experiment, there were slight but significant changes in the amount of SOM (C and N). Initially (1981) total soil C in the upper 10 cm was equal across systems, yet by 1990 total C was greatest in LIP-CG, intermediate in LIP-A, and lowest in the conventionally managed (CONV) system (Table 2). Total soil N in 1981 was greatest in CONV, intermediate in LIP-A, and lowest in LIP-CG, but 10 years later soil N was greatest in LIP-A, intermediate in LIP-CG, and lowest in CONV. Another soil sampling was taken in 1991 of the entire plow layer (approximately 0 to 25 cm) using the same analytical procedure (i.e., dry combustion with Carlo Erba C/N analyzer). As expected, C and N levels were lower for the whole Ap layer samples than for the 0- to 10-cm samples, yet the same trends were found, i.e., the low input systems had greater concentrations of C and N than the conventional after 11 years of different cropping history.

Table 2 Total Soil C and N Contents (g kg^{-1}) of FST Soils 1981 and 1990 (0 to 25 cm)

Cropping System	Carbon		Nitrogen		Carbon	Nitrogen
	1981[a]	1990[b]	1981	1990	1991[c]	1991
LIP-A	22.7a[d]	23.4ab	3.3ab	3.5a	21.4a	3.1a
LIP-CG	23.6a	24.5a	2.9b	3.4ab	22.3a	3.1a
CONV	22.3a	21.3b	3.4a	3.3b	19.8b	2.9b

[a] Analyzed archived soil from October, 1981 (0 to 10 cm).[8]
[b] Soil collected in October, 1990 (0 to 10 cm).[8]
[c] Soil collected in June and July, 1991 (0 to 25 cm).
[d] Within columns, different letters following means indicate values significantly different at 0.05 level.

Management-induced changes in soil C and N contents were difficult to document in these soils because much of the observed difference in SOM levels was due to the heterogeneity of texture and drainage characteristics. In addition, any newly accumulated SOM (after 1981) represented only a small fraction (≤10%) of the total organic matter pool. The value of this newly added organic matter must not be overlooked, however, because it did contribute to improving soil physical and biological properties (see below). Total C and N levels indicate gross changes of SOM, but provide little information about soil quality because they do not indicate turnover of organic matter or distinguish among the various active and passive "pools" of SOM that control nutrient cycling dynamics.

B. MICROBIOLOGICAL PROPERTIES

While changes in total soil C and N occurred very slowly among the cropping systems, changes in microbial biomass and activity were greater. As early as 1982, microbial biomass and activity levels were greater in the LIP-A soils, and to a lesser degree the LIP-CG soil, than in the CONV soil.[9] This trend became more pronounced in later years. Results suggested that the crop being grown at the time of sampling and the most recent tillage events had the greatest short-term effects on the soil biomass.

In 1985, after the first 5 years of the study, seasonal trends of soil microbial biomass and potentially mineralizable N were measured.[9] Only the corn-producing plots were studied. The corn entry point in the LIP-CG farming system included a hairy vetch cover crop at the time of the April sampling. This was plowed down 1 week prior to the May sampling. At the two early dates, the LIP-CG soil had much greater quantities of microbial biomass than the other systems. It had the equivalent of an additional 33 kg N ha^{-1} in soil biomass, assuming a microbial biomass C:N ratio of 8.5. At the later sampling dates, microbial biomass levels were equal in the low-input systems and only slightly lower in the CONV system soil.

A study conducted in 1988 suggested differences in soil biological activity similar to those found in the 1985 study.[10] Following the 1988 corn harvest, net N mineralization and microbial biomass C and N (based on the chloroform fumigation-incubation method), in 0 to 15 cm soil, were measured throughout a 200-day incubation of the soils at 25°C brought to 55% of water-holding capacity. The amount of N mineralized (N_m) was measured in KCl extracts and then the N mineralization potential (N_o) and N mineralization rate constant (k) were calculated ($N_m = N_o(1\text{-}e^{-kt})$: t = time) (Figure 3). Soil from the LIP-A system mineralized the greatest amount of N. Therefore, the microbes in this soil under optimal incubation conditions had a higher metabolic rate (activity) or they were more numerous or both conditions existed.[10] The cause of this microbial response may be that the LIP-A soils had a larger labile (active) organic N pool. Nitrogen mineralization rates were slightly greater in LIP-CG cropping system soils than in CONV system soils. The corresponding microbial biomass data (not shown) indicated that larger biomass pools existed in the LIP-A and LIP-CG soils than in the CONV soil.

In 1990, average seasonal soil respiration rates, available soil N, and N mineralization rates (measured under 2-week aerobic laboratory incubations) again indicated that the size and rate of biological activity was greatest in the LIP-A soil, intermediate in the LIP-CG soil, and least in the CONV cropping system soil.[8]

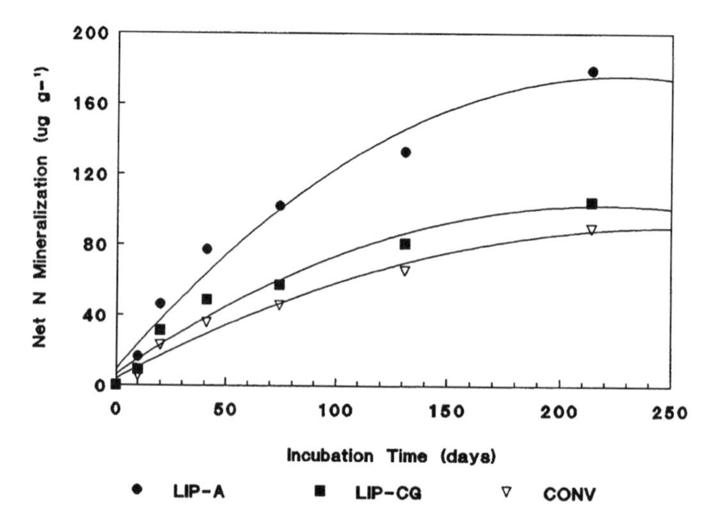

Figure 3 Mineralized nitrogen during long-term soil incubation.

C. PHYSICAL PROPERTIES

Differences in soil physical properties were also noted during the course of the FST experiment. More rapid water infiltration rates were recorded in the soils from both low-input systems relative to the infiltration rates in the conventionally managed soil. Water infiltration measurements were taken in corn plots just prior to harvest in 1990 and 1991 for in-row and between-row positions in each farming system. A 15-cm diameter steel ring was sunk 8 cm into the soil. A plexiglass cylinder, 112 cm high and 10 cm in diameter, was filled with water, placed on top of the steel ring, and the rate at which the water infiltrated the soil was then measured over a 10-minute period.[7] Results are reported as millimeters of water (within the plexiglass cylinder) infiltrated per unit time (minutes). For the in-row positions, in particular, the soil in the low-input plots absorbed water at a faster rate than did the CONV soils during both years (Figure 4). The large amount of soil "hilled-up" from cultivation was largely the cause; however, the addition of cover crops in the low-input systems was probably also a factor. For between-row positions, where all the wheel traffic was concentrated, the infiltration rates for all systems were much slower (data not shown). At these positions, however, the CONV soils again had the slowest infiltration rates in both years. Environmental factors also influenced infiltration characteristics since overall infiltration rates, regardless of system, were much more rapid in 1990 when soil moisture contents were higher.

The management practices influenced soil aggregation; yet the crop and temporal effects were also important. In 1985, the amount of water-stable aggregates was determined in the corn-producing plots of all farming systems.[11] Measurements were collected monthly, from May to November, of eight

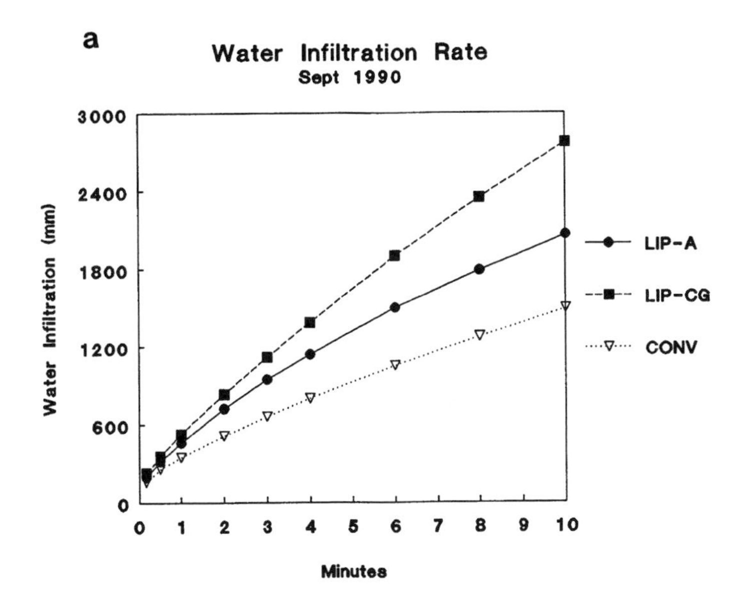

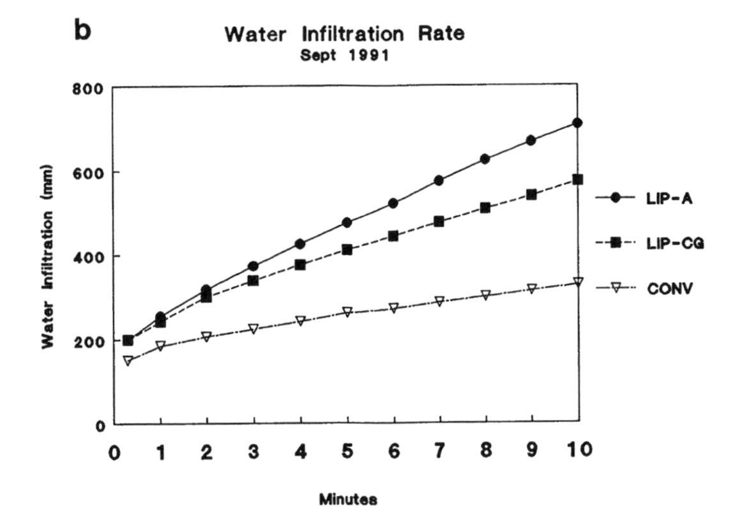

Figure 4 (a) Water infiltration rate, September 1990 and (b) water infiltration rate, September 1991.

aggregate size classes ranging from <0.15 to >8 mm. An early summer plowdown of hairy vetch in the LIP-CG led to the greatest amount of macroaggregation (>0.25 cm). Throughout the season, aggregate stability fluctuated the least in the LIP-CG and the most in CONV. Despite the differences in fluctuation patterns, all the farming systems had identical water-stable aggregate arrays (proportions of each size class) at the end of the growing season. This suggests that the particular crop being grown (in this case corn) had a stronger overall influence on water-stable aggregates than the additional input of organic residues (manure, hairy vetch).

Similar results were observed in 1990.[12] Water-stable aggregates were characterized in June and October of 1990 in all plots (Figure 5). There were no differences among the farming systems in the aggregate sizes smaller than 2 mm (data not shown). When rotational entry points were combined, results indicated that the low input systems had a greater amount of larger aggregates (i.e., >2 mm) than the CONV soils at both sampling dates.

Separating out the effect of crop-type indicated that the crops grown only in the low input plots (red clover hay, interseeded red clover and oats, and relay-cropped soybeans into spring barley) fostered the greatest amount of soil aggregation. These plots may have contained a greater density of roots than either the corn or the soybean monoculture plots.

The corn-producing plots had intermediate levels of aggregation. The results are similar to the results from the 1985 corn plots. In June the ranking was LIP-CG > LIP-A > CONV, while in October they all had similar aggregation.

Within the same cropping system, soil aggregation in the soybean-producing plots was sometimes greater (LIP-A in June), sometimes equal (LIP-A in October, CONV in June), and sometimes less (CONV in October) than in the corn-producing plots. Aggregation, however, was greater at both dates in the soybean-producing plots in the LIP-A system than in the CONV system. Therefore, while the crop present certainly affected aggregation, the influence of other management practices included in the low-input cropping systems, such as additional residue inputs (i.e., manure or cover crops) and more diverse crop rotations, must also be considered.

IV. THE IMPACT OF SPECIFIC MANAGEMENT PRACTICES

The observed changes in soil physical, chemical, and biological properties, some of which are summarized above, are primarily a direct or indirect result of the management practices represented in the FST experiment. While it is impossible to determine precisely cause and effect relationships between specific management practices and corresponding soil changes, the net effect of several management practices is clearly evident.

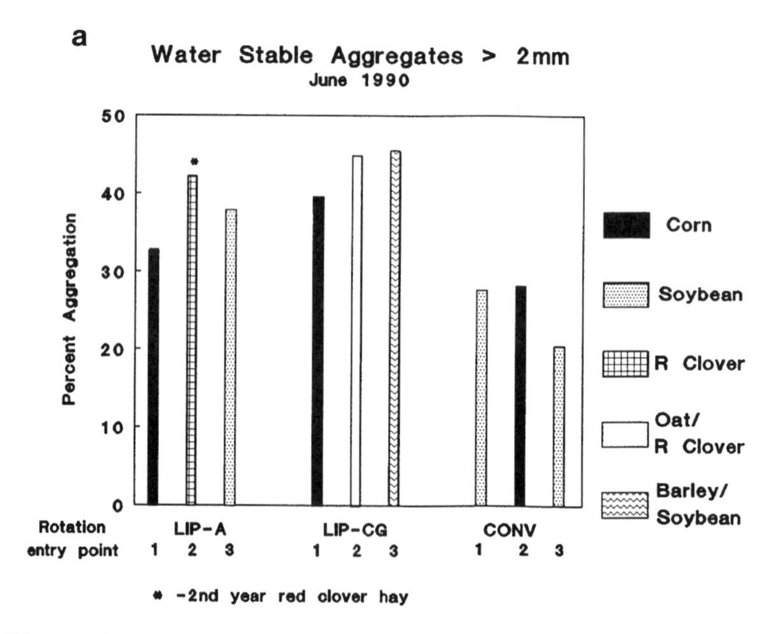

Figure 5 (a) Water stable aggregates >2 mm, June 1990 and (b) water stable aggregates >2 mm, October 1990.

A. FERTILIZATION AND CROP RESIDUE INPUTS

The kinds and amounts of soil amendments applied and plant residues returned impact SOM and related characteristics by their effect on microorganisms, on the supply of plant-available nutrients, and on soil structure. The specific amendment regimes of the three farming systems were as follows:

- All three cropping systems received potassium sulfate in 1989. P was not applied except in starter fertilizer in the CONV. Dolomitic limestone was applied to the CONV in 1982 and high calcium limestone was added to the LIP-A and LIP-CG in 1989.
- The LIP-A system was amended with partially composted cattle manure applied immediately before plowing the ground for corn (2 out of 5 years). Additional N was derived from the unharvested portion of leguminous hay stands incorporated prior to corn production. Both of these organic amendments provided energy for soil fauna; the manure contributed additional P, S, cations, and micronutrients.
- The LIP-CG system included a leguminous green manure cover crop of red clover or hairy vetch once every 3 years, which was plowed under at least 1 week before planting corn. Less inorganic N was released during the growing season from the cover crop than from the cattle manure applied to the LIP-A soil. Nitrogen additions from the legumes, however, supported grain yields of corn comparable to the LIP-A corn throughout the experiment. Even though the legumes did not add other nutrients, the resulting increase in root activity may have increased P and K availability to plants. Soluble root exudates have been found to influence potassium supply and rhizosphere microorganisms.[13]

Changes in soil C and N levels cannot be explained by simply accounting for the quantity of amendments applied and residues returned to the soil. During the 11-year period (1981 to 1991), an estimate of the net N balance between inputs and outputs indicates that the LIP-CG had an N deficit (output exceeded input) of about 50 kg N ha^{-1} y^{-1}, while the LIP-A and CONV systems had more N applied than was removed (Table 3). Therefore, assuming a good correlation exists between SOM and total N levels, the LIP-A, and to a lesser extent the CONV, soils should accumulate organic matter or at least maintain initial levels (excluding losses via leaching or erosion). In contrast, the LIP-CG SOM level should decrease. In fact, total soil N levels did increase slightly in the LIP-A, yet they declined in the CONV soil. The LIP-CG soil, despite the N deficit based on the input/output calculations, actually accumulated C and N (Table 2). An accurate account of changes in total C and N, however, must include sampling to a depth below the rooting zone (i.e., $\cong$ 1 m) and these data are currently not available.

The CONV system returned the most plant residue to the soil as unharvested crop stubble and weeds (Figures 6 and 7). This was primarily because corn was grown more often in the CONV rotation (3 out of 5 years). Corn produced more biomass than any of the other crops in the study, and in the CONV

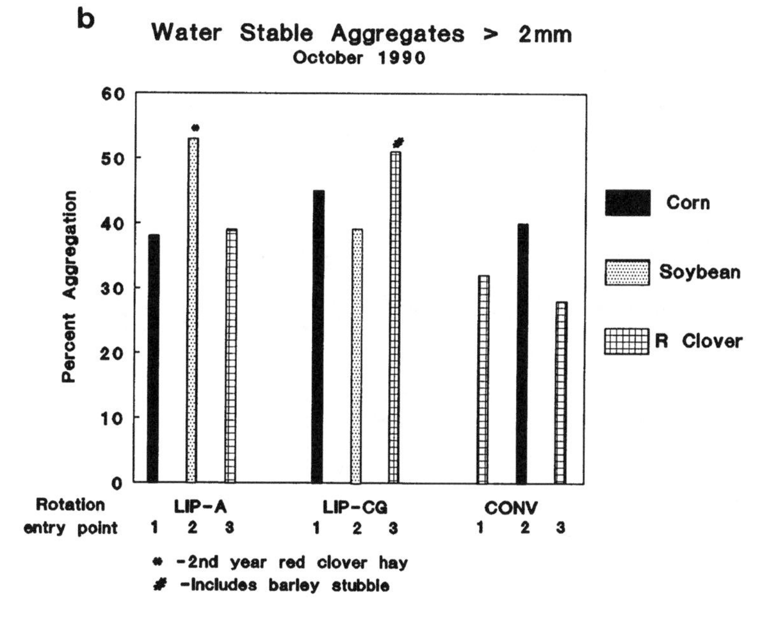

Figure 5 (continued)

Table 3 Nitrogen Budget of FST Cropping Systems 1981 to 1991

Cropping System	N Source/Sink	Nitrogen (kg ha^{-1})[a] Input	Output	Net
LIP-A	Manure	1106	0	1106
	Hay	807	646	161
	Soybean	460	460	0
	Grain corn	0	276	–276
	Silage corn	0	314	–314
	Small grain	0	262	–262
	Total	2373	1958	415
LIP-CG	Green manure[b]	530	144	386
	Soybean	524	524	0
	Grain corn	0	458	–458
	Small grain	0	443	–443
	Total	1054	1569	–515
CONV	Fertilizer	984	0	984
	Soybean	932	932	0
	Grain corn	0	871	–871
	Total	1916	1803	113

Note: The following factors were excluded from the analysis:

a. Losses of N from leaching, denitrification, volatilization.
b. Losses of N from soil erosion.
c. Additions of N from precipitation.
d. Additions of N from asymbiotic fixation.
e. Additions of N mineralized from SOM.

The following assumptions were made based on both literature values and actual measurements:

a. The amount of N in the above-ground biomass of legume hays and green manures is equivalent to the amount of N fixed by these crops.
b. First- and second-year hay (LIP-A) and green manures (LIP-CG) contain 36 g kg^{-1} N. Third-year plow-down hay (LIP-A) contains 30 g kg^{-1}.
c. A soybean crop results in neither a net gain or loss of N (i.e., the amount of N removed at harvest equals the amount of N fixed by these soybeans).
d. The N content of harvested crops (dry-weight basis) is as follows: grain corn 20 g kg^{-1}; silage corn 10 g kg^{-1}; oats and barley 25 g kg^{-1}; wheat 30 g kg^{-1}; soybean 80 g kg^{-1}.

[a] Eleven-year total (averaging the three rotational entry points each year).
[b] Harvested for hay in 1981 to 1985 and 1987.

system virtually all of the stover remained on the field after grain harvest. Despite this, the total soil C contents in this system decreased (Table 2).

The LIP-A returned to the soil a similar amount of organic residues as the CONV but only half was plant material while the rest was animal manure (Figure 7).

The LIP-CG returned the least amount of organic residues (excluding below-ground residues) since no animal manure was applied and corn was grown only once every 3 years. Green manures, however, were an important residue input in this system.

During the initial years the CONV system had the highest productivity. Even after the transition period, however, this system continued to produce 10 to 20% more biomass annually and return to the soil 10 to 20% more plant residues than the LIP-CG system.

The estimates of N and residue budgets are crude since measurements were based on above-ground inputs and outputs only. Variations among the farming systems of the factors not included in the analysis, such as root residues and gaseous and leaching losses, likely had a significant influence on SOM levels. The amount of N fixed by the red clover, hairy vetch, and soybeans may have been underestimated, which would account for the lower values reported in the LIP-CG farming system. Also, estimates of N fixed by the red clover and hairy vetch were based on spring biomass samples with no regard to N fixed the previous fall. This N fixed in the fall may have entered the soil N pool during the winter rather than having been recycled back into the plant in the spring.

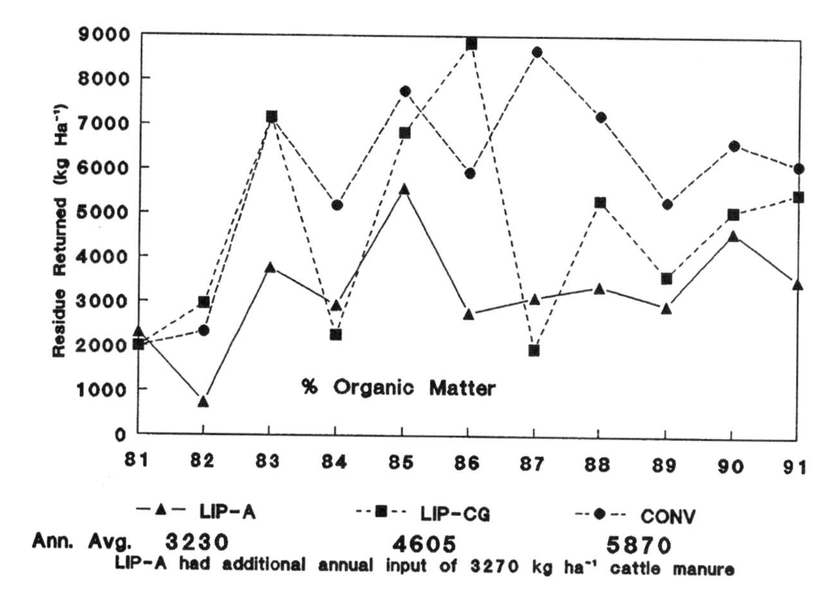

Figure 6 Annual above-ground residue returns.

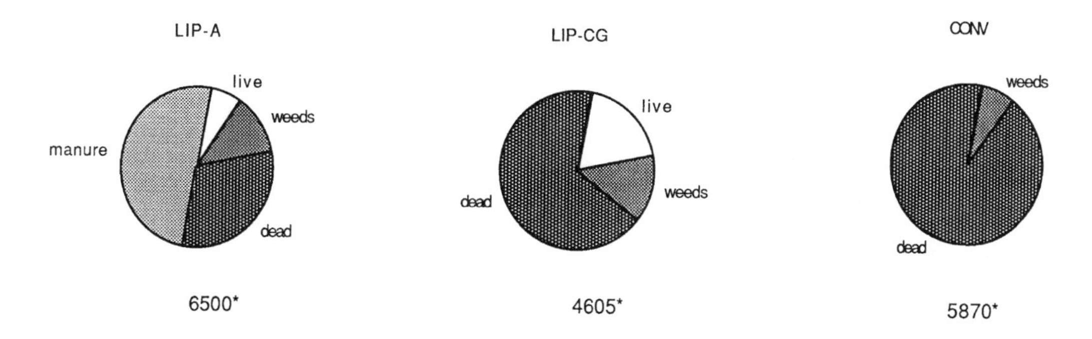

Figure 7 Proportion of organic residues returned to soil as live crop, dead crop, weeds, or cattle manure from 1981 to 1991.

The amount of carbon translocated to the roots of each plant species in a crop rotation and the structure of these root systems could be the most important factor affecting SOM dynamics. Forage legume and grass (besides corn) roots, for example, were important components of the low-input systems but not of the conventional system. The relative ranking of above-ground biomass produced among the systems (i.e., CONV > LIP-A > LIP-CG) might be expected to be similar for the below-ground portion. The allocation of plant biomass, however, to above- and below-ground structures can vary. The shoot to root ratios can be affected by the fertility and moisture status of plants.[14] The fine-root biomass of corn plants in this study, for example, was greater in the low-input farming systems than in the conventional system.[15]

The diversity of residue types incorporated could also affect SOM. Four types of residue were identified based on their chemical composition and addition characteristics (Figure 7). Residual hay and legume green manures (red clover or hairy vetch) were included in the "live plant residue" category. These materials generally had a C:N ratio between 11 and 13 when they were plowed into the soil. Corn and soybean stover made up most of the "dead plant residue" and had a C:N ratio of 40 to 60 when they were incorporated. Some of the N from these residues may have moved into the soil by leaching, since the residues remained on the soil surface for 5 to 6 months during the winter before being plowed under. The weeds had a C:N ratio of about 30 when they were incorporated before late-summer-planted hairy vetch or fall-planted winter wheat. Weed C:N ratios were probably at least 50 when plowed under in the spring. The cattle manure came from a confinement pen, was only partially composted, and included a substantial amount of straw bedding. It contained about 50% C and 3% N.

In the LIP-A, cattle manure was the major input, while dead residues were one half and one third of those in the LIP-CG and CONV, respectively (Figure 7). In the CONV, dead residues made up most of the returned material. Like the CONV, the LIP-CG had no cattle manure inputs, but, unlike the CONV, live plant residues constituted nearly 20% of the total returned. The presence of these live residues could be a key factor for the LIP-CG system's higher levels of soil C and N.

The cropping system crop rotations governed the length of time over a growing season that photosynthetically active plants were present. The presence of hay, green manure, and small grain crops all lengthened the active cropping season. As this time period was increased, so too was the time in which carbon was added to the soil environment. The low-input systems maintained live plant cover (and hence live roots) for a considerably longer period than the CONV system (Figure 1, Table 1). The LIP-A hay crops and LIP-CG legume green manures began growing in late March and did not cease growing until sometime in November. Winter wheat, overseeded with red clover, also provided winter cover. In contrast, the CONV system raised only corn and soybeans, which emerged in mid- to late May and reached physiological maturity by mid- to late September. During the winter months the CONV soil was devoid of living plants.

B. TILLAGE

Primary and secondary tillage (plowing, disking, harrowing, etc.) and cultivation practices can have a major impact on SOM since these kinds of disturbances can increase SOM decomposition and release of nutrients. All three cropping systems used the moldboard plow; yet, there were subtle differences in the timing and frequency of tillage events.

Over its 5-year crop rotation cycle, the LIP-A was plowed in April three times (for corn and soybean production), once in October (prior to planting wheat), and not tilled the year the field was in hay. Fall plowing for wheat increased potential for N leaching by stimulating mineralization at a time when young wheat plants would not take up appreciable amounts of soluble N. During hay cropping the soil was not tilled, so this was the most N- and C-conserving part of the rotation.

The LIP-CG farming system was plowed annually. Over its 3-year crop rotation cycle, the soil was tilled in May for corn and soybean production and in October before wheat was sown. It was also plowed in August prior to hairy vetch establishment. During the corn-producing entry point, the LIP-CG soil was plowed later than the other two systems. As a result, soil temperatures were warmer and early corn growth was more rapid, which may have reduced N leaching losses.

The CONV treatment soil was plowed in April every year for corn and soybean production. The soil in this system was never plowed in the fall. Also, there were no winter small grains or cover crops grown to prevent leaching of nutrients during the winter.

C. WEED CONTROL

The low-input cropping systems used a variety of cultural and mechanical techniques for controlling weeds. These included a diverse crop rotation, cover-cropping, relay-cropping, and cultivation. The conventional system relied principally upon herbicides. These divergent sets of practices resulted in different amounts of soil disturbance and different levels of weed growth.

A greater amount of soil disturbance occurred in the low-input systems. The soil was rotary hoed and cultivated twice during corn and soybean production. Furthermore, the second cultivation was rather aggressive with substantial mounding of soil around plants.

The CONV system included only one light soil cultivation in 3 out of 5 years, when corn was sidedressed with N fertilizer. There was no apparent soil physical disturbance (other than a little compaction from the wheels of the spray rig) when herbicides were applied. Weed control was better in the herbicide-amended CONV system than in the low-input systems.

The LIP-A system tended to be weedier than the LIP-CG system. This was partly due to the introduction of annual weed seeds in the cattle manure and to the regrowth of perennial weeds established in hay stands.

Maintaining a low level of weed pressure with herbicides, as in the CONV system, is generally considered desirable, but there are tradeoffs. Curbing weed growth assures the farmer that weeds will not compete with the crop plants for nutrients and water. In addition to the high cost of herbicides, however, the near elimination of weeds reduces substrate (energy) for soil fauna, which could potentially slow down biological activity and nutrient cycling. Further, the roots of weeds can improve soil structure and help reduce erosion.

V. SUMMARY AND CONCLUSIONS

An integral component of a sustainable agroecosystem is the presence of a continual supply of SOM to provide substrate for an active, healthy soil biological community, an environment for vigorous plant-root development which resists soil erosion, and adequate nutrients for plant growth. Some of the factors controlling the amount and turnover time of SOM, such as parent material and soil texture, are beyond the farmer's control. Other factors, however, including soil moisture, temperature, and fertility can be influenced by specific management practices such as the kinds and amounts of organic residue inputs.

The three cropping systems compared in this study were each defined by a specific set of cultural techniques including crop rotation, weed control, fertilization, and tillage practices. These techniques influenced the soil environment and hence the quantity, quality, and dynamics of SOM.

The low-input systems can be distinguished from the conventionally managed cash grain system in that they include

1. A greater diversity of crops
2. More N-rich residues returned to soil
3. Less above-ground plant residue returned to soil
4. A longer period of active plant growth
5. Greater organic matter inputs from roots
6. The elimination of pesticides and ammonia-based fertilizers
7. Cattle manure as a major input (animal system only)

Within 5 years after the experiment began, several soil biological and physical parameters, including microbial biomass, N-supplying potential, water infiltration rates, and aggregate stability, were all greater in the low-input systems than in the conventional, and these differences increased with time. Total C and N levels, however, were only slightly higher for the low-input systems after 11 years and some of this observed difference may be attributed to inherent soil textural variability.[11] Recording changes in total soil C and N may be useful for assessing the stability of a farming system after 50 to 100 years, but other physical and biological tests are needed for characterizing short term (1 to 10 years) changes in the cycling dynamics of SOM.

The low-input systems led to SOM accumulation, while the CONV soil degraded slightly. This occurred despite lower apparent plant inputs in the low-input systems. The LIP-A system accumulated the most N and was the most biologically active.[8,10] In contrast, the LIP-CG soil accumulated the most total C, suggesting that this system was the most resource-conserving of the three considered.

The increased infiltration rates in both low-input systems compared to the conventional made them less prone to erosion and water loss from surface runoff (which does occur at this site after heavy rains) and may have increased plant-available rainfall. Furthermore, the more aggregated nature of the low-input soils creates a more ideal environment for healthy plant roots and soil fauna by improving aeration and water-holding capacity.[16]

The conventionally grown crops sustained reasonably high yields throughout the 11 years, despite declining soil quality characteristics relative to low-input managed soils. Therefore, the quality and quantity of SOM present in the CONV soil has not yet reached a low enough level to cause a decline in crop yields and these soils may support good plant growth for a long time.

Results from the long-term experiments at Rothamsted and Woburn in England indicated that, if organic matter levels were increased after a few years of green or animal manure additions to soils previously cropped continuously for over 100 years, yields of several crops improved.[17] Much of the SOM-derived benefits were attributed to non-N effects such as increased plant-available water. The significant finding was that, while a crop may perform adequately for many years in soils for which SOM is declining, yields would eventually decrease without *high-quality* organic inputs (i.e., green or animal manures rather than only C-rich crop residues). Therefore, detecting a deterioration of soil structure or a decrease in microbial biomass could be an early warning of poor crop response in the future.

The slight, but observable, increases of SOM in the low-input systems is an indication that, if present practices are continued, SOM will eventually attain a new equilibrium at a higher level in accordance with climatic and soil conditions. Whether such increases result in correspondingly higher crop yields is not known, but it is likely that greater yield stability from year to year will be realized. In like manner, the slight declines of SOM in the conventionally managed soils may not lead to decreased crop yields (at least for some time), but yield stability will probably decline.

Two basic questions remain concerning the dynamics of SOM. What time period is required for SOM levels to change and what is the relative importance of each particular management practice in affecting this change? Addressing these questions will improve our ability to manipulate SOM turnover to enhance soil fertility and sustain the physical soil resource.

REFERENCES

1. U.S. Department of Agriculture, Report and Recommendations on Organic Farming, U.S. Government Printing Office, Washington, DC, 1980.
2. National Research Council, *Alternative Agriculture*, National Academy Press, Washington, DC, 1989.
3. Schlesinger, W.H., Changes in soil carbon storage and associated properties with disturbance and recovery, in *The Changing Carbon Cycle: A Global Analysis*, Trabalka, J.R. and Reichle, D.E., Eds., Springer-Verlag, New York, 1986, 194.
4. Oades, J.M., The retention of organic matter in soils, *Biogeochemistry*, 5, 35, 1988.
5. Bolin, B., Changes of land biota and their importance for the carbon cycle, *Science*, 196, 613, 1977.
6. Liebhardt, W.C., R.W. Andrews, M.N. Culik, R.R. Harwood, R.R. Janke, J.K. Radke, and S.L. Rieger-Schwartz, Crop production during conversion from conventional to low-input methods, *Agron. J.*, 81(2), 150, 1989.
7. Peters, S., R. Janke, and M. Bohlke, Rodale's Farming Systems Trial 1986–1990, Rodale Institute Research Report, 1992.
8. Wander, M.M., S.J. Traina, B.R. Stinner, and S.E. Peters, The effects of organic and conventional management on biologically active soil organic matter pools, *Soil Sci. Soc. Am. J.*, 58, 1130, 1994.
9. Doran, J.W., D.G. Fraser, M.N. Culik, and W.C. Liebhardt, Influence of alternative and conventional agricultural management on soil microbial processes and nitrogen availability, *Am. J. Altern. Agric.*, 2(3), 99, 1987.
10. Harris, G.H., Nitrogen Cycling in Animal-, Legume, and Fertilizer-Based Cropping Systems, Ph.D. dissertation, Michigan State University, East Lansing, 1993.
11. Werner, M.R., Impact of Conversion to Organic Agricultural Practices on Soil Invertebrate Ecosystems, Ph.D. dissertation, SUNY College of Environmental Science and Forestry, Syracuse, 1988.
12. Friedman, D.B., Carbon, Nitrogen and Aggregation Dynamics in Low-Input and Reduced Tillage Cropping Systems, M.S. thesis, Cornell University, Ithaca, NY, 1993.
13. Kraffczky, I., G. Trolldenier, and H. Beringer, Soluble roots exudates of maize: influence of potassium supply and rhizosphere microorganisms, *Soil Biol. and Biochem.*, 4, 315, 1984.
14. Klepper, B., Root-shoot relationships, in *Plant Roots, The Hidden Half*, Waisel, A. and Kafkafi, U., Eds., Marcel Dekker, New York, 1991.
15. Pallant, E., D.L. Lansky, J.B. Rio, L.D. Jacobs, G.E. Schuler, W.G. Whimpenny, and B.R. Churchwell, Corn Root Growth under Low Input and Conventional Farming Systems, in preparation.
16. Elliot, E.T. and D.C. Coleman, Let the soil work for us, *Ecol. Bull.*, 39, 23, 1988.
17. Johnston, A.E., Soil fertility and soil organic matter, in *Advances in Soil Organic Matter Research*, Wilson, W.S., Ed., Royal Society of Chemistry, Cambridge, 1991, 299.

Chapter 14

Crop Rotation, Manure, and Agricultural Chemical Effects on Dryland Crop Yield and SOM over 16 Years in Eastern Nebraska

Gary W. Lesoing and John W. Doran

CONTENTS

I. BACKGROUND

In the mid-1970s, emphasis was placed on increasing corn production in the United States. Marginal land, much of which was low in fertility and highly erodible, was put under cultivation. A continuous corn monoculture was common throughout the Midwest. Farmers applied large amounts of synthetic fertilizer to increase yields and pesticides for weed and insect control. With these intensive farming practices, there was a potential for problems of soil erosion, increased weed and insect pests, high external input costs, decreased economic productivity, and groundwater contamination to develop and create problems for farmers in future years.

Nebraska, with its large number of feedlots and beef-processing plants, has the potential to utilize feedlot and paunch manure as alternative fertility sources. In 1975, Dr. Warren Sahs and a committee of researchers, extension personnel, and farmers planned and initiated an experiment at the University of Nebraska Agricultural Research and Development Center (ARDC) near Mead, NB. Research on the use of manure and crop rotations was common throughout the U.S. prior to the development of synthetic fertilizer. However, in 1975 information was limited on the use of alternative farming systems, particularly with improved crop varieties and other modern production techniques. Research was needed to compare conventional and alternative farming systems because of the uncertain future availability and economics of synthetic fertilizer and pesticide use.

The objectives of this study were to (1) evaluate the effectiveness of using crop rotations and manure as substitutes for fertilizer and (2) evaluate the effect of crop rotations that include legumes on soil characteristics and crop yields.

II. SITE DESCRIPTION AND EXPERIMENTAL APPROACH

This experiment was initiated in 1975 at the University of Nebraska ARDC, Latitude 41:10:00, longitude 096:25:00. The experiment was conducted on a Sharpsburg silty clay loam (fine, montmorillonitic, mesic, typic Argiudoll) under dryland conditions. The site is gently sloping (0 to 5%) with a previous cropping history of alfalfa the 4 years prior to this experiment initiation. The long-term (30 year) average precipitation is 737 mm/y, 65% of which occurs between May and September. The research site received an average of 680 mm of precipitation per year during the 16 years of study. Mean annual temperature is 10°C with a mean air temperature between May and September of 21°C. Climatic data, precipitation, and air temperature for the 16-year period since 1975 are shown in Appendix Table 5.

The experiment was designed to compare (1) chemical-free farming practices that use crop rotation and manure, (2) crop rotation practices that use chemicals, and (3) continuous corn with a chemical-based cropping system. These plots will be hereafter referred to as the organic/conventional research plots. Plots were arranged in a randomized complete block design, with 4 replications and 13 plots (12.2 × 38.1 m) per replicate. A 4-year crop rotation of corn-soybean-corn-oat/clover was planted on 12 plots of each replication. Each crop in the rotation sequence was grown every year and was managed under one of three systems: (1) rotation-feedlot manure, designated R/Man; (2) rotation-synthetic fertilizer, designated R/F; and (3) rotation-synthetic fertilizer and herbicide, designated R/FH treatment. The fourth treatment was continuous corn which received fertilizer, herbicide, and insecticide and is designated as the CC/FHI system.

A disk-plant tillage system was the primary method used for all crops in the rotation. The exception was the use of a Noble Blade to undercut clover in the spring of the year prior to planting corn. Nebraska Certified Hybrid 611 corn, a full-season drought-tolerant hybrid, was planted in mid-May in 0.76-m rows at a population of 43,500 to 47,500 plants per hectare, based on soil moisture and expected growing season climatic conditions. Yield goals for corn were 5.41 Mg ha^{-1} from 1976 to 1982, 6.36 Mg ha^{-1} from 1983 to 1986, and 7.0 Mg ha^{-1} from 1987 to 1991.

From 1976 through 1984, Woodworth soybeans (Group III maturity category) were planted in mid- to late May in 0.76-m rows at a rate of 68 kg ha^{-1}. In 1985 Pella soybeans (Group III maturity) were substituted; improved Pella 86 has been used since 1987.

Oats/clover were drilled as early as possible in late winter or early spring (March 15). Lang oats were used the first 8 years of the experiment. Ogle oats, a higher-yielding improved variety was substituted the last 8 years. Oats were seeded at a rate of 91 kg ha^{-1} along with 9.1 kg ha^{-1} of sweet clover and 4.5 kg ha^{-1} of red clover.

Herbicides were used on the R/FH treatment for soybeans and corn on the CC/FHI treatment. For soybeans, alachlor and metribuzin were broadcast after planting. In corn, alachlor and cyanazine were broadcast pre-emergence. In 1988 to 1991, atrazine was added to the CC/FHI treatment to improve weed control. The insecticide Terbufos was banded in the corn rows for root worm control in the CC/FHI treatment. Corn and soybeans were also rotary hoed and cultivated for weed control, with this being the primary method of weed control for the R/Man and R/F treatments. All soybean plots were hand rogued, primarily to remove velvetleaf. Table 1 lists the tillage operations for each crop on the organic/conventional research plots. Sahs and Lesoing[1] and Andrews et al.[2] detail the management schedule, fertilizer, and manure application and crop yields.

III. RESULTS AND DISCUSSION

Corn grain yields as influenced by organic/conventional management systems are reported in Table 2. Average corn grain yields were low the first 4 years of the experiment, particularly for the R/Man management system. During this period the rotation effects were becoming established and effective weed control practices were developed. Yields of the other management systems, particularly the CC/FHI treatment, were influenced by the previous cropping history of alfalfa and periodic occurrence of drought stress. The first 4 years of the rotation were included in the subsequent yield averages because it is important to include the transition period for establishment of crop rotations to determine their overall impact on crop yields and the cost of establishment. The last 12 years of the experiment were

Table 1 Tillage Operations on the Organic/Conventional Research Plots at Mead, NE

Crop	Tillage Operation[a]
Continuous corn	Disk twice to 10–15 cm in spring, harrow
Corn following soybeans	Disk twice to 10–15 cm in spring, harrow
Corn following oats/clover[b]	Undercut with Noble blade at about 20 cm in spring, or shallow plow, disk, harrow
Oats/clover[c]	Fall disking, light spring disking, harrow
Soybeans	Disk twice spring, harrow

[a] A Noble blade was used to undercut clover in early spring if plots were dry enough. In 1987 to 1989, oats/clover plots were shallow moldboard plowed (to 15 cm).
[b] Plots planted in oats/clover the following spring were fall disked, weather permitting, to expedite early spring tillage and oat seeding.
[c] Plots usually were disked twice to 10 to 15 cm for weed control and incorporation of fertilizer or manure. All corn and soybean plots were rotary hoed (to 5 cm) two times and cultivated (to 10 cm) at least once for weed control. Treatments without herbicide (R/Man, R/F) were cultivated three times.

characterized by intermittent drought but excellent growing conditions in some years. The 16-year average indicates the yield advantage for the rotational management systems, as corn yields were increased 11% by the R/Man, 21% by R/FH, and 15% by R/F compared with the CC/FHI. Yield results from individual years (Table 3) indicate that corn yields for the R/Man systems are equal to or higher than the other rotational management systems when timely weed control was achieved, particularly in years of drought stress. These data also show that the yield stability is improved for corn in rotation compared with continuous corn. An illustration of the variability of corn yield for the CC/FHI management system compared to the rotational management systems was in 1983, 1984, and 1985. In 1983 and 1984, two drought stress years, continuous corn yields were only 1.72 and 1.27 Mg ha^{-1}, respectively, compared to 3.03 and 3.95 Mg ha^{-1} for the rotational management systems. In 1985, a normal year, continuous corn yielded 6.55 Mg ha^{-1} compared to an average of 7.17 Mg ha^{-1} for the rotational management systems. These results demonstrate that corn grown in rotation is affected less by drought stress than continuous corn, but has greater yield potential under more favorable environments. A portion of these results are reported by Lesoing et al.[3] Peterson and Varvel[4] observed similar results on a different rotation experiment at the ARDC.

Table 2 Corn Grain Yields as Influenced by Organic/Conventional Management Systems (1976 to 1991)

	Management System (Mg/ha)			
Years	CC/FHI	R/MAN	R/FH	R/F
4-y average (1976–1979)	3.88[a,b]	3.44[b]	4.26[a]	4.01[a]
8-y average (1976–1983)	4.26[a]	4.20[c]	4.90[a]	4.65[b]
12-y average (1976–1987)	4.26[c]	4.77[b]	5.28[a]	5.03[a]
16-y average (1976–1991)	4.20[c]	4.65[b]	5.09[a]	4.84[a]

Note: CC/FHI — continuous corn, fertilizer, herbicide, insecticide; R/MAN — crop rotation, manure only; R/FH — crop rotation, fertilizer, herbicide; R/F — crop rotation, fertilizer only. [a,b,c] Means within each row with unlike superscripts differ significantly ($P<.05$).

The effects of previous crop on corn grain yield are shown in Table 4. Average grain yield increases were 10% for corn following oat/clover and 21% for corn following soybean, compared with corn following corn. Above-ground clover was harvested, analyzed for nitrogen, and dry matter and nitrogen yields per hectare of clover as influenced by the different management systems were determined (Table 5). Clover yields were not influenced by management system, but primarily by the previous year's clover establishment and climatic conditions (precipitation and temperature) in the spring. High clover dry matter yields did not necessarily result in higher corn grain yield the following year. Managing clover cover crops under conditions of variable precipitation was problematic. In wet years, incorporation of clover was difficult, resulting in poor seed beds and, consequently, lower than optimum corn plant

Table 3 Corn Grain Yields as Influenced by Organic/Conventional Management Systems (1976 to 1991)

	Management System (Mg/ha)			
Year	CC/FHI	R/MAN	R/FH	R/F
1976	2.80	2.74	3.50	3.56
1977	2.10	0.89	2.10	1.78
1978	8.15	8.46	8.91	8.40
1979	2.35	1.72	2.48	2.35
1980	3.88	4.96	4.58	4.71
1981	7.19	6.24	7.64	7.38
1982	5.66	5.85	6.75	6.11
1983	1.72	2.99	3.37	2.86
1984	1.27	4.01	3.95	3.95
1985	6.55	7.45	7.13	6.94
1986	5.22	6.49	7.25	7.00
1987	3.75	5.66	5.66	5.60
1988	3.69	4.07	4.33	3.75
1989	4.01	1.97	3.18	2.99
1990	4.26	5.09	4.71	4.58
1991	4.65	5.41	5.22	5.22

Note: R/Man — crop rotation, manure only; R/F — crop rotation, fertilizer only; R/FH — crop rotation, fertilizer, herbicide; CC/FHI — continuous corn, fertilizer, herbicide, insecticide.

Table 4 Corn Grain Yield for the Organic/Conventional Plots at Mead, NE as Influenced by Previous Crop (1976 to 1991)

	Previous Crop (Mg/ha)		
Year	C(C)	O/CL	S
3-year average (1976–1978)	4.33[a]	4.77[a]	4.20[a]
7-year average (1976–1983)[a]	4.52[b]	4.96[a]	4.90[a]
11-year average (1976–1987)	4.39[c]	5.15[b]	5.41[a]
15-year average (1976–1991)	4.33[c]	4.77[b]	5.22[a]

Note: C(C) — continuous corn; O/CL — oats/clover; S — soybeans. [a,b,c] Means within each row with unlike superscripts differ significantly ($P<0.05$).

[a] Data unavailable for 1979.

populations and corn grain yields. In dry years, the clover cover crop depleted soil water, resulting in a soil water deficit for the succeeding corn crop which reduced corn grain yields significantly. These results agree with those of Peterson and Varvel.[4]

Soybean yield as influenced by management system is reported in Table 6. Soybean yield was substantially lower during the first rotation cycle for the R/Man system compared to the R/FH and R/F management systems. This relationship remained consistent throughout the 16-year period, although yield differences were less between the R/Man system and the other systems. The lower soybean yields for the R/Man system were attributed to poorer weed control; as the rotation became established and weed control improved, there was a greater improvement of soybean yields of the R/Man management system in comparison to the other management systems.

Oat yields were similar for all management systems (Table 7). By seeding a high population of oats along with clover, weeds were not a yield-limiting factor in the R/Man management system in this study. Nitrogen application rates for oats, based on soil tests, were minimal for all management systems.

The average application rate (4.9 Mg/ha) and analysis of beef feedlot manure applied to the organic/conventional plots over the 16-year period are given in Table 8. A blanket application rate of manure was applied on all R/Man corn plots during the first 5 years of the experiment. From 1981 to 1991, beef feedlot manure was applied according to soil tests of residual mineral N for each plot to meet the nitrogen requirements of the succeeding corn crop. In some years the beef feedlot manure was

Table 5 Above-Ground Dry Matter and N Yields of Sweet and Red Clover as Influenced by Organic/Conventional Management Systems (1985 to 1987 and 1990 to 1991)

	Dry Matter (Mg/ha)			Nitrogen[1] (kg/ha)		
Year	R/Man	R/FH	R/F	R/Man	R/FH	R/F
1985	3.39[a]	3.22[a]	3.53[a]	128[a]	114[a]	128[a]
1986	3.61[a]	3.68[a]	3.04[b]	119[a]	117[a]	95[b]
1987	1.76[a]	1.67[a]	1.78[a]	65[a]	59[a]	63[a]
1990[2]	1.48[a,b]	1.89[a]	1.38[b]	24[a]	40[a]	34[a]
1991	0.20[b]	0.33[a]	0.23[b]	7[b]	11[a]	7[b]

Note: R/MAN — crop rotation, manure only; R/FH — crop rotation, fertilizer, herbicide; R/F — crop rotation, fertilizer only. [a,b] Means within each row with unlike superscripts differ significantly (P<0.05).

[1] Nitrogen yields were determined by harvesting above-ground dry matter analyzing clover for nitrogen content, and calculating kilograms per hectare of nitrogen in the clover.

[2] Dry matter and N yields were not measured in 1988 and 1989 due to very low yields.

Table 6 Soybean Yield as Influenced by Management System (1976 to 1991)

	Management System (Mg/ha)		
Years	R/MAN	R/FH	R/F
4-y average (1976–1979)	1.64[b]	1.91[a]	2.05[a]
8-y average (1976–1983)	2.18[b]	2.45[a]	2.45[a]
12-y average (1976–1987)	2.18[b]	2.45[a]	2.45[a]
16-y average (1976–1991)	2.25[b]	2.39[a]	2.39[a]

Note: R/MAN — crop rotation, manure only; R/FH — crop rotation, fertilizer, herbicide; R/F — crop rotation, fertilizer only. [a,b] Means within each row with unlike superscripts differ significantly (P<0.05)

Table 7 Oat Yield as Influenced by Management System (1976 to 1991)

	Management System (Mg/ha)		
Years	R/MAN	R/FH	R/F
4-y average (1976–1979)	1.42[a]	1.35[a]	1.35[a]
7-y average (1976–1983)[1]	1.67[a]	1.64[a]	1.64[a]
11-y average (1976–1987)	2.29[a]	2.15[b]	2.15[b]
15-y average (1976–1991)	2.40[a]	2.29[b]	2.22[c]

Note: R/MAN — crop rotation, manure only; R/FH — crop rotation, fertilizer, herbicide; R/F — crop rotation, fertilizer only. [a,b,c] Means within each row with unlike superscripts differ significantly (P<0.05).

[1] Oats not harvested for grain in 1980 due to poor stand.

composted before application. This resulted in the inclusion of large amounts of soil in the manure as indicated by the high ash content in 1978 (Part IV of this volume, Table 2). This apparently increased the sand content in the surface 0- to 7.5-cm layer of the R/Man plots. The average N applied to the crops as manure or fertilizer for the organic/conventional plots (1976 to 1991) is given in Table 9. Nitrogen application rates were similar for all rotational management systems (Part IV, Table 2). Averaged across these management systems, corn after soybeans required 8 kg/ha less N, based on residual soil mineral N contents, than corn after oat/clover. The nitrogen application rate for continuous corn needed to meet N requirements was 6 kg/ha higher than corn following oat/clover and 13 kg/ha higher than corn following soybean. Peterson and Varvel[4] also reported higher nitrogen requirements for continuous corn.

Table 8 Average Application Rate and Analysis of Beef Feedlot Manure Applied to Rotation/Manure Plots at Mead, NE (1976 to 1991)

Application rate (Mg/ha)	N (g/kg)	P (g/kg)	K (g/kg)	Moisture (%)
4.90	16.14	5.22	16.44	50.71

Note: Application rate (wet weight) and total N, P, and K expressed on a dry matter basis.

Table 9 Average Nitrogen Applied to Crops as Manure or Ammonium Nitrate (34-0-0) for the Organic/Conventional Plots at Mead, NE (1976 to 1991)

Management System[a]	Crop[b] (kg/ha/y)			
	C(S)	C(O/CL)	C(C)	O
R/MAN	75	86	—	9
R/F	83	88	—	12
R/FH	81	88	—	12
CC/FHI	—	—	93	—

[a] R/MAN — crop rotation, manure only; R/F — crop rotation, fertilizer only; R/FH — crop rotation, fertilizer, herbicide; CC/FHI — continuous corn, fertilizer, herbicide, insecticide.

[b] C(S) corn after soybeans; C(O/CL) corn after oats clover; C(C) continuous corn; O oats.

Soil characteristics of the organic/conventional plots as influenced by the four different management systems for 1980, 1987, and 1990 are shown in Table 10. These results indicate that some changes in organic C, Kjeldahl N, P, K, and pH from 1980 to 1990 have occurred in the top 15 cm of soil. Differences in years may be attributed to sampling method and date of sampling. There is a consistently higher soil organic C, Kjeldahl N, P, K, and pH for the R/Man management system. This, however, appears to result mainly from initial applications of manure prior to 1980. The continued application of manure in R/Man system has maintained these levels of nutrients and pH. The soil pH in the continuous corn plots has been consistently lower than that in the other rotational management system plots. This was a result of the continual annual application of commercial fertilizer over the 16-year period for the continuous corn compared to less-intense fertilizer application for the R/F and R/FH treatments.

Manure also appeared to act as a buffer against drought stress and improved water utilization under favorable environments. Corn yield results (Table 3) indicate that, under drought stress conditions of 1980, 1983, 1984, and 1988, the R/Man system yielded comparable or higher than the other systems. Yields were also similar or higher for the R/Man system under the more favorable environments of 1985, 1987, 1990, and 1991. These results may be explained by a study conducted in 1981 on these plots in which Hussain et al.[7] reported higher saturated hydraulic conductivity for the R/Man treatment compared to the R/FH and continuous corn treatments. For the 0- to 76-mm depth, saturated soil hydraulic conductivity was 30, 6, and 2 mm h^{-1} for corn in the R/Man, R/FH, and continuous corn treatments, respectively. The greater saturated hydraulic conductivity for the R/Man treatment would allow for a higher infiltration rate and improved utilization of available water. Lower corn yields for this system in the other years of the study were attributed to severe weed pressure which would negate any yield advantage received from improved water utilization. Results of this experiment indicate that, following an initial adaptation period (4 years), corn yields produced from a crop rotation that receives manure only are comparable to crop rotations that receive commercial fertilizer.

There was some question of whether the results obtained were influenced by differences in soil bulk density and depth of sampling, which were not considered in our reported values for organic C.

There were slight differences in soil bulk density between management systems with surface soils of manured plots being slightly less dense. In another study, soils from these plots were sampled at three intervals to a depth of 30 cm and bulk density measurements were also made.[5] As shown in Table 11, comparisons of total organic C, which accounted for differences in soil bulk density and deeper sampling depths, confirmed the findings of the present study, namely, that only the application of manure resulted in a significant increase in soil organic C content. As with gravimetric concentration estimates of soil organic C, this increase in soil organic C was confined to the surface 0- to 15-cm layer of soil.

Table 10 Soil Chemical Characteristics in 0- to 15-cm Depth of Organic/Conventional Plots at Mead, NE, in 1980, 1988, and 1990 as Influenced by Four Different Management Systems

Management System	Organic C (g/kg)	Kjeldahl N (g/kg)	P[1] (mg/kg)	K[1] (mg/kg)	pH
	1980 — Two Samples/Plot Taken in November				
R/MAN	23.7[a]	1.99[a]	92[a]	693[a]	7.06[a]
R/FH	20.0[b]	1.75[b]	15[b]	375[b]	6.93[a,b]
R/F	19.5[b]	1.74[b]	12[b]	381[b]	6.69[b]
CC/FHI	19.1[b]	1.67[b]	18[b]	365[b]	5.75[c]
	1988 — Five Samples/Plot Taken in March				
R/MAN	21.3[a]	1.97[a]	109[a]	564[a]	7.03[a]
R/FH	17.3[b]	1.65[b]	15[b]	352[b]	6.70[b]
R/F	17.1[b]	1.64[b]	13[b]	326[b]	6.82[a,b]
CC/FHI	17.0[b]	1.61[b]	20[b]	327[b]	6.38[c]
	1990 — Five Samples/Plot Taken in November				
R/MAN	22.6[a]	1.95[a]	90[a]	586[a]	6.96[a]
R/FH	18.5[b]	1.68[b]	15[b]	358[b]	6.72[b]
R/F	19.4[b]	1.71[b]	17[b]	363[b]	6.72[b]
CC/FHI	17.1[b]	1.68[b]	16[b]	322[b]	6.18[c]

Note: R/MAN — crop rotation, manure only; R/FH — crop rotation, fertilizer, herbicide; R/F — crop rotation, fertilizer only; CC/FHI — continuous corn, fertilizer, herbicide, insecticide. [a,b,c] Means within each column with unlike superscripts differ significantly ($P<0.05$) for each year sampled.

[1] Bray extractable P and exchangeable K (1 *M* ammonium acetate).

Table 11 Organic C Content of Conventional/Organic plots at Mead, NE, for Several Sampling Depths (Values Averaged across Crops)

	Organic C Content for Respective Depths (Mg C/ha)				
Management System	0–7.5 (cm)	7.5–15 (cm)	15–30 (cm)	0–15 (cm)	0–30 (cm)
R/MAN	22.0[a]	17.8[a]	27.5[b]	39.8[a]	67.3[a]
R/F	16.1[b]	16.2[b]	26.2[b]	32.3[b]	58.5[b]
R/FH	16.3[b]	16.0[b]	27.5[b]	32.3[b]	59.8[b]
CC/FHI	15.4[b]	15.4[b]	28.0[b]	30.8[b]	58.8[b]

Note: Arithmetic means of samples taken in fall of 1981 and 1982. R/MAN — rotation, manure only; R/F — rotation, fertilizer only; R/FH — rotation, fertilizer, herbicide; CC/FH — continuous corn, fertilizer, herbicide; CC/FHI — continuous corn, fertilizer, herbicide, insecticide. [a,b] Means within each column with unlike superscripts differ significantly ($P<0.05$).

Data from Fraser, D. G., Doran, J. W., Sahs, W. W., and Lesoing, G. W., *J. Environ. Qual.*, 17, 585, 1988.

Under the conditions of this experiment, none of the variables such as manure application, fertility level, use of synthetic chemicals, crop rotation, tillage, or soil microclimate had a significant effect on surface soil (0 to 15 cm) organic C levels for the 10-year period from 1980 through 1990. The application of manure in the R/Man system during the first 5 years, however, did maintain higher organic C and Kjeldahl N levels in the top 0- to 15-cm layer of soil. Rotation management systems had higher corn yields than continuous corn; however, crop rotation did not significantly influence soil organic C and Kjeldahl N levels. A N balance for the years 1980 through 1988 for the organic/conventional management systems confirmed these results (Table 12). Initial and final soil N levels in the top 0- to 15-cm depth have remained unchanged for each management system. Each rotational management system resulted in a negative N balance ranging from 108 to 151 kg N/ha. However, the contribution of the legume roots was not accounted for and may have provided an additional N input of up to 50 kg/ha, thus reducing the actual N balance deficits from 58 to 101 kg/ha. Continuous corn, on the other hand, resulted in a positive N balance of 305 kg/ha. However, this additional N with continuous corn is unaccounted for and presumably lost from soil through leaching, volatilization, or erosion. Other researchers working on this experiment have confirmed this finding, but have also shown that crop rotation affects soil microbial and physical properties which are important to nutrient cycling and crop growth.[5–7]

Table 12 N Balance for Organic/Conventional Management Systems and Changes in Soil N for 0- to 15-cm Depth (1981 to 1988)

Management System	Initial Soil N (1980)	Final Soil N (1988)	N app.[a] (kg/ha) 1981–1988)	Legume N[b] (kg/ha) 1981–1988)	N removed[c] (kg/ha[c]) 1981–1988)	Net N Bal[d]
R/MAN	3731	3694	338	168	630	–124
R/F	3426	3347	387	165	660	–108
R/FH	3315	3200	375	160	686	–151
CC/FHI	3181	3200	765	—	460	+305

Note: R/MAN — crop rotation, manure only; R/F — crop rotation, fertilizer only; R/FH — crop rotation, fertilizer, herbicide; CC/FHI — continuous corn, fertilizer, herbicide insecticide.

[a] Nitrogen applied as fertilizer or manure.

[b] Average yearly above-ground legume N contribution for 1985, 1986, 1987, and 1990 multiplied by 2.

[c] Nitrogen removed as grain.

[d] N balance = (N applied in fertilizers, manures and legumes) – (N removed).

REFERENCES

1. Sahs, W.W. and Lesoing, G.W., Crop rotation and manure versus agricultural chemicals in dryland grain production, *J. Soil Water Conserv.*, 40, 511, 1985.
2. Andrews, R.W., Peters, S.E., Janke, R.R., and Sahs, W.W., Converting to sustainable farming systems, in *Sustainable Agriculture in Temperate Zones*, Francis, C.A., Flora, C.B., and King, L.D., Eds., Wiley-Interscience, New York, 1990, 281.
3. Lesoing, G.W., Francis, C.A., and Sahs, W.W., Rotation and precipitation effects on dryland crop yield stability, *Am. Soc. Agron. Abstr.*, 57, 1988.
4. Peterson, T.A. and Varvel, G.E., Crop yield as affected by rotation and nitrogen rate, III, *Agron. J.*, 81, 735, 1989.
5. Fraser, D.G., Effects of Conventional and Organic Management Practices on Soil Microbial Populations and Activities, M.S. thesis, University of Nebraska, Mead, 1984.
6. Fraser, D.G., Doran, J.W., Sahs, W.W., and Lesoing, G.W., Soil microbial populations and activities under conventional and organic management, *J. Environ. Qual.*, 17, 585, 1988.
7. Hussain, S.K., Mielke, L.N., and Skopp, J., Detachment of soil as affected by fertility management and crop rotation, *Soil Sci. Soc. Am. J.*, 52, 1463, 1988.

Chapter 15

Sanborn Field: Effect of 100 Years of Cropping on Soil Parameters Influencing Productivity

G.A. Buyanovsky, J.R. Brown, and G.H. Wagner

CONTENTS

I. INTRODUCTION

Sanborn Field, on the University of Missouri campus at Columbia, was established as the Rotation Field by Dean J. W. Sanborn in the fall of 1888. His original plan called for demonstration of the value of

0-8493-2802-0/97/$0.00+$.50

manure and crop rotation in retarding the loss of soil productivity due to cropping land formerly in tall grass prairie. The field was renamed Sanborn Field in 1924 by the University Board of Curators. Sanborn wrote that the field was

> *designed to show the value of various rotations; of constant tillage crops on soil fertility; of chemicals fed according to crop analysis to the necessity of rotation; and of various crops unmanured to soil fertility, etc.; etc.*[1]

From his experiments, he hoped to obtain practical answers to questions farmers were asking about the use of different cropping systems.

One hundred years is a very long period for plot maintenance and data collection from a field experiment. During that period, new technologies of cropping and soil treatments evolved, improved varieties of crops were introduced, and the use of mineral fertilizers changed dramatically. Naturally, the initial design of the field required adjustment in order to incorporate modern practices which were being used in commercial farming and to delete those systems proven to be impractical. Despite the efforts of different managers of the field to interject their own ideas, several treatments maintained their integrity throughout the first 100 years. Those systems provide us with invaluable data for determining long-term effects of crop and soil management on productivity and properties of the soil. Organic matter in soils under such systems are near to equilibrium. Some soluble amendments, such as phosphorus and potassium, require shorter time to equilibrate within the soil, and this field provides the resources for evaluating numerous changes in nutrient status.

Each long-term field is unique in the soil resource, the management practices, and the climatic setting. In spite of the differing results between long-term sites, there should be some discernible underlying principles concerning the soil-plant-environment continuum. This is especially true if Jenny's 1941 assessment of forming factors[2] has any validity.

This paper describes the setting and management of Sanborn Field and presents summaries of 100 years of results to illustrate the impact of long-term culture on nutrient status and organic matter level in the soil.

II. SITE INFORMATION

The site for Sanborn Field, when selected in 1888, was in an uninhabited area southeast of the campus of the University of Missouri in Columbia. Since 1888, the area around the field has been built up with private housing across the city street to the north and campus buildings, green areas, and parking lots across the streets on the other three sides.

Columbia is located at 93°20′ west longitude and 38°57′ north latitude. The area is subjected to a continental climate.[3] The average monthly temperatures, with July as the hottest month of the year, are shown in Table 1. Forty-six percent of the 916 mm of annual precipitation falls between May 1 and September 30. Potential evapotranspiration increases rapidly from April to July and periods of severe moisture stress are expected from mid-June through early September. Corn (*Zea mays* L.) is especially sensitive to these midsummer stresses. The increases in temperature from late April through May hasten winter wheat (*Triticum aestivum* L.) maturity so that harvest usually occurs during the last 10 days of June. The average date of the last killing frost in Columbia is April 8 and of the first killing frost in autumn is October 24.

The site occupied by Sanborn Field was described in early handwritten notes as being in grass with scattered woody species.[4] The first soil survey of Boone County, MO, made in the early 1960s provided the basis for soil description.[5] The field consists mostly of Mexico silt loam grading at the east into Lindley loam. The current taxonomic classifications of these two soils are

- Mexico — fine, montmorillonitic, mesic Udollic Ochraqualfs
- Lindley — fine-loamy, mixed, mesic Typic Hapludalfs[6]

The Mexico soils are somewhat poorly drained, formed in loess under tall prairie grasses, and the Lindley soils are well drained, formed in glacial till under hardwood forest. The east series of plots on the field are actually on the boundary between the two soils.

The soils of Sanborn Field are in the claypan resources area for which climate has been one of the most influential soil forming factors.[7] Environmental factors have facilitated the downward movement of chemical components and clays within a highly weatherable soil-forming material. The soils possess

Table 1 Weather Parameters — Columbia, MO, 1951 through 1980

Parameter	January	February	March	April	May	June	July	August	September	October	November	December
Temperature (°C)												
Maximum	2.2	5.6	11.1	18.3	23.3	28.3	31.7	30.6	26.7	20.0	11.7	5.0
Minimum	–7.2	–5.0	0.0	6.7	12.2	16.7	19.4	18.3	18.3	7.8	1.1	–4.4
Precipitation (mm)	39.9	47.2	81.0	97.2	1.4	95.5	89.1	74.4	92.5	84.8	51.3	49.5

From National Oceanic and Atmospheric Administration, Climatological Data, Annual Summary, Missouri: National Climatic Data Center, Asheville, NC, 1991.

well-expressed horizons. The parent materials are of two sources: Kansan Age glacial till and loess. The field is in the intermediate loess thickness zone (1 to 2 m thick) within Boone County; thus, loess is the dominant parent material. Most of this loess is probably of Peorian Age.[5] Parts of the lower solum may be directly formed in the Kansan Age glacial till or there has been basal mixing of the loess within the till.

The field is 80 m (NS) × 251 m (EW) or 2.0 ha in area. The summit near the north fence is 235 m above mean sea level and the lowest point at the southwest corner of the field is 232 m. The southwest slope is 1.9% from the summit and the southeast aspect slope is 2.2%. The slope on a few individual plots ranges up to 5% for short distances.

The individual plots are 30.6 × 9.5 m in area and are separated by a 1.5-m wide grass border. A diagram showing a plot layout appears in Figure 1. Over the years, erosion has occurred on each plot except those in continuous meadow. This has been limited by the plot borders which serve to prevent run-on of surface water. Each border is breached at the low point of the plot to allow excess surface water to run off the plot. A crude estimate of erosion can be made using the top of the argillic horizon as the reference point. Unfortunately, Dean J. W. Sanborn did not take soil samples when the field was established so such indirect measurements are necessary.

III. FIELD MANAGEMENT

Emphasis in the initial experiment was given to practical approaches in crop production that could be adopted readily by the farmers of the state without the need for off-farm inputs, which were not available to them. All farms had an array of domesticated livestock that required feed and were sources of manure.

Nine cropping practices, using corn (*Z. mays* L.), oats (*Avena sativa* L.), winter wheat (*T. aestivum* L.), red clover (*Trifolium pratense* L.), and timothy (*Phleum pratense* L.) under three soil treatments, were originally used in the experiment (Table 2). The practices were continuous cropping systems of each of the five crops and rotational cropping sequences of 2, 3, 4 and 6 years. Winter wheat, corn, and oats were the major grain crops, and clover and timothy were the main forage components of rotations around 1888. Wheat and corn have been grown continuously and in seven systems of crop rotation. Only one plot with continuous wheat was designed to receive "complete fertilizer for maximum crop". The rate was determined as "annual application of enough sodium nitrate, dissolved bone or acid phosphate, and muriate of potash to replace all of the nitrogen, phosphorus, and potassium removed in a 40 bushel crop of wheat with 2 tons of straw."[8] In accord with plant composition data, this rate approximated today's full treatment for wheat of 50-20-20 (N-P-K in kilograms per hectare). Timothy has been grown continuously for 100 years with and without manure. Varieties for each crop were selected to represent the highest yielding ones at a given time. Attempts were made to use the same variety for several consecutive years.

Barnyard manure was used as fertilizer. The manure is thought to have been from a horse barn for the first 50 years or so. Later, it has been from a dairy barn, and, at times, has undergone some composting. No analyses of the manure were made. During the first 50 years, nitrogen was applied as Chile niter (sodium nitrate that occurs naturally) and later (since 1950) as ammonium nitrate (34-0-0). Fertilizer P was applied as crude grade acid phosphate or normal superphosphate (16 and 20% P_2O_5, respectively) for the first several decades. After the mid-1940s, P was applied as superphosphate. Potassium was applied as muriate of potash (potassium chloride).

All plot operations were performed using field-size equipment, with changes throughout the period dictated by the development of agricultural machinery, including the disappearance of the horse and the introduction of the tractor. A moldboard plow set for about 20-cm depth has been used for conventional tillage in the fall. Seedbed preparation just prior to planting was performed with two passes of a disk and one pass of a finishing harrow.

The planting of crops was usually done at recommended times with minor adjustments due to weather conditions. Winter wheat was usually planted in the middle of October and corn in late April or early May. All crops were seeded at the recommended seeding rates. Wheat was sown at 100 kg ha^{-1} in 15-cm rows. Corn seeding rates were variable, reflecting the nutrient status and expected yield goal of each plot. The population was 30,000 plant per hectare for continuous corn (40,000 plants for fertilized corn from 1950 onward); in rotations, the population was 47,000 plants per hectare with manure, and 32,000 plants per hectare with no treatment. Forage crops that followed a small grain crop in rotation were usually frost seeded into the grain crop during early February. Plant vigor and persistence have been problems in continuous timothy with no treatment. Therefore, reseeding of both continuous timothy plots was done in the fall about every 4 to 5 years in a prepared seedbed.

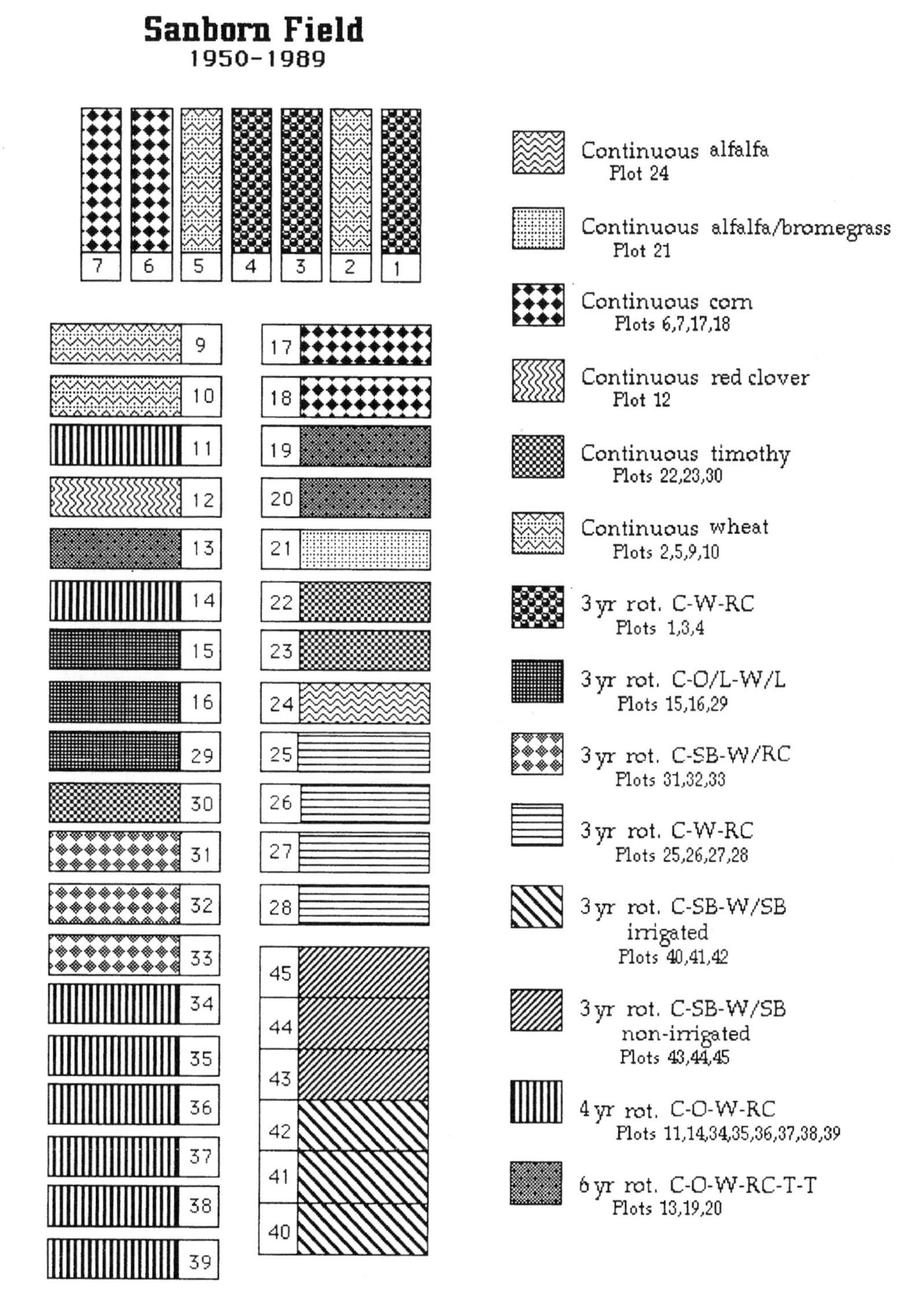

Figure 1 Schematic diagram of Sanborn Field experimental plots for 1950 to 1989.

The grain crops were harvested at maturity either by hand or using current farming technology. In normal weather years, wheat was harvested in late June to early July and corn in late September. Initially, the forage yields were cut with a field mower, allowed to dry, and the entire plot yield of hay was weighed. Perennial grasses and red clover were normally harvested twice each year. In recent years, forage yields have been estimated from strips cut with a plot mower. The plot was mowed and the remaining forage removed. Up to the early 1950s, corn stover was weighed and removed from plots

Table 2 Cropping Practices and Soil Treatments on Sanborn Field, 1888 to 1913

Crop Sequence	Treatment (number of plots)		
	None	Manure	Fertilizer
Continuous corn	1	1	0
Continuous oat	1	1	0
Continuous wheat	2	6	1
Continuous clover	1	1	0
Continuous timothy	1	1	0
Wheat, clover	1	2	0
Corn, wheat, clover	1	3	0
Corn, oat, wheat, clover	1	4	0
Corn, oat, wheat, clover, timothy, timothy[a]	1	6	1

[a] Also one plot had 1/2 rate of manure and fertilizer.

under continuous cropping systems. However, during the last 40 years stover has been spread uniformly over the plots.

Periodically during the past 100 years, a number of revisions was made in the experimental plan on some plots. The changes were prompted by a lack of practicality of some systems of management, by a desire to include new crops and cropping practices, and to add more variation of fertilization in the experiment. Evaluations of management of the field that led to revisions were made in 1914, 1928, 1940, and 1950. Additional information concerning the management of the field is presented in Section IV of this volume. Historically, the changes in Sanborn Field management paralleled the development of understanding of the nature and properties of soils and particularly its translation into practical approaches of liming and fertilization.

IV. SAMPLING AND ANALYSES

There were no records of sampling of soils on the field prior to 1914.[4] Since then detailed soil samples have been taken about every 25 years. Limited details concerning sampling are given in Table 3. In addition, surface soil samples have been taken at various times. Crushed, air-dried soil samples are archived in glass jars.

In 1963, soil samples were taken at 12 locations on each plot in a systematic fashion. Two cores at each sampling station were taken from the soil surface extending well into the argillic horizon (approximately 90 cm). One of the two cores was divided by diagnostic horizons and the other divided into 10-cm increments. In 1988, the sampling with depth was repeated to document changes at the end of 100 years of cropping. The sampling stations were slightly offset from the 1963 locations and only the central four stations on each plot were sampled. As in 1963, two cores were taken. One was used for

Table 3 Outline of Archived Soil Samples Taken from Sanborn Field

Year Sampled	Sampling Depths, cm	Measures
1914	0–10, 10–20, 20–30, 30–46, 46–61, 61–91, 91–122	Total C, N, P, K, and % volatile matter
1928	0–20, 20–46, 46–61	Total N, P, Ca, Mg, S, and % volatile matter
1938	0–10, 10–20, 20–30, 30–46, 46–61, 61–91, 91–122	Total N, CEC, exchangeable Ca, H, pH (0–10 and 10–20 cm only)
1949	0–10, 10–20	Total N
1963	Same as 1938, also Ap, A_1, A_2, and top 30 cm of B_2	Routine soil tests and total N (0–10, 10–20 cm)
1988	To 120- × 10-cm increments	Routine soil tests
	By diagnostic horizons to 120 cm	Analyses by National SCS Lab characterization methods

Note: In addition, there have been several sets of plow layer samples taken for testing to provide a basis for modification of P, K, and limestone treatments.

morphological characterization and the other cut into 10-cm increments for chemical analyses. Samples were taken to 120-cm depth. Each hole was filled with a core taken from the adjacent border.

Sampling of plant material and grain has been sporadic. Prior to 1950, crop residues were removed from the plots but no analyses were made. No plant samples have been stored.

There have been no comprehensive analyses of soil samples taken on the different dates using the same methodology. The methods of analysis were those in vogue at the time of the sampling. The most consistent methodology has been the use of the Kjeldahl total N analysis. That method, however, has seen several changes since it was introduced, particularly in catalysts.

Ammonium acetate (neutral 1 normal) was used on the 1963 and 1988 soil samples to extract cations and estimate cation exchange capacity. Flame emission spectroscopy has been used for K and Na analysis since 1963. The Ca and Mg in the 1963 extracts were measured by EDTA titration with Eriochrome Black T. Atomic absorption spectroscopy was used for Ca and Mg analysis of the 1988 extracts.

Total C, when done, has been estimated by some form of dry combustion in a purified stream of oxygen. Soil organic matter content has been estimated by quick-test chromic acid methods. Some measurements of bulk density have been made.

V. YIELDS

Crop yields obtained from Sanborn Field over the 100 years of culture have been impacted by cropping systems, technology, and management. All these factors have been influenced, in turn, by the natural variability in weather. Other factors that have had a significant impact on yield variability are changes in varieties, disuniformity of manure applications, and infestations of plant pests. Early references are made to open pollinated corn and to Reid's Yellow Dent. During and after World War II, US-13 was the corn hybrid of choice. After 1950, the corn seeded on the field has been a widely grown commercial hybrid which was used for several consecutive years. Similar contrasts over time have influenced yields of clover and timothy due to changes in varieties. Wheat varieties have had a similar impact on yields of this crop. Early varieties were characterized by long straw and a low grain-to-straw ratio. In the mid-1970s, a shift was made to high-yielding varieties which have much shorter straw and are less subject to lodging under high fertility levels.

The yields from the manured plots illustrate another trend. Yields on manured plots tended to decline prior to 1950 and then increase (Table 4). Soil data show that soil P declined from 1914 through 1949, then showed a dramatic increase through 1989.[9] This shift reflected a change in the quality of manure from the use of copious quantities of straw for bedding to minimum bedding for animals with higher nutrient level. An added factor to the soil P shift could have been return of crop residues to the plots after 1950.

The decline in soil productivity among plots between 1888 and 1913 is reflected in the comparison of yields, particularly between untreated and manured plots (Table 4). Intense cultivation during this period drew heavily on the nutrient reserves present in the soil at the outset of the experiment. A further difference among untreated plots is the effect of crop rotation in delaying the decline in soil productivity. Nutrient levels in the various plots will be covered in more detail later in this chapter.

In evaluating the data in Table 4, one must be aware of certain problems related to the effect of plant pests. Wheat, when grown continuously under full treatment, has become infected with fungal diseases, especially take-all (*Gaeumannomyces graminis*). Thus, fully fertilized wheat barely outyields continuous wheat that receives only manure. The timothy yields from the untreated plots would better be called weed yields. The low fertility of plot 23 does not support a stand of timothy even though frequent frost seeding is used to encourage timothy. Red clover has also invaded the plot.

A. WHEAT

Wheat grown continuously for 100 years without fertilizer showed a gradual decrease in productivity (Figure 2). Grain yield for the first 50 years (1888 to 1938) averaged 0.7 Mg ha^{-1} but in some years was as high as 1.3 Mg ha^{-1}. Yields were significantly lower during the 1940s and 1950s (0.25 to 0.43 Mg ha^{-1}).

Nonfertilized wheat in a 3-year rotation with corn and clover had about 50% higher yields than nonfertilized continuous wheat (Table 4). A 4-year rotation, which included corn, oats, and a legume, gave a significant increase in wheat yield compared with other rotations. This was probably due to the different crop sequence: in the 3-year rotation, wheat followed corn; in the 4-year rotation, it followed oats. Due to the short time period between corn harvest (late September to early October) and wheat planting (late October), there is little opportunity for replenishment of soil water storage and the recycling

Table 4 Crop Yields from Sanborn Field, 1888 to 1990

Plot	Crop	Treatment	Management Period (Mg ha^{-1})							
			1888–1913	1914–1927	1928–1939	1940–1949	1950–1962	1963–1969	1970–1990	100-y average
					Continuous Cropping					
2	Wheat	Full[a]	1.32	1.36	1.43	1.29	1.98	3.16	2.67	1.89
9	Wheat	None	0.67	0.71	0.72	0.26	0.43	1.01	0.63	0.63
10	Wheat	Manure[b]	1.18	1.42	1.28	1.17	1.67	3.05	2.56	1.76
6	Corn	Full	—	—	—	—	4.80	6.67	7.51	6.33
17	Corn	None	1.27	1.20	1.06	1.04	0.64	0.78	0.53	0.93
18	Corn	Manure[b]	2.18	2.19	1.99	2.63	2.63	3.19	3.75	2.65
23	Timothy	None	3.34	2.15	2.46	2.06	2.60	1.90	2.22	2.39
22	Timothy	Manure[b]	5.60	6.00	5.44	6.50	7.19	6.07	6.07	6.12
					3-y Rotations					
27	Corn	None	2.14	2.26	2.16	2.17	3.73	3.40	3.56	2.77
	Wheat	None	0.94	0.83	0.74	0.44	0.61	1.77	2.09	1.06
	Clover	None	2.17	2.04	0.96	1.99	1.57	5.62	4.86	2.74
25	Corn	Manure[b]	2.43	3.59	3.71	3.69	4.79	7.70	7.81	4.82
	Wheat	Manure[b]	1.72	1.87	1.41	1.21	1.61	2.94	3.26	2.00
	Clover	Manure[b]	4.17	7.19	2.24	1.88	3.39	10.14	9.18	5.46

[a] "Full treatment": the quantity of N, P, K, and limestone recommended according to soil tests and yield goals.

[b] 13.4 Mg ha^{-1} annually.

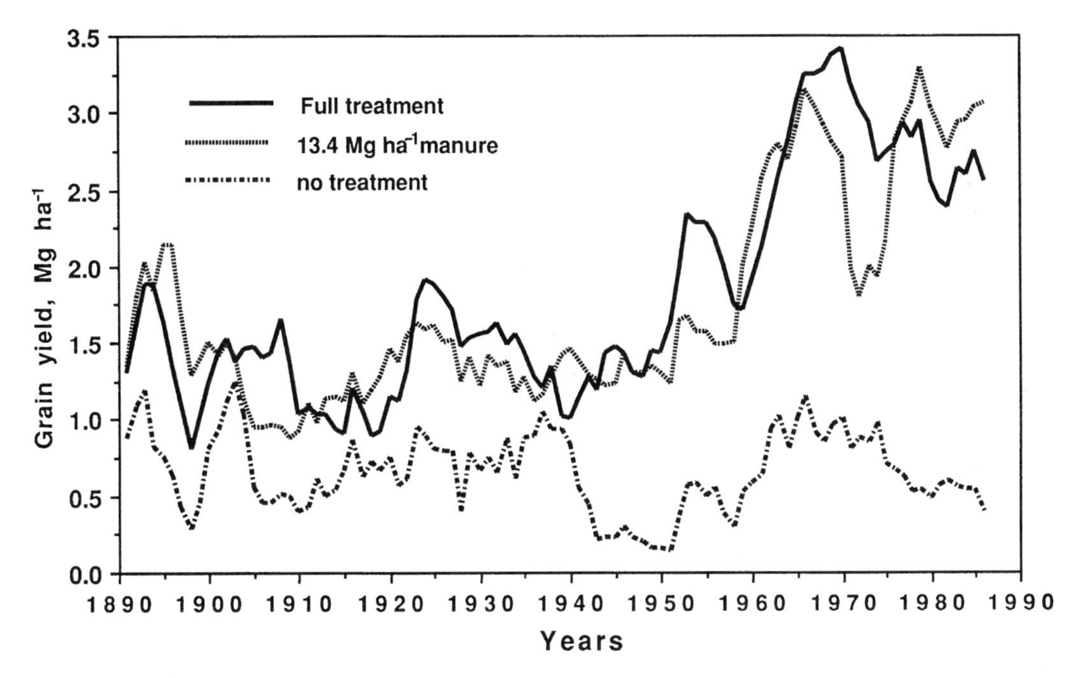

Figure 2 Yields of winter wheat grown continuously with no treatment, with annual treatment of 13.4 Mg ha^{-1} manure, and full fertilizer as 5-year moving averages.

of nutrients from residues. This could cause a negative effect on wheat development. In contrast, in the 4-year rotation, oats are harvested in July, allowing a fallow period during which the soil accumulates water and available nutrients. Alternatively, corn may be more efficient in using some nutrients in the nonfertilized soil than is oats. Thus, when wheat is planted following corn, it is subjected to a different residual supply of nutrients. Likely, both factors are involved and contribute to the "rotation effect". Adding 2 years of timothy to the rotation decreased wheat yields to about 75% of those in the 4-year rotation.

Application of barnyard manure or chemical fertilizers was very effective in increasing yields of continuously grown wheat (Figure 2). Manure at 6.7 Mg ha^{-1} wet weight increased average yield for the 100 years to 1.48 Mg ha^{-1}, or more than double that of nonfertilized wheat. With 13.4 Mg ha^{-1} manure, average yield was 1.76 Mg ha^{-1}. The trend of 5-year moving averages (Figure 2) consistently indicates a greater yield from the 13.4 Mg ha^{-1} rate. Much of the increase in yield during the past 30 years was presumably due to improved varieties, which were more responsive to fertilizer inputs, and to higher quality manure.

Mineral fertilizers were not used widely in 1888, and they were applied only on two plots (2 and 3). For more than 60 years, "chemicals for 2.7 Mg ha^{-1} of wheat grain" were applied to plot 2. In 1950, the plot was switched to so-called "full treatment", which was based on soil test and yield goals, with N-P-K rates ranging up to 60-22-22 kg ha^{-1}. It should be noted, however, that, during the first 60 years, yields never reached 2.7 Mg ha^{-1} and averaged only 1.32 ± 0.38 Mg ha^{-1}. For the 1950 to 1988 period, average yield was close to the initial goal. In some years it reached 3.7 to 4.0 Mg ha^{-1}, with a 39-year average of 2.35 Mg ha^{-1}. In general, yields of continuously grown wheat with full treatment (60-22-22 kg ha^{-1} applied annually) were slightly higher (though not significantly) than yields of wheat with 13.4 Mg ha^{-1} manure. The exception was the 6-year rotation with full treatment, where the highest average wheat yield (2.53 Mg ha^{-1}) was attained. Under this treatment, corn received 157-27-27 kg ha^{-1}, oats/wheat received 67-22-0 kg ha^{-1}, timothy received 74-0-0 kg ha^{-1}, and clover received P and K according to the amounts removed in forage.

B. CORN

Several systems involving corn have been maintained since 1888 (Table 2). Nonfertilized, continuous corn had the lowest productivity (Table 4). The decline in yield continued throughout the second 50 years despite the introduction of new more productive varieties (Figure 3). The average yield for 1888 to 1938

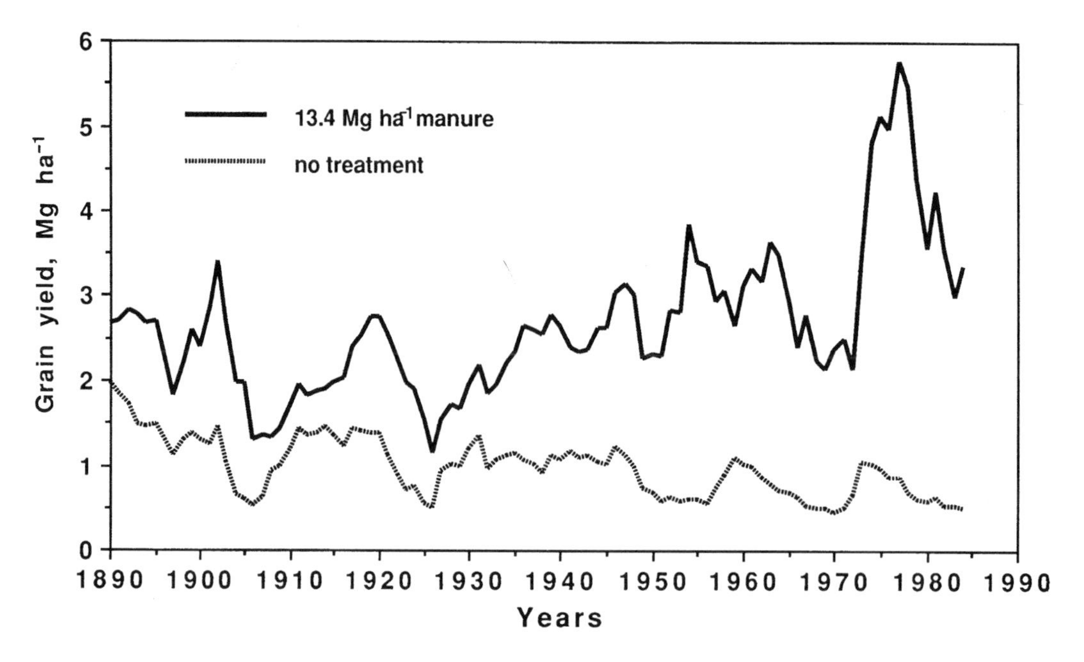

Figure 3 Yields of corn grown continuously with no treatment and 13.4 Mg ha^{-1} manure annually, as 5-year moving averages.

was 1.2 Mg ha^{-1}. For five consecutive decades since 1938, yields gradually declined to 0.5-0.6 Mg ha^{-1} during the last 20 years. In contrast, manure application (13.4 Mg ha^{-1}) increased average yield of continuous corn to 2.1 Mg ha^{-1} for 1888 to 1938. Average yield was significantly higher for the last 50 years (3.0 Mg ha^{-1}), and even more during the last 15 years (4.25 Mg ha^{-1}).

Corn yield from nonfertilized rotations was two to three times higher than that from nonfertilized continuous corn for the whole period, and the difference was greatest during the last 15 years (Table 4). Since 1975, corn in rotation yielded five to six times more than that grown continuously. Productivity of corn in all nonfertilized rotations (3-, 4-, and 6-year rotation) was similar for the last 15 years, averaging between 3.13 and 3.76 Mg ha^{-1}. For the first 50 years of the Sanborn Field experiments, productivity of these systems varied between 1.9 and 2.5 Mg ha^{-1}.

Manure application was effective for improving corn yield in 3-year rotation (Table 4). In the 4-year rotation (corn-oats-wheat-clover), with 13.4 Mg ha^{-1} of manure applied annually, the yield during the earlier period (1888 to 1938) varied from 2.5 to 3.0 Mg ha^{-1}, and increased for the whole period (to 1990) to 3.7 to 4.0 Mg ha^{-1}. A significant increase was obtained when manure treatment was supplemented with nitrogen (4.6 Mg ha^{-1}).

C. TIMOTHY

Timothy was grown continuously and in a 6-year rotation (corn-oats-wheat-clover-timothy-timothy). Timothy grown continuously without fertilizer decreased steadily in productivity (Figure 4). After high yields in 1891 to 1895 (4.3 to 6.0 Mg ha^{-1}), yields decreased to nearly half, and, for the 100 years, the average was only 2.4 Mg ha^{-1} (Table 5). The stand of nonfertilized continuous timothy deteriorated rather rapidly, and the plot had to be reseeded almost every 4 to 5 years.

Timothy grown in a 6-year rotation without fertilizer gave the same amount of forage as nonfertilized continuous timothy (average for the first 50 years was 2.9 Mg ha^{-1} and for 100 years was 2.5 Mg ha^{-1}). Since the late 1940s, this system could no longer sustain vigorous timothy. Second-year timothy was usually choked by weeds, mainly summer annual grasses, which covered nearly all of the plot.

Application of manure at 13.4 Mg ha^{-1} doubled forage yield in continuous timothy (average for the first 50 years was 5.6 Mg ha^{-1}, for the whole period 6.12 Mg ha^{-1}). In some years, up to 7.8 Mg ha^{-1} of hay was collected. The effect of manure on hay yield in the 6-year rotation was similar (5.13 Mg ha^{-1} of forage on average).

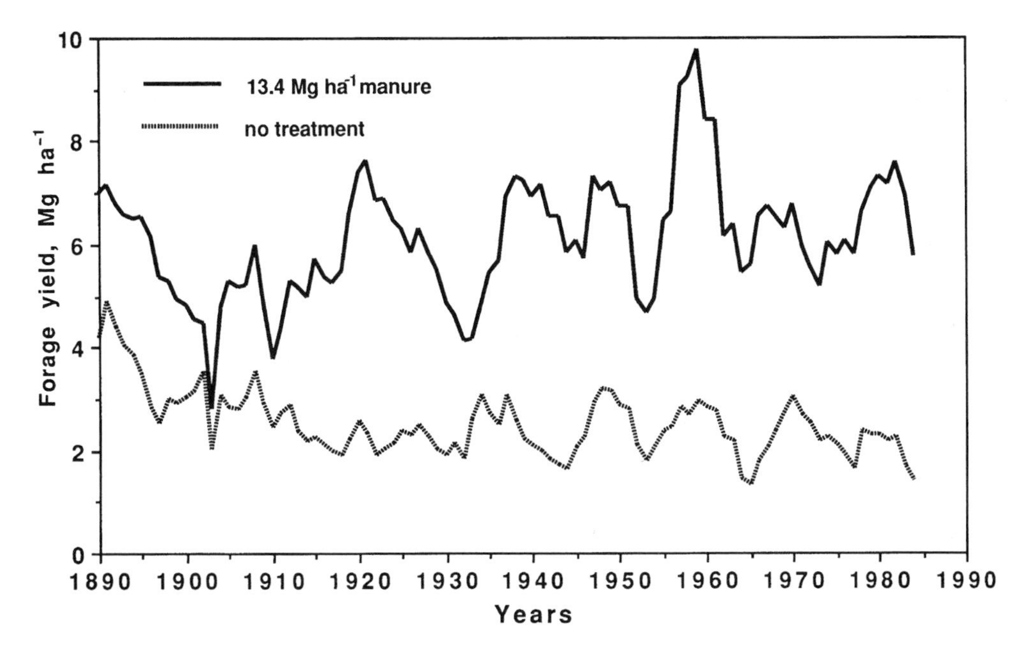

Figure 4 Forage yields of timothy grown continuously with no treatment and 13.4 Mg ha^{-1} manure annually, as 5-year moving averages.

VI. SOIL CHARACTERISTICS

A. SOIL MORPHOLOGY

The morphology and horizon sequence of the soil reported by Miles and Hammer[7] suggests that a A-E-Bt-Bg horizon sequence representative of past Putnam and Mexico soil series designations is appropriate for Sanborn Field. The Ap and E horizons are very silty (Figure 5). The Bt horizon exhibits a large increase in translocated clay evident from a silty clay texture and evidence of translocated clay on horizonal and vertical ped surfaces. The color of the lower solum and the presence of concretions reflect anaerobic conditions which could be both relict and contemporary. The Ap and E horizons contain concretions, which are probably a reflection of the perching of water above the claypan.

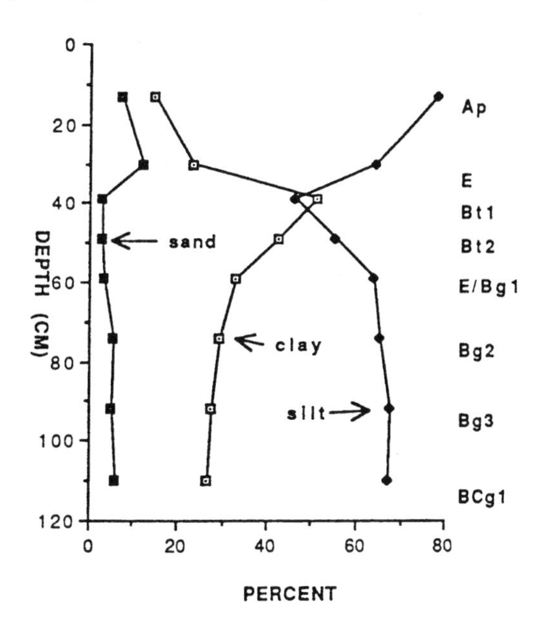

Figure 5 Particle size distribution with depth for the soil of Sanborn Field.[7]

The examination of various profiles among the plots led Miles and Hammer to conclude that morphological properties of the soil resource of Sanborn Field represent well-expressed pedogenic horizonation, specifically with respect to translocation of the clay. They stated that the summit area of the field tends to be more like the typical A-E-Bt horizon of the Missouri claypan area. Many of the shoulder, backslope, and footslope areas do not possess typical subsurface horizons (i.e., AB, BA, and E). Because of the lack of initial soil data, it is not known whether these hillslope positions originally contained these soil features. However, they feel it is plausible that the cultural practices (i.e., tillage and subsequent erosion) have had a strong influence in changing subsurface horizonation if it was originally present.[7]

B. PARTICLE SIZE DISTRIBUTION

The influence of landscape position, erosion, and parent materials are reflected in the particle size depth distributions of the plots on Sanborn Field (Figure 5). All of the plots possess a dominant silty matrix which is due to the influence of the parent loess material.[7] Clay which is less than 20% in the silt loam Ap and E horizons abruptly increases to over 50% in the silty clay Bt horizon and then decreases to <30% in silty clay loam Bg horizons. A higher clay content in Ap horizon of the plots that have experienced erosion is most likely due to incorporation of the silty clay Bt by tillage. Upchurch et al.[4] reported that in 1961 the surface soil material deposited by erosion on the east end of plots 1 through 7 was moved back to the west end (upslope) to restore uniformity of topsoil depth.

C. SOIL MINERALOGY

Miles[7] has accumulated limited data on the mineralogy of the soil of Sanborn Field. Semiquantitative estimation of X-ray diffraction analysis of the clay size and silt size components from a pit on the south edge of plot 12 (near the field summit) showed that montmorillonite was the most abundant clay size mineral in both the Bt and Bg horizons. Hydroxy-inter-layered montmorillonite and vermiculite were the next most-abundant minerals. Moderate amounts of vermiculite and mica exist in both horizons, whereas kaolinite is present in small quantities in the Bt horizon with a slightly greater quantity in the Bg horizon.

The silt in the samples from plot 12 was composed primarily of quartz, plagioclase feldspars, and potassium feldspars. Quartz was dominant. Greater potassium feldspar contents (two or three times) were found in the Bt horizon in comparison to the Bg horizon. A slightly greater plagioclase feldspar component was found in the Bg than the Bt horizon.

D. SOIL PHYSICAL PROPERTIES

In terms of enhancing soil physical properties, the best soil fertility management is the one involving annual additions of manure and the worst is that without treatment.[13] Annual additions of manure for 100 years significantly decreased bulk density by an average of 0.06 g cm^{-3}. Saturated hydraulic conductivity of manured plots was significantly increased by an average of one order of magnitude (i.e., from 0.119 m h^{-1} for fertilized corn to 1.089 m h^{-1} for manured corn). Even though use of inorganic fertilizers did not add as much organic matter to the soil, this treatment was not significantly different from that of manuring with regard to the soil properties monitored.

There were no significant differences in soil physical properties measured between the continuous corn, continuous wheat, continuous timothy, or the 3-year corn-wheat-red clover rotation treatments. Even with erosion and an increase in clay percent which were apparent for the untreated corn and wheat plots, the surface soil can apparently maintain good tilth for corn and wheat since the silt loam texture is a very easily managed material.

E. SOIL EROSION

One of the strongest effects of crop rotations has been the demonstrated reduction in the amount of soil erosion. Using topsoil depth data from the 75th-year sampling, average erosion rate was estimated.[14] Assuming that soil was neither lost from nor added to the timothy plots, the difference in topsoil depth between the rotation plots and the timothy plots was used to determine that 12.4 cm of soil was lost from the rotation plots in 74 years. This equals about 21 t ha^{-1} y^{-1} of soil erosion in the rotation plots. Similar calculations were performed for the difference between the topsoil depth in the timothy minus the corn plots. Because it was observed that subsoil had been mixed in the plow layer of the corn plots as early as 1948 (Woodruff, personal communication), it was concluded that 22 cm of soil was lost from the corn plots in 60 years. This approximates an average annual soil loss of 50 t ha^{-1} y^{-1}. This is confirmed by the fact that in the continuous corn plot the A and E horizons have been completely removed and the argillic horizon became a part of the "plow layer". Productivity estimates indicate that 60% reductions in corn yield of that expected in an uneroded soil might be anticipated for a claypan, even when managed with high fertility.

F. SOIL NUTRIENT STATUS

The effect of long-term cultivation on soil nutrient status can be seen in progressive changes in several major nutrients and in the general level of the fertility in various plots today, after 100 years of cropping. When first established, the principal treatment on Sanborn Field was 13.4 Mg ha^{-1} of manure applied annually. In subsequent years, this amount was considered to be larger than practical. For crop varieties used in the early period of Sanborn experiments, such large applications of manure prompted lodging in wheat and oat crops and injured clover stands either through smothering or through microbial competition for plant nutrients. In 1913, the annual manure application on some plots was reduced to 6.7 Mg ha^{-1} and on others was omitted. On plots where manure was discontinued after 25 years of use, its residual effect on crop yields decreased rapidly. Crops sustained by any residual nutrients from manure produced yields that within a few years were no better than yields from plots that had not received a manure treatment.

Table 5 Estimated Amounts of Nutrients Removed from the Soil by Continuous Cropping during 100 years[9]

Cropping System	Mg C ha^{-1}		kg ha^{-1}			
	Grain	Residue/Hay	Ca	P	N	K
Continuous wheat	53.5	88.2	154	217	1419	1265
Continuous corn	92.5	228.0	1250	445	3428	3665
Continuous timothy	—	95.7	355	134	1014	1521

Note: Data are from plots that received no fertilizer. Removal was calculated from weight of grain and residue, based on data on average composition of grain and forage.

Comparisons made between the use of manure and commercial fertilizers on rotation plots during the period of 1888 to 1942 showed that fertilizer produced higher yields for wheat, oats, and timothy. Corn and clover production, in contrast, was higher on manured plots. Annual application of manure provides an additional input of organic material to the soil beyond that provided by the crop residues. Analyses of soil collected periodically over the history of Sanborn Field from the 25th to 75th year showed that organic C in soil treated annually with 13.4 Mg ha^{-1} was about one half a percentage unit higher than the level in an equivalent cropping system without manure.

The removal of major nutrients during 100 years (Table 5) and the annual removal rates for these elements based on their contents in the forage[10] removed for each quarter of the century (Table 6) have been calculated.[11] Amounts of P and K removed from the soil over 100 years are much higher than the soil-available storage pools under native vegetation (Table 7). As the pools of these elements declined, their rate of uptake decreased and the productivity of the soil deteriorated dramatically. Blanchar and Conkling[12] analyzed differences in P concentration in the lower B and upper C horizons in fertilized and nonfertilized plots and came to the conclusion that crops have the potential to remove 2 to 3 kg ha^{-1} y^{-1} of P from the subsoil.

Continuous wheat had the least detrimental effect on the soil when 13.4 Mg ha^{-1} of manure was applied annually. Under this treatment, soil pH increased substantially, organic matter content stabilized at 2.8%, and soil accumulated available P_2O_5 and K_2O to very high levels (Table 7). One hundred years of continuous nonfertilized corn dramatically changed the nutrient level of the soil. Most of the organic matter and nutrients were lost from the soil partly as a result of erosion. A lower level of fertility likely decreased biomass yields which in turn offered less soil protection, especially since the stover also was removed. In contrast, application of 13.4 Mg ha^{-1} manure maintained pH, organic matter, and built up P and K reserves. The lack of corn yield response to manure noted previously may be more closely associated with soil properties other than fertility status.

The general loss of productivity demonstrated on Sanborn Field for continuous nonfertilized farming systems has proven them to be impractical. Further research on these plots might be to split them in

Table 6 Estimated Yearly Rates of Nutrient Uptake by Continuous Nonfertilized Crops[11]

Cropping System	Years	Ca (kg ha^{-1})	P (kg ha^{-1})	N (kg ha^{-1})	K (kg ha^{-1})
Continuous wheat	0–25	2.2	2.5	17.3	18.0
	25–50	2.8	3.0	20.6	22.1
	50–75[a]	1.0	1.5	9.7	8.1
	75–100	0.2	1.6	9.2	2.4
Continuous corn	0–25	12.4	5.5	39.0	37.1
	25–50	12.5	4.8	35.7	37.0
	50–75	13.2	4.4	34.3	38.4
	75–100	11.9	3.4	28.2	34.1
Continuous timothy	0–25	3.8	1.5	11.5	17.3
	25–50	3.3	1.3	10.0	15.0
	50–75	3.4	1.4	10.3	15.5
	75–100	2.9	1.2	8.7	13.1

[a] Large differences in nutrient uptake by wheat in years 0 to 50 compared with years 50 to 100 are due to the change in residue treatment. During the first period, straw was removed with the grain; during the last 40 years, straw was returned to the plot.

Table 7 Nutrient Levels of Various Plots of Sanborn Field (0- to 20-cm layer)[11]

Cropping System		Organic Matter (%)	P Bray 1 (kg ha^{-1})	K (kg ha^{-1})	pH
Continuous wheat	Nonfertilized	1.4	19	231	4.5
	6.7 Mg ha^{-1} manure	2.2	103	711	5.8
	13.4 Mg ha^{-1} manure	2.8	213	885	5.8
	Full treatment	2.2	109	350	5.8
Continuous corn	Nonfertilized	1.0	11	277	4.5
	13.4 Mg ha^{-1} manure	2.4	183	657	6.2
Corn-wheat-clover	Nonfertilized	2.0	7	132	5.1
	13.4 Mg ha^{-1} manure	3.2	94	246	5.6
Corn-oats-wheat-clover	Nonfertilized	1.9	8	116	5.0
	13.4 Mg ha^{-1} manure	2.4	68	263	5.6
Corn-oats-wheat-clover-timothy-timothy	Nonfertilized	1.9	8	113	4.6
	13.4 Mg ha^{-1} manure	2.9	57	186	5.3
	Full treatment	2.0	68	260	6.6
Continuous timothy	Nonfertilized	2.1	7	118	4.8
	13.4 Mg ha^{-1} manure	3.4	176	268	6.1
Tucker Prairie[a]		4.2	417	346	6.7

[a] Tucker Prairie is a mixed-grass prairie that is managed to maintain a vegetation diversity similar to native prairie. Soil is Putnam-Mexico silt loam, similar to that of the original Sanborn Field site. The P extractant in this case was Bray 2 (0.1 *N* HCl in 0.03 *N* NH_4F)

order to determine how long it would take to restore, at least partly, the productivity of exhausted soil by introducing rotations, manure, or mineral fertilizers.

VII. SOIL ORGANIC MATTER

Soil organic matter (SOM) has been significantly influenced by the crop and soil management practices applied to Sanborn Field during the last 100 years. When cultivation was introduced on the field, SOM declined from the virgin level.[15] The exact level prior to cultivation has not been firmly established but is thought to be nearly double that in most plots today and similar to that in Tucker Prairie, a virgin tallgrass prairie located 30 km east of Sanborn Field. The rate of loss of SOM was initially rapid and progressive until 25 to 45 years later, at which time a new equilibrium level presumably was reached. The new level was characteristic of the crop and the management practices followed on each respective plot. Annual input of carbonaceous residues, their humification, and the yearly mineralization of the organic matter existing in the soil were factors defining the equilibrium level. Several of the plots demonstrated a further adjustment in SOM at a later date when subsequent changes in management were introduced. Of particular interest are plots in which the annual residue input increased after 1950 because straw and stalks were incorporated into the soil rather than removed. This additional residue input resulted in a progressive increase in SOM until a new equilibrium level was reached (Figure 6). The response after 1950 for plot 2 cultivated to wheat and plot 6 cultivated to corn clearly shows the positive effect due to additional residues.

The level of SOM is reflected in measurements of both soil organic C and total N. These determinations were made at intervals of 25 years or more frequently during the 100-year history of the field; each parameter is discussed separately.

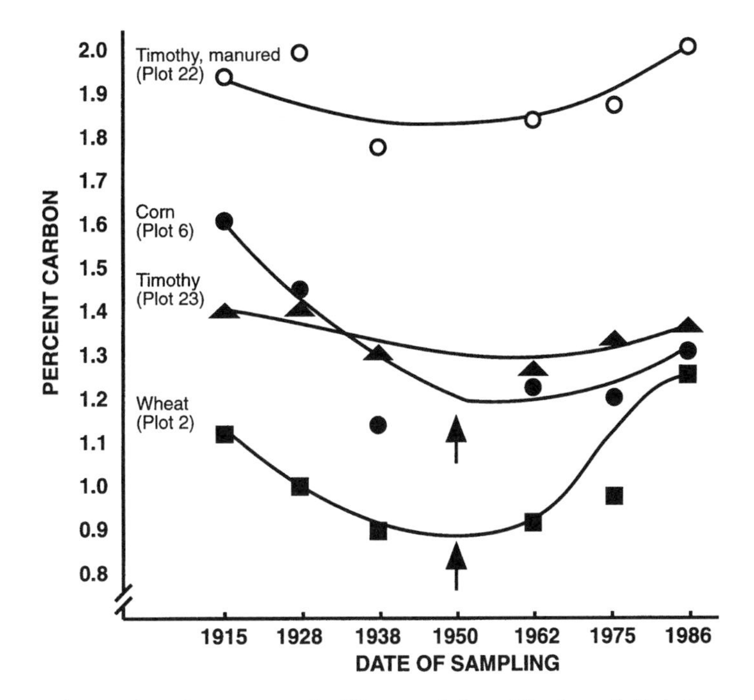

Figure 6 Changes in organic carbon content of soil in several plots of Sanborn Field during cropping of wheat, corn, and timothy. Arrows indicate date after which all above-ground residues on particular plots were returned to the soil.

A. SOIL CARBON

The soil organic C content for selected plots of Sanborn Field during the first 100 years of the field (Figure 6) serves to show the variability among plots and the magnitude of changes over time. Soil C level was highest under timothy which received annual additions of manure (plot 22). Timothy without manure had an organic C level, over the course of study, that was 0.5 percentage points less than that receiving manure annually. A similar effect of manure applied to corn and wheat also was found.[15]

Soil cultivated to various forages until 1949 demonstrated a progressive decline in soil C up to that time (plot 6). From 1950 onward, the plot was cultivated continuously to corn with the residue incorporated into the soil each year. This resulted in an increase in the level of soil C and it is probable that the level in 1986 represents a new equilibrium for this management system at about 1.3% C.

Continuous wheat, chemically fertilized to optimize yields, demonstrated an equilibrium level of soil C at 40 years of about 0.9% (plot 2). This was characteristic of the management system where straw was removed from the plot each year so that the annual input of residues was only that of stubble and roots. From 1950 onward, the straw was added back and incorporated into the soil of this plot. The additional residue input resulted in a significant increase in soil C over the next 40 years after which a probable equilibrium level is developing of at least 1.2% C, the level in 1986. Residue humification and SOM mineralization as they influence these levels has been previously reported.[15] Confirmation of achieved equilibrium of soil carbon for corn (plot 6) and for wheat (plot 2) will require another 10+ years with the same management practices in place. The further sampling and measuring of C in these plots is an objective to be achieved by following the 2nd-century plan for Sanborn Field which does not alter treatments for these plots.

1. Natural Abundance of ^{13}C in SOM

A unique approach to characterizing the soil humus complex employs ^{13}C natural abundance to distinguish the origin of soil C.[16] This has been especially useful in the study of long-term changes in SOM. The long persistence of a major portion of the soil organic C in plots on Sanborn Field has been demonstrated by this approach. The technique allows examination of changes in origin of SOM over time beginning with native prairie grass residues which were humified before the land was brought under cultivation. The "natural label" permitted us to distinguish the prairie C and its loss along with total C changes over the subsequent history of the field. At the same time, the method permitted the estimation of new soil organic carbon arising from residues of the cultivated crop that was introduced.

Total C and ^{13}C data demonstrated that, under wheat culture, much SOM of prairie origin was rapidly lost and very slowly replaced in part by new SOM of crop origin,[16] as illustrated in Figure 7. The initial loss of prairie-origin C from partially humified plant litter comprised a very labile component. Early phases of the breakdown curve are not precisely defined, but it was estimated that this labile fraction had a half-life of 11 years. Only part of the prairie SOM possessed this labile character. A sizable pool of SOM was resistant to the degree that it will continue to persist for hundreds of years. This resistant pool was estimated to have a turnover time approaching 1000 years.[17] After 100 years of cultivation, the resistant pool of prairie origin still comprised at least 50% of current SOM under all systems of cultivation studied.

The natural abundance method also was adapted to show the process of replacement of the soil C under long-term cultivation following change from one crop to another[18] (Figure 8). Some of the SOM that had previously accumulated to an equilibrium level after 60 years of C-3 forages progressively decreased when cropping shifted to continuous C-4 corn. The decay curve for the soil C originating from C-3 plants was evaluated for the next 25 years. Even though the total SOM level remained about the same during this period under corn, the tracer method showed the decline in C-3 SOM at a rate demonstrating a half-life of 12 years or a value similar to that for the labile fraction of prairie SOM. The study also showed that the SOM of C-3 crop origin included a resistant pool, relatively small in size, but possessing a persistence like that for the stable prairie SOM.

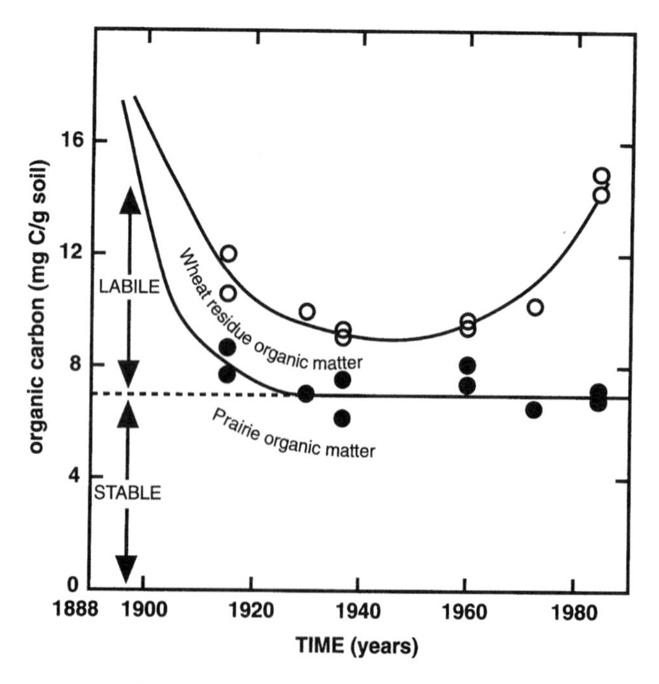

Figure 7 Changes in amount and origin of soil organic carbon with cropping to winter wheat on formerly virgin prairie soil. Open circles indicate total soil carbon; closed circles indicate that from prairie vegetation. Each data set includes observations for 0- to 10- and for 10- to 20-cm sampling depths.

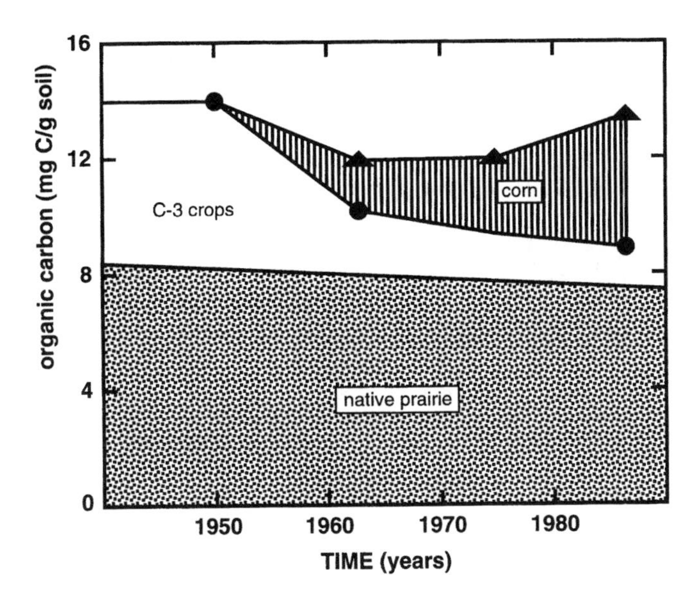

Figure 8 Changes in amount and origin of organic carbon in soil that has been cropped to forages for 60 years and then to corn for 36 years. (From Balesdent, J., Wagner, G. H., and Mariotti, A., *Soil Sci. Soc. Am. J.*, 52, 118, 1988. With permission.)

The decay curves for labile SOM are imperfect representations of first-order kinetics. Calculating a decay constant (k) at different dates shows that the value becomes progressively smaller with time. In the case of the C-3 forage SOM, k changed from 0.089 to 0.049 y^{-1} for observations beginning at 10 and extending to 36 years.

A well-established premise of SOM dynamics concerns the decline in SOM with cultivation to a level defined by the particular cultivation system. Using the ^{13}C technique to better understand this phenomenon, we partitioned current levels of total soil C between the sources of origin for several cultural practices.[18] The partitioning among the sources of origin was done after an equilibrium level had been achieved for a particular system of cultivation. Several of these relationships are shown in Figure 9. Under timothy, the extent of persistent prairie SOM was greater than that under cultivation of corn or wheat. The amount of SOM attributable to the wheat crop was related to the quantity of wheat residue returned to the land annually. With straw from wheat included as part of the input, the crop-origin SOM was approximately doubled as compared with only the stubble and roots. In addition to input from crop residues alone, annual manure additions demonstrated appreciable new SOM and, in the case of timothy culture, suggested that the rate of loss of prairie-origin SOM also was slightly reduced.

2. Humification and Mineralization Balances

Turnover of SOM partitioned by the tracer method was evaluated by calculating a balance sheet for C using annual inputs reconciled against known mineralization rates for the native SOM pool (Table 8). Assuming a mineralization rate of native SOM under wheat at 1.7% y^{-1}, from a pool of 2300 g m^{-2} C to a depth of 20 cm in the fertilized plot, the annual release of CO_2 would equal 39 g of C. Annual replacement by stubble and roots totaling 185 g of C m^{-2} is expected to contribute soil C, adjusted by the appropriate coefficient of humification of 0.2,[19] at an input of 37 g C m^{-2} y^{-1}. For the particular plot on Sanborn Field where these data were obtained during the period 1939 to 1950, total soil C was apparently at equilibrium level.

After 1950, when straw was not removed from the aforementioned plot, annual residue input was significantly increased so that net input was estimated at 65 g C m^{-2} y^{-1}. This annual input of C to humus then exceeded net mineralization by 26 g C m^{-2} y^{-1}. The positive balance should result in an accumulation of crop origin C, which stood at a low of 645 g C m^{-2} in 1950. When measurements were taken in 1986, the level of crop-origin C had risen to 1665 g m^{-2} and total soil organic C was 3275 g m^{-2}, with the latter value assumed to be at or approaching a new equilibrium level characterized by a balance between humification and mineralization.

Using equilibrium levels and quantities of SOM attributable to the crop, along with data on annual input of residues, we were able to calculate mean turnover times. For wheat stubble in two contrasting

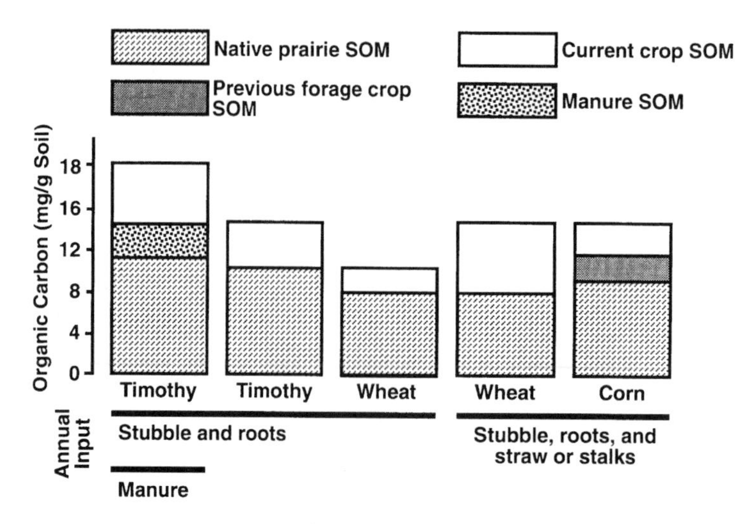

Figure 9 Amount and origin of soil organic C after an equilibrium level was attained under long-term cultivation of different crops and inputs. (From Wagner, G. H., *Int. Symp. Use Stable Isotopes in Plant Nutrition, Soil Fertility and Environmental Studies,* (IAE-SM-313), International Atomic Energy Agency, Vienna, 1991, 261. With permission.)

Table 8 Soil Organic Carbon Relationships in the Ap Horizon under Different Crop Management Systems (g m^{-2})

	End of Period		Annual		
Crop Residue Input and Period	Total Organic C	Crop-Origin C	Carbon Mineralized from SOM	Fresh Residue Carbon	Net Input of Humus-C
Wheat stubble, 1938–1950	2300	645	39	185	37
Wheat stubble and straw, 1980–1986	3275	1665	65	370	65
Corn stubble, 1975–1986	3375	1147	84	917	91
Timothy stubble, 1915–1986	3360	988	40	156	39
Timothy stubble plus manure, 1915–1986	4945	1997	77	157[a] 375[b]	77

[a] Attributed to timothy residues.
[b] Attributed to manure.

From Wagner, G.H., *Int. Symp. Use Stable Isotopes in Plant Nutrition, Soil Fertility and Environmental Studies,* (IAE-SM-313), International Atomic Energy Agency, Vienna, 1991, 261. With permission.

situations, one without the bulk of the straw and the other with all straw added back, the turnover times were 4.5 and 4.7 years, respectively. Because the ancient, prairie-origin C was excluded from the calculation, the turnover times for the two wheat plots were essentially the same, apparently controlled by the kind of residue and cultivation practiced but not the quantity of residue added. If the calculations had been made using total organic C in the soil as has been the classical approach, the conclusion would have been that wheat stubble without the bulk of the straw was characterized by a slower turnover than was the case when all the straw was incorporated into the soil.

B. SOIL NITROGEN

Changes in total soil N over time for corn, wheat, and timothy plots with and without manure are shown in Figure 10. Without treatment, the soil demonstrated a progressive decline in N throughout the 100 years of experimentation for both corn and wheat. In the case of timothy, soil N was stable after 1915. Scattered legumes in the timothy plot or the occurrence of nonsymbiotic N fixation is probably keeping N at a uniform level even though the annual harvest of N in the forage removed was 11 kg ha^{-1} y^{-1}.

The timothy plot receiving an annual manure treatment to supply an estimated 110+ kg N ha^{-1} y^{-1} had a soil N level over time that was 0.05 percentage points higher than that for the unmanured plot. Differences in total soil N between manured and nonmanured timothy appear to be particularly significant at the 1986 sampling. Efforts after 1980 to be more precise in defining the N added in manure may have been the reason for the latter observation. The N level in the corn and wheat plots receiving manure annually appears to be unchanged during the past 75 years.

Total N levels for manured and nonmanured plots cultivated to two different rotations are reported in Figure 11. These rotations show equilibrium for N achieved at 50 years. Again, the manure treatment displayed a total N level of about 0.05 percentage points above that for the unmanured corresponding rotation.

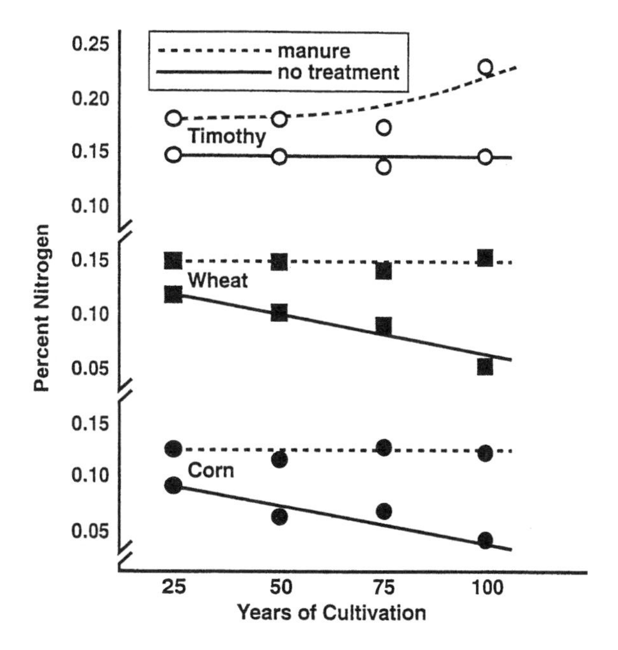

Figure 10 Changes in total soil nitrogen over time for plots of Sanborn Field cultivated continuously for 100 years to various crops without treatment and with 13.4 Mg ha^{-1} of manure annually.

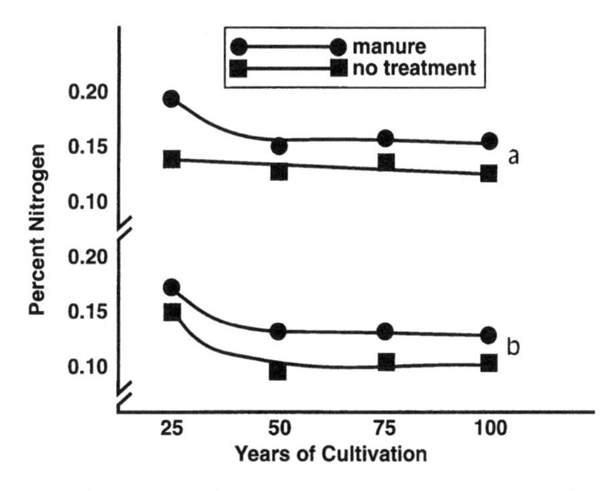

Figure 11 Changes in total soil nitrogen during 100 years of cropping to corn, wheat, red clover rotation (a), and corn, oats, wheat, red clover rotation (b), each with manure and without treatment.

VIII. CONCLUSIONS

The data reported in this paper demonstrated close interaction of soil nutrient status and crop performance. Near-complete mineral depletion and loss of SOM have been shown under continuous grain crops, and loss of mineral nutrients occurred under continuous timothy when no manure or fertilizer was applied. The content of organic matter under corn and wheat decreased to one third to one fourth that in the native prairie in our geographical area.

Yields of crops grown without manure or fertilizers for 100 years were one half to one third those with these treatments because the essential nutrients were depleted. Visual observation today allows one to distinguish the plot with continuous wheat and the plot with continuous corn even in winter when no crop is growing by the soil color and structure. Continuous timothy did not have such a deleterious effect on SOM, but the level of organic matter in this plot is significantly lower than that under native prairie. This may be due to less extensive rooting of timothy, generally lower yield potential, and periodical stand deterioration requiring that it be reseeded regularly. Tillage kills the roots, mixes the organic matter in the plow layer, and accelerated the loss of organic carbon.

Rotations showed less negative effects over the 100 years compared with nonfertilized continuous cropping. In some cases, rotations had the same influence as the annual application of manure or mineral fertilizers. Adding 2 years of timothy to a 4-year rotation to make a 6-year rotation had little influence on soil properties when plots were nonfertilized. When 13.4 Mg ha^{-1} manure was applied annually, the addition of timothy to the rotation caused an increase in organic matter, but phosphorus, potassium, and pH were lower than for the 4-year rotation. This again suggests that 2 years of timothy in a rotation was not sufficient to maintain soil productivity.

The management decision not to return crop residues to the plots on Sanborn Field during the first 60 years clearly accelerated the decline in SOM. Our data suggest that an equilibrium level was reached sometime between 1935 and 1945. Following the return of crop residues after 1950, SOM on treated plots increased toward a new equilibrium level. Retaining the residue did not increase yields or soil nutrient accumulation with continuous corn or continuous wheat. A short-lived increase in yield was observed in continuous wheat after a new, more productive variety was introduced. Wheat yield subsequently returned to the previous low productivity.

Manure maintained SOM at higher levels than where not applied. The fewer the grain crops in the rotations, the greater the level of SOM. Soil effects attributable to rotations are reduced soil erosion and a depletion in soil nutrients, especially phosphorus.

Crop yields over the 100 years did not necessarily parallel the level of SOM. The improvement in varieties especially after 1939, the change in manure characteristics, and the return of crop residues to the plots after 1950 tended to confuse the direct relationship of SOM and yield. A straightforward comment would be that the greatest difficulty with long-term studies such as Sanborn Field is in maintaining a focused direction by several successive field managers.

In the current era of overproduction, environmental concerns, and the rising popularity of "organic agriculture", we must consider historical, current, and future attitudes regarding continuous cropping and the value of simple rotations. On Sanborn Field, yields of wheat from a nonfertilized 3-year rotation (corn-wheat-clover) were not different from continuous wheat with full treatment for the last 15 years. Corn, with high requirements for nutrients, does not benefit from rotation as well as wheat (corn yields in the same rotation are one half that in the full mineral fertilizer system). However, we cannot preclude that, under the pressure of environmental concerns, some producers in the near future will opt for lower yields of "organically grown" corn and other grain crops. There is growing evidence that the public will be willing to pay higher prices for products that have been produced using fewer chemicals.

REFERENCES

1. Missouri Agricultural Experiment Station Records, Bound Field Notebook, 1888–1902, Rotation Series Notes, UMC Agronomy Dept. Vault, 30.
2. Jenny, H., *Factors of Soil Formation*, John Wiley & Sons, New York, 1941.
3. National Oceanic and Atmospheric Administration, Climatological Data, Annual Summary, Missouri: National Climatic Data Center, Asheville, NC, 1991.
4. Upchurch, W.J., Kinder, R.J., Brown, J.R., and Wagner, G.H., Sanborn Field Historical Perspective, Missouri Agric. Exp. Stn. Res. Bull. 1054, University of Missouri, Columbia, 1985.
5. Krusekopf, H.H. and Scrivner, C.L., Soil Survey — Boone County, Missouri Series 1951, No. 12. Soil Conservation Service, USDA, Washington, DC, 1962.

6. Held, R.J., Soil survey of Montgomery and Warren Counties, Missouri, USDA Soil Conservation Service and Missouri Agricultural Experiment Station, U.S. Government Printing Office, Washington, DC, 1978.
7. Miles, R.J. and Hammer, R.D., One hundred years of Sanborn Field: soil baseline data, in Proc. of the Sanborn Field Centennial, Missouri Agric. Exp. Stn. Spec. Rep. 415, 1989, 100.
8. Miller, M.F. and Hudelson, R.R., Thirty years field experiments with crop rotation, manure and fertilizers, Missouri Agric. Exp. Stn. Bull. 182, 1921.
9. Brown, J.R. and Breight, S.G., Changes in plant nutrients, a century of change, in Proc. of the Sanborn Field Centennial, Missouri Agric. Exp. Stn. Spec. Rep. 415, 1989, 124.
10. Morrison, F.B., *Feeds and Feeding. A Handbook for Student and Stockman*, 22nd ed., Morrison Publishing, Ithaca, NY, 1956.
11. Buyanovsky, G.A., Nelson, C.J., and Breight, S., 100-year-old cropping systems: yields and soil properties, in Proc. of the Sanborn Field Centennial, Missouri Agric. Exp. Stn. Spec. Rep. 415, 1989, 140.
12. Blanchar, R.W. and Conkling, B.L., Differences in subsoil phosphate in P-fertilized and unfertilized Mexico soil, in Proc. of the Sanborn Field Centennial, Missouri Agric. Exp. Stn. Spec. Rep. 415, 1989, 92.
13. Anderson, S.M. and Gantzer, C.J., Soil physical properties after 100 years of cropping on Sanborn Field, in Proc. of the Sanborn Field Centennial, Missouri Agric. Exp. Stn. Spec. Rep. 415, 1989, 71.
14. Gantzer, C.J., Anderson, S.H., and Thompson, A.L., Estimating soil erosion after 100 years of cropping on Sanborn Field, in Proc. of the Sanborn Field Centennial, Missouri Agric. Exp. Stn. Spec. Rep. 415, 1989, 83.
15. Wagner, G.H., Lessons in soil organic matter from Sanborn Field, in Proc. Sanborn Field Centennial, Missouri Agric. Exp. Stn. Spec. Rep. 415, 1989, 64.
16. Wagner, G.H., Using natural abundance of ^{13}C and ^{15}N to examine soil organic matter accumulated during 100 years of cropping, in *Int. Symp. Use Stable Isotopes in Plant Nutrition, Soil Fertility and Environmental Studies*, (IAEA-SM-313), International Atomic Energy Agency, Vienna, 1991, 261.
17. Hsieh, Y., Pool size and mean age of stable soil organic carbon in cropland, *Soil Sci. Soc. Am. J.*, 56, 460, 1992.
18. Balesdent, J., Wagner, G.H., and Mariotti, A., Soil organic matter turnover in long-term field experiments as revealed by ^{13}C natural abundance, *Soil Sci. Soc. Am. J.*, 52, 118, 1988.
19. Buyanovsky, G.A. and Wagner, G.H., Crop residue input to soil organic matter on Sanborn Field, in *Soil Organic Matter in Temperate Agroecosystems in North America*, Paul, E.A., Paustian, K., Elliott, E.T., and Cole, C.V., Eds., CRC Press, Boca Raton, FL, 1997, chap. 4.

Chapter 16

Soil Organic Matter under Long-Term No-Tillage and Conventional Tillage Corn Production in Kentucky

W.W. Frye and R.L. Blevins

CONTENTS

I. INTRODUCTION

Soil organic matter improves the agricultural quality of soil and helps mitigate certain environmental problems. It improves the soil's physical characteristics, plant-water relations, fertility status, and overall productivity. It decreases soil erosion and, in doing so, enhances water quality.

Recently, soil organic matter has been ascribed another important role in environmental enhancement. It has the potential to contribute to decreasing the CO_2 emissions associated with global warming. To achieve that potential, substantial quantities of CO_2 must be sequestered as soil organic C for long periods of time. This would involve long-term increases in soil organic matter while still using the soil for intensive, economic crop production.

A report of the Council for Agricultural Science and Technology estimated that conservation tillage could decrease the current levels of CO_2 emissions from U.S. agriculture by one third.[1] Their estimate was based on a scenario established by U.S. Environmental Protection Agency that conservation tillage

0-8493-2802-0/97/$0.00+$.50

could change soil organic matter from a net loss of about 0.01 Gt y^{-1} of CO_2 to a net sequestering of about 0.04 Gt annually[2] (Gt or gigaton = 10^9 metric tons or 10^{12} kg).

No-tillage and other forms of conservation tillage may result in higher soil organic matter contents than conventional moldboard plow tillage under similar soil, climate, and cropping conditions. A common misconception, however, is that no-tillage always increases soil organic matter. In cases where the soil organic matter content is initially relatively high, such as grasslands, conventional tillage accelerates the decomposition rate more than no-tillage, but soil organic matter is likely to decrease under both tillage systems.[3] Conversely, if the soil organic matter content is initially low because of previous intensive cropping with conventional tillage, no-tillage usually will substantially increase the level of soil organic matter.[4]

A number of factors affect the relationship between tillage and soil organic matter. Chief among these are amount of soil disturbance, distribution of plant residues and soil amendments, soil erosion, and effect of tillage on yields of crop and cover crop. The soil disturbance and manipulation by plow-tillage and the secondary tillage (mostly disking) normally associated with it increase the rate of decomposition of organic matter by increasing aeration, exposing heretofore protected organic matter, and mixing organic residues from the surface into the soil where conditions are usually more favorable for rapid decomposition.[5,6]

In no-tillage on the other hand, little soil is moved or mixed. Plant residues, fertilizers, lime, manures, and agricultural chemicals are applied to the soil surface and are not mechanically mixed into the soil. Plant roots tend to concentrate near the soil surface. The surface few centimeters of soil usually are cooler, wetter, less oxidative, and more acidic than in conventional tillage soil.[3,5,6] These conditions in no-tillage tend to cause the soil organic matter content to increase, or at least decrease at a slower rate, relative to conventional tillage. Furthermore, less soil erosion occurs with no-tillage than with conventional tillage.[7,8] Loss of topsoil through soil erosion disproportionately diminishes the soil's organic matter, nutrient supply, water infiltration capacity, rooting environment, and inherent productivity because the topsoil is the seat of these qualities.

Crop yields are not only affected by soil organic matter; in turn, they affect soil organic matter content. Soil organic matter supplies plant nutrients, particularly N, through mineralization and improves soil structure, thereby increasing infiltration of water and increasing the capacity of the soil to supply water to a crop during droughty weather. As crop yields increase, more residue is returned to the soil, increasing soil organic matter content. To a point, this becomes a self-perpetuating relationship between soil organic matter and soil productivity.

Our objective was to determine the interrelationships between tillage systems, N fertilizer rates, corn (*Zea mays* L.) grain yields with time, and soil organic C and N contents in a long-term (20 year) no-tillage and conventional plow-tillage field experiment with continuous corn.

II. EXPERIMENTAL SITE AND PROCEDURES

A. SITE CHARACTERISTICS

The field experiment, which was the site of this study, was started in 1970 on the Kentucky Agricultural Experiment Station Spindletop Farm at Lexington. The experiment is located about 38° 07′ N, 84° 29′ W. The soil is a Maury silt loam (fine, mixed, mesic Typic Paleudalfs). The Maury soil is deep, well drained, and formed in residuum of phosphatic limestone. Mineralogy is mixed, but predominant clays are vermiculite, kaolinite, and illite, usually in that order. The soil contains no free $CaCO_3$; however, exchangeable Ca^{2+} is usually in the range of 8 to 10 $cmol_c$ kg^{-1} throughout most of the root zone. The site is on a 1 to 3% slope with south aspect. Erosion class is none-to-slight.

The native vegetation in the area was mostly hardwood forest. The Maury soil, however, has a thicker and darker colored surface horizon than is normal for soils formed under forest vegetation, indicating influence from grass vegetation. Bluegrass (*Poa pratensis* L.) is native to the area. The site had been in a bluegrass pasture for horses for about 50 years before the start of the experiment. Thoroughbred horse pastures are typically characterized by high maintenance and low grazing intensity. However, we have no records of specific prior management on this site.

Climate is temperate with udic rainfall and mesic temperature regimes. Average annual rainfall is about 1140 mm with about 40% of it occurring during the period May through September. Table 1 shows the monthly and seasonal rainfall amounts for May through August in 1970 through 1990. The average annual temperature is about 13°C. The growing period averages about 175 days.

Table 1 Monthly Rainfall Amounts during the Growing Season for 1970 through 1990 at Spindletop Farm, Lexington, KY

	Rainfall (mm)				
Year	**May**	**June**	**July**	**August**	**4-month total**
1970	82	90	53	33	258
1971	175	95	177	20	467
1972	88	80	123	118	409
1973	155	160	112	38	465
1974	162	237	78	300	777
1975	93	138	85	105	421
1976	78	90	140	60	368
1977	45	105	88	183	421
1978	130	60	120	200	510
1979	110	75	165	108	458
1980	49	59	196	82	386
1981	117	61	134	54	428
1982	50	103	91	126	370
1983	85	270	26	41	422
1984	153	121	34	37	345
1985	113	125	167	98	503
1986	80	22	131	73	306
1987	43	149	74	19	285
1988	64	14	97	85	260
1989	126	140	111	99	476
1990	163	116	121	130	530

B. DESIGN AND TREATMENTS

The experimental design was a split block with four replications.[3] Each 11.0- × 48.8-m block was split horizontally for randomized N fertilizer treatments of 0, 84, 168, and 336 kg N ha^{-1} as ammonium nitrate (330 g N kg^{-1}) each year. The blocks were split vertically, and conventional tillage and no-tillage treatments were assigned randomly to the two halves of each block. The tillage and fertilizer treatments were maintained on the same plots throughout the experiment. Each subplot was 5.5 × 12.2 m and contained six rows of corn 0.916 m apart. The subplots were split horizontally in 1973 into 5.5- × 6.1-m sub-subplots for installation of limed (6.7 Mg ha^{-1} agricultural lime) and unlimed treatments. Agricultural lime was applied again in 1975 at the rate of 11.2 Mg ha^{-1}. In 1983, all plots were limed according to soil test recommendation to achieve uniform pH across treatments.

C. EXPERIMENTAL PROCEDURES

The conventional tillage plots were plowed to a depth of 20 to 25 cm in late April or early May each year, about 1 to 2 weeks before planting corn. Secondary tillage of the conventional tillage plots consisted of two trips over the plots with a tandem disk harrow cutting to a depth of about 8 cm. Corn was planted about 10 May each year with a no-tillage planter equipped with fluted coulters. The seeding rate was aimed at obtaining a final stand of about 50,000 plants ha^{-1}. Cultivars were as follows: P.A.G. SX-29 in 1970; Pioneer 3369A in 1971 to 1982; B73XPA91 in 1983 to 1988; and Pioneer 3165 in 1989 and 1990. All plots were sprayed at planting with a mixture of herbicides described by Blevins et al.[3] Muriate of potash (KCl, 500 g K kg^{-1}) was applied broadcast at the rate of 100 kg K ha^{-1} when needed to maintain soil test K at a very high level. No P was applied because the soil is naturally high in phosphate.

Corn was harvested from the two center rows of each sub-subplot about 1 October each year, except 1983 when a severe drought decimated yields and the corn was not harvested. Grain yields were corrected to a water content of 155 g kg^{-1}. Rye (*Secale cereale* L.) seed was broadcast on all plots in mid-September (2 weeks before harvesting corn) at a rate of about 188 kg ha^{-1}. Shredded corn stalks were left on the soil surface following corn harvest.

Soil samples were collected 23 August 1989 from depths of 0 to 5, 5 to 15, and 15 to 30 cm of each sub-subplot, dried, ground, and analyzed for soil organic C and N. Separate samples (about 400 to 500 g each) were collected from each depth using a 1.9-cm diameter Oakfield soil probe to determine soil bulk density. The samples of known volume were oven dried at 110°C for 48 hours, cooled in a desiccator, and weighed.

For all statistical analyses, we accepted the 5% level of probability ($P \leq 0.05$) as significant.

III. RESULTS

A. GRAIN YIELDS

Corn grain yields, averaged for 5-year periods, are shown in Table 2. Average yields were greater for conventional tillage than no-tillage during the first 5 years, about the same for both tillage systems for the 5- to 10- and 10- to 15-year periods and greater for no-tillage during the 15- to 20-year period. Overall grain yields were not significantly different for the two tillage systems when compared for the entire 20 years of the experiment.

Table 2 Corn Grain Yields as Affected by N Fertilizer, Tillage System, and Time

N Rate (kg ha^{-1})	Tillage	Years (Mg ha^{-1}) 1970–1974	1975–1979	1989–1985	1986–1990	Tillage Mean[a] (Mg ha^{-1})	N Mean[a] (Mg ha^{-1})
0	CT[b]	7.11	4.84	3.91	3.23	4.77a	4.54c
	NT[b]	5.79	3.68	3.73	3.98	4.30b	
84	CT	8.98	7.05	6.15	5.08	6.81a	6.79b
	NT	8.67	6.81	5.93	5.66	6.77a	
168	CT	8.78	7.16	6.68	5.83	7.11b	7.31a
	NT	8.89	7.19	7.46	6.45	7.50a	
336	CT	8.96	7.42	7.71	5.59	7.42a	7.58a
	NT	9.04	7.66	7.39	6.24	7.58a	
Tillage mean[a]	CT	8.46a	6.62a	6.11a	4.93b		
	NT	8.10b	6.34a	6.13a	5.58a		
Years mean[a]		8.28a	6.48b	6.12b	5.26c		

Note: Yields averaged for each 5-year increment (1970 through 1990).

[a] Tillage means (for each N rate and for each time period), N means, and years means followed by different letters are significantly different ($P<0.05$) by LSD.

[b] CT = conventional tillage; NT = no tillage.

The yield response to N fertilizer was largest for the first 84-kg N ha^{-1} increment, as would be expected, and was small and nonsignificant for the last increment, i.e., 336 kg N ha^{-1}. Yields without N fertilizer were the lowest, and yields with the 84-kg N ha^{-1} rate were lower than with the 168- or 336-kg N ha^{-1} rates, which were not significantly different. With no N fertilizer, conventional tillage outyielded no-tillage; the trend was the reverse at the higher N rates where no-tillage yields were higher than conventional tillage yields with 168 kg N ha^{-1} (Table 2).

Finally, grain yields have clearly decreased with time under both tillage systems and at all N rates. The decrease was greater with conventional tillage than with no-tillage. We are uncertain about the reason or reasons for this decline. Growing seasons with unfavorable rainfall distribution occurred frequently during the 1980s (Table 1) and weed problems have erupted periodically during the course of the study. These factors have undoubtedly decreased grain yields during certain years, but could hardly explain a significant decreasing trend with time. The only significant ($P<0.05$) correlation coefficient obtained between grain yields and rainfall amounts, however, was $r = 0.55$ between the 4-month total rainfall and overall average corn yields.

B. SOIL ORGANIC C

Soil organic C in the 0- to 5-cm depth, averaged across N rates, was significantly greater in no-tillage than in conventional tillage soil after both 5 (1975) and 20 years (1989) (Table 3). Conventional tillage resulted in more organic C than no-tillage at the 5- to 15-cm depth in 1975 and at the 15- to 30-cm depth in 1989. Our results were generally consistent with results obtained by other researchers at several locations.[10–13]

Organic C apparently increased between 1975 and 1989 with both tillage systems (Table 3). With no-tillage, the increase was greatest in the 5- to 15-cm depth, but was also sizable in the 15- to 30-cm depth. Under conventional tillage, soil organic C remained constant in the 0- to 5-cm depth, but increased

Table 3 Soil Organic C Distribution to 30-cm Depth after 5 (1975) and 20 Years (1989) of Continuous Conventional Tillage and No-Tillage Corn

N Rate	Soil Organic C							
	1975				1989			
(kg ha^{-1})	CT[a]	NT[a]	Average	Sod	CT	NT	Average	Sod
				Mg ha^{-1} in 0–5 cm				
0	8.6	13.3	11.0	18.8	8.3	14.6	11.5d	17.9
84	9.4	14.4	11.9		9.4	15.5	12.4c	
168	10.1	14.9	12.5		9.8	16.9	13.4b	
336	10.1	16.4	13.2		11.0	19.0	15.0a	
Average	9.5b[b]	14.8a			9.6b	16.5a		
				Mg ha^{-1} in 5–15 cm				
0	16.2	16.1	16.2	17.9	19.7	21.1	20.4c	18.6
84	18.3	16.5	17.4		20.2	23.0	21.6b	
168	18.8	15.6	17.2		21.9	22.6	22.2ab	
336	18.3	17.8	18.0		23.1	23.7	23.4a	
Average	17.9a	16.5b			21.2a	22.6a		
				Mg ha^{-1} in 15–30 cm				
0	15.0	17.4	16.2	16.7	20.9	19.6	20.5c	19.0
84	20.1	17.5	18.8		26.6	19.8	23.2b	
168	18.8	15.9	17.4		24.7	19.1	21.9bc	
336	17.5	18.5	18.0		27.2	23.5	25.4a	
Average	17.8a	17.3a			24.9a	20.5b		

[a] CT = conventional tillage; NT = no-tillage.

[b] Average values followed by different letters are significantly different P<0.05 for a given year and depth increment by Duncan's multiple range test.

Adapted from Blevins, R.L. et al., *Agron. J.*, 69, 1977 and Ismail et al., *Soil Sci. Soc. Am.*, 58, 194, 1994.

below that depth between 1975 and 1989. The greatest increase with conventional tillage occurred at the 15- to 30-cm depth. Plowing generally ranged from 20- to 25-cm depth, thus mixing much of the corn residue and rye cover crop into the 15- to 30-cm depth increment.

Nitrogen fertilizer significantly increased organic C at all depths in 1989 (Table 3) and at the 0- to 5- and 5- to 15-cm depths in 1975.[3] The most likely reason for this is that higher N rates produced greater biomass, adding more plant residue to the soil. The corn residue added an estimated 7.4 Mg ha^{-1} of dry matter annually and a rye cover crop added about 2.0 Mg ha^{-1} of dry matter each year on the plots receiving 168 kg N ha^{-1}.

Organic C content of the bluegrass sod plots was about the same in 1989 as in 1975.

C. SOIL ORGANIC N

As expected, organic N distribution in the soil followed a similar pattern to organic C (data not shown). The C:N ratios to a depth of 15 cm tended to be higher for no-tillage than conventional tillage, especially in 1989 (Table 4). The reverse tendency appeared in the 15- to 30-cm depth. Moreover, values were lowest at that depth for both tillage methods, but were especially apparent with no-tillage.

Table 4 Organic C:N Ratios to 30-cm Depth after 5 (1975) and 20 Years (1989) of Continuous Conventional Tillage and No-Tillage Corn Averaged across N Fertilizer Rates

Depth	1975		1989	
(cm)	CT	NT	CT	NT
0–15	9.7	10.4	10.6	11.4
5–15	9.5	9.5	9.5	10.4
15–30	9.1	8.0	8.2	7.3

Note: CT = conventional tillage; NT = no-tillage.

D. BULK DENSITY

The bulk density of the 0- to 8-cm depth of soil in March 1975 was reported by Blevins et al.[3] to be nearly identical at 1.25 Mg m^{-3} for all treatments, including the undisturbed bluegrass sod plots. The soil organic C calculations to a depth of 15 cm in 1975 were based on that bulk density; but, for the 15- to 30-cm depth in this article, we used 1.47 Mg m^{-3}, the value obtained in 1989.

The 1989 calculations were based on the bulk density values in Table 5. Bulk density increased significantly with soil depth in both of the tillage treatments and for bluegrass sod plots. There were, however, no significant differences in bulk density among the tillage and bluegrass sod treatments.

Table 5 Bulk Density of Soil under Bluegrass Sod and after 20 Years of Continuous Conventional Tillage and No-Tillage Corn

	Bulk Density (Mg m^{-3})		
Depth (cm)	**Bluegrass Sod**	**CT[a]**	**NT[a]**
0–5[b]	1.15c[c]	1.27b	1.17c
5–15[b]	1.28b	1.27b	1.36b
15–30	1.41a	1.47a	1.47a

[a] CT = conventional tillage; NT = no-tillage.
[b] Sampled as single increment in CT.
[c] Values in a column followed by different letters are significantly different ($P<0.05$) by Duncan's multiple range test.

Adapted from Ismail et al., *Soil Sci. Soc. Am.*, 58, 194, 1994.

IV. DISCUSSION AND CONCLUSIONS

The soil on which this experiment was established in 1970 was in bluegrass sod and had not been tilled for at least 50 years. It had a high soil organic matter equilibrium level. When the soil was cropped in continuous corn, the soil organic C decreased rather rapidly during the first 5 years, even with no-tillage. However, soil organic C under no-tillage did not decrease as low as under conventional tillage. Several factors may contribute to this. Plant residues accumulate at the soil surface in no-tillage and are not mechanically mixed throughout the plow layer as they are in the case of conventional tillage. When mixed into the soil, the residues are in a more favorable environment for decomposition, humification, and mineralization than when left on the soil surface. Moreover, soil disturbance associated with tillage generally accelerates the rate of organic C mineralization by increasing aeration and oxidation and exposing heretofore protected soil organic matter.[5,6]

The pattern of organic C distribution in the soil can also be explained by the soil disturbance in conventional tillage and the lack of disturbance in no-tillage. We were surprised, however, by the small increase in organic C in the 0- to 5-cm depth under no-tillage between 1975 and 1989. With most crop residues added at the soil surface, the greatest accumulation of organic C might be expected there. Surprising also was the larger than expected increase in organic C in the 5- to 15- and 15- to 30-cm depths under no-tillage. Clearly, this represents incorporation of organic matter without mechanical disturbance. Crop roots, along with natural disturbance, such as freezing and thawing and earthworm activity, could account for the incorporation.

When summed for the entire soil profile from the surface to 30 cm (Table 6), the soil organic C differences between tillage systems are less dramatic than when shown on an incremental-depth basis (Table 3). Further, the summed values are somewhat misleading in that they do not reveal the current organic matter distribution nor its redistribution with time as are shown by Table 3. This points out a deficiency with using only the total soil organic matter value for the soil profile without taking into account its distribution in the profile. However, where the objective is to sequester CO_2, total organic C accumulated is more important than its distribution within the soil profile.

We believe that the tendency for the C:N ratio to be higher for no-tillage than conventional tillage in the upper 15 cm of soil is related to the age of the organic matter as proposed by Jannsen.[14] Soil organic matter would be expected to be younger the nearer it is to the surface in a no-tillage soil, and young organic matter would likely have higher C:N ratio than would old organic matter. The plowing in conventional tillage mixes the young and old organic matter, thereby diminishing much of the effect of age of organic matter on C:N ratio.

Table 6 Soil Organic C to 30-cm Depth after 5 (1975) and 20 Years (1989) of Continuous Conventional Tillage and No-Tillage Corn

N Rate	1975 (Mg ha^{-1})			1989 (Mg ha^{-1})		
(kg ha^{-1})	CT[a]	NT[a]	Sod	CT	NT	Sod
0	39.7	46.8	53.4	48.9	55.3	55.5
84	47.8	48.4		56.2	58.3	
168	47.7	46.3		56.4	58.6	
336	45.9	52.8		61.3	66.2	
Average	45.3	48.6		55.7	59.6	

[a] CT = conventional tillage; NT = no-tillage.

Corn grain yields tended to parallel soil organic C content; however, yields have continued to decline even though soil organic matter has been replenished. Grain yields with no-tillage have surpassed yields with conventional tillage at all N fertilizer levels. This may be an effect of the much-higher organic matter content in the 0- to 5-cm depth of the no-tillage soil (Table 3).

Our results indicate that soil organic matter can be restored after being depleted, even with continuous corn production. Organic C content of the surface 30 cm of both conventional tillage and no-tillage soil returned to at or above that of the bluegrass sod within 15 years following an initial drop of 15% with conventional tillage and 9% with no-tillage during the first 5 years of the experiment. Clearly, with such practices as a cover crop, adequate N fertilization (excess N is not necessary), crop residues left in the field, and erosion kept under control, organic matter can be replenished in a soil where it has been depleted by cropping. Appreciable CO_2 is sequestered in the process. Our results suggest that these practices are more important than the tillage system per se.

In our experiment, from 1975 to 1989, organic C in the soil profile increased by 10.4 Mg ha^{-1} for conventional tillage and 11.0 Mg ha^{-1} for no-tillage when averaged across all N rates (Table 6). At the 336-kg N ha^{-1} rate, the increases were 15.4 Mg C ha^{-1} for conventional tillage and 13.4 for no-tillage. Assuming the soil had reached equilibrium concentration of organic C in 1989, the potential total sequestration for this soil is about 200 to 220 Mg CO_2 ha^{-1}. The further a soil is from its equilibrium organic C level for the existing conditions, the greater the potential to sequester CO_2. For example, in our case, an average of 40 Mg CO_2 ha^{-1} was sequestered from 1975 to 1989 by no-tillage compared to 38 by conventional tillage. The sequestration would likely occur more rapidly with no-tillage than with conventional tillage. Thus, the greatest potential for rapidly sequestering the most CO_2 would be when maximum biomass is produced using no-tillage on a soil with a history of being deprived of organic matter. Once the new, higher-level equilibrium has been achieved, the management conditions would have to be maintained to keep that amount of CO_2 sequestered. That could be done far more easily with no-tillage, or some other effective form of conservation tillage, than with conventional moldboard plow tillage.

REFERENCES

1. Council for Agricultural Science and Technology, Preparing U.S. agriculture for global climate change, Task Force Report No. 119, Council for Agricultural Science and Technology, Ames, IA, 1992.
2. U.S. Environmental Protection Agency, The Impact of Conventional Tillage Use on Soil and Atmospheric Carbon in the Contiguous United States, EPA/660/3-91/056, U.S. Environmental Protection Agency, Washington, DC, 1991.
3. Blevins, R.L., Thomas, G.W., and Cornelius, P.L., Influence of no-tillage and nitrogen fertilization on certain soil properties after five years of continuous corn, *Agron. J.*, 69, 383, 1977.
4. Frye, W.W., Bennett, O.L., and Buntley, G.J., Restoration of crop production on eroded or degraded soils, in *Soil Erosion and Crop Productivity*, Follett, R.L. and Stewart, B.A., Eds., Soil Science Society of America, Madison, WI, 1985, chap. 20.
5. Doran, J.W., Soil microbial and biochemical changes associated with reduced tillage, *Soil Sci. Soc. Am. J.*, 44, 765, 1980.
6. Rice, C.W., Smith, M.S., and Blevins, R.L., Soil nitrogen availability after long-term continuous no-tillage and conventional tillage corn production, *Soil Sci. Soc. Am. J.*, 50, 1206, 1986.
7. Triplett, G.B., Jr. and Van Doren, D.M., Jr., Agriculture without tillage, *Sci. Am.*, 236, 28, 1977.
8. Blevins, R.L., Frye, W.W., Baldwin, P.L., and Robertson, S.D., Tillage effects on sediment and soluble nutrient losses from a Maury silt loam soil, *J. Environ. Qual.*, 19, 683, 1990.

9. Phillips, R.E., Blevins, R.L., Thomas G.W., Frye, W.W., and Phillips, S.H., No-tillage agriculture, *Science*, 208, 1108, 1980.
10. Baeumer, K. and Bakermans, W.A.P., Zero tillage, *Adv. Agron.*, 25, 77, 1973.
11. Elliott, L.F., Papendick, R.I., and Bezdicek, D.F., Cropping practices using legumes with conservation tillage and soil benefits, in *The Role of Legumes in Conventional Tillage System*, Power, J. F., Ed., Soil Conservation Society of America, Ankeny, IA, 1987, 81.
12. Giddens, J., Rate of loss of carbon from Georgia soils, *Soil Sci. Soc. Am. Proc.*, 21, 513, 1957.
13. Utomo, M., Frye, W.W., and Blevins, R.L., Sustaining soil nitrogen for corn using hairy vetch cover crop, *Agron. J.*, 82, 979, 1990.
14. Jannsen, B.H., A simple method for calculating decomposition and accumulation of 'young' soil organic matter, *Plant Soil*, 76, 297, 1984.

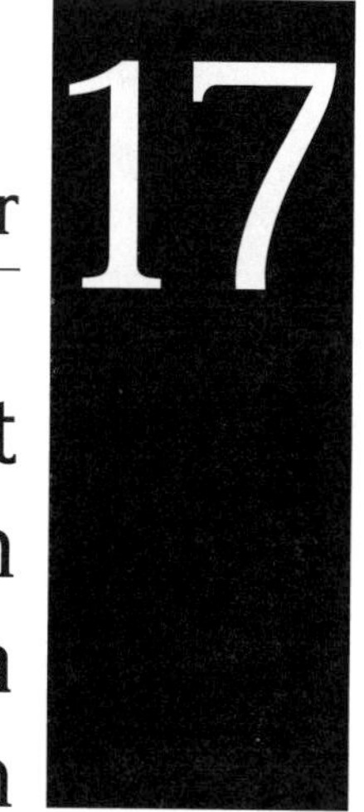

Chapter 17

Long-Term Patterns of Plant Production and Soil Carbon Dynamics in a Georgia Piedmont Agroecosystem

Paul F. Hendrix

CONTENTS

I. INTRODUCTION

The Horseshoe Bend (HSB) Agroecosystem Project began in 1978 with the establishment of a set of no-tillage and intensively plowed research plots. A variety of crop rotation and ecosystem process studies has been conducted within these plots over the past 15 years but the tillage management has been maintained continuously. Together with plots at Bledsoe Farm, University of Georgia, Agricultural Experiment Station, Griffin, GA,[1] and the P-1 Watershed at the USDA-ARS, Southern Piedmont Conservation Research Center (SPCRC), Watkinsville, GA,[2] these are among the longest-running, continuous no-tillage research agroecosystems on the Southern Appalachian Piedmont. As such, they represent an important resource for assessing the capacity of agricultural soils in the region to accumulate organic C.

Bruce et al.[3] describe soil C research at the SPCRC. This paper presents the site history of the HSB project, along with research results on plant production and soil organic C (SOC) dynamics in the experimental agroecosystems at the site. Previous reports give additional results from this site.[4–7]

0-8493-2802-0/97/$0.00+$.50

II. SITE DESCRIPTION AND HISTORY

A. GEOGRAPHY AND CLIMATE

The HSB site is located in the Southern Appalachian Piedmont Plateau physiographic region. Originally a broad, smooth plain, the region has been dissected into landscapes ranging from moderately to strongly sloping uplands to nearly level or short, steep-sloped bottomlands near stream channels. Soils are shallower to bedrock in strongly sloping to steep areas than on broad sloping ridges.[8]

Climate in the region is humid subtropical,[9] moderated to the southeast by the Atlantic Ocean, to the south by the Gulf of Mexico, and to the north and northwest by the Appalachian Mountains, which act as a partial barrier to winter cold air masses flowing from those directions. Climatic data have been recorded for the past 100 years at Ben Epps Airport in Athens, GA, approximately 3 km from the HSB site. A portion of this data set is used as the source of climatic data for the HSB site (Table 1).

Table 1 Long-Term Climatic Data at Ben Epps Airport, Athens, Clarke County, GA, 3 km from the Horseshoe Bend Site

Month	Precipitation (mm)	Temperature (°C)		
		Mean	Minimum	Maximum
January	119.4	5.94	0.44	11.38
February	113.3	7.72	1.72	13.72
March	134.9	11.66	5.33	17.94
April	100.6	16.50	9.88	23.11
May	107.9	20.94	14.55	27.27
June	98.3	24.83	18.77	30.88
July	123.7	26.33	20.66	32.00
August	88.6	25.83	20.22	31.44
September	89.4	22.77	17.16	28.33
October	76.7	16.88	10.55	23.22
November	90.7	11.44	5.27	17.61
December	102.1	7.05	1.50	12.61

Note: Values are averages for each month over the years 1962 to 1991.

Mean annual air temperature over the past 30 years is 16.3°C (25.5°C in summer and 6.7°C in winter). Average frost-free days number 220. Precipitation is relatively evenly distributed throughout the year averaging a total of 1245 mm (average minimum of 77 mm in October and maximum of 135 mm in March). Mean annual potential evapotranspiration is 1564 mm.[8,10,11]

B. SITE HISTORY

The HSB site is situated on a bottomland terrace within a meander loop on the flood plain of the North Oconee River in Athens, Clarke County, GA (33° 54′N, 83°24′W) (Figure 1). The soil is a fine loamy, siliceous, thermic Rhodic Kanhapludult (Hiwassee taxajunct, 66% sand, 12% silt, 22% clay) derived from alluvial material under mixed hardwood and pine forest (presently dominated by *Quercus* spp., *Carya* spp., and *Pinus* spp.). The last flood to innundate the area occurred in 1967. Selected soil properties measured in 1983 and in 1990 are shown in Table 2.

At least since 1938 (when aerial photographs were first made) and probably for much longer, the site has been under agricultural management. Bottomlands were preferred areas for cultivation by the first settlers in the region in the early 1800s and probably also by native Americans.[12] For 30 years prior to 1966 the site was used by the University of Georgia dairy for pasture and forage production[4] (E. P. Odum, personal communication).

In 1964, HSB was acquired by the Institute of Ecology and in 1966 the site now occupied by the agroecosystem plots was divided into two 0.4 ha plots, each enclosed by sheet-metal fencing. These plots were treated with 568 kg ha^{-1} 6-12-12 NPK fertilizer, planted in millet (*Panicum ramosum* L.), and one was sprayed with the carbamate insecticide carbaryl (2.3 kg a.i. ha^{-1}).[13] After 6 months of study, both plots were allowed to undergo secondary succession which was studied until 1970. In that year one half of each plot was subjected to a late winter burn which temporarily (6 months) reduced the standing stock of surface litter relative to the unburned areas.[14] During the subsequent 8 years, old field succession continued without disturbance.

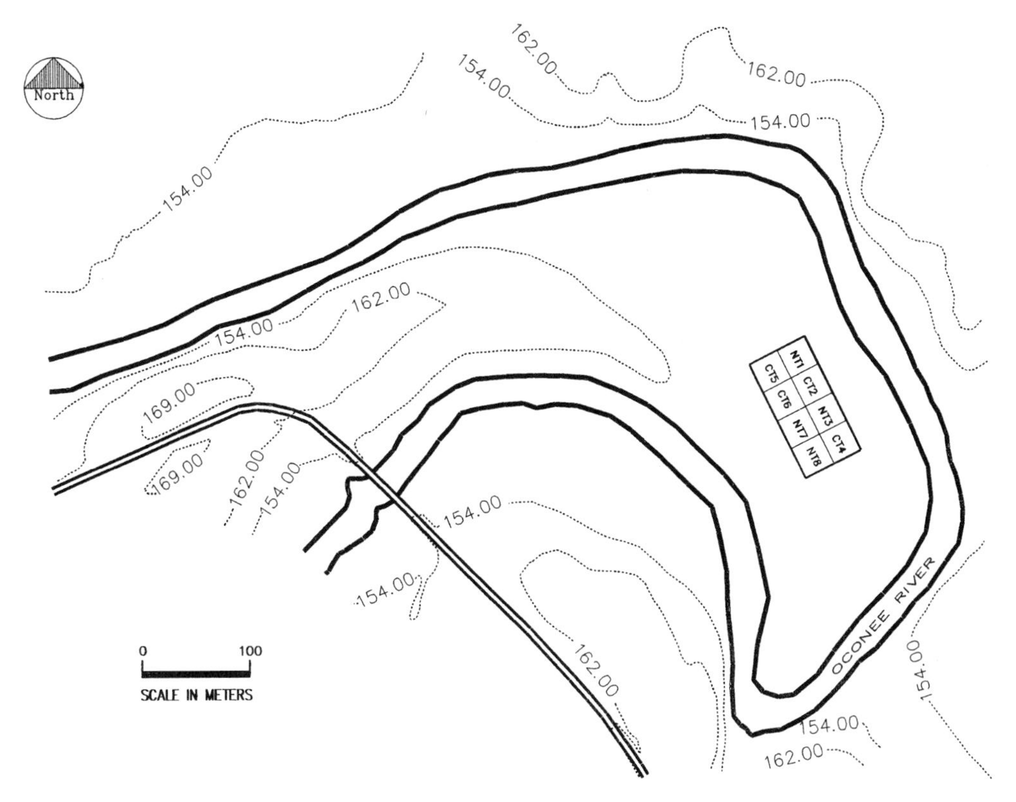

Figure 1 The Horseshoe Bend Experimental Area in Athens, Clarke County, GA, showing the long-term agroecosystem research plots. NT indicates no-tillage and CT indicates conventionally mold board-plowed plots.

Table 2 Selected Physical and Chemical Properties of Soil from the Long-Term Horseshoe Bend Agroecosystem Plots on Two Dates

		Organic C (g/kg)		Organic N (g/kg)		Bulk Density (Mg/m^3)		CEC (cmol (+) / g)		pH	
Date	Depth (cm)	CT	NT	CT	NT	CT	NT	CT	NT	CT	NT
1983[6]	0–5	13.0	23.4	1.2	2.0	1.40	1.21	4.86	7.77	5.9	6.4
	5–13	9.6	8.8	1.1	1.0	1.49	1.51	4.70	3.30	6.0	5.5
	13–21	5.7	4.0	0.08	0.07	1.57	1.53	3.36	2.13	5.6	4.8
1990[11]	0–5	13.5	24.6	1.11	2.32	1.38	1.20	5.32	8.72	6.8	7.1
	5–15	11.1	10.5	0.90	0.87	1.51	1.52	5.67	5.31	7.0	6.8

In spring, 1978, the two 0.4-ha plots were hand cleared of woody vegetation, rotary mowed, and each was divided into four, 0.1-ha plots. The resulting eight plots were placed under either no-tillage (NT) or conventional tillage (CT) management in a completely randomized design with four replicates per treatment (Figure 1). These tillage treatments have been maintained to the present time under a variety of crop management regimes.

III. PRIMARY PRODUCTION

Cropping systems at HSB have consisted of summer grain crops [*Sorghum bicolor* (L.) Moench or *Glycine max* (L.) Merr.] followed by winter cover crops (*Secale cerale* L. or *Trifolium incarnatum* L.) as described previously.[4–7] The plots were fallowed during summer 1988 and winter 1992/1993, and crop failures occurred in winter 1982 due to severe cold and in summer 1986 due to severe drought.

Nearly continuous measurements have been made of above-ground net primary production (ANP), including crop and weed biomass and summer grain yields. Measurements consist of end-of-season above-ground standing dry-weight (40°C) biomass from three to five replicate 0.25 or 1.0 m^2 quadrats for weeds and winter crops or from five replicate 0.5-m lengths of row for summer crop biomass and grain yield.

Over the 1978 to 1990 period of observation, several trends in ANP have emerged. First, winter weed ANP has been consistently and usually significantly ($p<0.05$) greater in NT than in CT. This has resulted from limited herbicide use (only in spring prior to planting summer crops in both CT and NT) and from fall tillage in CT. In summer, weed ANP is typically greater than in winter but has shown no consistent tillage effect. Despite higher weed growth in NT in winter, cover crop ANP has been fairly uniform across tillage treatments, although in recent years it has been slightly but not significantly higher in CT. Finally, summer crop ANP and grain yields have usually been higher in NT than in CT, sometimes significantly ($p<0.05$) so.

Two features of crop management at HSB are relevant to soil C dynamics. First, as already mentioned, herbicide use in most years has been limited to a single application per year of a contact herbicide (glyphosate 6 L a.i. ha^{-1}). As a result, weed biomass and probably total ANP have been higher than in typical row crop systems in the region. Second, in many years, after fall crops were sampled for grain yield and ANP, the entire crop was mowed down and returned to the soil surface in NT or incorporated into the soil in CT. This, combined with high weed biomass and winter cover crops, which are also mowed down as green manure in spring, has resulted in large organic inputs to the soil.

Table 3 shows mean C inputs to the soil in CT and NT over the 7-year period from 1984 to 1990. These values were calculated as dry-weight biomass × 0.45, the mean proportion of C in sorghum and rye grown at HSB. Inputs were nearly evenly distributed between summer (4 months) and winter (7 months) growing seasons. Weeds constituted approximately 50% of summer ANP in CT and 40% in NT. On average, the higher total ANP over the year in NT resulted in a 20% higher above-ground input of C to the soil than in CT (7.1 vs. 5.8 Mg C ha^{-1}). Previous studies at HSB estimated below-ground (crop plus weed) production to be 33 and 24% of total production in CT and NT, respectively, during the 1986 summer growing season.[15] If these values are applied to the mean annual ANP values in Table 3, total C inputs to the soil are estimated to be 9.3 Mg C ha^{-1} y^{-1} in NT and 8.6 Mg C ha^{-1} y^{-1} in CT.

Table 3 Mean Seasonal and Annual ANP (as End of Season Dry Biomass) and Inputs of C to Soil in the Long-Term Horseshoe Bend Agroecosystem Plots between 1984 and 1990

	CT	NT
Mean Annual Biomass (Mg ha^{-1})		
Winter ANP		
Crop	6.12	5.50
Weed	0.25	1.61
Total	6.37	7.11
Summer ANP		
Crop	3.42	5.14
Weed	3.19	3.48
Total	6.61	8.62
Annual ANP		
Total	12.98	15.73
Mean Annual C Input (Mg C ha^{-1})		
Above-ground[a]	5.84	7.08
Below-ground[b]	2.86	2.24
Total	8.60	9.32

[a] Annual total ANP × 0.45.
[b] Calculated from above-ground/below-ground production data from Reference 15.

Higher inputs of organic matter in NT are consistent with higher annual CO_2 efflux in soil respiration observed previously in NT at the site.[7,16] Total annual CO_2-C effluxes of 11.7 Mg C ha^{-1} in NT and 10.1 Mg C ha^{-1} in CT were estimated for 1983/1984.[16] The excess of CO_2-C loss (2.9 Mg C ha^{-1} y^{-1} in NT and 2.3 Mg C ha^{-1} y^{-1} in CT) over total C input represents the amount of CO_2-C in root respiration minus C lost in runoff and leaching or sequestered in the soil.

IV. SOIL ORGANIC CARBON

Numerous studies have documented losses of soil organic carbon (SOC) following cultivation.[17–20] In the Southern Appalachian Piedmont region, intensive cultivation over the past 100 to 150 years has resulted in widespread degradation of soils through erosion,[12,21,22] but changes in SOC have not been adequately studied. Conservation practices (e.g., double cropping and NT management), which reduce erosion and supply organic matter to the soil from plant residues, are now widely used. However, rates and magnitudes of SOC formation resulting from these practices are not well known for this region.[1,3,23] In this section, data are utilized from the literature and from the HSB site to explore SOC dynamics in the Southern Piedmont region.

A. POTENTIAL CARBON STORAGE

Various estimates are available for SOC content of soils in the Southern Piedmont prior to and following agriculture (Table 4). Climatic and life-zone models estimate pre-agriculture values between 92 and 137 Mg C ha^{-1}. Using data from soils under virgin forests, fescue sods, and old fields, values from 57 to 78 Mg C ha^{-1} to a depth of 1 m were calculated, depending on assumed bulk density and depth profile distribution of C.[24,25] This range of values may represent upper limits to C storage in Piedmont soils receiving inputs from *in situ* plant production. Limits at any particular site will depend on soil texture, landscape position, primary productivity, and management practices.

Table 4 Estimates of Potential Organic C Levels in Southern Piedmont Soils

Type of System	Mg C ha^{-1}	Type of Estimate	Ref.
Natural ecosystems			
Natural environment	137	Climate model	29
Warm-temperate moist forest	93	Life-zone model	30
Subtropical moist forest	92	Life-zone model	30
Piedmont forests	57–78	Historic data	24, 25
Agroecosystems			
Disturbed environment	47	Climate model	29
Field cropland	29–33	Geographic database	31
Horseshoe Bend	30	Field data	

Following establishment of agriculture, SOC levels are estimated to decline (Table 4). The climate model gives the highest prediction of 47 Mg C ha^{-1} in "disturbed" environments. Recent geographic data base estimates as well as measurements at HSB indicate SOC levels of approximately 30 Mg C ha^{-1} in agricultural soils. The actual measured values (i.e., historic data and field data) represent declines in C content of approximately 50 to 60% from estimated pre-agricultural levels.

B. LIMITS TO CARBON STORAGE

Carbon concentrations in the Ap horizon (0 to 15 cm) of soils under continuous sod or forest vs. continuous plowing and row cropping for various lengths of time are shown in Figure 2. These chronosequences suggest upper as well as lower limits of C storage in Piedmont soils.

The aggrading curve begins with sod establishment (fescue or alfalfa) on C-depleted soils (5.8 g C kg^{-1}). Carbon concentrations appear to approach an equilibrium value of approximately 23 g C kg^{-1} within 30 years. The degrading curve begins with plowing of a virgin forest and shows a 31% decline in C concentration in the first 3 years. The degrading curve may reach equilibrium in 30 years or less. The apparent lower limit of approximately 6 g C kg^{-1} in these heavily cultivated soils may be a function of soil texture (clay and silt content) and associated residual C.

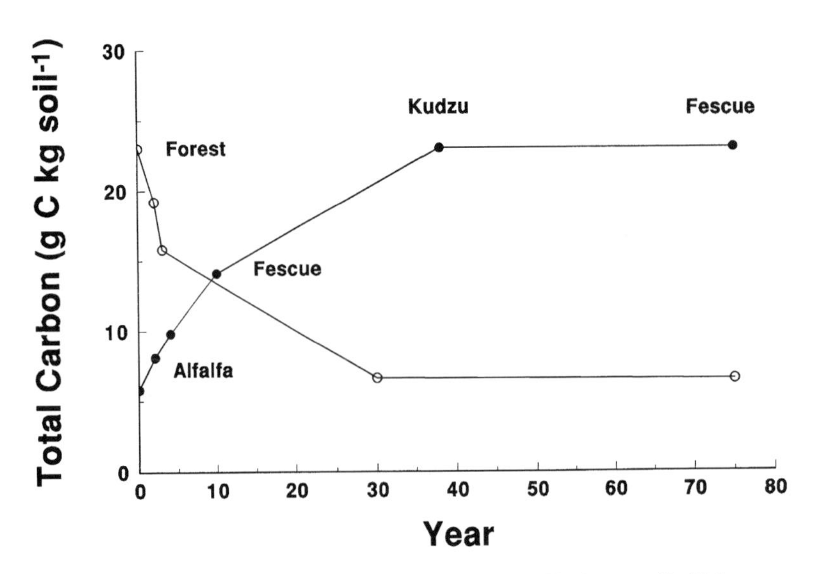

Figure 2 Chronosequences of organic C concentration in Georgia Piedmont soils. Values represent C concentrations in the plow layer (0 to 15 cm) of a variety of soils under different vegetation for various lengths of time.

C. CHANGES IN SOC AT HORSESHOE BEND

SOC to a depth of 21 cm has been measured periodically on air-dried, sieved (2-mm) soils from the long-term plots at HSB. During the first 2 years of the project (1978 to 1980), SOC was determined by mass loss on ignition at 500°C and calculation of C using 0.58 as a conversion factor.[26] Thereafter, SOC was determined on archived or recently collected soils via dry combustion and infrared gas analysis using a LECO Carbon Determinator or a Carlo Erba Automated C/N Analyzer.

In 1989, new short-term CT and NT plots (four each; 3 × 10 m) were established in a grass sod adjacent to the long-term plots in order to measure rapid changes in SOC immediately following the start of agricultural management. Each spring after the CT plots were plowed, all of the short-term plots were planted with grain sorghum. In fall sorghum residues were returned to the soil at 2.5 Mg dry weight ha^{-1}, the NT plots were surface-broadcast with rye, and the CT plots were plowed and left in bare winter fallow. Soil sampling was conducted in the short-term plots at the end of each cropping cycle in spring and fall. Grass sod plots were also sampled as references. Soil C was measured with a Carlo Erba Automated C/N Analyzer.

Changes in SOC concentrations in the long-term plots at HSB are shown in Figure 3. Initial SOC concentrations in 1978 at the beginning of the long-term study are not known, but soils analyzed in fall 1978 after the first spring and fall plowing indicate values of approximately 30 g C kg^{-1} soil in the upper 5 cm and approximately 15 g C kg^{-1} at 5- to 13-cm depth (data not shown). The continuous long-term data set begins in 1982 (Figure 3), with SOC concentrations reduced from the 1978 values. At 0 to 5 cm, both CT and NT showed net declines in SOC during the 8 years of observation, but values remained significantly higher in NT. At 5 to 13 and 13 to 21 cm, SOC concentrations were usually higher in CT than in NT, but showed a net decline only in CT at 5 to 13 cm. These data indicate a redistribution and homogenization of SOC in the plow layer of CT and vertical stratification of SOC in NT. A similar pattern was seen in SOC concentrations in the short-term plots (Figure 4), which began with lower initial values than those in the long-term plots.

Changes in vertical distribution of SOC in the long-term plots at HSB were accompanied by overall declines in amounts of C within the soil profile of both CT and NT, after correction for bulk density (Figure 5). Soil C content was not significantly different between CT and NT on any sampling date over the 8 years of observation. However, considering initial and final values in Figure 5, soil C content showed net declines of approximately 25% (9 Mg C ha^{-1}) in CT and approximately 15% (5 Mg C ha^{-1}) in NT; total losses since 1978 are probably somewhat higher. In the short-term plots (Figure 6), soil C showed no net change during the 2-year study. Redistribution of particulate organic C into smaller size fractions, homogenization of SOC in the soil profile, but no net loss of C during the first few years following cultivation have been reported previously.[27,28]

It is interesting that declines in C content of soils in the long-term plots occurred under both CT and NT management. These losses may be due to a combination of (1) high sand content and low stability

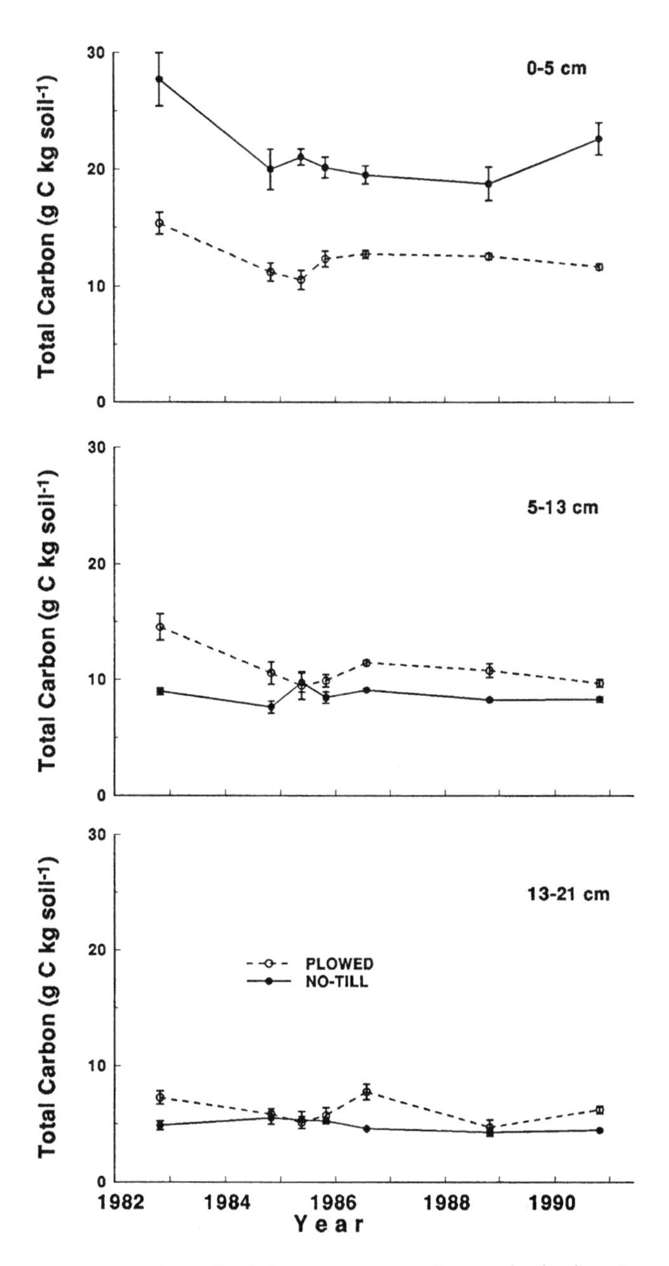

Figure 3 SOC concentrations at three depth increments over 8 years in the long-term plots at Horseshoe Bend. Points represent means with standard error bars.

of soil aggregates which may be important in protecting SOC in these soils[11] and (2) conversion from old field and forest vegetation to annual crop plants with more readily decomposable tissues. Although equilibrium values as high as 60 to 80 Mg C ha^{-1} may be approached in forest- or sod-based systems in the Southern Piedmont region,[24,25] these levels appear unlikely under row-crop production, even with conservation management (Hargrove et al., 1982). Our data and that from Bruce et al.[3] suggest that equilibrium values around 30 Mg C ha^{-1} are more realistic. Higher levels might be attained in finer-textured soils, with sod in the rotations, or with applications of animal manures or organic wastes.

V. SUMMARY

Data from a long-term agroecosystem research site and from available literature were used to assess patterns in plant production and soil organic matter dynamics under different types of agricultural

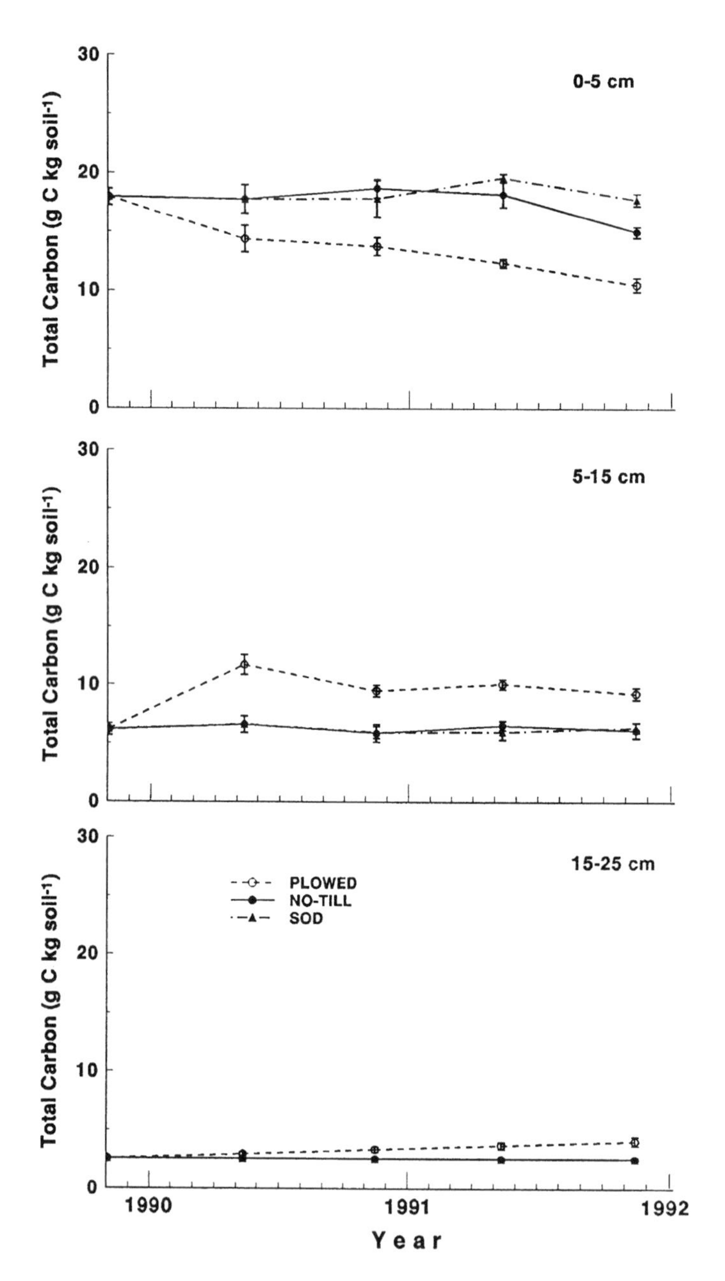

Figure 4 SOC concentrations at three depth increments over 2 years in the short-term plots at Horseshoe Bend. Points represent means with standard error bars.

management on the Southern Appalachian Piedmont of Georgia. The study focused on effects of tillage, residue management, and cropping systems on ANP and SOC concentrations and standing stocks over a 12-year period. Crop management consisted of double cropping with summer grain crops (*Sorghum bicolor* (L.) Moench or *Glycine max* (L.) Merr.) followed by winter green-manure cover crops (*Secale cerale* L. or *Trifolium incarnatum* L.) or winter bare fallow, all grown under NT or CT management.

Due to minimal use of herbicides (equally in CT and NT), winter weed ANP was significantly higher ($p<0.05$) under NT, while winter cover crop and summer weed ANP were similar under the two tillage treatments. Summer crop ANP and grain yields were usually higher under NT. Return of most of the annual ANP to the soil, with higher average quantities in NT (15.73 Mg ha^{-1} y^{-1}) than in CT (12.98 Mg ha^{-1} y^{-1}) resulted in an average of 20% higher inputs of organic matter into NT soil. Considering possible below-ground contributions, mean annual inputs of carbon to the soil may approach 9.32 Mg C ha^{-1} y^{-1} in NT and 8.60 Mg C ha^{-1} y^{-1} in CT.

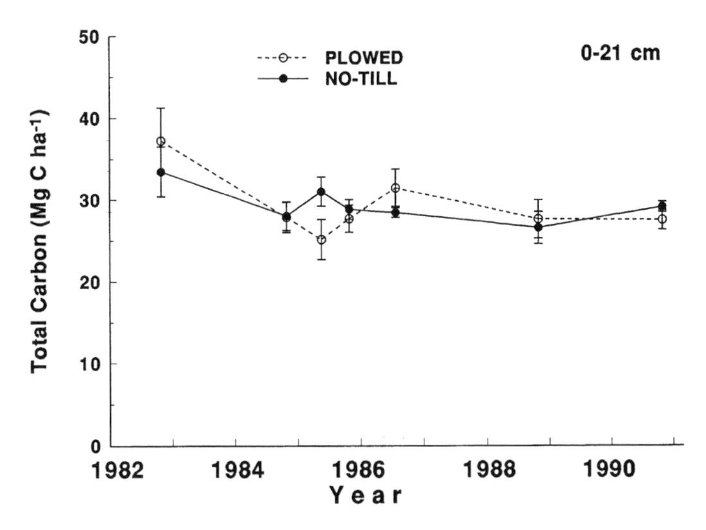

Figure 5 Total C content in the soil profile to 21 cm over 8 years in the long-term plots at Horseshoe Bend. Points represent means with standard error bars.

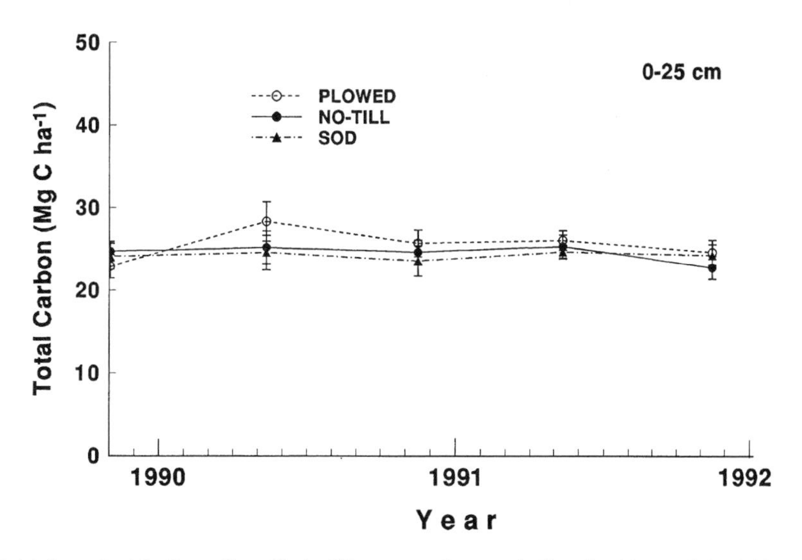

Figure 6 Total C content in the soil profile to 25 cm over 2 years in the short-term plots at Horseshoe Bend. Points represent means with standard error bars.

During 8 years of observation, SOC concentrations in the 0- to 5-cm layer were consistently higher in NT but were higher in CT at 5 to 21 cm, indicating vertical redistribution of C; similar patterns were observed after 2 years in nearby short-term plots. After 8 years, total C content in the soil profile showed a slight but not significant tillage effect, reflected as a decline of approximately 25% (9 Mg C ha^{-1}) in CT and approximately 15% (5 Mg C ha^{-1}) in NT. Total C content showed no change after 2 years in the short-term plots. Equilibrium levels of 60 to 80 Mg C ha^{-1} may be approached in forest- or sod-based systems, but levels of approximately 30 Mg C ha^{-1} may be more realistic for continuous row-crop systems in this region.

ACKNOWLEDGMENTS

This work has been supported over the past 12 years by the Institute of Ecology, University of Georgia, and by grants DEB-8207206, BSR-8506374, and BSR-8818302 from the National Science Foundation to the University of Georgia Research Foundation, Inc. Mike Beare and two anonymous reviewers provided helpful comments on earlier drafts of the manuscript. Appreciation is also expressed to David Coleman and D.A. Crossley, Jr.

REFERENCES

1. Hargrove, W.L., Reid, J.T., Touchton, J.T., and Gallaher, R.N., Influence of tillage practices on the fertility status of an acid soil double-cropped to wheat and soybeans, *Agron. J.*, 74, 684, 1982.
2. Langdale, G.W., Mills, W.C., and Thomas, A.W., Use of conservation tillage to retard erosive effects of large storms, *J. Soil Water Conserv.*, 47, 257, 1992.
3. Bruce, R.R. and Langdale, G.W., Soil carbon level dependence upon crop culture variables in a thermic-udic region, *Soil Organic Matter in Temperate Agroecosystems: Long-Term Experiments in North America*, Paul, E.A., Elliott, E.T., and Paustian, K., Eds., CRC Press, Boca Raton, FL, 1996, chap. 18.
4. Stinner, B.R., Crossley, D.A., Jr., Odum, E.P. and Todd, R.L., Nutrient budgets and internal cycling of N, P, K, Ca and Mg in conventional tillage, no-tillage, and old-field ecosystems on the Georgia Piedmont, *Ecology*, 65, 354, 1984.
5. House, G.J., Stinner, B.R., Crossley, D.A., Jr., and Odum, E.P., Nitrogen cycling in conventional and no-tillage agroecosystems: analysis of pathways and processes, *J. Appl. Ecol.*, 21, 991, 1984.
6. Groffman, P.M., House, G.J., Hendrix, P.F., Scott, D.E., and Crossley, D.A., Jr., Nitrogen cycling as affected by interactions of components in a Georgia piedmont agroecosystems, *Ecology*, 67, 80, 1986.
7. Beare, M.H., Parmelee, R.W., Hendrix, P.F., Cheng, W., Coleman, D.C. and Crossely, D.A., Jr., Microbial and faunal interactions and effects on litter nitrogen and decomposition in agroecosystems, *Ecol. Monogr.*, 62, 569, 1992.
8. Robertson, S.M., Soil Survey of Clarke and Oconee Counties, Georgia, U.S. Department of Agriculture, U.S. Government Printing Office, Washington, DC, 1968.
9. Bailey, R.G., Explanatory supplement to ecoregions map of the continents, *Environ. Conserv.*, 16, 307, 1989.
10. N.O.A.A, Local Climatological Data: Annual Summary with Comparative Data, Athens, Georgia, National Oceanic and Atmospheric Administration, National Climatic Data Center, Asheville, NC, 1992.
11. Beare, M.H., Hendrix, P.F., and Coleman, D.C., Water-stable aggregates and organic matter fractions in conventional and no-tillage soils, *Soil Sci. Soc. Am. J.*, 58, 718, 1994.
12. Trimble, S.W., Man-Induced Soil Erosion on the Southern Piedmont: 1700–1970, Soil Conservation Society of America, Ankeny, IA, 1974.
13. Barrett, G.W., The effects of an acute insecticide stress on a semi-enclosed grassland ecosystem, *Ecology*, 49, 1019, 1968.
14. Odum, E.P., Pomeroy, S.E., Dickinson, J.C., III, and Hutcheson, K., The effect of late winter litter burn on the composition, productivity and diversity of a 4-year old fallow-field in Georgia, *Proc. Annu. Tall Timbers Fire Ecol. Conf.*, 399, 1973.
15. Cheng, W., Coleman, D.C., and Box, J.E., Jr., Root dynamics, production and distribution in agroecosystems on the Georgia Piedmont using minirhizotrons, *J. Appl. Ecol.*, 27, 592, 1990.
16. Hendrix, P.F., Han, C., and Groffman, P.M., Soil respiration in conventional and no-tillage agroecosystems under different winter cover crop rotations, *Soil Tillage Res.*, 12, 135, 1988.
17. Dalal, R.C. and Mayer, R.J., Long-term trends in fertility of soils under continuous cultivation and cereal cropping in Southern Queensland. II. Total organic carbon and its rate of loss from the soil profile, *Aust. J. Soil. Res.*, 24, 281, 1986.
18. Havlin, J.L., Kissel, D.E., Maddux, M.M., Claassen, M.M., and Long, J.H., Crop rotation and tillage effects on soil organic carbon and nitrogen, *Soil Sci. Soc. Am. J.*, 54, 448, 1990.
19. Chan, K.Y., Roberts, W.P., and Heenan, D.P., Organic carbon and associated soil properties of a red earth after 10 years of rotation under different stubble and tillage practices, *Aust. J. Soil Res.*, 30, 71, 1992.
20. Roth, C.H., Waldemar, W., and de Castro Filho, C., Effect of tillage and liming on organic matter composition in a Rhodic Ferralsol from Southern Brazil, *Z. Pflanzenernahr. Bodenk.*, 155, 175, 1991.
21. Langdale, G.W., Hargrove, W.L., and Giddens, J., Residue management in double-crop conservation tillage systems, *Agron. J.*, 76, 689, 1984.
22. Little, C.E., *Green Fields Forever: The Conservation Tillage Revolution in America*, Island Press, Washington, DC, 1987.
23. Wood, C.W. and Edwards, J.H., Agroecosystem management effects on soil carbon and nitrogen, *Agric. Ecosystems Environ.*, 39, 123, 1992.
24. Giddens, J., Rate of loss of carbon from Georgia soils, *Soil Sci. Soc. Proc.*, 21, 513, 1957.
25. Jones, L.S., Anderson, O.E., and Stacy, S.V., Some effects of sod-based rotations upon soil properties, *Georgia Agric. Exp. Stations*, 166, 5, 1966.
26. Page, A.L., Miller, R.H., and Keeney, D.R., *Methods of Soil Analysis. Part 2. Chemical and Microbiological Properties*, 2nd ed., American Society of Agronomy, Madison, WI, 1982.
27. Tiessen, H. and Stewart, J.W.B., Particle-size fractions and their use in studies of soil organic matter. II. Cultivation effects on organic matter composition in size fractions, *Soil Sci. Soc. Am. J.*, 47, 509, 1983.
28. Angers, D.A., Pesant, A., and Vigneux, J., Early cropping-induced changes in soil aggregation, organic matter, and microbial biomass, *Soil Sci. Soc. Am. J.*, 56, 115, 1992.

29. Meentemeyer, V., Gardner, J., and Box, E.O., World patterns and amounts of detrital soil carbon, *Earth Surface Proc. Landforms*, 10, 557, 1985.
30. Post, W.M., Pastor, J., Zinke, P.J., and Stangenberger, A.G., Global patterns of soil nitrogen storage, *Nature*, 317, 613, 1985.
31. Kern, J.S. and Johnson, M.G., Conservation tillage impacts on national soil and atmospheric carbon levels, *Soil Sci. Soc. Am. J.*, 57, 200, 1993.

Chapter 18

Soil Carbon Level Dependence upon Crop Culture Variables in a Thermic-Udic Region

R.R. Bruce and G.W. Langdale

CONTENTS

I. INTRODUCTION

Our intention is to use about a decade of research that has been carried out near Watkinsville, GA (83°24′W, 33°54′N) as a data base from which to discuss soil carbon. This region is known as the Southern Appalachian Piedmont or just Southern Piedmont land resource area in the U.S.A. A large proportion of the land area in this region has been subjected to nearly 200 years of rather intense cropping. Trimble[1] concludes that the frontier expanded into Georgia by 1810 when clean cultivation of cotton began to spread. He reports that a dramatic increase in erosive land use occurred between 1810 and 1860 and that a medium to high level of erosive land use continued until about 1920. Since 1920 land use has changed dramatically and the state of the land resource today represents the influence of federal government programs interacting with economics.

It is important to recognize the contributions of Giddens and Garman,[2] Giddens,[3] and Giddens et al.[4] They characterized the distribution of C in soils under forests, undisturbed by man, and in soils cultivated 45 to 150 years. Although their data must be adjusted for differences in procedures for C determination, very sound interpretations were made. For example, C in the surface 0.15 m was reduced by 57% from forest to cultivated conditions and C:N ratios changed from 14 to 11.[2] The profile distributions of C showed no difference between forest and cultivated below 0.4 m and not much below 0.2 m.[4] Their data showed a highly significant, positive correlation (r = 0.515) between % soil organic matter, Y, and % clay plus silt, X, where

$$Y = 0.233 + 0.04648X \quad (1)$$

Studies by Adams et al.[5] and Barnett et al.[6] at Watkinsville, GA, and by Jones et al.[7] at Griffin, GA, focused on cropping rotations and associated culture as a means of modification of soil conditions including organic matter content (SOM). In a 17-year experiment including both annual and perennial species the aggradation and degradation of SOM was shown. SOM was more than doubled in the topsoil by using corn (*Zea mays* L.)-oats (*Avena sativa* L.)-lespedeza [*Lespedeza cuneata* (Dum.Cours) G. Don]-lespedeza, and cotton (*Gossypium hirsutum* L.)-oats-lespedeza-lespedeza rotations for 9 years. A subsequent 8 years of clean-tilled, continuous cotton reduced the SOM to the initial level. Complementary measurements of soil N were also made.

II. VARIABLES AFFECTING SOIL CARBON

In a bulletin entitled "Factors affecting organic matter in Georgia soils" Giddens et al.[4] introduced the topic of soil C levels relative to cropping practice. They assumed that organic materials were soil incorporated and that climate variables were known and not worthy of particular consideration. In our list of variables in Table 1 no such assumptions are made. If one intends to deal in a global context with soil C sequestration, climate and soil characteristics are primary. Only after climate and soil characteristics at a site are known can we treat crop cultural variables. The crop production variables that determine effective utilization of the climate and soil resource including soil carbon status are shown in Table 1.

Table 1 Crop Production Variables Affecting Soil C Level and Soil State

1. Climate — annual temperature and rainfall regime, solar radiation, evapotranspiration
2. Soil characteristics — biological, physical, and chemical state — prior culture effects
3. Crop culture — crop sequence, soil water, fertilization, import and export of organic materials, tillage, energy import
4. Plant species and cultivar — quality and quantity of plant material
5. Annual production of plant material
   Above ground — stover; grain
   Below ground — affected by species and cultivar selection
6. Crop residue handling
   Incorporated
   Soil surface

Elements of the climate for Watkinsville, GA, that have influenced the experimental approach as well as the data to be reported are shown in Table 2. On the average there are 264 days annually with daily minimum temperature greater than –4.4°C; rainfall deficiency occurs only in the summer, i.e., 21 mm (ET-Rainfall). Although there are a number of days each year when the probability of significant biomass production is not very high, a photosynthetically active crop canopy must be maintained over the maximum land area throughout the year to realize the potential of the climate.

The relation of soil C level to climate reflects upon soil chemical, biological, and physical processes, but also reflects the quantity and kind of biomass produced in situ or imported and how it is managed. This is the premise that has influenced our research for at least 20 years. However, political, economic, and cultural forces have been a deterrent to the effective application of physical and intellectual resources in research that might determine the actual response of soil systems to the variety of inputs. The culture associated with cotton production in the 19th and most of the 20th century in the southeastern U.S.A. was introduced by settlers from western Europe. It was inappropriately applied in a climate and on soils with consequent disastrous results.

The horizons of soils which dominate the landscape and represent field research sites are shown in Figure 1. These soils are in the Cecil and Pacolet soil series of the clayey, kaolinitic, thermic Typic Kanhapludults family. The Ap horizon varies from 0.08 to 0.25 m in depth and 60 to 400 g kg^{-1} clay. This variation frequently represents a range in soil erosion, from severe to slight. The Bt horizon has a clay content from 300 to 600 g kg^{-1}. This lies immediately below the Ap or is separated by a transition BE or B/A as shown for the slightly eroded pedon. The mean pH at 0.5 m is 5.2. Many published data exist describing these soils.[8,9]

Table 2 Rainfall, Temperature, Solar Radiation, and Evaporation Patterns for Watkinsville, GA (33°54′N, 83°24′W)

Mean annual rainfall	1252 mm			
Mean annual temperature	17°C			
Mean annual pan evaporation	1564 mm			
	Winter		**Summer**	
	November–March	**April**	**May–September**	**October**
Rainfall (mm/month)	116	122	96	69
Potential ET (mm/month)	36	94	117	64
Solar radiation (ly/day)	542	875	902	610
Air temperature	above 0°C	above –2.2°C	above –4.4°C	
Mean number of days	213	240	264	
Soil temperature				
at 0.05 m 0 to 36°C				
at 0.5 m 5 to 25°C				

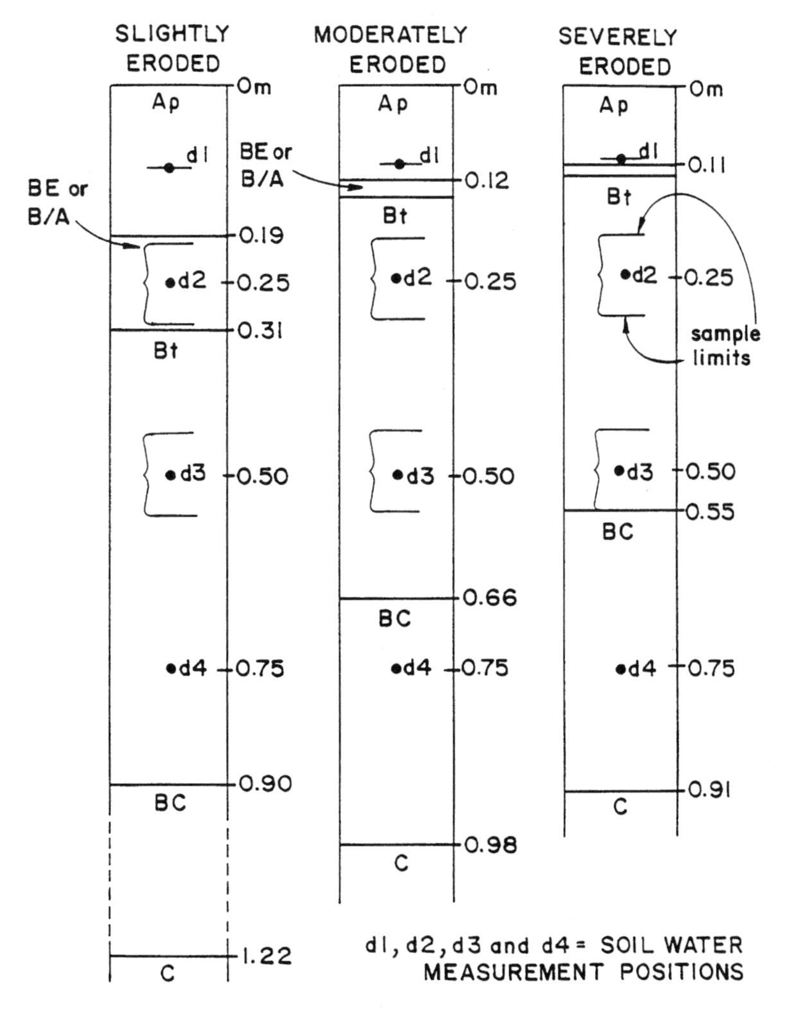

Figure 1 Soil horizons in the average pedons on slightly (E1), moderately (E2), and severely (E3) eroded Cecil-Pacolet soils in 40 farm fields near Watkinsville, GA.

Winter and early spring rainfall is usually sufficient to completely recharge the soil water. By early summer, perennial crops can deplete the soil water to 1.5 m and thereafter are dependent on rainfall. Rainfall infiltration rates that minimize runoff are essential for summer crop production. Our studies during the past decade include measurement of soil water at 0.10, 0.25, 0.50, and 0.75 m at least twice per week.

III. FARM FIELD VARIABILITY ACROSS EROSION CLASSES

During the 1970 to 1980 decade, large areas were devoted to continuous soybean production without a winter crop. Preplant tillage consisted of disc harrowing at least twice. This incorporated nearly all residues, volunteer winter vegetation, and a herbicide. Occasionally chisel plowing preceded disc harrowing. After planting, a combination of tillage and herbicides was used for weed control. Although some terrace systems were maintained which afforded some soil erosion control, many were only remnants and considered to be obstacles to freedom in cultivation. In both 1982 and 1983, 24 farm fields were selected, with the aid of the Soil Conservation Service, USDA, to determine dry matter production and associated soil conditions. All fields were on similar landscapes and dominated by Cecil or Pacolet soils and had been growing soybean for more than 10 years. In each field, slightly, moderately, and severely eroded sites were designated for study (Figure 1). Other methodology is reported by White et al.[10] and Bruce et al.[11,12]

There was 100% more grain and stover yield on slightly eroded than on severely eroded sites (Table 3). Yield was significantly negatively correlated with clay content in the surface 0.1 m. The mean soil carbon level in the surface 0.1 m was 7.3 g kg^{-1}; the range was 2.3 to 14.3 g kg^{-1}; and the difference among erosion classes was insignificant. The mean soil carbon value was similar to values associated with continuous cotton under similar culture and residue production.[5,6] From factor analyses, including over 40 variables, 67% of crop yield variance was explained by clay, soil carbon, and pH in the surface 0.1 m plus soil water at 0.1 m and rainfall during critical period for fruiting.[11,12] The components regression equation was

$$Y = -0.2001 - 0.0366Cl_{0.1} + 0.0041R + 0.1074C_{0.1} + 0.2513pH_{0.1} - 0.0643IW_{0.1} \quad (2)$$

where Y = predicted soybean yield; $Cl_{0.1}$ = clay content at 0 to 0.1m; R = rainfall between day 214 and day 280; $C_{0.1}$ = soil carbon at 0 to 0.1 m; $pH_{0.1}$ = pH at 0 to 0.1 m; and $IW_{0.1}$ = integrated soil water tension at 0.1 m from 214 to 280 days

It was concluded that the crop yield variability was associated with available soil water which was a function of rainfall infiltration. Soil carbon and clay content were proxy variables reflecting the soil water deficit problem and only soil carbon can be potentially managed.[12] It was also interesting that soil carbon and silt content were positively correlated at probability of 0.0001 and carbon and clay or (clay + silt) were not.

IV. SOIL CARBON REGULATION BY CROP CULTURE

A. TWO CROPS PER YEAR AND REDUCED TILLAGE

Langdale et al. [13] reported the results of a field experiment on Cecil sandy loam in which grain sorghum was grown after winter wheat under three tillage treatments for 4 years. The mean production of stover

Table 3 Soybean Yield of Stover and Grain on Three Erosion Classes with Designated Silt, Clay, and Soil Carbon Content in 0- to 0.1-m Depth

Erosion Class	Ap Depth (m)	Silt (g kg^{-1})	Clay (g kg^{-1})	Stover (Mg ha^{-1}) 1982	Stover (Mg ha^{-1}) 1983	Grain (Mg ha^{-1}) 1982	Grain (Mg ha^{-1}) 1983	Soil Carbon (g kg^{-1})
Slightly	0.19	166	82	3.27	2.17	2.78	2.00	6.98
Moderately	0.12	149	221	2.36	1.72	1.89	1.50	7.36
Severely	0.11	135	371	1.57	1.12	1.33	1.00	7.56

From White, A.W., Jr., Bruce R.R., Thomas A.W., Langdale, G.W., and Perkins, H.F., *Erosion and Soil Productivity*, (Publ. 8-85), ASAE, St. Joseph, MO, 1985, 83. With permission.

was 10.7 Mg ha^{-1} y^{-1}. In November of the final year soil C levels in the surface 10 mm were 18.6 g kg^{-1} for no-tillage and 11.8 g kg^{-1} for fall and spring disc harrowing (Table 4). Disc harrowing only in the fall (minimum tillage) and no-tillage produced similar mean grain and stover yields for grain sorghum which were significantly greater than disc harrowing both in fall and spring. The soil carbon levels of fall disc harrowing and spring plus fall disc harrowing were not statistically different. As pointed out by Bruce et al.[14] the soil carbon level does not always reflect the soil condition that determines yield. In fact, Langdale et al.[13] found that carbon level determined on samples of 0- to 75-mm depth did not differentiate among tillage treatments.

Table 4 Effect of Tillage Treatments Applied for 4 Years (1976 through 1980) to Winter Wheat-Grain Sorghum Sequence upon Mean Grain and Stover Yield and Soil Carbon

	Grain Sorghum (Mg ha^{-1} y^{-1})		Wheat (Mg ha^{-1} y^{-1})		Soil Carbon[a]
Tillage	**Grain**	**Stover**	**Grain**	**Stover**	**(g kg^{-1})**
Conventional[b]	2.88 A	5.32 A	3.08 A	4.69 A	11.8 A
Minimum[c]	3.72 B	6.94 B	3.08 A	4.54 A	13.7 A
No-tillage	3.73 B	6.67 B	3.13 A	4.53 A	18.6 B

[a] 0 to 10-mm depth.
[b] Fall and spring disc harrowing.
[c] Fall disc harrowing only.
From Langdale, G.W., Hargrove, W.L., and Giddens, J., *Agron. J.*, 76, 689, 1984. With permission.

B. SURFACE SOIL AGGRADATION AND DEGRADATION

As previously shown (Figure 1, Table 3) the soil surface of the Southern Piedmont varies greatly in clay content and may alter the response to cultural treatment. Slightly, moderately, and severely eroded areas within a radius of about 0.3 km were selected with the assistance of the Soil Conservation Service, USDA, for use in evaluating the impact of imposed treatments upon soil characteristics. The Ap horizons had a thickness of 0.25, 0.15, and 0.09 m and clay contents of 60, 110, and 200 g kg^{-1}, respectively. The treatments imposed on each were (see Section IV of this volume, Appendix B and C)

1. Grain sorghum no-till planted into crimson clover, (NT+CL)G.
2. Grain sorghum planted into a disc-harrowed seedbed (conventional tillage) from untilled winter fallow, CTG.
3. Soybean planted into a disc-harrowed seed bed from untilled winter fallow, CTS.

Each of the treatments had an irrigated (W2) and a nonirrigated (W1) component. The quantity of stover and grain produced annually on each treatment was measured for 5 years (1983 through 1987). Soil water was measured at 0.1-, 0.25-, 0.50-, 1.0-, and 1.5-m depths three times per week throughout the summer growing season. In the fall of 1986 and spring of 1988 soil samples were taken at 0 to 15, 15 to 30, and 30 to 80 mm for soil C and water-stable aggregate determinations. Infiltration was measured by rainfall simulation in March and April 1988. Daily rainfall was recorded. Further details of methodology were reported by Bruce et al.,[15] West et al.,[16] and Langdale et al.[17]

The stover yields for each treatment on each erosion class show a range from 4.9 Mg ha^{-1} y^{-1} for CTG W1 to 15.8 Mg ha^{-1} y^{-1} on CTS W2 (Table 5). Irrigation greatly increased biomass production on CTS and CTG compared to (NT+CL)G as shown by (W2-W1)/W1. The (NT+CL)G culture produced about 6 Mg ha^{-1} y^{-1} of crimson clover on both irrigated and nonirrigated treatment.

The CTS treatment was imposed on all treatments for an additional three years (1988, 1989, and 1990) in order to observe degradation of soil carbon, aggregate stability, and infiltration. Across the original treatments an irrigation, I, and no-irrigation, NI, split was imposed to provide two levels of biomass production and consequent degradation. In the spring of 1989 and 1991 soil samples were taken at 0 to 15, 15 to 30, and 30 to 80 mm for soil carbon and water-stable aggregate determinations.

The mean soil carbon level across erosion classes in the 0- to 15-mm depth of no-till grain sorghum into crimson clover treatment increased from 7.3 to 24.9 g kg^{-1} over 4 years (Figure 2A, Table A1). At 15- to 30- and 30- to 80-mm soil carbon levels were 13.0 and 10.1 g kg^{-1}, respectively. Similar C levels were measured in the spring of 1988 before the culture was changed to soybean under conventional tillage. By the fall of 1988 the soil C level was about 12 g kg–1 at all depths compared to about 8 g kg–1 in fall 1982. (Table A2 and A3). During the 1988 through 1990 period the annual stover levels

Table 5 Cumulative Stover Yields (Mg ha^{-1}) as a Function of Erosion Class and Crop Culture over 5 Years (1983 to 1987, Inclusive)

Crop Culture[a]	Water Regime[b]	E1[c]	E2[c]	E3[c]	Cumulative Mean	Annual Mean	(W2-W1) /W1
CTS	W1	43.3	33.7	53.3	43.4	8.7	
	W2	74.1	80.9	81.3	78.8	15.8	0.816
CTG	W1	22.5	24.5	26.2	24.4	4.9	
	W2	35.1	43.6	38.0	38.9	7.8	0.594
(NT+CL)G	W1	54.5	66.1	58.9	59.9	12.0	
	W2	66.5	78.3	70.5	71.7	14.3	0.197

[a] CTS = soybean into disc-harrowed seedbed from winter fallow; CTG = grain sorghum into disc-harrowed seedbed from winter fallow; (NT+CL)G = grain sorghum no-till planted into crimson clover.

[b] W1 = rainfall dependent; W2 = irrigated when soil matric potential reached –30 J kg^{-1} at 0.22-m depth.

[c] E1 = slightly eroded, E2 = moderately eroded, E3 = severely eroded.

From Bruce, R. R., Langdale, G. W., and West, L. T., *Trans. 14th Int. Cong. Soil Sci.*, 6, 4, 1990. With permission.

were 5.7 Mg ha^{-1} instead of 12.0 Mg ha^{-1} and soil mixing to 100 to 120 mm occurred. Soil C data for the CTG and CTS crop cultures during the entire period do not show the stratification with depth that occurred in no-till (Figure 2A to C). Under irrigation mean stover levels were higher and soil carbon levels were higher but patterns over time were similar to the rainfall-dependent situations (Figure 3A to C). The effect of crop culture, erosion class, and irrigation upon soil carbon is shown for three soil depths and 3 years in Tables A1, A2, and A3.

In the period 1983 through 1987 regulation of soil carbon was restricted to a depth of about 0.08 m in the no-till treatment. The quantity of carbon in this soil volume for each treatment at four dates of measurement was calculated (Table 6). An increase of about 60% soil carbon was produced by the 4- and 5-year (NT+CL)G culture as reflected in 1986 and 1988 data, i.e., 9.9 to 15.8 Mg ha^{-1}. This is close to the magnitude of a value given by Giddens et al.[2] for decrease from forest to cultivated of 57%.

V. SOIL AGGREGATE STABILITY AND SOIL CARBON

In the soil aggradation and degradation experiment the quantity of water stable soil aggregates >250 μm was determined on a portion of the samples that were also subsampled for the C determinations. The relation between quantity of water-stable soil aggregates and soil C for the March 1988 sampling of the severely eroded site is shown in Figure 4. A logarithmic relation as suggested by Kemper and Koch[18] poorly fits these data as well as the data from slightly and moderately eroded sites.[19] All sites show a very rapid increase in soil aggregate stability over a small range in soil C and then show very little effect of increasing soil C after aggregate stability reaches approximately 0.85 kg kg^{-1}. The soil carbon content, at which this threshold value of aggregate stability was reached, increased from 8 g kg^{-1} for slightly eroded sites to about 15 g kg^{-1} for severely eroded sites with respective clay contents of 60 and 200 g kg^{-1}. The quantity of water-stable soil aggregates was very strongly related to 1n (cumulative stover) in this experiment (Figure 5). These data also show that soybean stover has been less effective than grain sorghum stover in generating water-stable soil aggregates.

The quantity of water-stable soil aggregates increased from about 0.5 to >0.8 kg kg^{-1} in 4 years of no-till grain sorghum no-till planted into crimson clover with 12.0 Mg ha^{-1} y^{-1} of stover. Although the greatest soil C level increase occurred at a depth of 0 to 15 mm, aggregate stability increased dramatically to a depth of 80 mm (Figure 6A). After 3 years of conventional till soybean with 5.7 Mg ha^{-1} y^{-1} of stover, the quantity of water-stable aggregates was about 0.6 kg kg^{-1} and rapidly decreasing. Irrigation of the summer crop throughout the 8 years significantly increased stover production, and this is reflected in a greater quantity of water-stable soil aggregates for all crop cultures. In 1988, 1989, and 1990 the conventionally tilled soybean with 11.6 Mg ha^{-1} y^{-1} of stover did not decrease the quantity of water-stable soil aggregates after no-till grain sorghum into crimson clover as rapidly as in the case of no irrigation (Figure 6B).

The quantity of water-stable soil aggregates was unchanged for the conventionally tilled grain sorghum and soybean that were rainfall dependent and planted into winter fallow (Figure 7A and B). Under irrigation, both conventionally tilled soybean and grain sorghum indicated a slow increase in water-stable soil aggregates until 1988 when a value near 0.8 kg kg^{-1} was reached, which then decreased in 1989 and 1990 (Figure 8A and B).

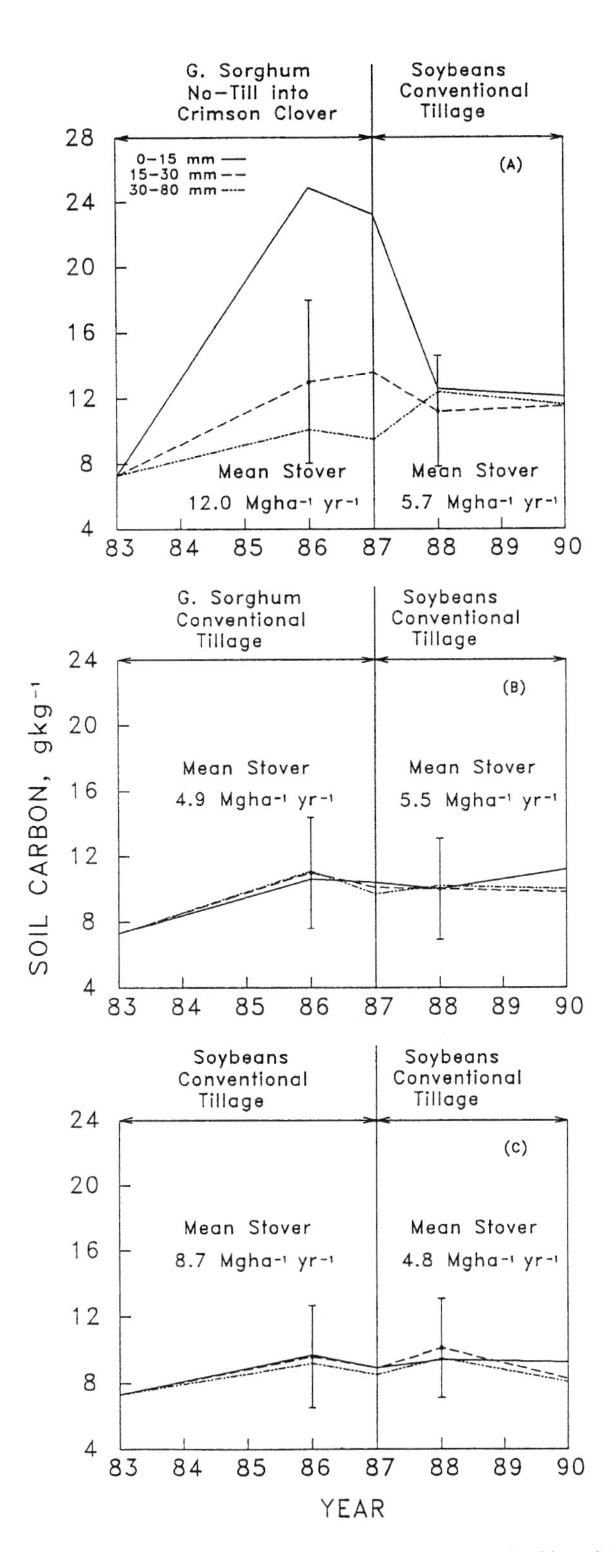

Figure 2 Soil carbon at three soil depths for eight years (1983 through 1990) *without irrigation.* (A) When grain sorghum was no-till planted into crimson clover, (NT+CL)G, from 1983 through 1987 followed by conventionally tilled soybean, CTS, for 3 years, with stover yields. (B) When grain sorghum was conventionally tilled after winter fallow, CTS, from 1983 through 1987 followed by conventionally tilled soybean, CTS, for 3 years, with stover yields. (C) When soybean was conventionally tilled after winter fallow, CTS, from 1983 through 1990, with mean stover yields.

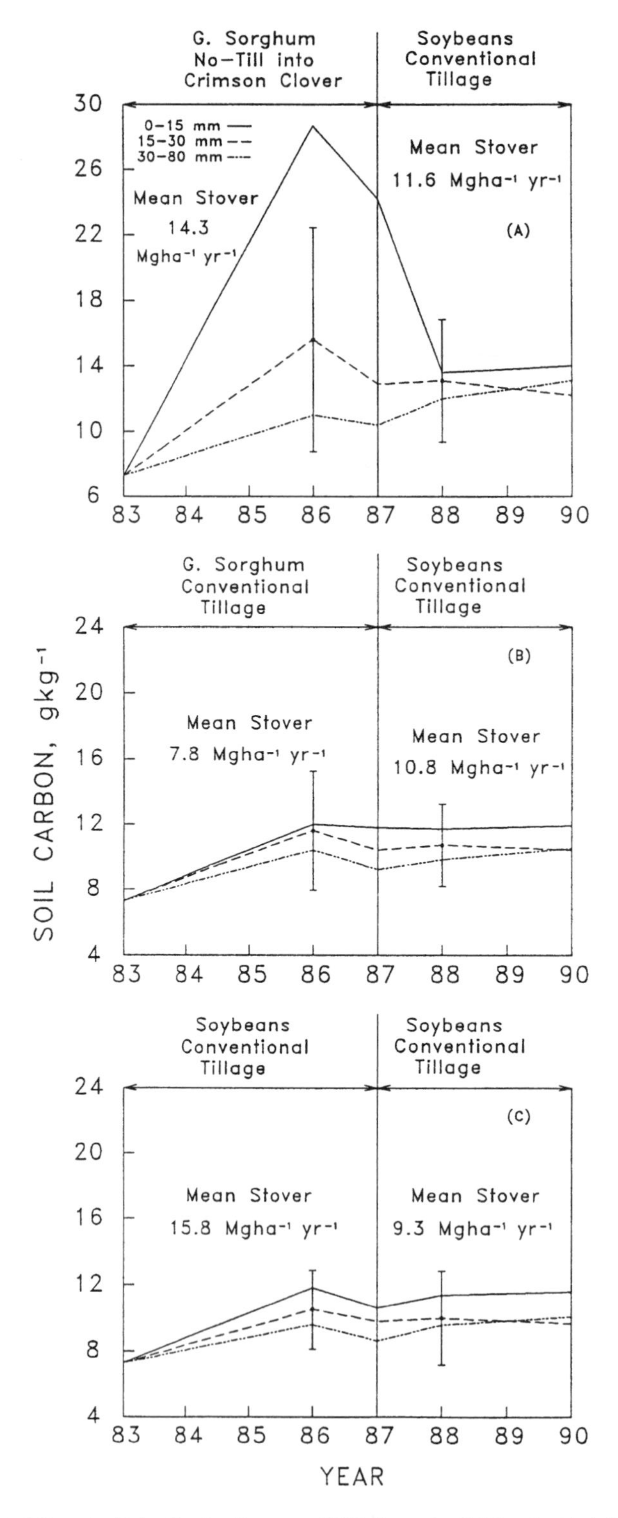

Figure 3 Soil carbon at three soil depths for 8 years (1983 through 1990) *all with irrigation.* (A) When grain sorghum was no-till planted into crimson clover, (NT+CL)G, from 1983 through 1987 followed by conventionally tilled soybean, CTS, for 3 years, with stover yields. (B) When grain sorghum was conventionally tilled after winter fallow, CTG, from 1983 through 1987 followed by conventionally tilled soybean, CTS, for 3 years, with stover yields. (C) When soybean was conventionally tilled after winter fallow, CTS, from 1983 through 1990, with mean stover yields.

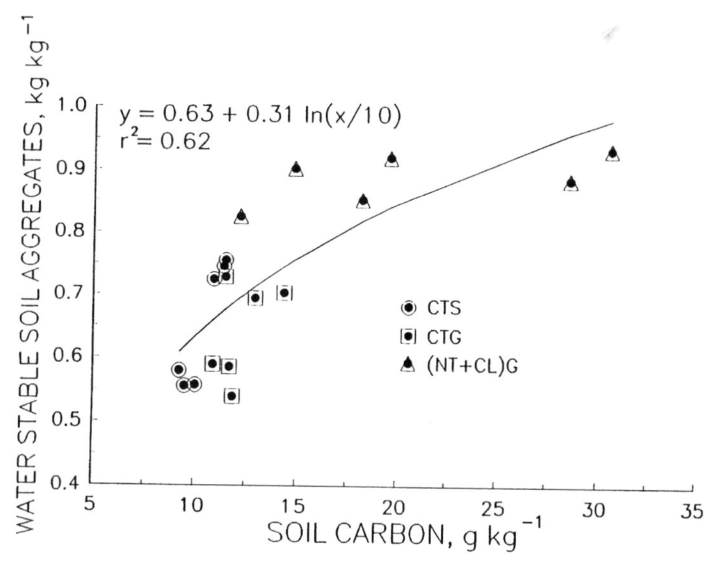

Figure 4 Water stability of soil aggregates in relation to soil carbon in 0- to 80-mm sampling depth for all treatments on severely eroded plots (E3) in March 1988.

Table 6 Soil Carbon (Mg ha^{-1}) in Surface 80 mm in Relation to Crop Culture, 1983 to 1987, and during Subsequent 1988 to 1990 CTS Culture

1983–1987 Culture[a]	Water Regime[b]	1986[c]	1988[c]	1989[d]		1990[d]	
				NI[e]	I[f]	NI	I
CTS	W1	10.34	9.54	10.60	10.40	9.22	10.24
CTS	W2	11.24	10.20	10.80	11.06	10.04	11.37
CTG	W1	12.10	10.95	11.22	11.25	11.29	11.24
CTG	W2	12.03	10.96	11.47	11.41	11.66	11.84
(NT+CL)G	W1	14.79	14.21	13.52	13.56	12.85	12.76
(NT+CL)G	W2	16.80	14.86	13.81	13.84	13.34	14.44

[a] CTS = soybean into disc-harrowed seedbed from winter fallow; CTG = grain sorghum into disc-harrowed seedbed from winter fallow; (NT+CL)G = grain sorghum no-till planted into crimson clover.
[b] W1 = rainfall dependent; W2 = irrigated when soil matric potential reached –30 J kg^{-1} at 0.22-m depth.
[c] Before CTS culture overall.
[d] After CTS culture overall.
[e] NI = rainfall dependent 1988, 1989, 1990.
[f] I = irrigation 1988, 1989, 1990.

Rainfall infiltration was significantly affected by crop culture and reflected the associated effects of crop culture upon water stability of soil aggregates (Table 7). With the residue removed, within 24 h of rainfall event, the infiltration rate on (NT+CL)G was about 100% greater than CTS and CTG, which were not different. Previously, the physical role of crop residues on the soil surface involving dissipation of raindrop energy, retardation of runoff, and consequent impedance to soil particle detachment, suspension, and transport has been emphasized.[20] These data indicate a modification of the soil characteristics at the surface, e.g., water stability of aggregates, and a stability that sustains the infiltration rate. In 1991, CTG was not different from (NT+CL)G treatment which had been significantly degraded (Table 8). Soil water data during critical growth and fruiting of summer crop confirm the presence of significantly greater quantities of water at >–0.1 MPa soil-water-pressure potential under (NT+CL)G treatment (Table 9). This additional water was found in the surface 0.5 m which is the summer rainfall recharge volume. No differences in soil water regime occurred below this depth.

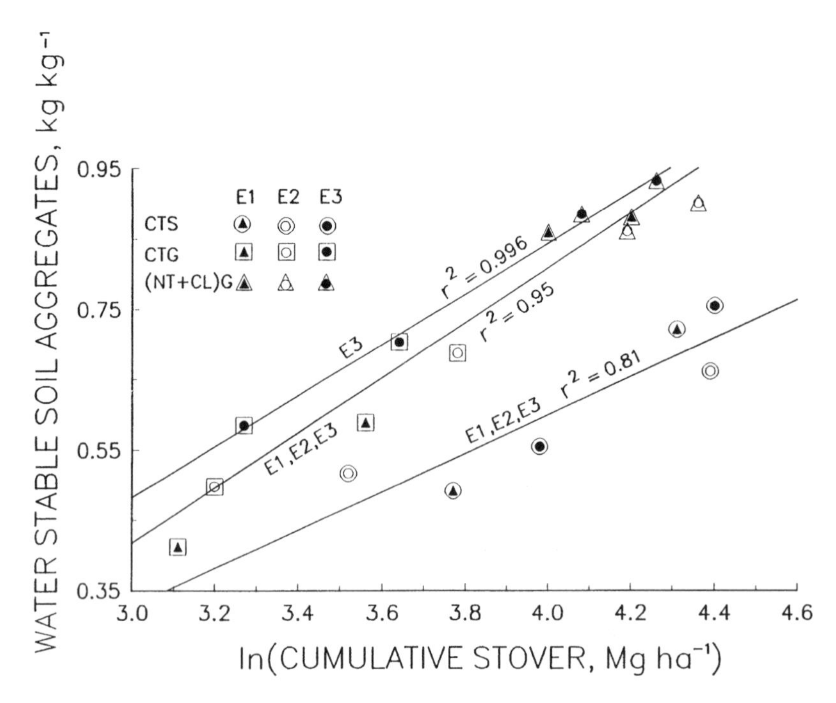

Figure 5 Water stability of soil aggregates at 0 to 15 mm, in March 1988, in relation to cumulative stover from each cultural treatment on three erosion classes.

VI. SOIL CARBON EQUILIBRIUM

Two small watersheds, P1 and P2, at Watkinsville, GA, have been in a mode of carbon sequestration for more than 16 years. Neither watershed has been tilled during the period. P1 has grown both a winter and summer crop, and an average of 9 Mg ha^{-1} y^{-1} of crop residue has been left on the soil surface. Wheat-soybean or crimson clover-grain sorghum have been most frequent crop sequences. A grid sampling at 29 sites on the 1.3-ha watershed has given the mean distribution of soil C in Figure 9, i.e., 2.5, 2.0, and 1.1% in the 0- to 15-, 15- to 30-, and 30- to 80-mm depths, respectively. There has been an increase in soil C from about 21.0 to 32.6 Mg ha^{-1} in the surface 0.3 m.

The P2 watershed was in unharvested common bermudagrass for 7 years followed by 7 years of no-till crimson clover-grain sorghum and recently crimson clover-summer forage (harvested by cattle grazing). Only at 0 to 15 mm is the soil carbon lower than P1, i.e., 22 g kg^{-1} compared to 25 g kg^{-1} (Figure 9). The carbon distribution of 1.5 m is identical. P2 and P1 have produced runoff only in cases of very high rainfall intensity and storms of considerable duration.[21]

VII. CONCLUDING REMARKS

Soil carbon levels approaching that under forest have been achieved in 4 years with appropriate crop culture, i.e., 16 Mg ha^{-1} top 80 mm. The distribution of soil C with depth in the two systems also was similar. The climate potential of growing both a winter and summer crop made it possible to produce 7 to 12 Mg ha^{-1} y^{-1} of crop residue without irrigation. The use of a no-tillage culture permitted the residue of both crops to serve as a mulch during decomposition and allowed soil carbon to accumulate in the surface 30 or 50 mm. The residue decomposition at the soil surface generated a nearly constant soil aggregate stability to a depth of 80 mm that assured a stable soil surface for rainfall infiltration. The increased rainfall infiltration rates increased the effectiveness of summer rainfall in meeting crop water needs.

A single crop per year without irrigation did not generate sufficient crop residue to achieve a stable soil surface or C sequestration. It was also a practice which failed to use the potential of the climate. Selection of crop species and cultivars with complementary characteristics in a two-crop-per-year system is required to achieve synchronization of crop biomass production activities with important climate characteristics and satisfy economic criteria. Tillage accelerates decomposition and has reduced soil

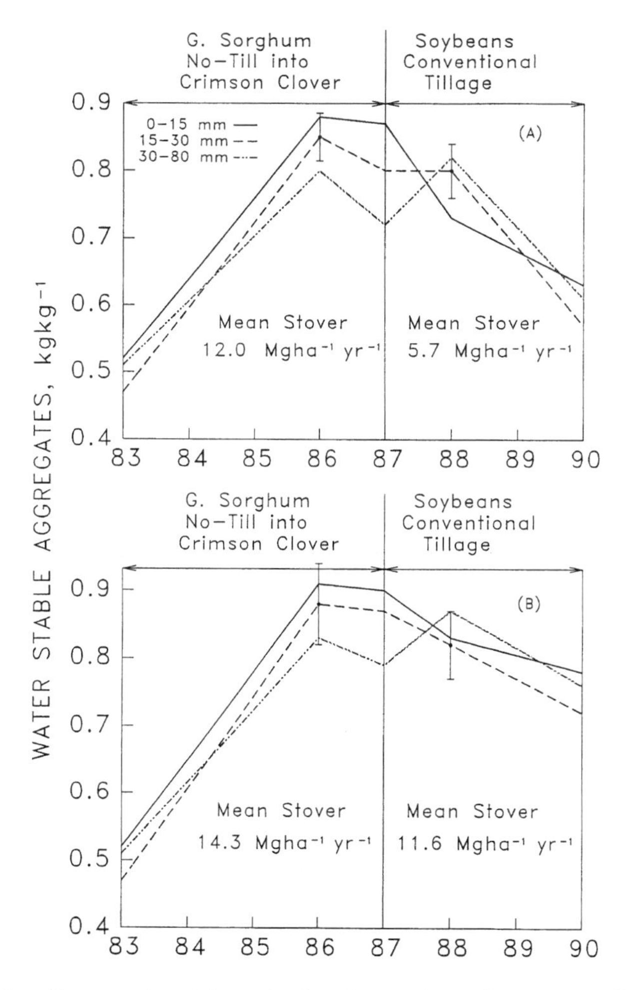

Figure 6 Water-stable soil aggregates at three depths when grain sorghum was no-till planted into crimson clover, (NT+CL)G, in 1983 through 1987 followed by conventionally tilled soybean, CTS, in 1988 through 1990 at associated stover yields. (A) All years *without irrigation.* (B) All years with irrigation.

surface stability by diluting the effect of decomposition products in a larger soil volume than occurs in a no-till culture. If tillage must be done it should be done in the fall when the erosion hazard is least.

Similar principles apply in the culture of perennial crops, e.g., crops for hay or grazing. In both annual and perennial crop cultures provision must be made to supply residue at the soil surface for decomposition and not incorporation by tillage. This means appropriate regulation of grazing and hay harvest. Although we do not have data to support the above statement it seems a reasonable extension of principles. Importation of organic materials may be substituted for in situ production of crop materials.

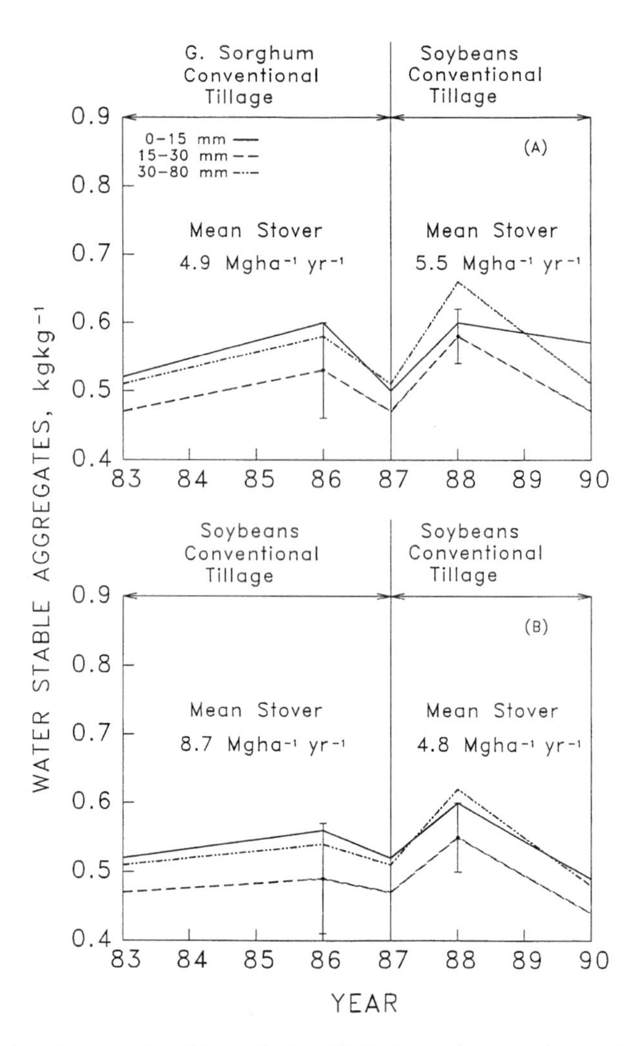

Figure 7 Water-stable soil aggregates at three depths. (A) Grain sorghum was conventionally tilled after winter fallow, CTG, in 1983 through 1987 followed by conventionally tilled soybean, CTS, in 1988 through 1990, at associated stover yields, all years *without* irrigation. (B) Soybean was conventionally tilled after winter fallow, CTS, in 1983 through 1990 at associated stover yields, all years *without irrigation.*

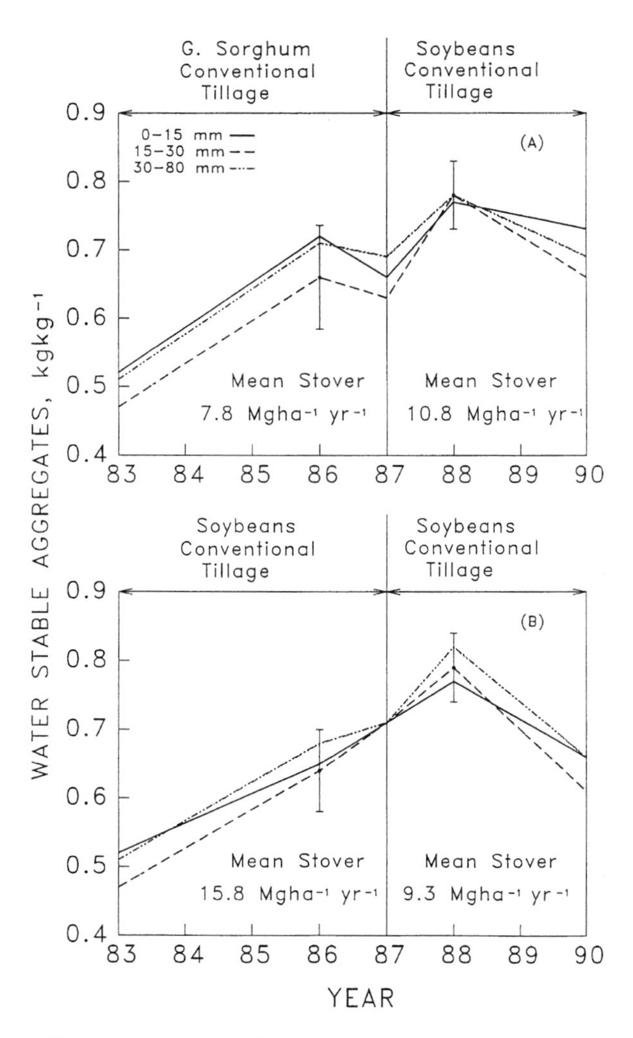

Figure 8 Water-stable soil aggregates at three depths. (A) Grain sorghum was conventionally tilled after winter fallow, CTG, in 1983 through 1987 followed by conventionally tilled soybean, CTS, in 1988 through 1990, at associated stover yields, all *with irrigation.* (B) Soybean was conventionally tilled after winter fallow, CTS, in 1983 through 1990 at associated stover yields, all years *with irrigation.*

Table 7 Infiltration Rate after 1 Hour of Simulated Rainfall at 60 mm h^{-1} for Three Crop Cultures with and without Surface Residue in March 1988

	CTS[a] ($mm\ h^{-1}$)	CTG[a] ($mm\ h^{-1}$)	(NT+CL)G[a] ($mm\ h^{-1}$)
With residue	32.0A[b]	35.7A	49.7B
Without residue[c]	23.8C	22.3C	46.3D

[a] CTS = soybean into disc-harrowed seedbed from winter fallow; CTG = grain sorghum disc-harrowed seedbed from winter fallow; (NT+CL)G = grain sorghum into no-till planted into crimson clover.

[b] Numbers fo wed by the same letter in a given row are not significantly different at the 5% level of probability.

[c] Residue removed within 24 hours of rainfall event.

From Bruce, R.R., Langdale, G.W., and West, L.T., Trans 14th Int. Cong. *Soil Sci.*, 6, 4, 1990. With permission.

Table 8 Infiltration Rate on Severely Eroded Site (E3) after 1 Hour of Simulated Rainfall at 60 mm h^{-1} on Two Crop Cultures in 1988 after 5 Years and in 1991 after 3 Years of CTS

	CTG (mm h^{-1})		(NT+CL)G (mm h^{-1})	
	1988	**1991**	**1988**	**1991**
With residue	34.5	21.5	46.0	23.2
Without residue	21.0	13.8	41.6	16.2

Note: CTS = soybean into disc-harrowed seedbed from winter fallow; CTG = grain sorghum into disc-harrowed seedbed from winter fallow; (NT+CL)G = grain sorghum no-till planted into crimson clover.

Table 9 Fraction of Measurement Period that Soil-Water-Pressure Potential Was Greater than –0.1 MPa in 0.5-m Depth on Three Soil Erosion Classes for Two Grain Sorghum Cultures

Crop Culture[a]	E1	E2	E3
CTG	0.35	0.18	0.35
(NT+CL)G	0.61	0.42	0.43
Difference	0.26	0.24	0.08

Note: Average period of measurement was 80 days between June 30 and September 17. Numbers are means for 5 years.

[a] CTG = grain sorghum into disc-harrowed seedbed from winter fallow; (NT+CL)G = grain sorghum no-till planted into crimson clover; E1 = slightly eroded; E2 = moderately eroded; E3 = severely eroded.

From Bruce, R.R., Langdale, G.W., and West, L.T., *Trans. 14th Int. Cong. Soil Sci.*, 6, 4, 1990. With permission.

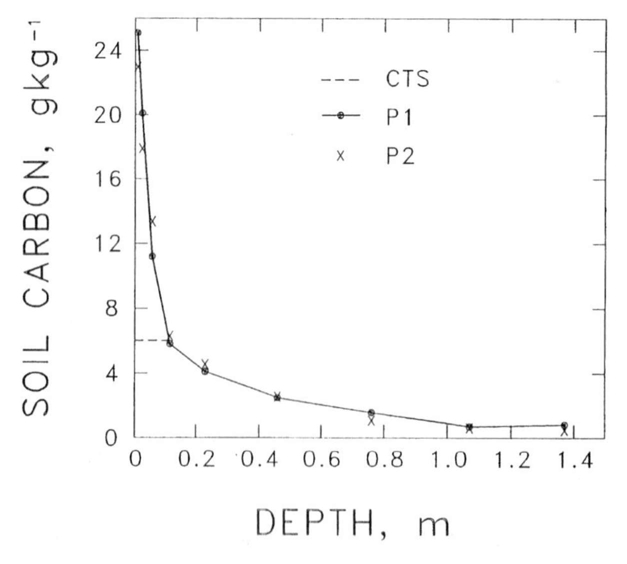

Figure 9 Soil carbon as a function of soil depth in P1 and P2 watersheds and, for CTS, conventionally tilled soybean, at shallow soil depths.

REFERENCES

1. Trimble, S.W., Man induced soil erosion on the Southern Piedmont — 1770-1970, *Soil Cons. Soc. Am.*, 1974, 180.
2. Giddens, J. and Garman, W.H., Some effects of cultivation on the Piedmont soils of Georgia, *Soil Sci. Am. Proc.*, 6, 439, 1941.
3. Giddens, J., Rate of loss of carbon from Georgia soils, *Soil Sci. Am. Proc.*, 21, 513, 1957.
4. Giddens, J., Morris, H.D., and Perkins, H.F., Factors Affecting Organic Matter in Georgia Soils, Georgia Agric. Exp. Stn. Tech. Bul. N.S. 10, 1957.
5. Adams, W.E. and Dawson, R.N., Cropping System Studies on Cecil Soils, Watkinsville, GA. 1943–62, Agric. Res. Serv, USDA, Watkinsville, GA, USDA-ARS 41, 1964.
6. Barnett, A.P., Roberts, J.S., Adams, W.E., and Welch, L.F., Cropping systems, organic matter and nitrogen, *Georgia Agric. Res.*, 3(1), 10, 1961.
7. Jones, L.S., Anderson, O.E., and Stacy, S.V., Some Effects of Sod-Based Rotation upon Soil Properties, Georgia Agric. Exp. Stn. Tech. Bul. N.S. 166, 1966.
8. Buol, S.W., Ed., Soils of the southern states and Puerto Rico, *Southern Coop. Series Bull.*, No. 174, 1973.
9. Bruce, R.R., Dane, J.H., Quisenberry, V.L., Powell N.L., and Thomas, A.W., Physical characteristics of soils in the region: Cecil, *Southern Coop. Series Bull.*, No. 267, 1983.
10. White, A.W., Jr., Bruce, R.R., Thomas, A.W., Langdale, G.W., and Perkins, H F., Characterizing productivity of eroded soils in the Southern Piedmont, in *Erosion and Soil Productivity*, (Publ. 8-85), ASAE, ST. Joseph, MO, 1985, 83.
11. Bruce, R.R., White, A.W., Jr., Thomas, A.W., Snyder, W.M., Langdale, G.W., and Perkins, H.F., Characterization of soil-crop yield relations over a range of erosion on a landscape, *Geoderma*, 43, 99, 1988.
12. Bruce, R.R., Snyder, W.M., White, A.W., Jr., Thomas, A.W., and Langdale, G.W., Soil variables and interactions affecting prediction of crop yield pattern, *Soil Sci. Soc. Am. J.*, 54, 494, 1990.
13. Langdale, G.W., Hardgrove, W.L., and Giddens, J., Residue management in double-crop conservation tillage systems, *Agron. J.*, 76, 689, 1984.
14. Bruce, R.R., Langdale G.W., and Dillard, A.L., Tillage and crop rotation effect on characteristics of a sandy surface soil, *Soil Sci. Soc. Am. J.*, 54, 1744, 1990.
15. Bruce, R.R., Langdale, G.W., and West, L.T., Modification of soil characteristic of degraded soil surfaces by biomass input and tillage affecting soil water regime, *Trans. 14th Int. Congr. Soil Sci.*, 6, 4, 1990.
16. West, L.T., Miller, W.P., Langdale, G.W., Bruce, R.R., Laflen, J.M., and Thomas, A.W., Cropping system effects on interrill soil loss in the Georgia Piedmont, *Soil Sci. Soc. Am. J.*, 55, 460, 1991.
17. Langdale, G.W., West, L.T., Bruce, R.R., Miller, W.P., and Thomas, A.W., Restoration of eroded soil with conservation tillage, *Soil Technol.*, 5, 81, 1992.
18. Kemper, W.D. and Koch, E.J., Aggregate Stability of Soils from Western United States and Canada, USDA-ARS Tech. Bull. No. 1355, U.S. Government Printing Office, Washington, DC, 1966.
19. Bruce, R.R., Langdale, G.W., West, L.T., and Miller, W.P., Soil surface modification by biomass inputs affecting rainfall infiltration, *Soil Sci. Soc. Am. J.*, 56, 1614, 1992.
20. Foster, G.R., Young, R.A., Romkens, M.J.M., and Onstad, C.A., Processes of soil erosion by water, in *Soil Erosion and Crop Productivity*, Follet, R.F., and Stewart, B.A., Eds., ASA, CSSA, and SSSA, Madison, WI, 1985, 137.
21. Mills, W.C., Thomas, A.W., and Langdale, G.W., Rainfall retention probabilities computed for different cropping-tillage systems, *Agric. Water Manage.*, 15, 61, 1988.

Chapter 19

Changes in Ecosystem Carbon 46 Years after Establishing Red Pine (*Pinus resinosa* Ait.) on Abandoned Agricultural Land in the Great Lakes Region

Kurt S. Pregitzer and Brian J. Palik

CONTENTS

I. INTRODUCTION

Reforestation of abandoned agricultural land may result in improved chemical and physical properties of degraded soil through the action of above- and below-ground litter input and increased root activity. Additionally, because forest ecosystems fix and store substantial amounts of C annually (Whitaker 1975; McClaugherty et al. 1982; Waring and Schlesinger 1985), expansive reforestation projects might be used to mitigate escalating atmospheric CO_2 levels (Cooper 1983).

The speed with which organic matter accumulates in mineral soil may influence changes in soil properties and the potential for long-term C storage in forest ecosystems (Olson 1963; Hillel 1982). Mineral soil can be globally massive and a relatively stable storage pool for carbon (Olson 1963; Schlesinger 1977; Post et al. 1982; Cooper 1983; Solomon et al. 1985) that may be little affected by above-ground dynamics associated with timber rotations (Cooper 1983). Decomposition and organic

0-8493-2802-0/97/$0.00+$.50

matter transfer to the soil is slow in litter with a wide C:N ratio and high lignin content (Gosz 1981; Staaf and Berg 1981; Melillo et al. 1982; Nadelhoffer et al. 1983; Post et al. 1982). Because litter chemistry varies by taxa, the choice of species for reforestation may influence the potential for soil to store C (Perala and Alban 1982; Nadelhoffer et al. 1983).

In 1942, foresters at Michigan State University initiated a long-term study of soil changes on an abandoned agricultural site. Bulk density, soil organic C, and total soil N concentrations were measured in 1942 and again in 1962, 12 years after the establishment of a red pine plantation. We re-examined the site in 1988 with the following objectives: (1) to quantify changes in soil C, N and bulk density over the 46-year period and (2) to determine the content and distribution of C and N in the red pine ecosystem.

II. METHODS

A. SITE DESCRIPTION

The study site is a 2.9-ha red pine plantation at Michigan State University's W.K. Kellogg Experimental Forest (42°22′N, 85°21′W). Average monthly climatic conditions are summarized in Table 1. The soil is classified as an Oshtemo sandy loam (coarse-loamy, mixed, mesic Typic Hapudalf). The site had been abandoned from agriculture for approximately 10 years prior to study initiation. In 1942, 11 permanent points were randomly established throughout the site. These points were to serve as loci around which soil would be sampled over the next 50 years at 10-year intervals. In 1950-1951 the site was planted with 2-0 red pine (*Pinus resinosa* Ait.) seedlings at a 3 × 3 m spacing. At the time of planting the site had a thin grass cover (original notes). Soil around the sampling points was not disturbed during planting.

B. SOIL

Soil was sampled in late October-early November of 1942, 1952 (bulk density only), 1962, and 1988. Litter was cleared to mineral soil in a 0.6 × 0.9 m area located 1.5 m from each permanent point. Soil sampling areas were spaced clockwise at 60° intervals around the center point, beginning at 0° north in 1942. A soil pit was excavated to 30 cm in a 0.3 × 0.45 m half of each area. In 1942 and 1962 three 9.5-cm deep volumetric samples were collected from 0- to 10- and 30- to 40-cm depths for bulk density determinations. In 1942 and 1962 three additional volumetric samples were composited by depth for chemical analysis. In 1988 forest floor was collected to mineral soil within a 0.10-m^2 template from each sampling area prior to soil sampling. Bulk density was determined from a composite of two 10-cm deep volumetric samples at each depth. Four additional samples were composited by depth for chemical analysis.

C. VEGETATION

In October 1988, five 500-m^2 circular plots were established around randomly chosen sampling points. In each plot, stem diameter (to the nearest 0.1 cm at 1.37 m) and height (to the nearest decimeter using a clinometer) of all individuals were recorded. Samples of live and dead branches, foliage, bole bark, and bolewood were collected from one individual in four other plots.

D. LABORATORY ANALYSIS

Soil samples collected for bulk density were oven-dried (105°C for 48 hours) and weighed. Samples collected for chemical analysis were air-dried and passed through a 2-mm sieve. Total soil N concentrations in 1942 and 1962 were determined on subsamples by the macro-Kjeldahl method and colorimetric analysis (Technicon 1977). Organic C concentrations in 1942 and 1962 were determined on subsamples by the Schollenberger method (Schollenberger 1927, 1931). We converted easily oxidizable C to total C using a multiplication factor of 1.15 (Allison 1965). Organic carbon concentrations in 1988 were determined on subsamples by the Walkley-Black method (Schulte 1980). Easily oxidizable C was converted to total organic C using a multiplication factor of 1.33 (Walkley 1947). Red pine and forest floor samples were oven-dried (70°C for 24 hours) and ground in a Wiley mill to pass a 0.5-mm sieve. Forest floor samples were weighed prior to grinding. Total N concentrations of tissues were determined as for soil. Ash-free percentages of tissue and forest floor subsamples were determined after dry combustion (550°C for 5 hours).

E. DATA ANALYSIS

Estimates of red pine biomass (Mg/ha) were determined for each plot using allometric equations based on mean stand height and basal area (Alban and Laidly 1982). Organic C in vegetation and forest floor

Table 1 Thirty-Year Mean Monthly Temperature and Precipitation Averages at Kellogg Forest

	January	February	March	April	May	June	July	August	September	October	November	December
Mean monthly temperature (°C)	–5.2	–3.9	1.3	8.6	14.8	20.0	22.3	21.4	17.6	11.4	4.3	–2.2
Mean monthly maximum temperature (°C)	–0.9	1.0	6.6	14.9	21.6	26.6	28.8	27.9	24.0	17.4	8.7	1.8
Mean monthly minimum temperature (°C)	–9.6	–8.9	–4.0	2.3	8.1	13.4	15.7	14.9	11.1	5.4	–0.2	–6.1
Precipitation (cm)	4.57	3.68	5.16	8.92	8.03	10.67	8.64	8.99	7.57	7.34	6.88	5.79

From NOAA, Climate Normals for the U.S. (Base: 1951–1980), 1st ed., National Oceanographic and Atmospheric Administration, Gale Research Company, Detroit, 1983.

was calculated as 48% of ash-free dry mass (Vitousek et al. 1982). Paired bulk density and C and N concentrations for each point were used to determine C and N content (Mg/ha). Total C accrual in the red pine ecosystem included that exported in experimental thinnings in 1975 and 1983. Carbon removed was calculated as total volume of sawlogs removed × density (481 kg/m^3; Panshi and de Zeeuw 1980) × ash-free percentage in 1988 (the mean of bark and wood) × 0.48. Soil C and N contents and concentrations and bulk density were compared separately by depth by one-way ANOVA grouped on year. Hypotheses of normality were accepted for all ten data sets (Shipiro-Wilk test; P>0.1). Hypothesis of homogeneous variances were rejected (Barlett's test; P <0.2) for all data sets except N concentration at 0 to 10 cm. A Brown-Forsythe modified f statistic was used to compare the means of groups with heterogeneous variances. When the overall test was significant, Bonferroni t-tests (modified for heterogeneous variances) were used to test hypotheses of increasing or decreasing means over time. Statistical methods follow Gill (1978). For comparison, we converted cited literature values of soil matter content to organic C using a multiplication factor of 0.58 (Broadbent 1965). Similarly, forest floor ash-free organic matter content was converted to organic C using a multiplication factor of 0.48 (Vitousek et al. 1982).

III. RESULTS

Bulk density decreased at both the 0 to 10- and 30- to 40-cm depths over the 46-year period (Figure 1). Differences were significant between 1942 and all subsequent dates at both depths (P <0.01) and between 1952 and 1988 at 0 to 10 cm (P<0.05).

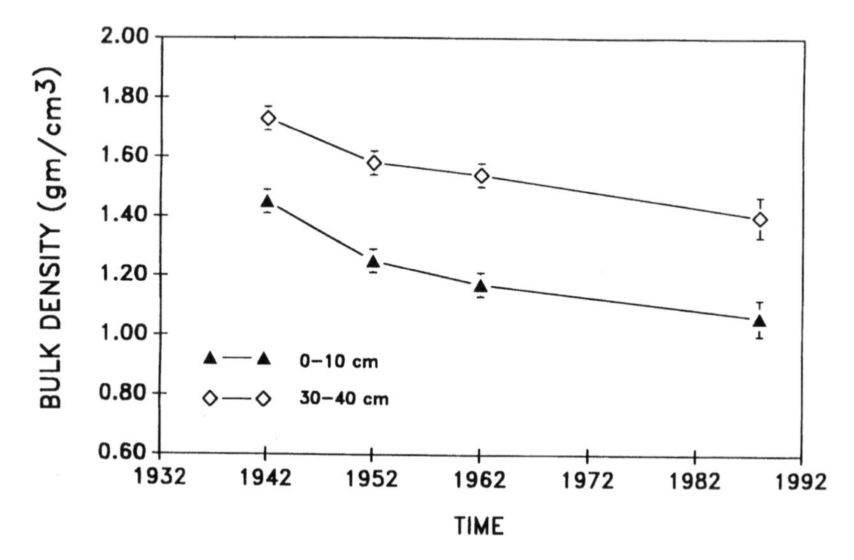

Figure 1 Bulk density changes over 46 years at two depths on an abandoned agricultural site. Red pine (*Pinus resinosa* Ait.) was planted in 1950/1951. Each point is the mean (± 1 SE) of 11 observations.

Soil organic C content decreased over 46 years at both sampling depths (Figure 2). Differences were not significant at 0 to 10 cm (P>0.1). The decrease from 1942 to 1988 was significant at 30 to 40 cm (P<0.05). Total soil N content at 0 to 10 cm decreased over time (Figure 2). Nitrogen content at 30 to 40 cm increased slightly from 1942 to 1962, followed by a slight decrease by 1988 (Figure 2). Differences were not significant at either depth (P>0.1). Carbon and N concentrations (Table 2) remained virtually unchanged at both depths (P>0.3). Changes in content at the measured depths were therefore more a function of decreases in bulk density.

The 1988 soil C content (11.3 Mg/ha) at 0 to 10 cm was 31 to 49% of values estimated from Perala and Alban (1982) and Nadelhoffer et al. (1983) for similar aged red pine stands on sandy soils with no history of cultivation. Total N content (726 kg/ha) was 38 to 65% of values reported in the aforementioned studies. Carbon content at 30 to 40 cm (3.6 Mg/ha) was 20% above and below two values estimated from Peral and Alban (1982) for a depth of 25 to 36 cm, while N content (363 kg/ha) was 34% higher.

The mass of C and N in the forest floor was 10.5 Mg/ha and 230.7 kg/ha, respectively (Table 3). Carbon content was 2 to 4 Mg/ha less than that estimated by Perala and Alban (1982) for two 40-year-old red pine

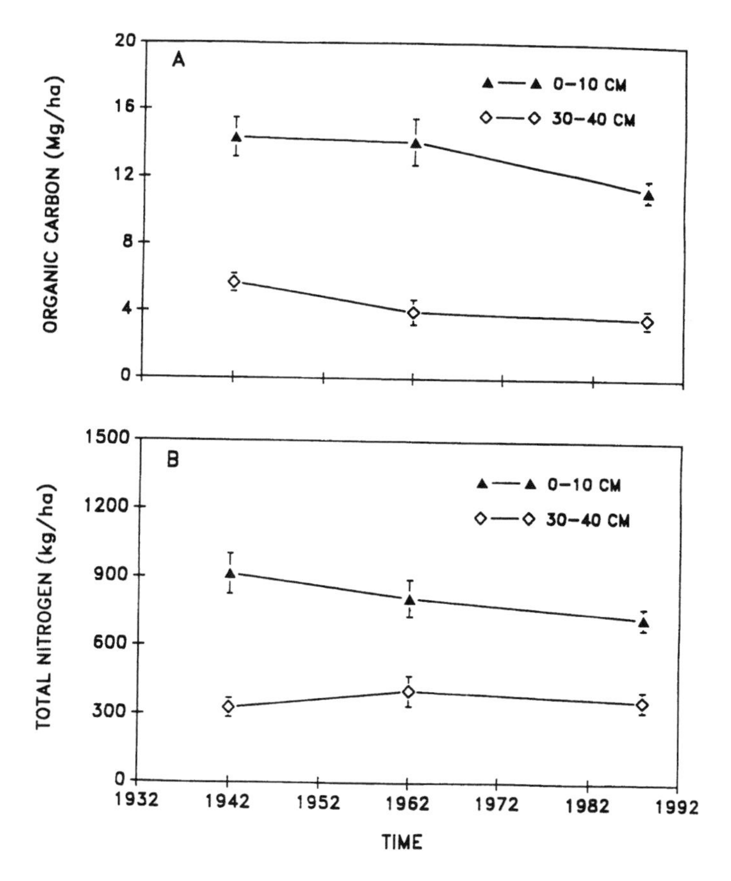

Figure 2 Changes in soil organic carbon content (A) and total nitrogen content (B) over 46 years at two depths on an abandoned agricultural site. Red pine (*Pinus resinosa* Ait.) was planted in 1950/1951. Each point is the mean (± 1 SE) of 11 observations.

Table 2 Soil Organic Carbon and Total Nitrogen Concentrations at Two Depths on an Abandoned Agricultural Site

	Organic Carbon [g (kg^{-1})]	Total Nitrogen [g (kg^{-1})]
0–10 cm		
1942	10.6 (1.1)	0.7 (0.08)
1962	13.1 (1.5)	0.8 (0.09)
1988[a]	11.0 (0.8)	0.7 (0.06)
30–40 cm		
1942	3.5 (0.2)	0.2 (0.03)
1962	2.7 (0.5)	0.3 (0.05)
1988	2.7 (0.5)	0.3 (0.05)

Note: Red pine (*Pinus reinosa* Ait.) was planted in 1950/1951.

[a] 1985 soil organic C and total N concentrations at 0 to 10 cm in an adjacent oak forest, with no history of cultivation, were 18.4% and 1.3 kg ha^{-1}. Each value is the mean (±1 standard error) of 11 observations.

stands and nearly 5 Mg/ha less than a 100+-year-old stand (Alway and Zon 1930). Nitrogen content was approximately 120 and 300 kg/ha less than in the sites studied by Peral and Alban (1982).

Organic C totaled 72,5 Mg/ha in above-ground biomass in 1988. Carbon was distributed in red pine components as bolewood > live branches > foliage > bole bark > dead branches (Table 3). An additional

Table 3 Organic Carbon and Total Nitrogen Content and Distribution in the Overstory and Forest Floor of a 38 Year-Old Red Pine Plantation

	Organic Carbon		Total Nitrogen	
	Mg/ha	%[a]	kg/ha	%[a]
Live branch	6.6 (0.1)	9.1	43.8 (1.3)	14.5
Dead branch	2.7 (0.1)	3.7	14.7 (0.6)	4.9
Foliage	5.3 (0.2)	7.3	128.1 (5.0)	42.5
Bolewood	52.8 (2.4)	72.8	85.5 (3.8)	28.4
Bole bark	5.1 (0.3)	7.0	29.4 (1.3)	9.8
Total Overstory	72.5		301.6	
Forest Floor	10.5 (1.2)		230.7 (33.8)	

Note: Each overstory and forest floor value is the mean (±1SE) of 4 and 11 observations, respectively.

[a] Percent of total overstory.

10 Mg/ha of accrued C was removed from the stand in thinnings. Nitrogen totaled 302 kg/ha in above-ground red pine biomass. Nitrogen was distributed as foliage > bolewood > live branches > bole bark > dead branches (Table 3). These distributions are similar to those reported in the literature (Perala and Alban 1982; Bockheim et al. 1986).

IV. DISCUSSION

A model for organic C accumulation in an aggrading forest ecosystem includes fixation and storage in vegetation, with redistribution of some fraction of fixed C to the forest floor and mineral soil. In our study site, C has accumulated in vegetation (standing crop + thinnings; 82.5 Mg/ha) and forest floor (10.5 Mg/ha) only. Over the 46-year period C content at the 0- to 10- and 30- to 40-cm depths has actually decreased slightly, concurrently with bulk density. Physical characteristics of the soil, including rates of infiltration and diffusion, have likely improved with the decrease in bulk density (Hillel 1982).

Perala and Alban (1982) and Nadelhoffer et al. (1983) both found higher soil organic matter and total N contents in red pine stands than in stands dominated by either jack pine, white pine, white spruce, aspen, or sugar maple. Nadelhoffer et al. (1983) suggested that low-quality red pine litter slows decomposition and net N mineralization, with a resultant increase in the recalcitrant soil organic matter and total N pools. Our data show that over 38 years little of this organic matter has accumulated in the mineral soil. Comparatively, the 1988 soil C and total N contents at 0 to 10 cm in our site were substantially lower than those of Peral and Alban (1982) and Nadelhoffer et al. (1983). The C:N ratio of the forest floor was wider in our site than either of the red pine sites studied by Perala and Alban (1982): 45 vs. 35 and 27, respectively. Relatively higher quality litter may have resulted in increased organic matter transfer to mineral soil and larger C and N pools in these sites.

Although we cannot be certain that soil C content was measurably greater at our site prior to cultivation, some indication that this was so is found in an adjacent oak-dominated forest, logged early in this century, but with no apparent history of cultivation. Soil organic C concentration at 0 to 10 cm was 1.84% in 1985, while total N concentration was 0.13% (D. Zak unpublished data).

Decreases in mineral soil C and N contents on this site were the result of decreases in bulk density. The near homeostasis in soil C and N concentrations over time reflects a negative feedback that is probably the result of poor litter quality and retarded organic matter dissolution and transfer.

Considerable C has accumulated in the forest floor and overstory of our site. Comparison of forest floor C mass with those of Alway and Zon (1930) and Perala and Alban (1982) suggests that additional C may yet be accrued before equilibrium mass is attained. Unfortunately, C stored in the forest floor is stable only as long as the overstory remains in place (Covington 1982; Mroz et al. 1985). Intensive management, to insure rapid reestablishment of leaf area, is required if rapid decomposition and loss of C stored in the forest floor is to be minimized. Above-ground storage is also temporary, with the length of sequestration dependent on species, rotation length, and the eventual use of the organic matter produced (Cooper 1983). Considerable C may be stored below ground in woody roots and its fate following harvest is poorly understood. Utilizing increased forest production to mitigate rising atmospheric CO_2 levels will be most successful if a portion of the fixed C can be transferred to mineral soil. The feedback mechanisms discussed above inhibit this process in the red pine ecosystem we studied.

Species composition, as it affects litter quality and organic matter transfer, may be an important consideration if landscapes are to be managed for C storage. Red pine is an important pulp and timber species in the Lake States (Bockheim et al. 1986). High nutrient use efficiency makes Red pine well suited for growth on poor sites, a benefit if biomass production is the primary management objective. Our data show that if concurrent or alternative objectives include maximization of C storage in the mineral soil and improved soil chemical properties, choice of this species would be inappropriate on sites similar to the one we studied.

ACKNOWLEDGMENTS

Support from the Michigan Agricultural Experiment Station and the McIntire-Stennis Cooperative Research Program is gratefully acknowledged. We thank the personnel of the W.K. Kellogg Experimental Forest, past and present.

REFERENCES

Alban, D.H. and P.R. Laidly, Generalized biomass equations for jack and red pine in the Lake States, *Can. J. Forest Res.*, 12, 13, 1982.

Allison, L.E., *Methods of Soil Analysis. Part 2*, Black, C.A., Ed., American Society of Agronomy, Madison, Wis., 1965, 1367.

Alway, F.J. and R. Zon, Quantity and nutrient contents of pine leaf litter, *J. Forest.*, 28, 715, 1930.

Bockheim, J.G., S.W. Lee, and J.E. Leide, Distribution and cycling of elements in a *Pinus resinosa* plantation ecosystem, Wisconsin, *Can. J. Forest Res.*, 13, 609, 1983.

Bockheim, J.G., J.E. Leide, and D.S. Travella, Distribution and cycling of macronutrients in a *Pinus resinosa* plantation fertilized with nitrogen and potassium, *Can. J. Forest Res.*, 16, 778, 1986.

Broadbent, F.E., Organic matter, *Agronomy*, 9, 1397, 1965.

Cooper, C.F., Carbon storage in managed forests, *Can. J. Forest Res.*, 13, 155, 1983.

Covington, W.W., Changes in forest floor organic matter and nutrient content following clearcutting in northern hardwoods, *Ecology*, 62, 41, 1982.

Gill, J.L., *Design and Analysis of Experiments in the Animal and Medical Sciences*, Vol. 1, Iowa State University Press, Ames, IA, 1978.

Gosz, J.R., Nitrogen cycling in coniferous ecosystems, in *Terrestrial Nitrogen Cycles*, (Ecological Bulletin 33), Clark, F.E. and T. Rosswall, Eds., Swedish Natural Science Research Council, Stockholm, 1981, 405.

Hillel, D., *Introduction to Soil Physics*, Academic Press, San Diego, 1982.

McClaugherty, C.A., J.D. Aber, and J.M. Melillo, The role of fine roots in the organic matter and nitrogen budgets of two forested ecosystems, *Ecology*, 63, 1481, 1982.

Melillo, J.M., J.D. Aber, and J.F. Muratore, Nitrogen and lignin control of hardwood leaf litter decomposition dynamic, *Ecology*, 63, 427, 1982.

Mroz, G.D., M.F. Jurgensen, and D.J. Frederick, Soil nutrient changes following whole tree harvesting on three northern hardwood sites, *Soil Sci. Soc. J.*, 49, 1552, 1985.

Nadelhoffer, K.J., J.D. Aber, and J.M. Melillo, Leaf-litter production and soil organic matter dynamics along a nitrogen-availability gradient in southern Wisconsin (U.S.A.), *Can. J. Forest Res.*, 13, 12, 1983.

NOAA, *Climate Normals for the U.S. (Base: 1951–80)*, 1st ed., National Oceanic and Atmospheric Administration, Gale Research Company, Detroit, 1983.

Olson, J.S., Energy storage and the balance of producers and decomposers in ecological systems, *Ecology*, 14, 322, 1963.

Panshin, A.J. and C. de Zeeuw, *Textbook of Wood Technology*, 4th ed., McGraw-Hill, New York, 1980.

Pastor, J., J.D. Aber, and C.A. McClaugherty, Aboveground production and N and P cycling along a nitrogen mineralization gradient on Blackhawk Island, Wisconsin, *Ecology*, 65, 256, 1984.

Perala, D.A. and D.H. Alban, Biomass, nutrient distribution and litterfall in *Populus*, *Pinus* and *Picea* stands on two different soils in Minnesota, *Plant Soil*, 64, 177, 1982.

Post, W.M., W.R. Emanuel, P.J. Zinke, and A.G. Stangengberger, Soil carbon pools and world life zones, *Nature*, 298, 156, 1982.

Schlesinger, W.H., Carbon balance in terrestrial detritus, *Annu. Rev. Eco. System.*, 89, 51, 1977.

Schollenberger, C.J., A rapid approximate method for determining soil organic matter, *Soil Sci.*, 24, 65, 1927.

Schollenberger, C.J., Determination of soil organic matter, *Soil Sci.*, 31, 483, 1931.

Schulte, E.E., Recommended chemical soil test procedures for the North Central Region, North Dakota Agricultural Experiment Station Bulletin Number 499, North Dakota State University, Fargo, 1980.

Solomon, A.M., J.R. Trabalka, D.E. Reichle, and L.S. Voorhees, The global carbon cycle, in *Atmospheric Carbon Dioxide and the Global Carbon Cycle*, J.R. Trabalk, Ed., (DOE/ER-0239), National Technical Information Service, Springfield, VA, 1985.

Staaf, H. and B. Berg, Plant litter input to soil, in *Terrestrial Nitrogen Cycles*, F.E. Clark and T. Rosswall, Eds., (Ecological Bulletin 33), Swedish Natural Science Research Council, Stockholm, 1981, 147.

Technicon Individual/Simultaneous Determination of Nitrogen and/or Phosphorous in BD Acid Digests, Industrial Methods Number 329-74W/B. Technicon Industrial Systems, Tarrytown, NY, 1977.

Vitousek, P.M., Nutrient cycling and nutrient use efficiency, *Am. Nat.*, 119, 553, 1982.

Vitousek, P.M., J.R. Gosz, C.G. Grier, J.M. Melillo, and W.A. Reiners, A comparative analysis of potential nitrification and nitrate mobility in forest ecosystems, *Eco. Monogr.*, 52, 155, 1982.

Walkley, A., A critical examination of a rapid method for determining organic carbon in soils — effect of variations in digestion conditions and of inorganic soil constituents, *Soil Sci.*, 63, 251, 1947.

Waring, R.H. and W.H. Schlesinger, *Forest Ecosystems: Concepts and Management*, Academic Press, Orlando, 1985.

Whittaker, R.H., *Communities and Ecosystems*, MacMillan, New York, 1975.

Part III

Site Management Effects on Productivity and Soil Organic Matter Characteristics in the Great Plains

Chapter

Crop Yield and Soil Organic Matter Trends over 60 Years in a Typic Cryoboralf at Breton, Alberta

N.G. Juma, R.C. Izaurralde, J.A. Robertson, and W.B. McGill

CONTENTS

I. INTRODUCTION

Most Cryoboralfs in Canada occur in the northern interior plains of Manitoba, Saskatchewan, and Alberta. The largest area occurs in Alberta (20 million hectares), of which about 5.7 million hectares are potentially arable.[8] About 3.2 million hectares are occupied and about 1.6 million hectares are cultivated.[8] Cultivated Cryoboralfs constitute about 15% of the total cultivated area in Alberta.[3]

Cryoboralfs have a thin A horizon which has a slightly acidic reaction (pH 6.0 to 6.5), low organic C content (10 to 20 g kg^{-1}), and nutrient-supplying capacity. The A horizons often develop poor tilth when cultivated. Most soil-related problems such as crusting, low water-holding capacity, low fertility, and low buffering capacity against pH change arise from the low organic matter content of the A horizon. Impeded water transmission and restricted root growth arise from the presence of a dense, very firm, and acidic B horizon. Problems in managing these soils were recognized early, and in 1930 Dr. F. A. Wyatt and Dr. J. D. Newton established plots on Cryoboralfs at Breton, Alberta.[15] Highlights of research conducted at the Breton Plots and a publication list have been prepared by Robertson.[12–14] These plots are the only continuous, long-term plots on Cryoboralfs in Canada and possibly in the world. This paper provides a summary of trends of crop yields and soil organic C over 60 years at this site.

0-8493-2802-0/97/$0.00+$.50

II. SITE DESCRIPTION, SOIL PROPERTIES, AND METEOROLOGICAL DATA

The Breton plots are located 110 km southwest of Edmonton (53° 07′ N, 114° 28′ W), 5 km from the town of Breton, on Typic Cryoboralfs.[11] Long-term climate data[1,2] show that Breton receives 547 mm of precipitation during the year with 405 mm as rain and 132 mm as snow. The months of greatest rainfall are June, July, and August and the greatest snowfall occurs during December and January. July is the warmest month with an average minimum temperature of 8.8°C and a maximum of 21.2°C. January is the coldest month with average minimum temperature of –19.5°C and average maximum of –8.6°C. Breton has an average of 80 frost-free days. Average precipitation and radiation data obtained from long-term (1951 to 1980) climate records collected at the town of Breton are presented in Figure 1. Soil properties of the Breton loam soil series are given in Table 1.

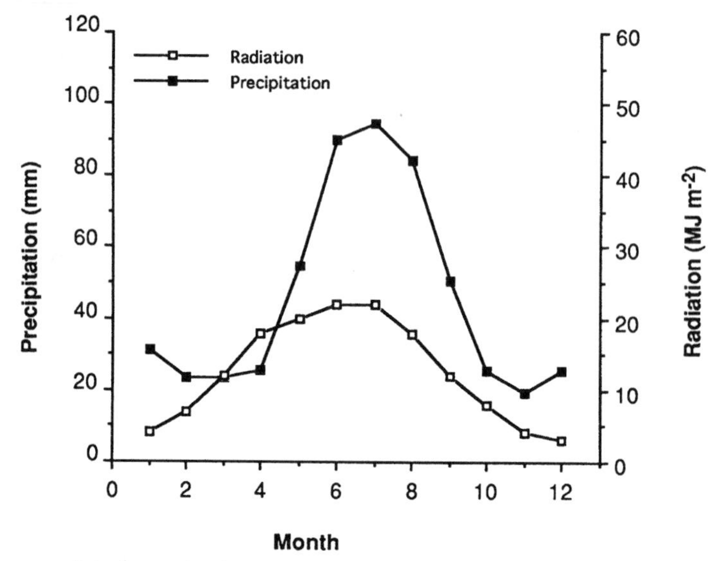

Figure 1 Average precipitation and radiation data obtained from long-term (1951 to 1980) climate records[1,2] collected at the town of Breton.

Table I Properties of Soil (Typic Cryoboralf) at Breton

Depth (cm)	Total C ($g\ kg^{-1}$)	Total N ($g\ kg^{-1}$)	pH[a]	Bulk Density ($Mg\ m^{-3}$)	Texture[b]
0–10	21.7	1.85	6.2	1.11	SiL
10–20	19.2	1.64	6.3	1.30	CL
20–30	11.3	1.04	6.1	1.55	CL

[a] 1:2 soil:water (mass:volume).
[b] Particle size was determined by the hydrometer method.
From Dinwoodie, G. D. and Juma, N. G., *Plant Soil*, 110, 111, 1988. With permission.

III. DESCRIPTION AND MANAGEMENT HISTORY OF SELECTED TREATMENTS OF THE BRETON PLOTS

The Breton Plots were established in 1930 to find "a system of farming suitable for the wooded soil belt."[12] The experiment was designed to compare two cropping systems and test several soil amendments (fertilizers, manure, lime) within the two rotations. Originally, the experiment consisted of 5 blocks of land (Series A to E) which accommodated the 2 cropping systems, across which ran 11 strips with the various soil amendments. In 1938, an additional block of land was added (Series F) to expand the 4-year rotation to a 5-year rotation. Further, in 1941 the continuous wheat system (Series E) was split in half to create the present-day 2-year rotation of wheat-fallow. The experimental unit is a series (i.e., a crop by soil amendment) which is not replicated. The crops for both rotations are grown every year but the individual treatments (crop by soil amendment) are not replicated.

The soil amendments included several combinations of the nutrients (nitrogen, phosphorus, potassium, and sulfur) as well as lime and farmyard manure.[15] In this paper, we discuss the results of three treatments [check (plot 1), manure (plot 2), and NPKS (plot 3)] (Table 2). Fertilizer application methods varied over the years. Initially, all fertilizers were annually broadcast. From 1946 to 1964, fertilizers were added every 2nd year. In 1964, annual applications were resumed and phosphate was drilled with the seed. Manure was applied once every 5 years. In 1972, lime was added to the east half of all plots of the 5-year rotation and to the entire area of the 2-year rotation. The check or control plots have not received fertilizers since 1930.

Major revisions were introduced in 1980.[5,15] Some of the fertilizer treatments were revised and the nutrient applications were brought into line with current recommendations to farmers. The annual applications of nitrogen, phosphorus, and potassium were also increased substantially (Table 2).

Table 2 Comparison of Original and Revised Fertilizer Application Rates (kg ha^{-1} y^{-1}) on the Long-Term Plots at Breton

	Former Treatment 1930–1979 inclusive				Current Cereal Crop Treatments 1980–present			
	N	P	K	S	N	P	K	S
Manure	76[a]	42	91	20	—[b]	—	—	—
NPKS	10	6	16	10	—[c]	22	46	5.5
Control	0	0	0	0	0	0	0	0

[a] Applied every 5th year, in later years at 44 t/ha. Nutrient rates shown are annual equivalents and are estimates based on manure applied from 1976 to 1986 inclusive.

[b] N application via manure depends upon the rotation. Wheat-fallow rotation: 90 kg N ha^{-1} for each wheat crop. Wheat, oat, barley, forage, forage rotation 175 kg N ha^{-1} every 5 years. Applied in two equal applications: after oat harvest and at the time of forage plowdown.

[c] N amounts depend on the crop and its place in rotation: wheat on fallow 90 kg N ha^{-1}; wheat after forage 50 kg N ha^{-1}; oats after wheat 75 kg N ha^{-1}; barley as nurse crop 50 kg N ha^{-1}; and legume-grass forages 0 kg N ha^{-1}.

The forage crop has varied over the years but has always included a legume. For the 1939 to 1954 period, the forage component consisted of "mixed legumes". For the period from 1955 to 1966, the forage component was a five-crop mixture (alfalfa, red clover, brome, creeping red fescue, timothy). In 1967, the forage mixture was changed to alfalfa and brome to reflect commonly recommended forage mixture for the area.

The grain crops are harvested in the fall, usually between August 20 and September 15, depending on the weather. The first-year forage crop is cut twice seasonally: the first cut is in early July and the second is in September. The second-year forage crop is harvested in July and the plot is clean cultivated for the remainder of the season. For all treatment combinations, all above-ground growth is removed (straw, grain, and hay) and no crop residues are returned to the soil. The plots are tilled once in the fall after harvest and again in the spring prior to planting. Planting is currently done with a press drill; planting usually occurs between May 1 and 15. The current seeding rates are 82 kg ha^{-1} for wheat; 89 kg ha^{-1} for oats and barley; 10 kg ha^{-1} and 15 kg ha^{-1} for alfalfa and bromegrass, respectively. All seeds are planted in 15-cm rows. Herbicides were used from 1960 onward, first to control broad leaf weeds and later to control wild oats. Standard recommended practices were used to select the types and rates of herbicide for particular cropping sequences. Weeds in the fallow plots in the 2-year rotation are controlled with tillage or herbicides. Lime was added to the east half of the 5-year rotation plots (series A to D and F) and to the complete plots of the 2-year rotation (series E) in 1972. The data for the unlimed (west half) of the plots in 5-year rotation are used in this paper.

IV. RESULTS AND DISCUSSION

A. CROP YIELD TRENDS

Wheat is the common crop in both the rotations. It is the first cereal crop in the 5-year rotation and benefits from 2 years of forages in rotation. In this section, we have first compared the performance of wheat in the two crop rotations, followed by the performance of other crops in the 5-year rotation.

The 5-year running averages for wheat grain yields in the check plot of the 2-year wheat fallow rotation (Figure 2A) were about 1 t ha^{-1}; however, the running averages of the crops in the 5-year rotation showed an upward trend over a period of 50 years particularly after 1960 (Figure 2B to F). The cultivars of wheat used have varied over time; however, this did not increase the yields in the 2-year wheat-fallow rotation compared to the yields of the 5-year rotation. Because all above-ground residues were removed from all the plots, the increases in wheat grain yields in the check plots of the 5-year rotation can be attributed to increased nutrient supply, especially N, through biological N fixation and increased root mass. It is possible that a greater amount of root residues and N could contribute more available nutrients for crops. This is supported by the observation that forage yields increased from 1960 onward (see below) and contributed to increases in wheat yields in the 5-year rotation but not in the 2-year wheat-fallow rotation.

The wheat grain yields in the manure-treated plots in the 2- and 5-year rotations were consistently greater than those obtained in the check plots (Figure 2A and B). We calculated that the addition of manure provided approximately 70 kg N ha^{-1} y^{-1} (Table 2) and resulted in the yields which were at least twofold greater than the check plots over the past 35 years (Figure 2A and B). The wheat grain yield from manured plots of the 5-year were consistently higher than the grain yields from the manured treatment of the 2-year rotation. This suggests a synergistic effect with the 5-year rotation on wheat yields.

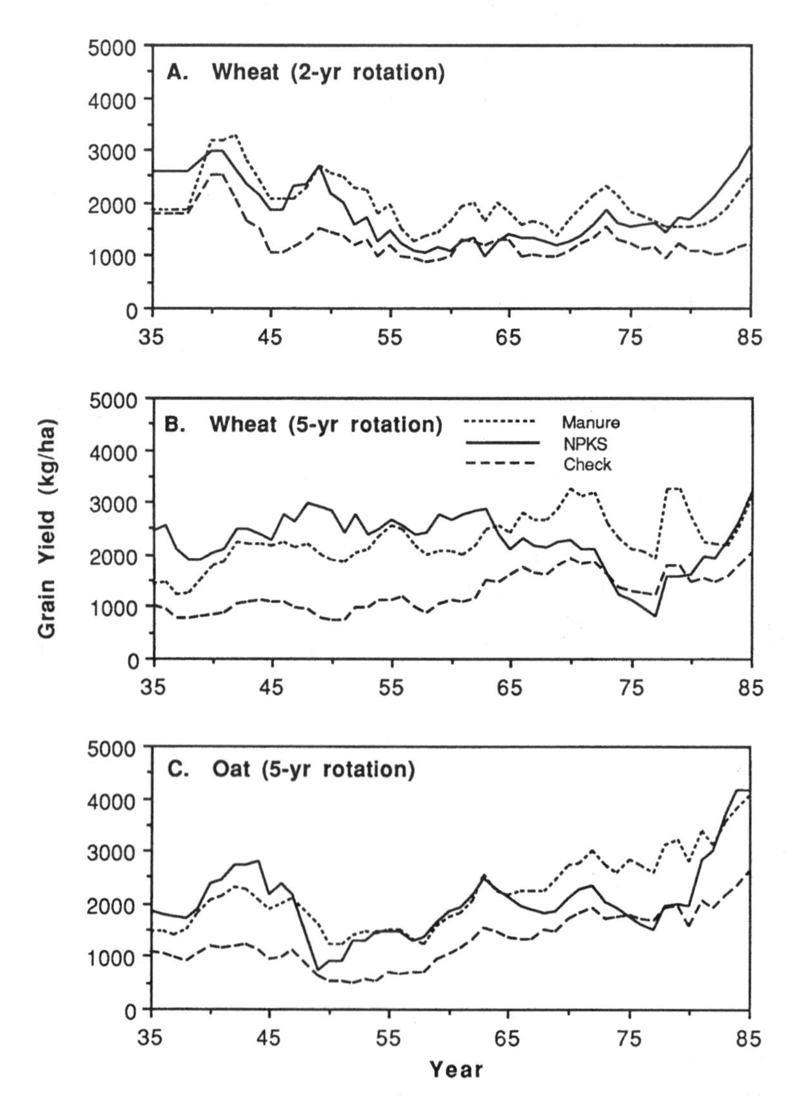

Figure 2 Five-year running averages of wheat grain yields for the 2-year rotation and for wheat, oat, and barley grain yields and dry matter yields of two forage crops for the 5-year rotation for the period 1935 to 1985. (From Juma, N.G., in *Impact of Macronutrients on Crop Responses and Environmental Sustainability on the Canadian Prairies*, Canadian Society of Soil Science, Ottawa, 1993, chap. 9. With permission.)

The wheat grain yields in the NPKS-treated plots of the 2-year rotation decreased after the first 20 years and became similar to those obtained in the check plots. These plots were N limited up to 1979 because fertilizer N had been applied at a rate of 10 kg ha^{-1}. In 1980, an adjustment in fertilizer application rate brought a balance between the N removed by harvest and that applied as fertilizer.[15] We observed an almost threefold increase in grain yields when fertilizer rates were adjusted. The grain yields in the NPKS-treated plots of the 5-year rotation were generally greater than those in the 2-year rotation up to 1970 which suggests that N supplied by legumes helped to maintain grain yields. In the late 1970s, there was a marked decrease in crop yields in the NPKS treatment of 5-year rotation. This can be partially explained by the type of legume being grown. In 1967, the legume combination of alfalfa and red clover was changed to alfalfa only. The plots were too acidic so the alfalfa yields were low (Figure 2E and F). The higher N fertilizer application rates of N fertilizer in the 1980s compensated for the lack of alfalfa growth in the forage in these plots. The higher fertilizer rates in the 1980s resulted in comparable yields in the NPKS and manure treatments in both rotations (Figure 2A and B).

In the 5-year rotation, the yield trends observed for oat (Figure 2C) and barley (Figure 2D) were similar to those for wheat. Generally, crop yields showed an upward trend with time. Yield ranking by treatment up to 1980 placed manure first followed by NPKS and check. On all plots, the average (1935 to 1979) wheat yields were greater than barley yields because the wheat crop benefited more from the nitrogen contributed by the legume in the preceding forage crop. By the time the barley crop was grown,

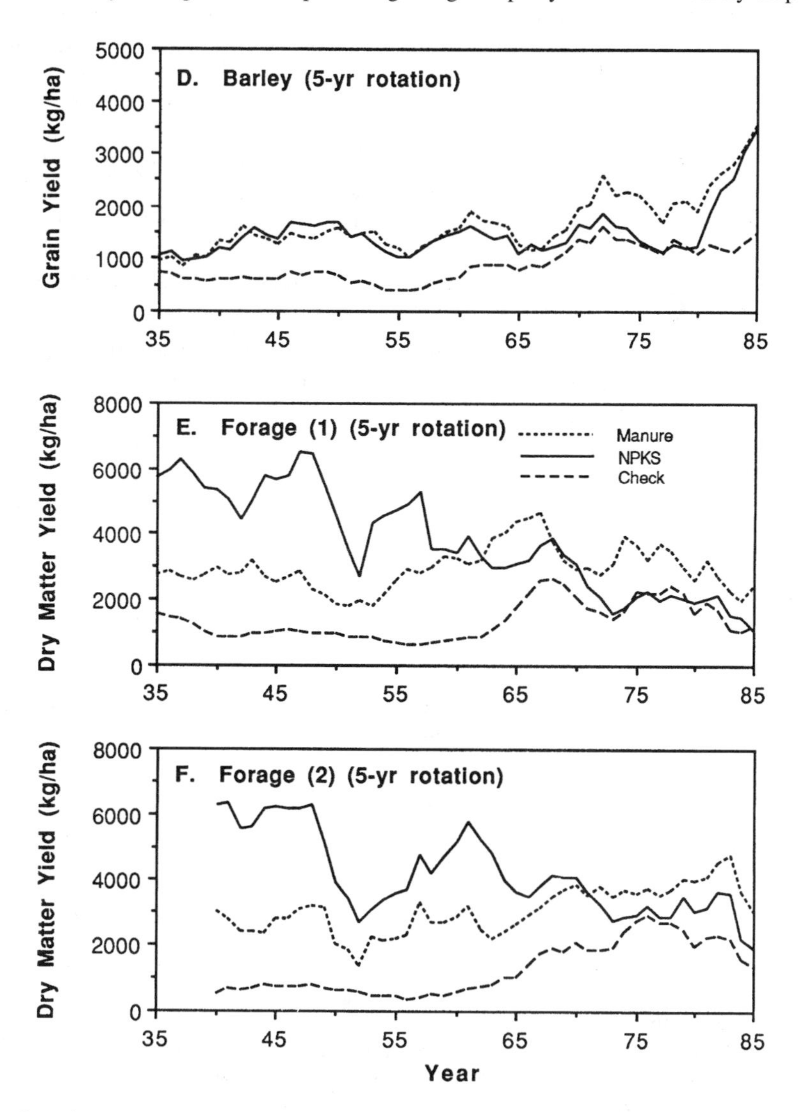

Figure 2 Continued.

much of the benefit of the legume in the forage crop was gone and the barley yields were considerably lower than the wheat yields. In these experiments the forages have been removed for hay in early to mid-July and the 2-year forages have been "plowed down" in late July without much regrowth. After the adjustment of fertilizer rates in 1980, the grain yields of manured and NPKS fertilized crops were similar (Figure 2A to D).

The running average of first-year forage yields obtained in the 4th year of the rotation in the check plots was around 1.0 t ha^{-1} from 1935 to 1960 (Figure 2E). After that time, the yield in the check plots rose to about 2 t ha^{-1}. We speculate that increased oil drilling in the area resulted in increased sulfur in the atmosphere and hence some sulfur added to the soil. The yields of manure-treated plots were almost two- to threefold higher and consistently greater than the check plots. Farmyard manure, added at about 44 t ha^{-1} every 5th year, produced a yield of about 3.0 t ha^{-1}. The nutrient content of the manure was not determined, but, based on values from the literature and some recent analyses, the amounts of nitrogen, phosphorus, and potassium added were several times larger than those added in the fertilizers (Table 2). The NKPS treatment had generally very high yields (4 to 6 t ha^{-1}) until about 1955, somewhat lower yields (about 3 t ha^{-1}) from 1955 to 1967, and still lower yields (about 2 t ha^{-1}) from 1967 to 1979. The likely explanation of the decline is the gradual acidification of the soil through application of ammonium-containing fertilizers (21-0-0 and 16-20-0). Such an effect would be particularly evident after 1967 when the legume crop was changed from red clover and alfalfa to alfalfa alone. Alfalfa is less tolerant of acidity than is red clover. The application of lime in other treatments of the Breton plots and in the east half of the NPKS plots resulted in higher forage yields (data not shown), indicating that the main effect was acidity and not the lack of S.

The running average of 2nd-year forage yields obtained in the 5th year of the rotation showed similar trends as the 1st-year forage yields (Figure 2F). The effect of increasing fertilizer rate in 1980 resulted in a marked increase of hay yields in 2nd-year forage in the early 1980s. The trend in N content of the crops in the 2- and 5-year rotations sampled in 1986 was NPKS > manure > check (Table 3). These data suggest that the adjustment of fertilizer application rates increased the N content of the plants.

Table 3 Nitrogen Analyses of Crops Sampled in 1986 for Selected Series in Each of the Check, NPKS, and Manure Treatments

Series	Plot	Crop	Legume N (g kg^{-1})	Grass N (g kg^{-1})	Grain N (g kg^{-1})	Straw N (g kg^{-1})
B	Check	Wheat	—	—	20.9	5.3
	NPKS	Wheat	—	—	22.6	6.5
	Manure	Wheat	—	—	21.9	5.4
A	Check	Oat	—	—	13.2	3.6
	NPKS	Oat	—	—	15.6	4.6
	Manure	Oat	—	—	13.4	3.0
F	Check	Barley	—	—	11.6	6.3
	NPKS	Barley	—	—	13.6	5.7
	Manure	Barley	—	—	13.1	6.0
D	Check	Forage 1	15.5	11.1	—	—
	NPKS	Forage 1	20.3	8.2	—	—
	Manure	Forage 1	19.2	9.3	—	—
C	Check	Forage 2	20.5	10.9	—	—
	NPKS	Forage 2	24.8	10.4	—	—
	Manure	Forage 2	21.0	9.8	—	—
E	Check	Wheat	—	—	18.2	4.6
	NPKS	Wheat	—	—	23.5	8.5
	Manure	Wheat	—	—	19.1	6.1

B. CARBON INPUT INTO SOIL

The above-ground crop residues are removed from both the rotations at the Breton plots. The input of C through the roots has not been measured for all the crops. However, the standing root mass has been measured for some crops. Fyles et al.[7] measured the standing root mass of two crops grown in plots which received manure. The average dry mass of roots of oat (263 g m^{-2}) was lower than that of 2nd-year alfalfa (483 g m^{-2}) in the 0- to 15-cm depth. Dinwoodie and Juma[6] measured the standing root mass of barley in plots adjacent to the long-term plots at Breton. The average dry mass of barley roots in the 0- to 10-cm depth was 45 g m^{-2}. The standing root mass estimates do not provide complete information on root inputs into soil. Van Veen et al.[16] summarized data from a wide variety of agroecosystems and reported that 60 to 90% of the total C assimilated by crops was stabilized in different pools of crop-soil system and 10 to 40% was released from the roots into soil. Estimations of annual inputs of C into soil by a growing crop ranged from 90 to 300 g m^{-2}. Large variations exist among plant species, cultivars, developmental stage of the cultivar, and environmental conditions. Campbell et al.[4] have estimated the above- and below-ground standing mass for different crops in Western Canada. The role of fertilizer nutrients and organic C inputs in rebuilding soil organic matter and the effect of C and N inputs on the dynamics of microbial biomass, faunal populations, and barley crop yields in plots adjacent to the long-term plots at Breton have been summarized by Juma.[9,10] As we have limited data on below-ground inputs of C for the Breton Plots, the total input of C has to be estimated from other various sources such as those mentioned above or through the use of simulation models.

C. SOIL ORGANIC MATTER TRENDS

Very few soil samples were taken or analyzed at the beginning of the Breton Plots so that soil changes from 1930 to 1979 cannot be assessed. Four samples were stored since 1936 and 1938, and their nitrogen contents were determined recently.[5] Various researchers over the years took samples and measured soil nitrogen content. Thus, for nitrogen we were able to follow the soil content over a 40-year period (Figure 3). Linear regression analyses showed that the slope of total N in the three treatments of the 2-year rotation were not significantly different from zero; therefore, there were no significant changes of total soil N

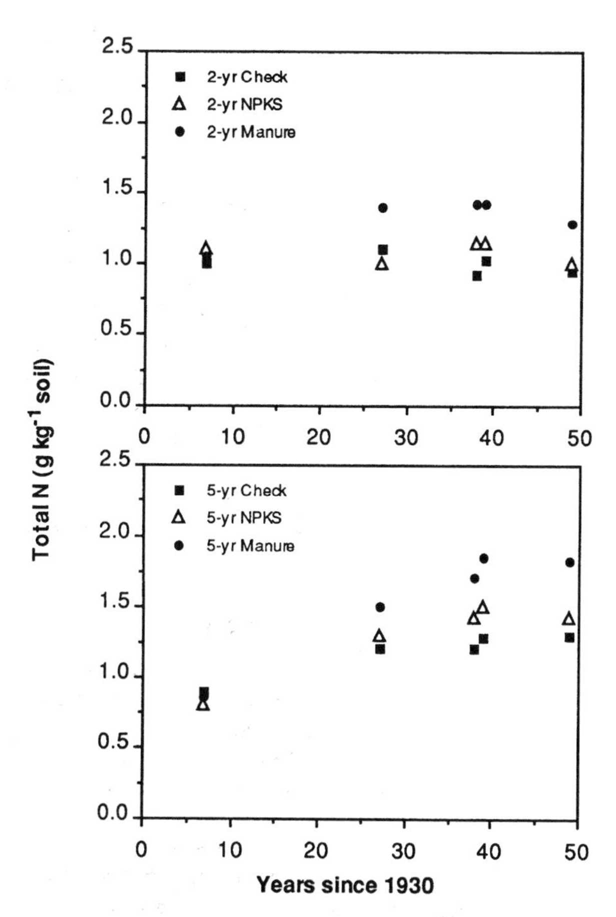

Figure 3 Total soil N in Ap horizon samples of the Check, NPKS, and Manure treatments of the two rotations over a period of 50 years. (From Juma, N.G., in *Impact of Macronutrients on Crop Responses and Environmental Sustainability on the Canadian Prairies*, Canadian Society of Soil Science, Ottawa, 1993, chap. 9. With permission.)

over 50 years (Table 4). In contrast, soil total N increased in all treatments of the 5-year rotation. The increase was the highest for the manure treatment followed by NPKS and check treatments (Table 4).

Cropping practices can greatly affect organic matter content of soils. After 50 years' cropping at the Breton Plots three important observations have been made. First, the soil organic matter content is about

Table 4 Trends of Total Soil N (g kg^{-1} soil) in the Ap Horizons (0 to 15 cm) of the Check, Manure, and NPKS Treatments of the 2- and 5-Year Rotation at Breton

Rotation	Treatment	Equation	R^2
2-year	Check	Total N = 1.04 (±0.0676) – 0.00182 (±0.0234) × Year	0.16 ns
	NPKS	Total N = 1.09 (±0.0732) – 0.000664 (±0.00254) × Year	0.02 ns
	Manure	Total N = 1.14 (±0.121) + 0.00687 (±0.00421) × Year	0.47 ns
5-year	Check	Total N = 0.940 (±0.0539) + 0.00942 (±0.00187) × Year	0.89**
	NPKS	Total N = 0.881 (±0.112) + 0.0163 (±0.00391) × Year	0.85**
	Manure	Total N = 0.926 (±0.117) + 0.0248 (±0.00407) × Year	0.92**

Note: ns = not significant; ** = significant at $P<0.01$. From Juma, N.G., in *Impact of Macronutrients on Crop Responses and Environmental Sustainability on the Canadian Prairies*, Canadian Society of Soil Science, Ottawa, 1993, chap. 9. With permission.

20% higher in the soil of the 5-year rotation plots than in that of the 2-year (wheat-fallow) rotation plots (Figure 3). This result can be explained in several ways. Crops are grown more continuously on the 5-year rotation and therefore more plant material (stubble and roots) is added than on the 2-year rotation plots. Further, the 5-year rotation plots are bare for a small portion of the time so that decomposition processes are probably slower and erosion losses are smaller. Finally, a greater amount of nitrogen is probably left by forage crop roots than cereal crop roots and this would be conducive to higher soil organic matter.

The second important observation is that soil organic matter is higher on the NPKS fertilized plots than on the control plots (Figure 3). The increase can be explained by the fact that crop yields (forages and grain) have been much higher on the fertilized plots. Thus the amount of stubble and root material added to the soil is greater on these plots although all forages are removed as hay and the straw of cereals is also removed.

Third, organic matter content is generally higher on the manured plots than on plots receiving commercial fertilizer. This result arises from the fact that manure is organic matter and the average annual addition was approximately 9 t ha^{-1}. Besides, manure served as a fertilizer material and improved crop growth and hence organic matter additions via roots and stubble. It should be noted, however, that the amount of manure added was greater than could be generated if all the crop produced were fed to livestock, i.e., it was not a sustainable application rate.

Thus, it is clear from these long-term experiments that appropriate management practices, including growing legumes, reducing fallowing, adding fertilizers, and returning manure, have resulted in increased soil organic matter.

V. CONCLUSIONS

Typic Cryoboralfs are low in organic matter. Cropping to a 5-year rotation has increased the organic matter content compared to the 2-year grain-fallow rotation. The effect is most noticeable on the manure plots but it has also occurred on fertilized plots which have produced good crop yields. Maintenance of soil organic matter requires continuous growth of crops (i.e., reduced fallow), return of crop residues and manures, and appropriate use of fertilizers to ensure good yields of crops. Forage crops appear to provide an added benefit.

Typic Cryoboralfs are low in nitrogen, sulfur, and phosphorus. Cereal crops following high-yielding legumes taken off as hay produce very good yields but the beneficial effect of the legumes is much reduced by the time of the third cereal crop. Nitrogen must be added for cereal crops to yield satisfactorily. The nitrogen can be obtained from legumes, manure, or commercial fertilizer.

REFERENCES

1. Atmospheric Environmental Service, *Canadian Climate Normals. 1951–1980*, Vol. 2, Temperature, Atmospheric Environment Service, Downsview, 1982.
2. Atmospheric Environmental Service, *Canadian Climate Normals. 1951–1980*, Vol. 3, Precipitation, Atmospheric Environment Service, Downsview, 1982.
3. Bentley, C. F., Peters, T. W., Hennig, A. M. F., and Walker, D. R., *Gray Wooded Soils and Their Management*, 7th ed., University of Alberta and Canada Department of Agriculture Bulletin Number B-71-1, Edmonton, 1971.
4. Campbell, C. A., Bowren, K. E., Schnitzer, M., Zentner, R. P., and LaFond, G. P., Effect of crop rotations and cultural practices on soil organic matter, microbial biomass and respiration in a thin Black Chernozem, *Can. J. Soil Sci.*, 71, 43, 1991.
5. Cannon, K. R., Robertson, J. A., McGill, W. B., Cook, F. D., and Chanasyk, D. S., Production Optimization on Gray Wooded Soils, Farming for the Future Project 79-0132 Report, University of Alberta, Edmonton, 1984.
6. Dinwoodie, G. D. and Juma, N. G., Factors affecting the distribution and dynamics of ^{14}C in two soils cropped to barley, *Plant Soil*, 110, 111, 1988.
7. Fyles, I. H., Juma, N. G., and Robertson, J. A., Dynamics of microbial biomass and faunal populations in long-term plots on a Gray Luvisol, *Can. J. Soil Sci.*, 68, 91, 1988.
8. Holmes, N. D., McNaughton, G. R., Phillips, W. E., Stothart, J. G., and Willman, J., *Alberta Farm Guide*, Alberta Agriculture, Edmonton, 1976, 18.
9. Juma, N. G., The role of fertilizer nutrients in rebuilding soil organic matter, in *Impact of Macronutrients on Crop Responses and Environmental Sustainability on the Canadian Prairies*, Rennie, D. A., Campbell, C. A., and Roberts, T. L., Eds., Canadian Society of Soil Science, Ottawa, 1993, chap. 9.
10. Juma, N. G., Dynamics of soil C and N in a Typic Cryoboroll and a Typic Cryoboralf Located in the Cryoboreal regions of Alberta, Lal, R., Kimble, J., Levine, E., Stewart, B. A., Eds., Lewis Publishers, Boca Raton, 1995, chap. 16.
11. Lindsay, J. D., Odynsky, W., Peters, T. W., and Bowser, W. E., Soil Survey of the Buck Lake and Wabamun Lake areas, Alberta Soil Survey Report No. 24, University of Alberta, Edmonton, 1968.
12. Robertson, J. A., Lessons from the Breton Plots, Agric. For. Bull., University of Alberta, Edmonton, 2 (2), 8, 1979.
13. Robertson, J. A., Publications from the Breton Plots, Agric. For. Bull., University of Alberta, Edmonton, 2 (3), 24, 1979.
14. Robertson, J. A., Sixty years' research at the Breton Plots, in 60th Annual Breton Plots Field Day and Soils/Crops Clinic Report, University of Alberta, Edmonton, 1990, 19.
15. Robertson, J. A. and McGill, W. B., New Directions for the Breton Plots, Agric. For. Bull., University of Alberta, Edmonton, 6 (1), 41, 1983.
16. Van Veen, J. A., Liljeroth, E., and Lekkerkerk, J. A., Carbon fluxes in plant-soil system at elevated atmospheric CO_2 levels, *Ecol. Appl.*, 12, 175, 1991.

Chapter 21

Soil Organic Matter Dynamics in Long-Term Experiments in Southern Alberta

H.H. Janzen, A.M. Johnston, J.M. Carefoot, and C.W. Lindwall

CONTENTS

I. INTRODUCTION

The soils in southern Alberta were first cultivated for agricultural production early in the 20th century. Developed under mixed prairie, they were generally fertile but productivity was limited by inadequate and sporadic precipitation. As a result, reliable crop production depended on the adoption of summer fallow and other moisture-replenishing practices. The frequent drought stress also prompted the early development of irrigation systems using streams originating in the Rocky Mountains.

0-8493-2802-0/97/$0.00+$.50

The evolution of farming practices in southern Alberta occurred concurrently and interactively with agricultural research activities. Agriculture and Agri-Food Canada established an experimental station at Lethbridge, Alberta, in August of 1906 when land in the area was first being cultivated. A series of large-scale experiments was established to compare the agronomic feasibility of proposed cropping treatments. The surviving plots from these early experiments span the entire time interval from initial cultivation to modern times. More recently, several supplementary experiments were established to permit more rigorous statistical analysis and evaluate cropping practices not yet conceived at the turn of the century.

We describe briefly the ongoing long-term experiments and highlight some findings with particular emphasis on soil organic matter dynamics. More comprehensive descriptions of these studies are provided in the references cited.

II. GEOGRAPHY

Lethbridge is situated in the southwest corner of Alberta, Canada (49°42′N, 112°50′W), approximately 110 km east of the Rocky Mountains. The region is characterized by a continental climate with short, warm summers and long, cold winters.[1] A prominent feature of the climate is the chinook (or foehn) winds that sweep down the eastern slopes of the Rockies. These warm, dry winds mitigate winter air temperature, but also increase evapotranspiration and result in soil erosion.

Average annual precipitation at Lethbridge is 402 mm (Table 1), considerably less than the potential evapotranspiration of 681 mm[2]. On average, about 40% of the total precipitation is received during the growing season (May through July) but distribution and amounts of rainfall are highly variable from year to year. Snow represents 30% of the annual precipitation, but frequent chinook events minimize the persistence of snow cover and its benefit for moisture replenishment.

Table 1 Long-Term Climatic Data (1902 to 1985) at the Agriculture and Agri-Food Canada Research Center, Lethbridge

Month	Precipitation (mm)	Pan Evaporation (mm)	Temperature (°C)			Wind Velocity (km h^{-1})
			Mean	Min	Max	
January	18	—	–9	–43	18	21
February	17	—	–6	–42	22	20
March	23	—	–2	–38	24	19
April	31	120	6	–27	34	21
May	54	186	11	–13	34	19
June	73	236	15	–3	36	18
July	43	222	18	1	39	15
August	42	194	17	–2	37	15
September	40	148	12	–16	36	17
October	22	86	7	–26	30	20
November	19	—	–1	–36	23	21
December	18	—	–6	–43	19	22

[a] Mean refers to mean daily temperature. Min and Max refer to daily temperature extremes (minimum and maximum, respectively).

The mean annual air temperature at Lethbridge is 5°C, with an average frost-free season of 118 days. On average, the site has 1689 cumulative growing degree days, calculated on a base temperature of 5°C. Chinook conditions cause significant daily temperature fluctuations, and changes of 15°C h^{-1} are common. Temperature fluctuations as high as 44°C in a 24-hour period have been recorded.[1]

Soils in the Lethbridge region were developed under native vegetation of short and mid grass species. At the Agriculture and Agri-Food Canada Research Center, the soil is an Orthic Dark Brown Chernozem (Typic Haploboroll) developed on alluvial lacustrine parent material. The soil typically has a 10- to 15-cm A_h, a 0- to 12-cm B, and a calcareous C horizon, and texture near the surface is usually loam to clay loam. When the land at the research site was first broken from grassland between 1907 and 1910, a small area of land was left uncultivated. This area has been maintained as a benchmark site for comparison with various cultivated treatments.

Spring wheat (*Triticum aestivum* L.) is the dominant crop grown on nonirrigated, arable land in southern Alberta. Other dryland crops of lesser importance include barley (*Hordeum vulgare* L.), canola (*Brassica napus* L. or *B. campestris* L.), flax (*Linum usitatissimum* L.), and winter wheat. Summer fallow is widely used to reduce economic risk;[2] in some areas almost 50% of the land is in fallow in a given year. Irrigation, which is now used on approximately 500,000 ha in southern Alberta, permits a much wider diversity of crops, including alfalfa (*Medicago sativa* L.), potatoes (*Solanum tuberosum* L.), sugar beets (*Beta vulgaris* L.), and various vegetables.

III. LONG-TERM EXPERIMENTS

The Agriculture and Agri-Food Canada Research Center at Lethbridge currently maintains seven long-term cropping systems studies, of which five have been cropped for 25 years or longer (Table 2). Few of these experiments were originally conceived as long-term studies; most were originally designed to provide short-term agronomic and economic evaluations of then innovative cropping systems. With increasing longevity, however, their emphasis has gradually shifted from agronomic considerations to concerns regarding resource and ecological sustainability. Today, the primary objective of the long-term cropping system experiments is to determine the capacity of selected management practices to conserve soil resources while maintaining environmental quality.

Table 2 Long-Term Cropping Systems Experiment Ongoing at the Agriculture and Agri-Food Canada Research Center at Lethbridge, Alberta

Experiment	Established	Variables	Comments
Rotation ABC	1911	Frequency of fallow; N, P fertilizer	Large, unreplicated plots; dryland
Rotation U	1911	P application	10-y rotation (alfalfa, sugar beets, cereals); manured; revised in 1988; irrigated
Rotation 116	1951	Frequency of fallow; inclusion of forages; N fertilizer; manure (green and animal)	Replicated (4); revised in 1985; dryland
Summer fallow Study	1967	Method of fallow (including no-till)	Replicated (6); split for wheat and barley; dryland
No-till study	1967	Tillage; seeding equipment	Large, unreplicated plots; dryland
Manure plots	1973	Rate of manure application	Barley; irrigated and dryland
Winter wheat rotation	1984	Rotation, tillage	Winter wheat, flax, canola; dryland; revised in 1995

Note: Unless indicated, the dominant crop is spring wheat

A. ROTATION ABC

1. Description

In 1911, a series of large plots was established to compare various prospective cropping systems for newly broken land in the region. Many of the treatments were abandoned over time, largely because they were found to be poorly adapted to climatic and edaphic conditions. Three of the original treatments, however, have been maintained: rotation A (continuous wheat or “W”), rotation B (fallow-wheat or “FW”), and rotation C (fallow-wheat-oats). The latter rotation was amended to fallow-wheat-wheat (FWW) in 1923.

The experiment consists of six 0.63-ha plots, representing each phase of every rotation without replication. In 1967, all plots were modified to include subplots receiving N fertilizer, and, in 1972, two additional subplots were added to yield four fertilizer treatments in each of the main plots: a factorial of two N rates (0 and 45 kg N ha^{-1}) and two P rates (0 and 20 kg P ha^{-1}). The original FW plots were abandoned in 1985 because of road construction. Since then, samples have been obtained from an adjacent pair of plots from one of the original rotations that had been reverted to fallow-wheat in the early 1950s.

A parallel set of smaller plots, one for each of the rotations, was also established in 1911 near the main experiment. These plots, “Chemist plots”, were originally designated for soil sampling and analysis.

Soil management practices have evolved appreciably over the years. The moldboard plow, extensively used in the first years, was replaced by disc implements in the late 1920s, which were replaced by the wide blade cultivator in 1939 when stubble mulch practices were adopted. In recent years, intensity of tillage has been further reduced by use of herbicides to control weeds. Prior to the 1940s, plots where wheat followed wheat required multiple cultivations for seedbed preparation and weed control, so that seeding was delayed by as much as 8 weeks relative to plots seeded after fallow. Herbicides were first used for control of broadleaf and grassy weeds in 1950 and 1961, respectively. All straw was removed from the plots until combine harvesters were introduced in 1943, after which all residues were retained on the plots.

2. Observations

Wheat yields in the rotation ABC experiment have fluctuated widely in response to changes in climatic, soil, and other factors over time (Figure 1). The lowest yields were observed in the droughts of the mid-1930s and the early 1960s. Annual precipitation (August through July) and growing season precipitation (May through July) accounted for 15 to 28% and 19 to 28%, respectively, of the variability in grain yield.

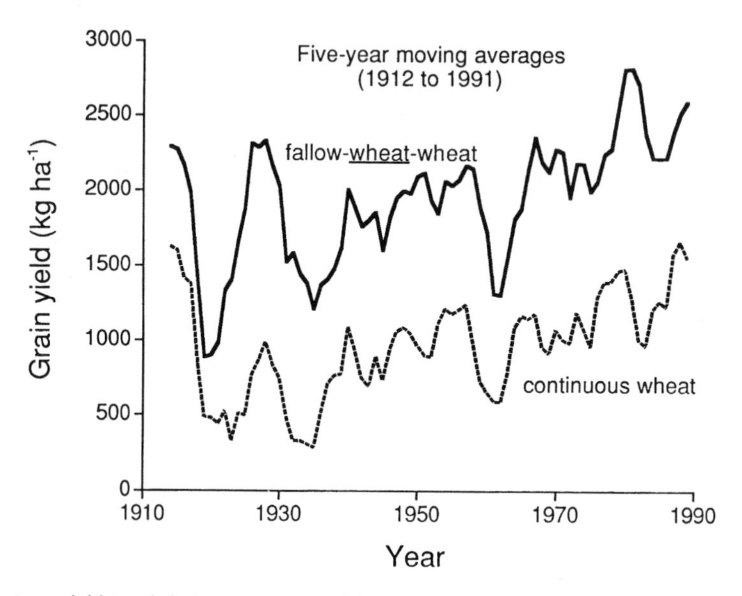

Figure 1 Long-term yield trends in two treatments of the historical ABC rotations. All yields are from unfertilized subplots. For fallow-wheat-wheat rotation, yields shown are from wheat after fallow. (From Janzen, H. H., *Can. J. Soil Sci.*, 75, 123, 1995. With permission.)

Although yields are highly variable among years, there was a trend toward increasing yields over time. Regression of yields against year demonstrated significant positive correlations for each of the four cropped treatments. The mean yield increase, as estimated from the slope of the regression line, ranged from 7 to 14 kg ha^{-1} y^{-1}. These findings imply that the productivity of the soils in the various rotation has been maintained for over 80 years, even without fertilizer application. In part, the apparent increase in yields over time is a reflection of improved crop cultivars, weed control methods, and soil management techniques. Despite the presence of these confounding effects with time, it is evident that the unfertilized soil has not yet approached irreversible loss of productivity, even in treatments with a high frequency of summer fallow.

Rotation yield (the average of yields in all phases including fallow) was similar for FW, FWW, and W in the absence of added fertilizer (Figure 2). When supplemented with N and P fertilizer, however, productivity in treatment W was 17 and 28% higher than that in the FWW and FW rotations, respectively. Nitrogen application elicited highest response in wheat seeded after wheat whereas most of the yield increases in fallow-seeded wheat were attributable to P application. These observations suggest that nutrients, and not water, are the primary constraints to long-term production in unfertilized soils. Only when nutrient constraints were removed by fertilizer application was the higher water use efficiency of continuous cropping[4] reflected in appreciably higher productivity.

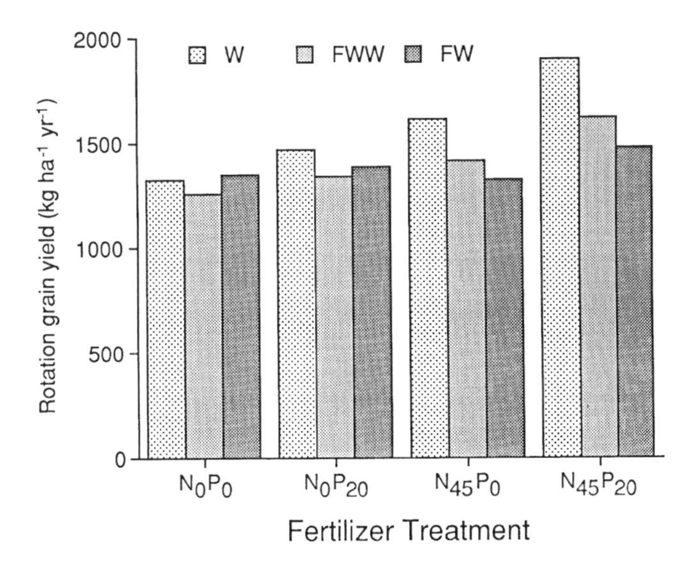

Figure 2 Mean rotation yields of three crop rotations from 1972 to 1991 as influenced by fertilizer application. W, FWW, and FW denote continuous wheat, fallow-wheat-wheat, and fallow-wheat, respectively. Subscripts following N and P in fertilizer treatment indicate application rates in kilograms per hectare.

The soil from the experimental site was first sampled in 1910, apparently soon after the land was broken from sod and the year prior to the establishment of the experiment. Thereafter, the soils were sampled approximately every 15 years. Archived and modern samples were analyzed for organic C in 1991 using an automated dry combustion technique with correction for carbonate content.[5] Organic C concentration fell sharply during the first several decades after cultivation and appears to have approached near-steady-state values inversely proportional to the frequency of fallow in rotation (Figure 3). The initial rate of organic C loss was more rapid in rotations with fallow but concentrations approached steady state earlier than in the continuous wheat treatment. Consequently, the comparatively large differences among rotations evident early in the experiment had diminished appreciably after 80 years of cultivation. This observation emphasizes that the long-term effects of cropping practices on organic matter concentration can be reliably determined only after the soils have approached steady state. The eventual convergence of organic C concentrations in the various unfertilized rotations is not surprising because cumulative residue inputs, as estimated from grain yields, were comparable.

Assuming a constant bulk density, the soils lost approximately 20% of C in the surface 15 cm from 1910 to 1990. Proportional decline in soil carbon in the 15- to 30-cm layer was similar, falling from 12.0 g kg^{-1} in 1910 to an average of 9.4 g kg^{-1} in 1990. These estimates are in agreement with an earlier report,[6] which estimated that typical organic C loss in the soils of western Canada ranged from 15 to 30%. Our values, however, do not include C lost immediately after cultivation (from 1908 to 1909).

Soil N concentrations have followed trends similar to those described for C (Figure 4). In all rotations, concentrations appear to have approached a new steady state after some 50 years of cultivation. If the soil N content is indeed constant, it would imply that there are appreciable inputs of N into the system from unknown sources. Estimated N removal in grain from unfertilized treatments averaged about 25 kg N ha^{-1} y^{-1} from 1967 to 1992. Since soil organic N was apparently near steady state, and therefore neither contributing nor withdrawing any N, this removal must be balanced by an equivalent net accretion. Our estimate of removal, furthermore, may be somewhat conservative because it assumes that there are no losses from the system other than through grain harvest. The input of appreciable N from unidentified sources has also been suggested in other long-term studies at Lethbridge[7] and elsewhere.[8,9] Possible extraneous sources of N include asymbiotic N_2 fixation, atmospheric N (e.g., ammonia), and nitrate from groundwater.

Physical and biochemical properties of soil also showed marked differences in the various rotation treatments.[10] For example, degree of aggregation and activities of selected enzymes in cultivated soil were consistently lower than those in adjacent uncultivated soil and were inversely proportional to frequency of fallow in the rotation (Table 3).

Nitrogen and P fertilizers, first applied in 1967 and 1972, respectively, have had significant effect on organic matter concentration (Table 4). In the continuous wheat plot, for example, N fertilization

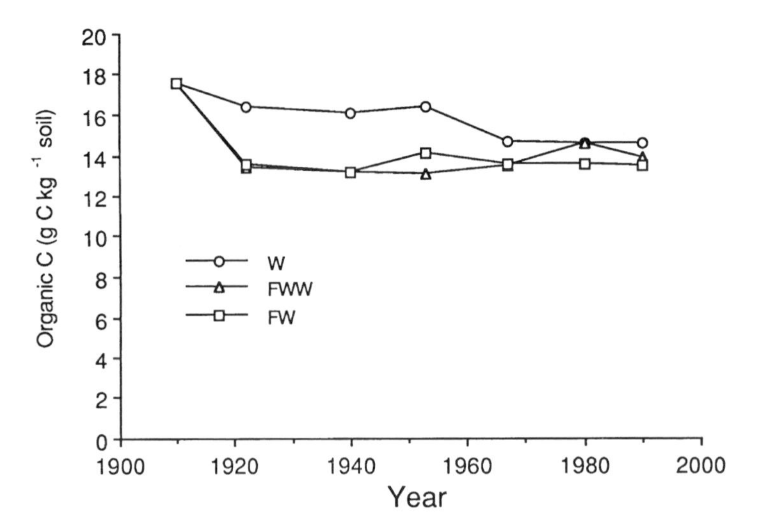

Figure 3 Organic C concentrations in surface soil (0 to 15 cm) as influenced by crop rotation in plots receiving no fertilizer. Values for 1953 through 1990 are from main rotation plots. Earlier samples were obtained from nearby chemist plots. Comparison of soils from both sets of plots in 1953 showed little apparent difference in concentration. Standard deviations (S.D.) for treatments W, FWW, and W in 1990 were 0.3, 0.5, and 1.2 g kg^{-1}, respectively (n = 4). Value for 1910 is average of values from the three plots (S.D. = 1.5 g kg^{-1}). Organic C for FW in 1990 was estimated from an adjacent plot. (From Janzen, H. H., *Can. J. Soil Sci.*, 75, 123, 1995. With permission. From data reported by Monreal, C. and Janzen, H. H.[5])

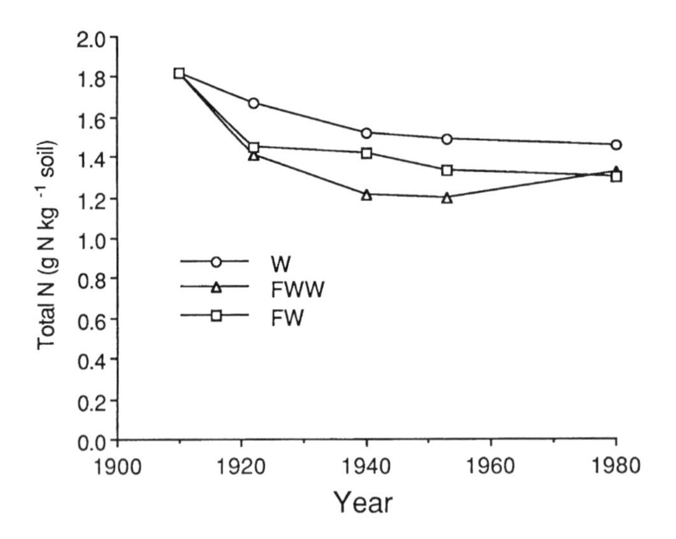

Figure 4 Organic N concentrations in surface soil (0 to 15 cm) as influenced by crop rotation in chemist plots. Value for 1910 is average of values from three treatments (1.97, 1.68, and 1.80 g kg^{-1} for W, FW, and FWW, respectively). (Drawn from data reported by Freyman, S., Palmer, C. J., Hobbs, E. H., Dormaar, J. F., Schaalje, G. B., and Moyer, J. R., *Can. J. Plant Sci.*, 62, 1982.)

increased organic C and N concentrations by approximately 14% relative to those in treatments receiving no N. Differences among fertilizer treatments are probably attributable to variable amounts of residue inputs, as estimated from yields (Figure 2). Nitrogen-fertilized treatments in the continuous wheat also had higher concentrations of labile organic C and N.[11]

B. ROTATION 116

1. Description

A rotation experiment was established in 1951 on land that had been used for a mixed crop rotation with "light" manure applications every 6th year since originally broken in about 1910.[12] Of the

Table 3 Effect of Long-Term Cropping Practices on Physical and Biochemical Properties of A_p at Lethbridge Relative to that of Uncultivated Soil (A_n)

	% of Uncultivated Soil			
Rotation	Water Stable Aggregates	Dehydrogenase Activity	Phosphatase Activity	Urease Activity
W	86	55	36	55
FWW	77	37	25	41
FW	73	29	20	27

Calculated from data reported by Dormaar, J. F., *Plant Soil*, 75, 51, 1983. With permission.

Table 4 Effect of Fertilizer Application on Organic C, Total N, and Total P in Surface Soils from the Continuous Wheat Plot (W)

Fertilizer (kg ha^{-1})		Organic C[29] (g kg^{-1})	Total N[29] (g kg^{-1})	Total P[30] (g kg^{-1})
N	p			
0	0	16.2	1.49	0.53
0	20	16.1	1.41	0.63
45	0	18.0	1.62	0.53
45	20	18.8	1.71	0.66

Note: Similar trends in C and N were also observed in the fallow-wheat-wheat plot (apparently sampled after fallow, not after wheat as originally reported). C and N data are from Reference 11; P data from Reference 29.

treatments originally established in 1951, five still remain: continuous wheat (W), fallow-wheat-wheat (FWW), fallow-wheat (FW), manured fallow-wheat-wheat (F_mWW), and fallow-wheat-wheat-hay-hay-hay (FWWAAA), where the hay is a mixture of alfalfa and crested wheatgrass (*Agropyron cristatum* L.). None of the original rotations receive any N amendments, except for the F_mWW rotation, where feedlot manure is applied during the fallow phase at a rate of 11.2 t ha^{-1} (approximately 67 kg N ha^{-1}).[13] The experiment was modified in 1985 to include annual legumes, fertilizer treatments, and a "native" grass treatment but these new rotations have not been in place long enough to permit reliable evaluation.

The treatments in rotation 116 are arranged in a randomized complete block design with four replicates. Each phase of a total of 13 rotations is represented every year on plots 3.2 × 36.6 m in size (total of 116 plots). Cultural practices have been relatively consistent over the course of the experiment, though there has been a slight shift toward reduced tillage over time.

2. Observations

Rotation grain yields from 1955 through 1991 ranged from 907 to 1107 kg ha y^{-1} (Table 5). Highest productivity was observed in the manured treatment, presumably because of response to added N and other nutrients. Wheat in the FWWAAA rotation also demonstrated relatively high productivity, though yields were often restricted by soil moisture stress induced by the deep-rooted forage crop.

The various rotations have exerted significant effect on soil organic matter characteristics[14,15] (Table 6). In unamended treatments, organic C and N concentrations in the surface soil layer (0 to 7.5 cm) were found to be inversely proportional to the frequency of fallow, with highest concentrations in the continuous wheat (W) treatment, and lowest in the fallow-wheat (FW) rotation. Manure amendment, however, mitigated the effects of summer fallow; organic matter in the manured rotation (F_mWW) was slightly higher than that in the unfertilized continuous wheat treatment. The inclusion of a forage crop (FWWAAA) did not increase surface organic matter above that observed in continuous wheat (W) (Table 6), but the FWWAAA rotation tended to have higher organic matter content in deeper soil layers (e.g., 15 to 30 cm).[15]

Table 5 Dry Matter and N yield of Spring Wheat Grain in Crop Rotations Established in 1951

	Grain Yield 1955–1991 (kg ha^{-1})		Grain N Yield 1978–1991 (kg ha^{-1})	
Rotation[a]	**Phase**[b]	**Rotation**	**Phase**	**Rotation**
(W)	1020	1020	27	27
F(W)	1813	907	44	22
F(W)W	1947	1025	46	25
FW(W)	1129		28	
F_m(W)W	2065	1107	52	28
F_mW(W)	1255		33	
F(W)WAAA	1792	1062	44	27
FW(W)AAA	1395		36	
Significance[c]	0.0001		0.0001	
LSD (0.05)	69		3	

Note: 'Dry matter yield' refers to yield at harvest moisture content (assumed to be 12.5% for calculation of N yield).

[a] F, F_m, W, and A denote summer fallow, manured summer fallow, spring wheat, and hay (alfalfa/crested wheatgrass), respectively. Parentheses indicate specific phase.

[b] Phase denotes mean yield of phase in parentheses; Rotation denotes mean yield of rotation including fallow; in FWWAAA, only the FWW phases were included in the calculation of rotation yields.

[c] Significance = *P* (as determined by analysis of variance for a randomized complete block design with four replicates)

Labile organic matter fractions were more responsive to crop rotation than were total C and N concentrations. In the continuous wheat treatment, light fraction organic matter (s.g. < 1.7 g cm^{-3}), N mineralization, and C mineralization was approximately 1.5 to 2 times that in the FW treatment. As for total concentrations, labile organic matter fractions in unmanured treatments were inversely related to frequency of summer fallow. Since estimated residue inputs were comparable among the various wheat rotations, differences in labile organic matter are probably attributable largely to variable decomposition rates. During a cropped phase, when moisture stress is prevalent and soil temperatures are often lower, labile organic matter may accumulate by virtue of inadequate opportunity for decomposition. Conversely, a fallow period, with higher soil moisture content and elevated temperature, may result in a flush of

Table 6 Effect of Crop Rotations Established in 1951 on Soil Organic Matter Characteristics in Surface Soil (0 to 7.5 cm) as Determined in 1992

Rotation[a]	**Organic C (g kg^{-1})**	**Organic N (g kg^{-1})**	**LF C**[b] **(g kg^{-1})**	**N min**[c] **(mg kg^{-1} 10 wk^{-1})**	**C min**[c] **(mg kg^{-1} 10 wk^{-1})**
W	18.7	1.81	3.3	52	960
FWWAAA	18.6	1.81	2.9	46	994
FWW	17.0	1.63	2.2	34	803
FW	15.9	1.58	1.6	33	632
F_mWW	21.7	2.06	3.7	35	997

Note: For further details, including statistical significance, bulk densities and properties of subsurface layers, see Reference 15. Trends similar to those in this table were reported in an earlier publication,[14] although absolute N min and C min values reported earlier may not be reliable.

[a] W, F, F_m, and A denote spring wheat, fallow, manured fallow, and alfalfa/crested wheat grass forage, respectively. Soils were sampled in September 1992. Values are means across all rotation phases.

[b] LF C denotes light fraction C (density < 1.7 g cm^{-3}).

[c] N and C mineralization was measured in a laboratory incubation at 25°C.

microbial activity and rapid decomposition of accumulated labile organic matter.[16] In part, the higher total organic matter content of soil from continuous wheat plots may simply reflect the accumulation of labile substrate (Figure 5). When these soils are incubated under favorable laboratory conditions, this labile material rapidly decomposes, resulting in a flush of mineralized C and N. The higher organic matter content of the continuous wheat, therefore, may be transient and subject to loss upon change in management.

Of the three rotations receiving no amendments (W, FWW, and FW), the former rotation has exhibited the highest removal of N in harvested grain (Table 5). This rotation has also retained the highest concentration of soil N, suggesting that it either receives higher external inputs or lower losses than the other systems. Based on an analysis of soil profile nitrate concentrations to a depth of 3 m,[17] nitrate leaching from these rotations appears to be minimal.

C. ROTATION U

1. Description

Established in 1911, rotation U is believed to be the oldest ongoing irrigated rotation experiment in North America.[18] The study originally consisted of a 10-year rotation: 6 years of alfalfa, 3 years of cereals, and 1 year of sugar beets or potatoes. Each phase of the rotation was present every year, resulting in a total of 10 plots, each 0.4 ha in size. Feedlot manure was applied at a rate of 27 t wet weight ha^{-1} once during each rotation from 1911 to 1942 and at a rate of 33.5 t wet weight ha^{-1} twice during each rotation from 1943 to 1986. Starting in 1943, each of the plots was split into two subplots, one of which received phosphate fertilizer (0-45-0 or 11-48-0, depending on year) at a rate of 110 kg fertilizer ha^{-1} three times during each 10-year rotation. While sugar beet tops were returned to the land, straw from cereal crops was removed after harvest. All plots were regularly plowed (moldboard) to a depth of 15 to 20 cm in fall. The plots were irrigated using surface flood methods until 1966 when line sprinklers were introduced.

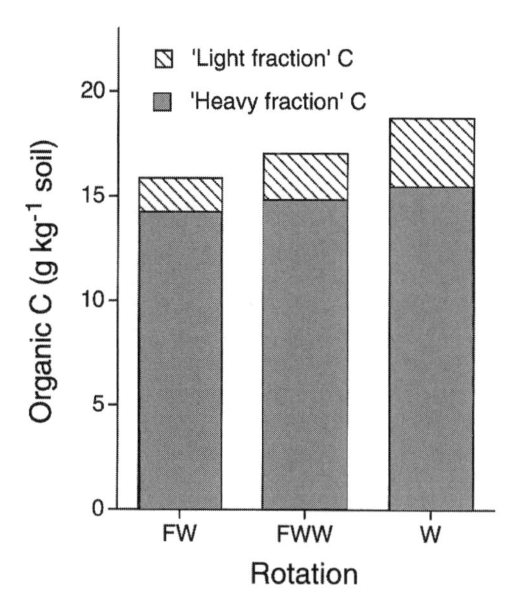

Figure 5 Organic C concentration and composition in surface soil (0 to 7.5 cm) after 42 years of cropping to various spring wheat rotations. Rotation designations are as follows: FW = fallow-wheat, FWW = fallow-wheat-wheat, W = continuous wheat. Values are averaged across rotation phases. (Drawn from data reported by Bremer, E., Janzen, H. H., and Johnson, A. M., *Can. J. Soil Sci.*, 74, 131, 1994.)

The experiment was modified in 1987 to introduce additional treatments and replication by reducing plot size and number of phases represented each year. The experiment now includes three 5-year rotations, one which approximates the original rotation (alfalfa-alfalfa-alfalfa-wheat-barley) and two which replace the alfalfa with wheat and either corn or faba bean. As well, a series of N fertilizer treatments have been superimposed as subplots over each of the main plots.

2. Observations

Crop yields in rotation U have consistently increased over time, even in subplots which have never received commercial fertilizer (Figure 6). These gradual yield increases are attributable to improved cultivars, improved soil and irrigation management, and the cumulative benefit of manure and legume growth on soil fertility. Despite the provision of adequate water through irrigation, the yields still show considerable variability, some of which has been linked to meteorological variables such as wind, solar radiation, and air temperature.[19]

Total soil N concentration, a reliable indicator of organic matter content, has shown consistent increases over time (Figure 7). The increases in the subsurface soil layer (15 to 30 cm) have been even more pronounced than those in the surface soil layer, probably because of residue and manure incorporation by plowing. These findings suggest that irrigation can significantly enhance soil organic matter by increasing the amount of organic substrate added to the soil.

D. TILLAGE STUDIES

1. Description

A summer fallow experiment was established in 1955 (revised in 1967) to evaluate the influence of various tillage and herbicide practices during the fallow phase on weed control, spring wheat yield, and soil properties. The experimental design currently includes six replicates of eight treatments including zero-tillage, several types of tillage practices, and combinations of herbicides and tillage for weed control.[20] Several of the treatments have evolved over time to take advantage of new herbicides and tillage practices.

A second experiment (long-term no-till study) was established in 1967 to compare tillage practices in several crop rotations. The experiment includes two treatments (no-till and conventional tillage using a heavy-duty cultivator) in a fallow-wheat rotation, a no-till fallow-wheat-flax rotation, and a minimum-till fallow-wheat-wheat rotation. Every phase of each treatment is represented on one relatively large 0.4-ha plot.

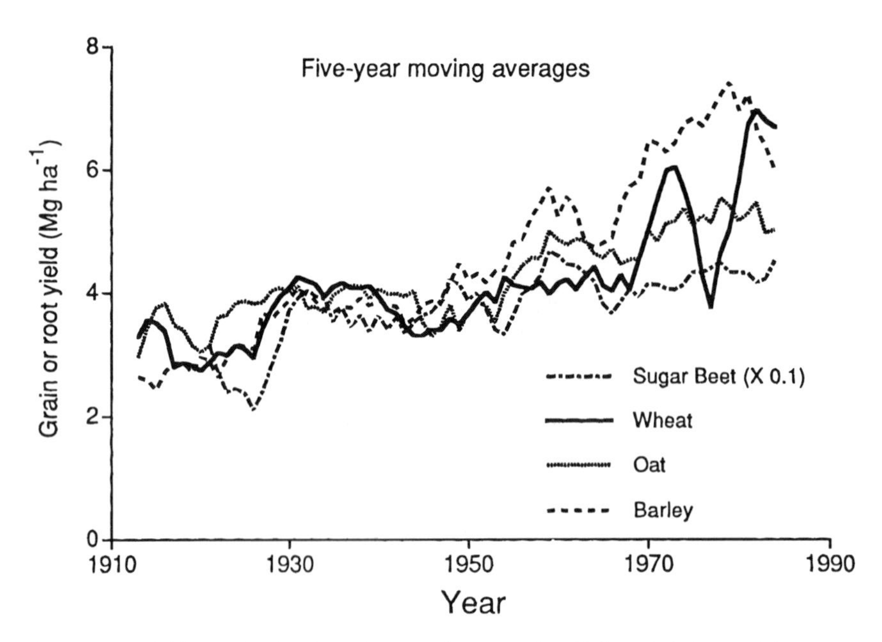

Figure 6 Long-term yield trends of selected crops from rotation U in subplots receiving no commercial fertilizer.

2. Observations

In the summer fallow study, long term-spring wheat yields have been only slightly affected by fallow treatment. Mean yields (1977 to 1991) ranged from 1913 to 2120 kg ha^{-1}, with highest yields occurring in a combination herbicide-tillage treatment.[21] Preliminary analysis of surface soil from this experiment

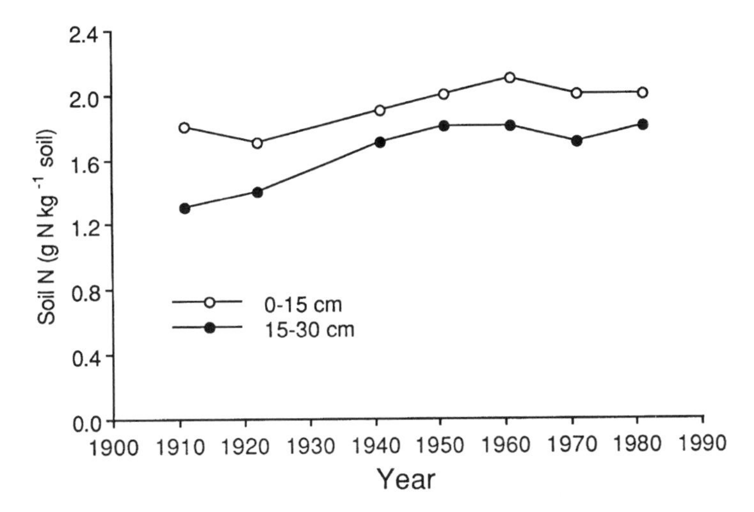

Figure 7 Changes over time of total N concentration in two soil layers from unfertilized plots in rotation U. (Drawn from data presented by Dubetz, S., Ten-Year Irrigated Rotation U. 1911–1980, Agriculture Canada Technical Bulletin 1983-21E, 1983.)

showed that intensive tillage reduced organic C content, but that no-till did not increase C above that in a stubble-mulch system.[22]

Carter and Rennie[23,24] compared the C and N characteristics of soil in the fallow-wheat rotation of the long-term no-till study. They found no significant effect of tillage on organic C or total N concentration in the soil. Furthermore, they concluded that tillage practices did not have a pronounced influence on the soil N cycle.

From preliminary results obtained to date, it appears that effects of tillage practices on soil organic C retention may not be as pronounced under the semi-arid conditions of southern Alberta as under more humid conditions. The comparatively small effect can perhaps be attributed to several factors: the use of conventional tillage practices that retain residues on the surface; similar (low) inputs of crop residues; and the presence of fallow in most rotations, which may negate the benefits of no-till for C conservation. Adoption of no-till may be more beneficial for soil C storage in continuous cropping systems.

E. RELATED STUDIES

A field experiment was established in Lethbridge in 1973 to determine the influence of annual feedlot manure applications on barley yield and soil properties.[25] After 11 years, soil organic C concentration had increased by as much as threefold, the magnitude of increase depending primarily on application rate.[26]

Agriculture Canada maintains a long-term grassland site in the Porcupine Hills near Stavely, in southwestern Alberta. An experiment was established at this site in 1949 to determine the carrying capacity of the Rough Fescue Prairie. The experiment consists of four fields (each 16 to 65 ha) with stocking rates of 1.2 to 4.8 animal unit months ha^{-1}.[27] High stocking rates resulted in lower concentrations of organic matter, total N, and mineralizable N[28] though total organic matter content (mass per unit area) may not have been affected because there was a concomitant increase in bulk density.

IV. PERSPECTIVE

The long-term experiments at Lethbridge have several inevitable limitations.[3] One of these is the absence of randomization and replication in the older studies. While this deficiency merits consideration in the interpretation of data, it does not necessarily minimize the value of the experiments for scientific investigation. Many of the conclusions first drawn from the unreplicated rotation ABC experiment, for example, were later confirmed in the replicated rotation 116. More troublesome than the rudimentary design, is the evolution of management practices and treatments over time. The confounding effect of these changes is particularly acute in the evaluation of economic or agronomic feasibility. The primary role of the long-term experiments, therefore, is to serve as model agroecosystems for studying nutrient dynamics that affect soil productivity and environmental quality.

Long-term experiments offer direct evidence for the sustainability of current agroecosystems. Based on results at Lethbridge, there is little evidence to suggest that current management practices will result in the eventual collapse of productivity under present conditions. Even in suboptimal systems such as fallow-wheat, for example, the soil appears to approach a new steady state specific to the system constraints and maintain productivity in accordance with that altered steady state.

Increasing concern over the predicted "greenhouse effect" has focused attention on the use of agroecosystems to store atmospheric carbon. Our results suggest significant potential for such storage, particularly under irrigated conditions, where the removal of drought stress greatly enhances primary production and the eventual conversion of photosynthesized carbon to soil humus. Under rainfed conditions, it appears that soil C storage can be encouraged by application of nutritive amendments and adoption of rotations that slow decomposition rates. Additional C storage may be achieved by a reduction in tillage intensity though only preliminary results are available in southern Alberta. In all cases, however, net C accumulation may be a short-term phenomenon and will presumably diminish as the soil approaches a new steady state.

The potential contributions of the long-term studies at Lethbridge to scientific and agronomic communities remain largely untapped. Current research activities seek to quantify the C storage potential of various cropping systems, elucidate missing pathways in the nitrogen cycle, examine the effects of management practices on soil biology, and quantify organic matter-yield relationships.

The current research objectives in the long-term experiments have evolved well beyond those posed at their inception. Similarly, if adequately maintained and nurtured, these studies may provide answers to critical scientific questions not yet envisaged.

ACKNOWLEDGMENTS

We gratefully acknowledge the foresight, perseverance, and commitment of those who established and maintained the experiments.

REFERENCES

1. Grace, B. and Hobbs, E. H., The Climate of the Lethbridge Agricultural Area: 1902–1986, LRS mimeo report 3 (revised), Agriculture Canada Research Station, Lethbridge, Alberta, 1986.
2. Campbell, C. A., Zentner, R. P., Janzen, H. H., and Bowren, K. E., Crop rotation studies on the Canadian Prairies, Publication 1841/E, Agriculture Canada, 1990.
3. Janzen, H. H., The role of long-term sites in agroecological research: A case study. *Can. J. Soil Sci.*, 75, 123, 1995.
4. Campbell, C. A., Zentner, R. P., and Steppuhn, H., Effect of crop rotations and fertilizers on moisture conserved and moisture use by spring wheat in southwestern Saskatchewan, *Can. J. Soil Sci.*, 67, 457, 1987.
5. Monreal, C. and Janzen, H. H., Soil organic carbon dynamics after eighty years of cropping a Dark Brown Chernozem, *Can. J. Soil Sci.*, 73, 133, 1993.
6. McGill, W. B., Dormaar, J. F., and Reinl-Dwyer, E., New perspectives on soil organic matter quality, quantity, and dynamics on the Canadian prairies, in Land Degradation and Conservation Tillage, Proceedings of the 34th Annual Canadian Society of Soil Science/Agricultural Institute of Canada Meeting, Calgary, Alberta, 1988, 30.
7. Bremer, E. and Janzen, H. H., unpublished data, 1994.
8. Cairns, R. R., Effect of long-term fallow on Solonetz soil and crop production, *Can. J. Soil Sci.*, 50, 449, 1970.
9. Powlson, D. S. and Jenkinson, D. S., Quantifying inputs of non-fertiliser nitrogen into an agro-ecosystem. in *Nutrient Cycling in Terrestrial Ecosystems. Field Methods, Application and Interpretation*, Harrison, A. F., Ineson, P., and Heal, O. W., Eds., Elsevier, London, 1990, 56.
10. Dormaar, J. F., Chemical properties of soil and water-stable aggregates after 67 years of cropping to spring wheat, *Plant Soil*, 75, 51, 1983.
11. Janzen, H. H., Effect of fertilizer on soil productivity in long-term spring wheat rotations, *Can. J. Soil Sci.*, 67, 165, 1987.
12. Pittman, U. J., Crop yields and soil fertility as affected by dryland rotations in southern Alberta, *Commun. Soil Sci. Plant Anal.*, 8, 391, 1977.
13. Anon., Guide to Farm Practices in Saskatchewan. Saskatchewan Agriculture Services Co-ordinating committee, Univ. of Saskatchewan, Saskatoon, 1987.
14. Janzen, H. H., Soil organic matter characteristics after long-term cropping to various spring wheat rotations, *Can. J. Soil Sci.*, 67, 845, 1987.

15. Bremer, E., Janzen, H. H., and Johnston, A. M., Sensitivity of total, light fraction, and mineralizable organic matter to management practices in a Lethbridge soil, *Can. J. Soil Sci.*, 74, 131, 1994.
16. Janzen, H. H., Campbell, C. A., Brandt, S. A., Lafond, G. P., and Townley-Smith, L., Light-fraction organic matter from long-term crop rotations, *Soil Sci. Soc. Am.*, 56, 1799, 1992.
17. Johnston, A. M., Janzen, H. H., and Lindwall, C. W., Nitrate-N accumulation under historic dryland cropping systems, *Can. J. Soil Sci.*, 72, 317, 1992.
18. Dubetz, S., Ten-Year Irrigated Rotation U. 1911–1980, Agriculture Canada Technical Bulletin 1983-21E, 1983.
19. Rao, U. M. B., Major, D. J., and Carefoot, J. M., Comparative risk of growing four irrigated crops in southern Alberta, *Can. J. Plant Sci.*, 68, 907, 1988.
20. Lindwall, C. W. and Anderson, D. T., Agronomic evaluation of minimum tillage systems for summer fallowing in southern Alberta, *Can. J. Plant Sci.*, 61, 247, 1981.
21. Johnston, A. M., Larney, F. J., and Lindwall, C. W., Spring wheat and barley response to long-term fallow management, *J. Prod. Agric.*, 8, 264, 1995.
22. Bremer, E., Janzen, H.H. and Lindwall, C.W. unpublished data. 1994.
23. Carter, M. R. and Rennie, D. A., Changes in soil quality under zero tillage farming systems: distribution of microbial biomass and mineralizable C and N potentials, *Can. J. Soil Sci.*, 62, 587, 1982.
24. Carter, M. R. and Rennie, D. A., Nitrogen transformations under zero and shallow tillage, *Soil Sci. Soc. Am. J.*, 48, 1077, 1984
25. Sommerfeldt, T. G. and Chang, C., Changes in soil properties under annual applications of feedlot manure and different tillage practices, *Soil Sci. Soc. Am. J.*, 49, 983, 1985.
26. Sommerfeldt, T. G., Chang, C., and Entz, T., Long-term annual manure applications increase soil organic matter and nitrogen, and decrease carbon to nitrogen ratio, *Soil Sci. Soc. Am. J.*, 52, 1668, 1988.
27. Willms, W. D., Smoliak, S., and Dormaar, J. F., Effects of stocking rate on a Rough Fescue Grassland vegetation, *J. Range Manage.*, 38, 220, 1985.
28. Dormaar, J. F. and Willms, W. D., Sustainable production from the rough fecue prairie, *J. Soil Water Conserv.*, 45, 137, 1990.
29. McKenzie, R. H., Mechanisms Controlling Soil Phosphorus Dynamics in Chernozemic and Luvisolic Soils, Ph.D. thesis, University of Saskatchewan, Saskatoon, 1989.
30. Freyman, S., Palmer, C. J., Hobbs, E. H., Dormaar, J. F., Schaalje, G. B., and Moyer, J. R., Yield trends on long-term dryland wheat rotations at Lethbridge, *Can. J. Plant Sci.*, 62, 609, 1982

OTHER REFERENCES FROM LONG-TERM SITES AT LETHBRIDGE

Beke, G. J., Janzen, H. H., and Entz, T., Salinity and nutrient distribution in soil profiles of long-term crop rotations, *Can. J. Soil Sci.*, 74, 229, 1994.

Campbell, C. A., Zentner, R. P., Janzen, H. H., and Bowren, K. E., Crop Rotation Studies on the Canadian Prairies, Publication 1841/E, Agriculture Canada, 1990.

Carter, M. R. and Rennie, D. A., Dynamics of soil microbial biomass N under zero and shallow tillage for spring wheat, using ^{15}N urea, *Plant Soil*, 76, 157, 1984.

Chang, C., Janzen, H. H., and Entz, T., Long-term manure application effects on nutrients uptake by barley, *Can. J. Soil Sci.*, 74, 327, 1994.

Chang, C. and Lindwall, C. W., Effect of long-term minimum tillage practices on physical properties of a Chernozemic clay loam, *Can. J. Soil Sci.*, 69, 443, 1989.

Dormaar, J. F., Monosaccharides in hydrolysates of water-stable aggregates after 67 years of cropping to spring wheat as determined by capillary gas chromatography, *Can. J. Soil Sci.*, 64, 647, 1984.

Dormaar, J. F. and Lindwall, C. W., Chemical differences in Dark Brown Chernozemic horizons under various conservation tillage systems, *Can. J. Soil Sci.*, 69:481, 1989.

Dormaar, J. F. and Pittman, U. J., Decomposition of organic residues as affected by various dryland spring wheat-fallow rotations, *Can. J. Soil Sci.*, 60, 97, 1980.

Dubetz, S., The fertility balance in a ten-year sugar beet rotation after forty-two years of cropping, *Proc. Am. Soc. Sugar Beet Technol.*, 8, 81, 1954.

Dubetz, S. and Dudas, M. J., Potassium status of a Dark Brown Chernozem soil after sixty-six years of cropping under irrigation, *Can. J. Soil Sci.*, 61, 409, 1981.

Dubetz, S. and Oosterveld, M., Effects of weather variables on the yields of sugar beets grown in an irrigated rotation for fifty years, *J. Am. Soc. Sugar Beet Technol.*, 19, 143, 1976.

Dubetz, S. and Oosterveld, M., Sixty-six year trends in irrigated crop yields — barley, wheat, and oats, *Can. J. Plant Sci.*, 59, 685, 1979.

Hill, K. W., Effects of forty years of cropping under irrigation, *Sci. Agric.*, 31, 349, 1951.

Hill, K. W., Wheat yields and soil fertility on the Canadian prairies after a half century of farming. *Soil Sci. Soc. Am. Proc.*, 18, 182, 1954.

Janzen, H. H. and Radder, G. D., Nitrogen mineralization in a green manure-amended soil as influenced by cropping history and subsequent crop, *Plant Soil*, 120, 125, 1989.

Janzen, H. H, Bole, J. B., Slinkard, A. E., and Biederbeck, V. O., Fate of N applied as green manure or ammonium fertilizer to soil subsequently cropped with wheat, *Can. J. Soil. Sci.*, 70, 313, 1990.

Johnston, A., To Serve Agriculture — 1906–1976, the Lethbridge Research Station, Historical Series No. 9 Research Branch, Agriculture Canada, Lethbridge, Alberta, 1977.

Johnston, A. M. and Janzen, H. H., Long-term rotation effect on crop production and soil quality, in Proc. Great Plains Soil Fertility Conference, Denver, 1992, 54.

Lutwick, S. J., The Effects of Long-Term Management on Soil Chemical and Physical Properties. M.Sc. thesis, University of Alberta, Edmonton, 1989.

McKenzie, R. H., Stewart, J. W. B., Dormaar, J. F., and Schaalje, G. B., Long-term crop rotation and fertilizer effects on phosphorus transformations. I. In a Chernozemic soil, *Can. J. Soil Sci.*, 72, 569, 1992.

Moyer, J. R. and Lindwall, C. W., Persistence and availability of paraquat in a Lethbridge clay loam soil, *Can. J. Soil Sci.*, 65, 523, 1985.

Olson, B. M. and Lindwall, C. W., Soil microbial activity under chemical fallow conditions: effects of 2,4-D and glyphosate, *Soil Biol. Biochem.*, 23, 1071, 1991.

Smith, E. G., Johnston, A. M., and Janzen, H. H., Influence of organic amendments on the economic performance of long-term rotations in southern Alberta, *J. Sustainable Agric.*, 4:31, 1994.

Sommerfeldt, T. G. and Chang, C., Soil-water properties as affected by twelve annual applications of cattle feedlot manure, *Soil Sci. Soc. Am. J.*, 51, 7, 1987.

Zentner, R. P. and Lindwall, C. W., Economic evaluation of minimum tillage systems for summer fallow in southern Alberta, *Can. J. Plant Sci.*, 62, 631, 1982.

Chapter

Crop Production and Soil Organic Matter in Long-Term Crop Rotations in the Sub-Humid Northern Great Plains of Canada

C.A. Campbell, G.P. Lafond, A.P. Moulin, L. Townley-Smith, and R.P. Zentner

CONTENTS

I. INTRODUCTION

There are about 30 million hectares of cultivated land on the Canadian prairies; of this about 50% is located in the Black soil zone (Udic Borolls). There are only two ongoing long-term (>30 years) crop rotation studies in this zone, both located on Agriculture Canada Research Stations at (1) Indian Head

0-8493-2802-0/97/$0.00+$.50

and (2) Melfort, Saskatchewan. The Black soils occur in the fescue prairie-aspen grove (parkland) and the prairie grassland areas. These soils are inherently very fertile, with surface horizons averaging 20 to 25 cm and organic matter averaging about 7%. Here, as in most areas of the prairies, spring wheat (*Triticum aestivum* L.) is the predominant crop, but, because of the favorable moisture regimes that normally prevail, we find a much greater diversity of crop types than elsewhere [e.g., barley (*Hordeum vulgare* L.) and canola (*Brassica napus* L. and *Brassica camprestis* L.) are commonly grown in this soil zone). Mixed farming (both beef cattle and hogs) is quite common in this region and the crop rotations tend to reflect this fact. Although moisture is not normally limiting, a significant proportion of the land is summer fallowed each year, mainly to control weeds and diseases. However, because of the associated effects on soil degradation, producers have been encouraged to reduce summer fallow frequency, and this has been the trend in the last decade.

In 1957, researchers at the Agriculture Canada Research Stations at Indian Head and Melfort initiated very similar crop rotation experiments in which they were primarily interested in production and economic considerations. Consequently, soil characterization was kept to a minimum. In 1987, Campbell and colleagues from the Agriculture Canada Research Station at Swift Current, Saskatchewan, initiated a series of studies to evaluate how these rotations and crop management practices had influenced various soil biochemical characteristics.[1–6]

This chapter summarizes the findings to date of the Indian Head and Melfort long-term rotation experiments as they relate to soil organic matter and crop production and examines the feasibility of managing crops on these soils so as to sequester C. According to the model proposed by Jenny[7] the potential for sequestering C in these soils is good due to high levels of biomass production and reduced mineralization associated with low temperatures and a short growing season.

II. GEOGRAPHY

Indian Head is located in southeastern Saskatchewan at 50°32′N, 103°31′ west, on the trans-Canada highway about 65 km east of Regina. Melfort is situated northwest of Indian Head at 52°49′N, 104°36′ west.

As was true of Swift Current,[8] the climate is continental with the warmest month in July and the coldest month in January. Mean annual temperature at Indian Head (2.5°C) is lower than at Swift Current (3°C) but greater than at Melfort (0.8°C) (Table 1). Mean annual precipitation at Indian Head (427 mm) is about 70 mm more than at Swift Current,[9] and slightly more than at Melfort (Table 1). However, the mean annual evapotranspiration (ET_p) at Indian Head is much higher than at Melfort (607 vs. 506 mm);[9] thus the moisture deficit is significantly lower at Melfort (96 mm) compared to Indian Head (180 mm). The main constraint to crop production at Melfort is its short growing season with frost-free days (>0°C) being only 93 days compared to 110 days for Indian Head and 117 days for Swift Current.[9]

The soil at Indian Head is classified as an Indian Head heavy clay. It was developed on clayey lacustrine deposits underlain by till and is situated on gently undulating topography showing slight evidence of wind erosion.[10] Virgin profiles of this soil had granular dark brown to grayish black Ah horizon (0 to 15 cm), a cloddy dark brown to blackish transitional A_2B_1 (15 to 30 cm) overlying a high lime, massive, gray B_2 horizon. Selected properties of this soil at the experimental site are shown in Table 2.

Prior to commencement of this experiment in 1957 (results started in 1958), the experimental site had been uniformly managed in a fallow-spring wheat or fallow-wheat-wheat rotation employing conventional tillage management to control weeds.

The experiment at Melfort is on land classified as a Melfort silty clay, developed on lacustrine deposits and situated on gently undulating topography.[11] The soil is moderately well drained with a black A horizon, 13 to 20 cm thick, underlain by a dark-brown to yellowish brown B horizon and a moderately calcareous C horizon. Selected properties of this soil are shown in Table 2.

III. DESCRIPTION OF EXPERIMENTS

The rotation experiment at Indian Head includes a series of spring-wheat-based rotations (Table 3) varying in frequency of summer fallow, fertilization, residue management, and includes legumes for green manure and forage (hay) crops.[1]

The experiment at Melfort (Table 4) includes treatments generally similar to those at Indian Head, except there are no residue management treatments.[2] At both sites commercial farm equipment was used

Table 1 Mean Monthly Air Temperatures and Precipitation and Long-Term Means for the Indian Head and Melfort Experimental Sites (1958 to 1991)

Period and Meteorological Characteristic		January	February	March	April	May	June	July	August	September	October	November	December	Mean[a]
Indian Head														
Temperature (°C)														
34-y mean	Max	–11.3	–7.9	–1.1	10.0	17.9	22.8	25.7	24.8	17.9	11.0	–0.7	–8.5	8.4
	Min	–21.8	–18.5	–11.4	–2.3	4.1	9.6	12.1	10.8	5.2	–1.0	–9.9	–18.2	–3.5
90-y mean	Max	–12.9	–8.4	–2.5	8.9	17.4	22.2	25.4	24.4	17.8	11.2	–0.5	–8.1	7.9
	Min	–23.0	–19.1	–13.1	–2.8	3.8	9.1	11.8	10.5	5.1	–0.7	–9.6	–17.8	–4.3
Precipitation (mm)														
34-y	Mean	19.1	15.3	20.9	26.7	50.7	67.9	62.0	51.0	41.6	24.4	16.5	21.7	418
90-y	Mean	21.0	17.9	21.8	28.3	49.9	73.9	53.1	55.9	42.3	24.7	17.1	21.5	427
Melfort														
Temperature (°C)														
34-y mean	Max	–14.2	–10.3	–3.8	7.7	17.4	21.9	24.2	23.2	16.3	8.9	–3.3	–11.2	6.4
	Min	–24.2	–20.4	–14.1	–3.1	3.7	8.8	11.1	9.7	4.4	–1.6	–11.4	–20.4	–4.8
55-y mean	Max	–14.8	–9.4	–3.7	8.1	18.0	22.1	24.9	23.5	17.1	10.2	–2.0	–10.1	7.0
	Min	–26.1	–22.1	–16.1	–4.2	2.9	8.0	10.5	8.7	3.6	–2.1	–12.1	–20.8	–5.8
Precipitation (mm)														
34-y	Mean	17.3	15.0	18.6	21.6	40.7	62.5	58.0	53.0	41.2	29.0	18.3	20.6	396
55-y	Mean	18.2	17.1	24.7	19.2	36.5	77.0	65.8	47.9	43.9	23.3	19.0	17.3	410

[a] Monthly temperatures or annual precipitation.

Table 2 Selected Properties of Indian Head and Melfort Soils (0- to 7.5-cm Depth)

Soil[a]	Order	pH $CaCl_2$	Organic C conc (g kg^{-1})	Particle Size Distribution (g kg^{-1})			Moisture Content at Potential (M Pa, g kg^{-1})	
				Sand	Silt	Clay	–0.03	–1.5
Indian Head heavy clay	Rego, thin Black Chernozem (Udic Boroll)	7.5	25	163	206	631	420	278
Melfort silty clay	Orthic, thick Black Chernozem (Udic Boroll)	6.0	55	166	420	414	380	213

[a] The Indian Head soil has a coarse granular to small cloddy structure in the surface horizon, and Melfort soil has a moderate, medium to coarse subangular blocky primary structure, with a fine to medium granular secondary structure. (From Campbell, C.A. et al., *Can. J. Soil Soc.*, 73, 579, 1993c. With permission.)

Table 3 Crop Rotations Sampled and Some Selected Soil Characteristics at Indian Head

Rotation Phase Sampled and Fertilization	Organic C (t ha^{-1})	Organic N (t ha^{-1})	Initial Potential Rate of N Mineralization (kg ha^{-1} wk^{-1})	Olsen-P (kg ha^{-1})
(F)-W	29.3	2.94	24.5	ND
(F)-W (N+P)	30.6	3.04	32.0	ND
(F)-W-W	29.3	2.97	28.3	37.0
(F)-W-W (N+P)	30.0	3.24	37.6	55.4
(F)-W-W-W (N+P) (straw removed)	27.6	3.07	32.0	ND
(GM)-W-W	34.0	3.37	44.1	25.6
GM-(W)-W	32.5	3.34	45.5	ND
(F)-W-W-H-H-H	33.4	3.48	48.7	23.5
F-W-W-(H)-H-H	34.5	3.68	47.0	ND
Cont W	29.2	3.17	36.1	30.8
Cont W (N+P)	34.2	3.46	55.4	42.1
Significance of F-ratio	*	**	***	**
LSD ($P<0.10$)	2.4	0.31	8.0	14

Note: F = fallow, W = hard red spring wheat, GM = sweetclover green manure, H = alfalfa-bromegrass cut for hay, Cont = annual cropping. Phases in parenthesis were the ones on which organic characteristics were measured. Organic N and initial potential rate of N mineralization measured in top 15 cm of soil sampled in 1987;[3] organic C determined on these same samples;[5] Olsen-P measured on samples taken from top 120 cm on 4 September, 1991. The previously published values for organic C were total C;[5] the present values are true organic C, corrected for inorganic C. ND = not determined.

Table 4 Effect of Rotations and Fertilizer on Soil Organic C and N in the top 15 cm of Melfort Soil

Rotation Phase Sampled and Fertilization	Total Soil Organic ($t\ ha^{-1}$)	
	C	N
(F) -W (N+P)	62.4	6.05
(F) -W-W	61.4	5.87
(F)-W-W (N+P)	61.2	5.98
(GM)-W-W (N+P)	62.0	6.18
GM-(W)-W (N+P)	66.1	6.23
Cont (W)	65.3	6.37
Cont (W) (N+P)	65.4	6.30
(F)-W-W-H-H-W	65.5	6.30
F-W-(W)-H-H-W	66.6	6.34
F-W-W-H-H-(W)	63.7	6.08
(F)-W-W-H-H-W (N+P)	65.9	6.47
F-W-(W)-H-H-W (N+P)	63.5	6.35
F-W-W-H-H-(W) (N+P)	61.3	6.15
Significance of F ratio	**	**
LSD (P<0.10)	2.5	0.24

Note: Samples taken in September 1987. H = bromegrass-alfalfa hay.

Adapted from Campbell, C.A. et al., *Can. J. Soil Sci.*, 71, 377, 1991b.

for cultural, tillage, and harvest operations, and a combination of tillage and herbicides were used for weed control up to 1989.[13] Green manure, and hay crops after the final harvest, were plowed using a rototiller. However, beginning in summer 1990, all treatments at Indian Head were changed to zero-tillage management and the crops direct seeded, while the sweetclover (*Melilotus officinalis* L.) green manure and alfalfa-bromegrass (*Medicago sativa* L.-*Bromus inermis* Leyss) hay crops were desiccated with herbicides rather than soil incorporated by plowing. Treatments were arranged in a randomized complete block design with four replicates.

The plots at Indian Head were 4.5 × 33.5 m with ample roadways; at Melfort plots were 3.7 × 38 m. At both sites all states of each rotation were present every year and each rotation was cycled on its assigned plots. The plots were seeded at recommended rates. In the case of green manure and forage crops, seeding was done at the same time as the preceding cereal crop (i.e., underseeded). Recommended varieties of crops were used each year but cultivars were changed as improved ones became available.

At Indian Head[14] and at Melfort,[15] rates of N and P fertilizer applied to rotations designated to receive fertilizer varied over the study period (Table 5). During 1958 to 1977, fertilizer N and P were applied at Indian Head according to rotation specifications and the generally recommended rates for spring wheat grown in this region.[16] Since 1978, N and P were applied based on soil test levels of N and P determined the previous fall. At Melfort, N and P fertilizers were applied in accordance with the general recommendations for the region[16] during the periods 1958 to 1971 and 1988 to 1990, while from 1972 to 1987 fertilizer was applied based on fall soil test levels and the recommendation criteria of the Saskatchewan Advisory Council on Soils.

At both experimental sites annual crops were swathed in early September at the full-ripe stage (i.e., Feekes scale 11.4).[17] Yield determinations were made by threshing the grain from the entire plot with a conventional combine. The straw was redistributed on the plots by a paddle-type straw spreader attachment on the combine except for one F-W-W rotation at Indian Head where about two thirds of the straw was baled and removed. Forage plots were cut at full-bloom (usually late June to early July), field-dried, baled, and the hay removed. The crowns after the final hay crop and the green manure crop were soil-incorporated by rototilling in mid-July. No measurements of N nor P in grain or crop residues were made.

Weed control on summer fallow areas was achieved by mechanical tillage. On average, five operations (range four to six) with a heavy-duty cultivator were required. In some years a rodweeder replaced one or more cultivation operations, and in the early years of the study a disc was also used.

Precipitation data and air temperature were collected at meteorological stations located about 1 km from each test site.

Table 5 Fertilizer Treatments over 34-Year Period

Year and Treatment	Average Annual Rates of Fertilizer Applied ($kg\ ha^{-1}y^{-1}$) N	P
At Indian Head		
	1960–1977 (General Recommendations)[a]	
Wheat on fallow	5.6	11.2
Wheat on stubble	22.4	8.4
	Since 1978 (Soil Test)[b]	
Wheat on fallow	10.1	7.8
Wheat on stubble	68.0	8.4
At Melfort		
	1958–1990 (Source and Criteria Variable)[c]	
Wheat on fallow (monoculture)	16.0	15.7
Wheat on fallow (6-y hay system)	13.0	15.7
Wheat after hay or green manure break	36.0	15.7
Wheat on stubble (monoculture)	55.0	14.6
Wheat on wheat stubble (legume system)	46.0	14.6
Hay crops	79.0	6.5
Wheat on fallow in 6-y hay-containing system	13.0	15.7

[a] Ammonium nitrate-phosphate (23-23-0) and monoammonium phosphate (11-48-0) used.
[b] Ammonium nitrate (34-0-0) and monoammonium phosphate used.
[c] In 1958 to 1971 and 1988 to 1990 general recommendations used; in 1972 to 1987, soil test criteria used. In 1958 to 1971 wheat on fallow received 11-48-0 while wheat on stubble received 27-14-0 or 23-23-0 seed placed. During 1972 to 1987 N rates were much higher than in the first period;[15] P was still applied with the seed at 11-48-0 while N as 34-0-0 was either broadcast prior to seedbed preparation or side banded at depth of planting. Forage crops received 34-0-0 and/or 11-48-0 broadcast in spring.

In early June 1987, soil samples were taken at Indian Head from 0- to 7.5- and 7.5- to 15-cm depths of 11 rotation phases of the 9 rotations shown in Table 3. Three subsamples were taken with a spade from near the central part of each plot and these composited by depth. Half of this soil was air-dried, sieved to <2 mm, crop residues remaining on the sieve discarded, and the soils stored until analyzed for total C by dry combustion, inorganic C by acid titration, and Kjeldahl N.[18] Separate soil samples (two cores per plot) were taken for bulk density (D_b) determination.[19] Bulk densities of the 0- to 7.5- and 7.5- to 15-cm depths were unaffected by treatment but differed by depth. Consequently, D_b were averaged within depths giving values of 0.95 and 1.22 Mg m^{-3} for the 0- to 7.5- and 7.5- to 15-cm layers, respectively. Bulk densities were used to convert concentrations to weight basis.

In mid-September 1987, immediately after wheat harvest, 13 rotation phases representing the 8 rotations were sampled at Melfort (Table 4). Soil organic C and N and bulk densities for the 0- to 7.5- and 7.5- to 15-cm depths were determined using similar methods as described for Indian Head. The bulk densities (D_b) were unusually low; values were 0.75 and 0.84 Mg m^{-3} for the 0- to 7.5- and 7.5- to 15-cm depths, respectively, but scientists at Melfort generally report 0.90 Mg m^{-3} for the 0- to 15-cm layer.

At each site, yield and soils data from the rotation phases sampled were subjected to analysis of variance and LSDs interpreted at either $P<0.10$ or $P<0.05$. Regression and correlation analyses were used to interrelate the various soil organic characteristics assessed and to relate these characteristics to the estimated crop residues and crop residue C and N returned to the land during the experiment.[1,2] Five-year running means of grain yields were calculated for selected cropping systems so as to minimize the influences of annual variability in weather on yield trends. Regression analysis was used to determine the time trends in yields using the year as the independent variable.[20]

IV. RESULTS AND DISCUSSION — INDIAN HEAD

A. WEATHER CONDITIONS

Over the 34-year period of the experiment, annual temperature and precipitation were similar to the long-term (90-year) average at Indian Head (Table 1). The 34-year average growing season precipitation

(1 May to 31 August) was 232 mm compared to the 90-year average of 233 mm. Annual and growing season precipitation were generally slightly below average from 1958 to 1972 and slightly above average in later years.[21]

B. YIELDS

Yields for wheat grown on fallow or partial fallow showed the positive influence of fertilizers and legumes (Table 6), but over the study period there were no consistent trends (data not shown). Similar results were reported for a long-term study conducted at Lethbridge, Alberta.[22] They suggested that the lack of marked fertilizer effects over time was because N and P mineralization during the fallow periods was sufficient to cover up any shortfalls due to the apparent gradual depletion in soil fertility.

The influence of N and P fertilizers on yields of wheat grown on stubble was positive and significant (Figures 1 and 2). The 34-year average yield increase of wheat grown on stubble in F-W-(W) was 916 kg ha^{-1} and for Cont W it was 931 kg ha^{-1} (Table 6). These results are similar to those obtained at Lethbridge, Alberta.[22] Yields increased at an average rate of 30 to 33 kg ha^{-1} y^{-1} for fertilized wheat grown on stubble (Figures 1 and 2). The widening difference between the unfertilized and fertilized systems was partly due to the major increase in N rate after 1978 when the criteria for fertilization was changed to one based on soil tests. However, the downward trend in check yields also suggests that the unfertilized system may be showing evidence of the depletion of soil nutrients. Corroborating this evidence was the lower organic matter and N-supplying power of the unfertilized F-W-W system when compared to unfertilized Cont W, although there was no difference in available P between these two systems after 34 years (Table 3). The yield decline was more apparent in F-W-(W), i.e., decline of 27 kg ha^{-1} y^{-1} (Figure 2), than in the Cont (W), i.e., decline of 11 kg ha^{-1} y^{-1} (Figure 1). This difference in rate of decline in yields probably reflects a greater incidence of erosion and lower quantity of residues being returned to the soil in the F-W-W system.[23] Soil loss from three of the Indian Head rotations was estimated[22] by comparing their $^{137}C_s$ concentration in the top 15 cm of soil to that of an adjacent grassed area as a control.[24] The soil loss over 30 years was estimated to be 933 Mg ha^{-1} from fertilized F-W system, 646 Mg ha^{-1} from fertilized Cont W, and 409 Mg ha^{-1} from the 6-year hay-containing rotation.[23] These results suggest that sizeable erosion can be expected to accompany tilled summer fallow systems and that the presence of hay crops may be more effective than Cont W in reducing erosion. The downward trends in yields of check treatments were not due to the amount of precipitation because, with the exception of the first decade, precipitation was generally above average.[21]

Legume (sweetclover) green manure increased yields of wheat grown on stubble compared to unfertilized F-W-(W) by an average of 6.7 kg ha^{-1} (Figure 2 and Table 6), likely a result of N fixation. Evidence of N fixation was seen in the 56% higher N-supplying power of the soil that included green manure compared to the unfertilized F-W-W rotation (Table 3). Even so, the yield trends of wheat grown on stubble in this green manure system was downward, similar to that of unfertilized F-W-(W) (i.e., decline of 26 kg ha^{-1} y^{-1}). This yield decline was probably due to the fact that legumes do not supply P, which is required for efficient crop production. In fact, available P in the top 120 cm of soil, measured in September 1991, showed 44% greater amounts in the unfertilized F-(W)-W rotation than in GM-(W)-W (Table 3). This was partly related to greater export of P in grain from the green manure rotation over the years (Table 6). There was an inverse relationship ($r^2 = 0.75^*$) between the apparent P deficit (i.e., P exported in grain minus fertilizer P applied over 34 years, Table 5) and Olsen-P (Table 3), for the six rotations in which P was measured.

The lower yields in the GM-W-(W) rotation compared to the fertilized F-W-(W) system was strongly related to available P (Table 3) and possibly to N fertilization, though it was not possible to estimate the relative contribution of each nutrient because the actual contribution of N from the green manure was not measured. Nevertheless, when F-W-(W) was fertilized based on soil tests, yield trends were superior to those for wheat grown on stubble in a sweetclover GM-W-W system that did not receive fertilizer, suggesting that the latter system may not be sustainable without fertilizer P additions in the long term.

Up to 1978, the unfertilized F-W-W-H-H-H system maintained higher yields of wheat grown on stubble than those of fertilized F-W-(W) (Figure 3). This was a period when the latter system was apparently underfertilized despite the use of the general fertilizer recommended for this particular soil. The greater yields of wheat grown on stubble in the 6-year rotation compared to F-W-(W), up to 1978, was partly because of much less erosion in the hay-containing system, as mentioned earlier. After 1978, however, wheat grown on stubble in fertilized F-W-(W) system, with its much higher N rates based on soil tests, was providing significantly greater yields than similar wheat in the hay-containing rotation.

Table 6 Estimated Amount of N and P Exported in the Grain and Hay from Rotations during the 34-Year Period (1958 to 1991) and Amount of Fertilizer N and P Applied at Indian Head

Yields and N and P Exported from System in Grain	Rotation Phase														
	F-(W)	F-(W) (N+P)	F-(W)-W	F-W-(W)	F-(W)-W (N+P)	F-W-(W) (N+P)	GM-(W)-W	GM-W-(W)	F-(W)-W-H-H-H	F-W-(W)-H-H-H	F-W-W-(H)-H-H	F-W-W-H-(H)-H	F-W-W-H-H-(H)	Cont (W)	Cont (W) (N+P)
Mean annual yield (kg ha^{-1})	2180	2480	2300	1100	2600	2010	2510	1440	2700	1850	1190	2640	2830	1030	1960
Mean annual N exported (kg ha^{-1})	54	62	58	27	65	50	63	36	68	46	25	56	59	26	49
N exported from rotation[a] in 34 y (kg ha^{-1})	918	1054	935		1265		1089		1397					884	1666
N applied as fertilizer in 34 y (kg ha^{-1})	0	134	0		542		0		0					0	1544
Mean annual P exported (kg ha^{-1})	9.1	10.4	9.7	4.6	10.9	8.5	10.5	6.0	11.3	7.8	3.1	6.9	7.4	4.3	8.2
P exported from rotation[a] in 34 y (kg ha^{-1})	155	177	157		213		182		200					147	280
P applied as fertilizer in 34 y (kg ha^{-1})	0	176	0		216		0		0					0	306

Note: Assumed %N in wheat grain = 2.5, in alfalfa = 2.6; in grass = 1.5; and %P in wheat grain = 0.42, in alfalfa = 0.28, in grass = 0.22. Assumed hay = 50% alfalfa and 50% grass.

[a] For the 2-year rotation this was the amount of N or P in grain for wheat on fallow × 17; for the 3-year rotations this was sum of N or P in grain for the two crop years × 11; for Cont W it was the amount for annual wheat × 34; and for the 6-year rotation it was the sum of N or P in wheat grain and hay in 5 crop years × 5.5.

Adapted from Campbell, C.A. et al., *Can. J. Soil Sci.*, 73, 555, 1993a and Campbell, C.A. et al., *J. Environ. Qual.*, 23, 195, 1994.

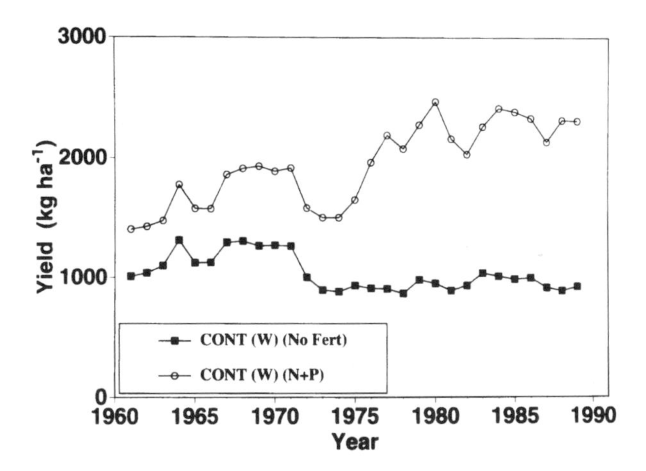

Figure 1 Yield trends of continuously grown hard red spring wheat at Indian Head, as influenced by fertilizer (the points are 5-year running means). (Adapted from Campbell, C.A. et al.)[21]

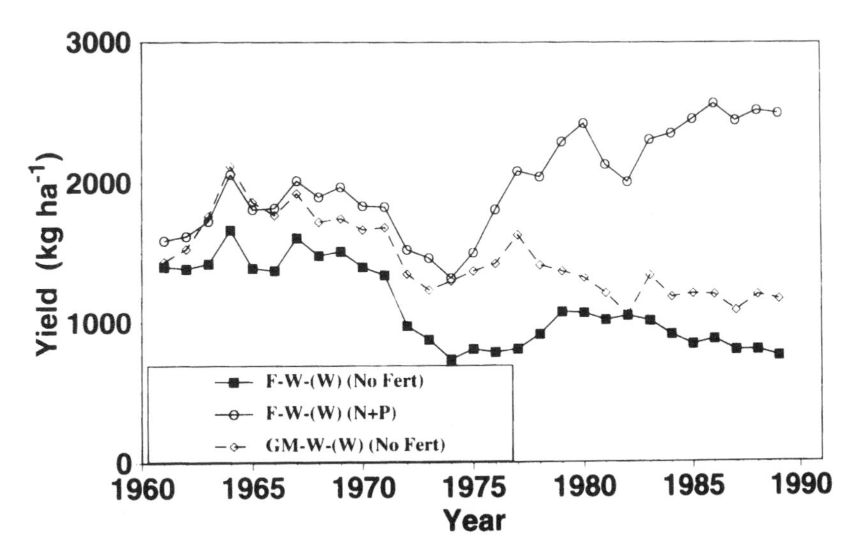

Figure 2 Yield trends of hard red spring wheat grown on stubble in fallow-wheat-wheat (F-W-W) rotation at Indian Head, showing the influence of sweetclover green manure (GM) and of N + P fertilizer (the values pertain to the rotation phase in parentheses; the points are 5-year running means). (Adapted from Campbell, C.A. et al.)[21]

Over the 34-year period, the yield trends of wheat in the 6-year rotation declined at a much slower rate (5 kg ha^{-1} y^{-1}) than rates of the other unfertilized systems, including the GM system. Both legume-containing rotations had similar, low amounts of available P in the top 1.2 m of soil when sampled in 1991 (Table 3), even though the hay-containing system exported slightly more P via harvest than the GM system (Table 6). Yields of wheat grown on stubble in the 6-year rotation were maintained at a higher level than in the GM rotation because of the significantly greater N-supplying power of the 6-year rotation (Table 3). The N-supplying power of the hay-containing rotation was significantly greater than that of the fertilized F-W-(W) system in 1987 (Table 3), but its much lower available P content compared to fertilized F-W-W likely contributed to a reduction in yield over time.

Removal of straw by baling did not reduce yields (Figure 4); in fact in some years yields were actually increased (data not shown), possibly due to better seed placement.[4]

Hay yields in the first year of utilization were low (Table 6), but this value included several years of poor crop establishment. The yields of 2nd- and 3rd-year hay crops were about equal and were more than twice as much as for 1st-year hay crop yields.

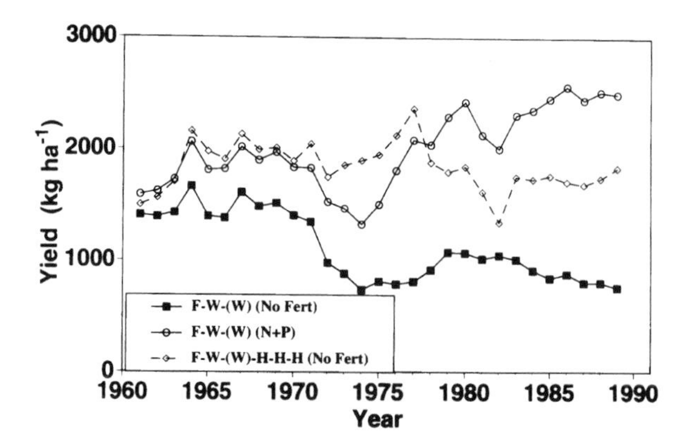

Figure 3 Yield trends of hard red spring wheat grown on wheat stubble in an unfertilized fallow-wheat-wheat-hay-hay-hay rotation at Indian Head, compared to N and P fertilized and unfertilized fallow-wheat-wheat systems (the points are 5-year running means). (Adapted from Campbell, C.A. et al.)[21]

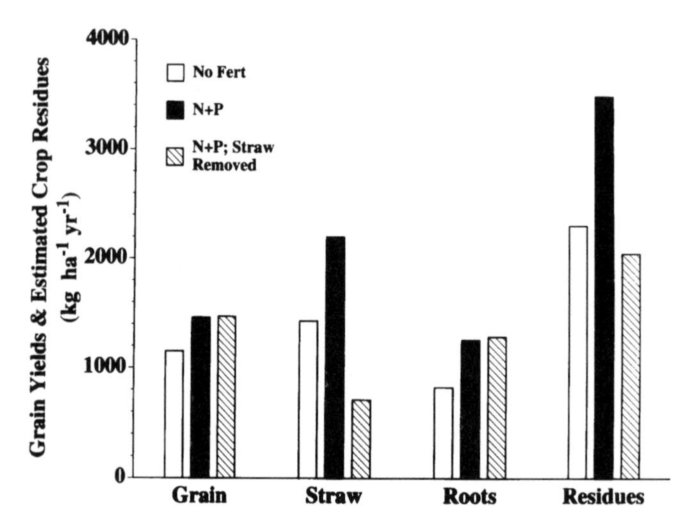

Figure 4 Effect of fertilizer and straw removal on grain yields and estimated crop residues returned annually to soil in fallow-wheat-wheat systems at Indian Head. (Adapted from Campbell, C.A. et al.)[4]

The amount of grain produced by wheat per rotation per year was calculated by summing the yields for each rotation phase and dividing by the number of rotation phases (including fallow) in the rotation. Thus, for Cont W the grain production would be the same as the unit area yield, while for F-W it would be one half the unit area yield. Production was directly proportional to cropping frequency when fertility was not limiting (Figure 5). Grain production per unit area was increased by green manure and increased even more by the inclusion of brome-alfalfa hay in the rotation. Straw removal did not influence grain production.

C. SOIL ORGANIC MATTER

Soil organic C in the 0- to 7.5- and 7.5- to 15-cm depths was influenced by treatments in a similar manner. Consequently, the results are discussed in terms of the 0- to 15-cm depth (Table 3).

The amount of soil organic C ranged between 27.6 and 34.5 t ha^{-1} and was directly related to cropping frequency (Figure 6). Although some workers in northwestern U.S.[25] found no effect of fertilization on organic C, we found a significant increase for Cont W though not for F-W or F-W-W in the Black Chernozem at Indian Head. The inclusion of sweetclover green manure (GM) or brome-alfalfa hay (H) in cereal rotations increased soil organic C compared to the 2- and 3-year fallow-cereal rotations, but they were similar to Cont W (fertilized or unfertilized). These results are generally similar to those reported for a Dark Brown Chernozem (Typic Boroll) in southern Alberta.[26] The 6-year hay-containing

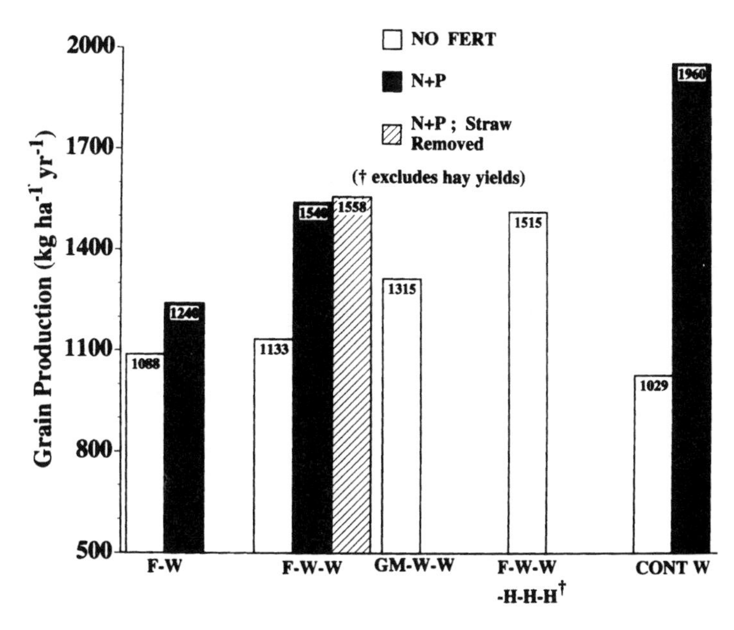

Figure 5 Influence of fertilizer, sweetclover green manure, alfalfa-bromegrass hay crop on grain production per rotation per year (including fallow) at Indian Head.

rotation had slightly more organic C than the 3-year GM-containing system. This may be due to the higher inputs of residue N relative to residue C,[4] which could have resulted in a C-conserving influence during decomposition.[27] Although straw removal had no influence on grain yield (Figure 4) it tended to decrease soil organic C (Table 3). In a previous assessment, Campbell et al.[4] suggested there was no effect on organic C but this discussion was erroneously based on total C. Nuttall et al.[28] found no effect of straw removal on yield or soil organic C on a thick Black Chernozem in north-central Saskatchewan, but Persson and Mattsson in Sweden[29] found results similar to ours.

Total soil N concentrations in the 0- to 7.5- and 7.5- to 15-cm depths were similar except for the 6-year mixed cereal-forage system and fertilized Cont W rotations where concentrations were higher in the upper layer (data not shown). No reliable values are available for the N concentrations of these soil layers at the start of the experiment in 1958.

Treatment effects on total N contents were greater in the top 7.5 cm than in the 7.5- to 15-cm layer, and there were more significant treatment effects ($P<0.10$) on total N in the top 7.5 cm as well.[5] Soil N content decreased with increasing fallow frequency (Figure 6); it increased with fertilization, with the inclusion of sweetclover green manure (GM) in the 3-year rotation and even more when the F-W-W system was lengthened to 6 years by including 3 years of hay (Table 3). The influence of fertilizers and fallow frequency on soil N is similar to the findings at Swift Current[30] and Lethbridge.[26,31] In contrast to results reported for a Dark Brown Chernozem at Lethbridge, Alberta,[26] the Indian Head results show that the 6-year mixed cereal-forage rotation increased the total N content compared with unfertilized Cont W. This difference in response in the two soils can be explained by the positive yield response of associated wheat crops to N fixed by the hay crop in the Black Chernozem where moisture was less limiting;[9] in contrast, moisture stress severely limited wheat yields in the drier, Dark Brown Chernozem.[26] Total soil N content in the 0- to 15-cm layer of the 6-year rotation was as high as that for fertilized Cont W, though it included 1 year of tilled summer fallow and even though hay was removed from the land in 3 of every 6 years.

Rotation phase in the green-manure system and in the mixed forage-cereal rotation system did not significantly ($P>0.10$) affect total soil N content. Baling and removing straw from fertilized F-W-W tended to decrease total N (by about 5.4%).[4]

Both amount and quality of soil organic matter are directly influenced by the amount and characteristics (C/N ratio, and possibly lignin and soluble carbohydrate content) of crop residues returned to the soil.[32,33] However, few field studies have attempted to quantify the relationship between crop residues and various biochemical characteristics that are used in assessing soil organic matter quality. We estimated the amounts of crop residues, and residue C and N returned to the soil over the first 30 years of study

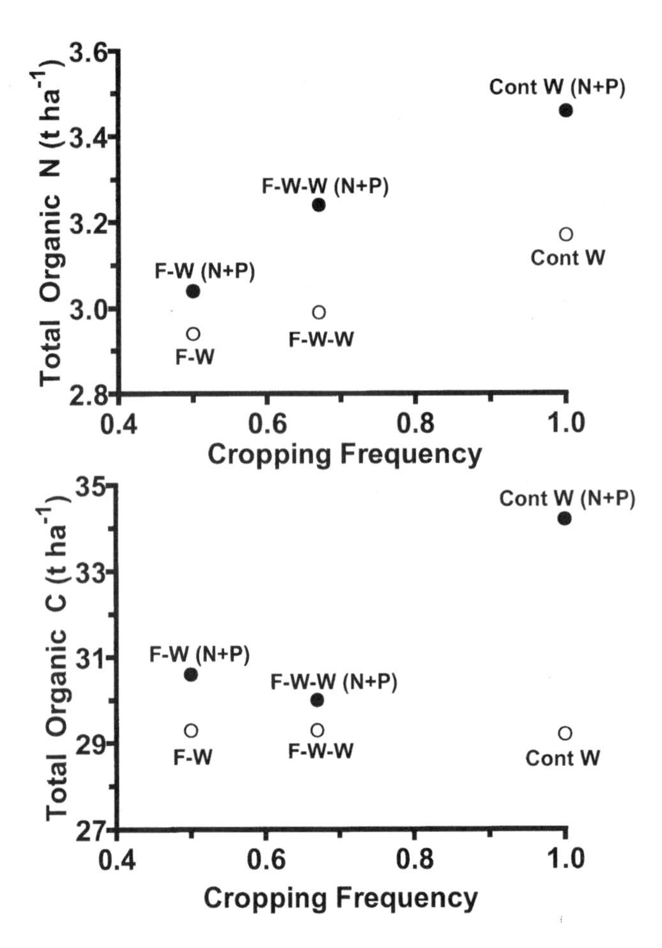

Figure 6 Effect of cropping frequency (e.g., fallow-wheat = 0.5; continuous wheat = 1.0) on soil organic matter at Indian Head. (Note: Values in original paper were total, not organic C as listed. The present values are organic C.) (Adapted from Campbell, C.A. et al.)[1]

by using the grain, green manure, and hay yields, recorded over the study period.[1,4] Estimates for the hay-containing system were less precise than those for cereal and green manure systems.

All biochemical indices examined were significantly and positively correlated with the estimated crop residue values.[1] Because of the approximate nature of the estimates of the independent variable, the equations in Table 7 are not statistically rigorous. However, they do allow us to estimate the changes in soil organic C that can be expected for this soil if we have values available for the crop residues (including roots) returned to the soil. For example, soil organic C and N in the top 15 cm of this soil can be expected to increase by about 83 and 7 kg ha^{-1}, respectively, for each tonne of crop residues returned per hectare (Table 7). These equations therefore allow us to approximate the rates of change in soil organic matter of this soil.

V. RESULTS AND DISCUSSION — MELFORT

A. WEATHER CONDITIONS

The 34-year mean temperatures during the study period were generally similar to the long-term means (Table 1). Mean annual precipitation during the study period was 14 mm less than the long-term mean while growing season precipitation (1 May to 31 August) was 13 mm less than the long-term mean (214 mm vs. 227 mm). Average growing season and total annual precipitation received during 1960 to 1971 were lower than the long-term averages.[15] In contrast, average growing season precipitation received during 1972 to 1986 was 8% higher than the long-term average and about 28% higher than that received during 1960 to 1971. The driest years were 1961, 1962, 1967, and 1969 when less than 124 mm of

Table 7 Relationships between Soil Organic C and N, in 0- to 15-cm Depth and Estimated Crop Residues (Straw and Roots), Crop Residue C, and N, Returned to Soil over 30 Years at Indian Head (t ha^{-1})

Soil Organic C = Y	Soil Organic N = Y
Y = 23.4 + 0.083 residues ($r^2 = 0.85^{**}$)	Y = 2.55 + 0.007 residues ($r^2 = 0.74^{**}$)
Y = 23.3 + 0.187 residue C ($r^2 = 0.81^{*}$)	Y = 2.56 + 0.016 residue C ($r^2 = 0.68^{**}$)
Y = 26.8 + 5.59 residue N ($r^2 = 0.72^{**}$)	Y = 2.80 + 0.56 residue N ($r^2 = 0.83^{**}$)

Note: The equations published previously [1] were total C (erroneously referred to as organic C). The present equations are the true relationships to organic C. *, ** = significant at P<0.05 and P<0.01, respectively.

Adapted from Campbell, C.A. et al., *Can. J. Soil Sci.*, 71, 363, 1991a.

growing season precipitation was received; the wettest years were 1973 and 1974 when growing season precipitation exceeded 345 mm.

B. YIELDS

Yields of stubble-wheat on fertilized and unfertilized rotations increased significantly over the first decade then leveled off or decreased slightly in later years (Figure 7) reflecting (1) higher fertilizer rates used with time; (2) the tendency for precipitation to be greater in years after 1971; and (3) improved wheat varieties used with time. For example, stubble-seeded wheat received on average 30 kg N ha^{-1} and 6.9 kg P ha^{-1} up to 1963; 26 kg N ha^{-1} and 11.2 kg P ha^{-1} from 1964 to 1971; variable but much higher rates of N (44 to 113 kg ha^{-1}) and P (11.6 to 21 kg ha^{-1}) from 1972 to 1981; and 65 kg N ha^{-1} and 14.6 kg P ha^{-1} to 1991. Yields of wheat grown on fallow and stubble were significantly correlated with growing season precipitation during 1960 to 1971 ($r = 0.60^{*}$), but not thereafter.

Fertilizer increased yields of stubble-wheat in the 3- and 6-year hay-containing and continuous wheat rotations (Figure 7 and Table 8). The yields of stubble-wheat in the fertilized GM system was less than that in the fertilized F-W-W system (Figure 7 and Table 8) but this difference could be accounted for by the lower amounts of N applied to the GM system over the years (Table 8). Presumably, N fixation in the GM system was suppressed by the frequent application of N or due to poor growth of sweetclover. Yields of unfertilized stubble-wheat in the 6-year hay-containing system were generally similar to those for the unfertilized F-W-W system (Figure 7) although greater amounts of N and P were removed in harvested material from the 6-year system (Table 8). Despite the much larger deficits in N and P (fertilizer applied minus N harvested) in the unfertilized systems compared to fertilized systems, we found little evidence to suggest that this very fertile soil is being degraded with time. The slight negative slope in yield trends over the last 15 years or so in the fallow-containing systems may indicate the occurrence of some degradation but, because the fertilized systems also show this trend, we doubt that there is a direct relationship with fertility. These results contrast with those shown earlier for Indian Head and are likely due to the high levels of organic matter present in this Melfort soil (see later discussion).

On a per unit area basis, wheat grain production at Melfort was generally directly related to cropping frequency, except for Cont W (Figure 8), which produced lower yields than those projected likely due to disease and weed control difficulties experienced with monoculture wheat in some years in this subhumid area.[15] Grain production was markedly increased by fertilizer, especially in Cont W. The hay crop increased grain production but the green manure crop did not, probably because in several years clover growth was impeded by weevil attack.[15]

C. SOIL ORGANIC MATTER

The thick Black Chernozem at Melfort had almost twice as much organic C in the 0- to 15-cm layer as in the thin Black Chernozem at Indian Head (61 to 67 t ha^{-1} vs. 28 to 34 t ha^{-1}, Tables 3 and 4). Similarly, there was almost twice as much organic N in the Melfort soil (5.9 to 6.5 vs. 2.9 to 3.5 t ha^{-1}).

In contrast to results obtained for Indian Head, neither fertilizer nor inclusion of a legume green manure crop significantly (P<0.10) affected soil organic C or N (Table 4). As found in a Dark Brown

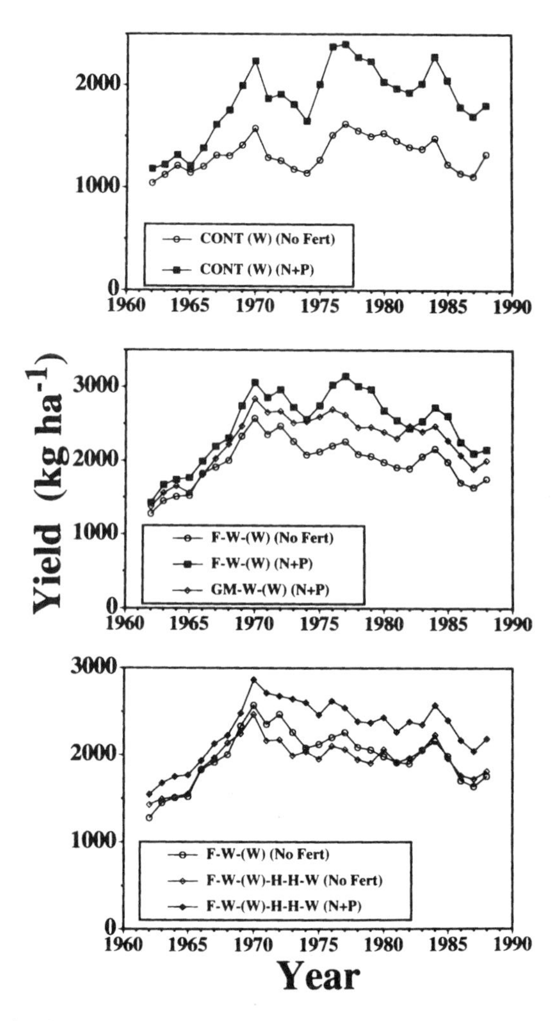

Figure 7 Top: Yield trends of hard red spring wheat grown continuously at Melfort as influenced by fertilizer. Middle: Yield trends of hard red spring wheat grown on stubble in fallow-wheat-wheat rotations at Melfort showing influence of sweet clover green manure and of N and P fertilizer. Bottom: Yield trends of hard red spring wheat grown on wheat stubble in unfertilized and fertilized fallow-wheat-wheat-hay-hay-wheat rotations at Melfort compared to an unfertilized fallow-wheat-wheat system. All relationships are 5-year running means, 1960 to 1990. (H = alfalfa-bromegrass.)

soil at Lethbridge, Alberta,[26] organic matter contents of the Cont W treatments were higher than those of F-W and F-W-W, but were similar to those of GM-containing or the 6-year hay-containing rotations. Further, the 6-year rotations had similar organic matter as the GM system. However, while the lack of difference in soil organic matter between the unfertilized 6-year F-W-W-H-H-H system and unfertilized Cont W in the Dark Brown soil may have been related to water deficit restricting cereal production in the 6-year system,[26] at Melfort, soil moisture rarely limits production. Presumably it is very difficult to enhance soil organic matter content of soils in which organic matter is already high, as was observed in a Black Chernozemic soil in Manitoba.[34]

Earlier we related changes in soil organic matter resulting from various crop rotations and cultural practices at Indian Head to the estimated amount of crop residues, residue C, and residue N that had been returned to the soil during the experimental period. A similar analysis was performed for the cropping systems at Melfort.[4] In contrast to the Indian Head results which showed significant relationships between soil organic matter characteristics and estimated crop residues returned to the soil,[1] the relationships for the Melfort soil were not significant (Figure 9). This was also true for the relationship between organic matter and residue C and N returned to the land (data not shown). The soil organic matter of the Melfort soil was much greater than that at Indian Head, but soil organic matter at Melfort

Table 8 Estimated Amount (kg ha^{-1}) of N and P Exported in Grain and Hay from Rotations during 31 years (1960 to 1990) and Amount of Fertilizer N and P Applied at Melfort

Rotation Phase and Fertilizer		Mean Annual Yield	Mean Annual N Exported	N Exported from Rotation[a] in 31 y	N Applied to Rotation in 31 y	Mean Annual P Exported	P Exported from Rotation[a] in 31 y	P Applied to Rotation in 31 y	Deficit[b]	
									N	P
F-(W)	(N+P)	2740	69	1065	295	11.5	180	240	–768	62
F-(W)-W		2400	60			10.1				
F-W-(W)		1940	49	1115	0	8.1	190	0	–1117	–187
F-(W)-W	(N+P)	2760	69			11.6				
F-W-(W)		2400	60	1330	773	10.1	225	305	–555	82
GM-(W)-W	(N+P)	2580	64			10.8				
GM-W-(W)		2190	55	1230	639	9.2	205	310	– 589	104
F-(W)-W-H-H-W		2450	61			10.3				
F-W-(W)-H-H-W		1920	48			8.0				
F-W-W-(H)-H-W		2200	45			5.5				
F-W-W-H-(H)-W		2860	59			7.1				
F-W-W-H-H-(W)		1990	50	1365	0	8.3	205	0	–1364	–204
F-(W)-W-H-H-W	(N+P)	2800	70			11.7				
F-W-(W)-H-H-W		2240	56			9.4				
F-W-W-(H)-H-W		2930	60			7.3				
F-W-W-H-(H)-W		3490	71			8.7				
F-W-W-H-H-(W)		2340	59	1640	1310	9.8	245	300	–332	58
Cont W		1320	33	1025	0	5.5	170	0	–1023	–171
Cont W	(N+P)	1820	46	1410	1705	7.6	236	440	294	204

Note: Assumed % N in wheat grain = 2.5; in alfalfa = 2.6; in grass = 1.5; and % P in wheat grain = 0.42; in alfalfa = 0.28 and in grass = 0.22. Assumed hay = 50% alfalfa and 50% grass.

[a] For the 2-year rotation this was the amount of N or P in grain of wheat on fallow × 15.5; for the 3-year rotations this was sum of N or P in grain for the two crop years × 10.3; for Cont W it was the annual wheat yield × 31; and for the 6-year rotation it was the sum of N or P in wheat grain and hay in 5 crop years × 5.2.

[b] Nutrient deficit = nutrient applied as fertilizer minus nutrients exported in harvested produce.

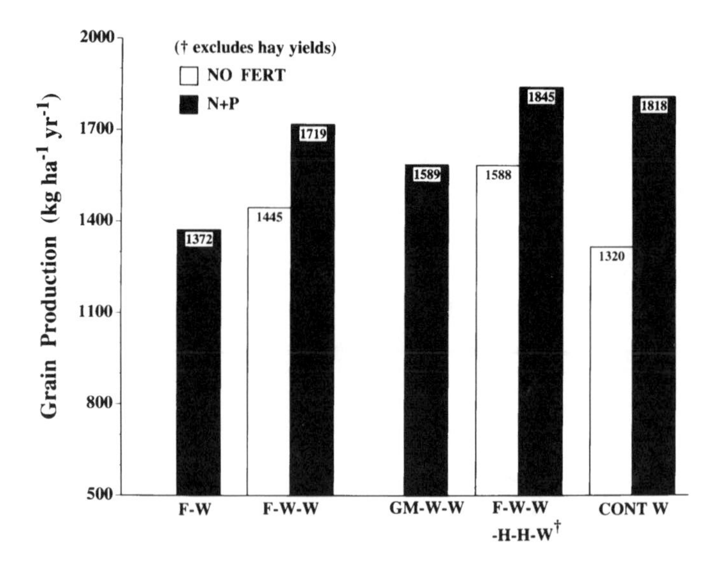

Figure 8 Influence of fertilizer, sweetclover green manure, alfalfa-bromegrass hay crop on grain production per rotation per year (including fallow) at Melfort.

was unaffected by the addition of crop residues while it was increased at Indian Head (Figure 9). The y-intercepts and regression coefficients for each pair of variables for the two soils were significantly different ($P<0.01$). The failure of the analysis to reveal marked changes in the total soil organic matter at Melfort, while these changes have been easily detected in other soils[1,5,26,27] may be related to the high organic matter content of the thick Black Chernozem. For a given input of crop residue C, the relative change in soil organic matter in the Melfort soil may be so small as to make such change easily masked by natural soil variability. For example, the range in organic C in the 0- to 15-cm depth was the same (6 t ha^{-1}) for Melfort and Indian Head soils, but Melfort had more than 60 t ha^{-1} and Indian Head about 30 t ha^{-1} organic C. Second, if organic matter decomposition obeys first-order kinetics, then we can expect the rate of organic matter decomposition to be greater in the higher organic matter soil at Melfort than at Indian Head. Replacing the decomposed organic matter through increased residue inputs resulting from improved management may be significant in absolute terms but in relative terms this might be undetectable. These results emphasize some of the difficulties in finding increases in the organic matter content of a soil that is already inherently rich in organic matter.[34]

VI. PERSPECTIVE

The experiments at Indian Head and Melfort were initiated by agronomists with the objective of providing yields and economic answers. Thus, neither organic matter nor nutrient use by the crop were monitored over the study period. Relative comparisons of the impact of the treatments on soil organic matter after 30 years, together with some deductive reasoning based on trends in crop production, can be used to estimate levels of C sequestered in soil in this subhumid region of the Canadian prairies.

The results emphasized that the amount of C sequestered in the soil will be first and foremost dependent on crop production per unit area. Soil C should increase more rapidly in the more productive Udic Boroll (Black Chernozem) soils compared to the Aridic or Typic Borolls (Brown and Dark Brown Chernozem). Further, we would expect soil C sequestration to be directly proportional to cropping frequency, to the use of fertilization (properly administered), and to the inclusion of legume green manure or legume-grass hay crops in rotations in subhumid or humid climates, but to be inversely affected by frequent use of tilled summer fallow. Tempering these general responses, however, is the initial level of C in the soil. At Melfort, where the soil organic matter is very high, it might be more difficult to increase soil organic C; in such circumstances all we may be able to achieve with good management is to reduce the rate of loss of soil C. On an absolute basis this may still be a positive outcome in terms of C sequestration since, without good management, considerable extra C would not have been captured. On the other hand, we must balance this picture with the fact that, even if we are able to maintain soil organic C, this might be at a net negative cost to C in the environment. This is because the fertilizer N

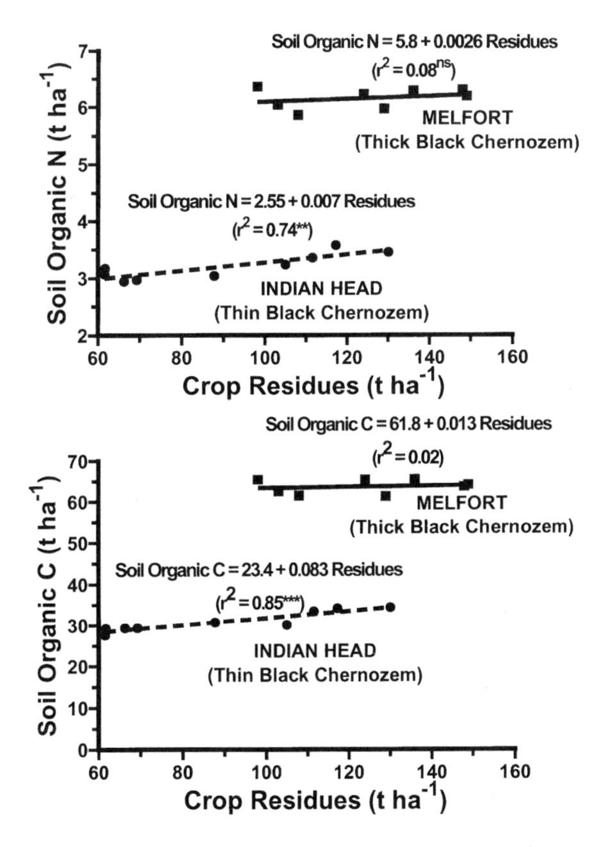

Figure 9 Relationships between soil organic C and N in the 0- to 15-cm depth and estimated crop residues returned to Melfort soil (31 year) and Indian Head soil (30 year). (Adapted from Campbell, C.A. et al.)[2]

and herbicides required, particularly if monoculture is practiced, may result in a greater expenditure in energy (C) in manufacture of these ingredients, and this may counterbalance any C sequestered by the system. Major increase in C sequestration may be anticipated in such regions as Indian Head with its moderately low organic matter content but very favorable weather conditions to promote corp and residue production.

These studies have been conducted using conventional tillage (the Indian Head study was changed to zero tillage in 1991). Zero tillage should increase soil organic matter[35] because it reduces erosion and because tillage will increase organic matter decomposition. Zero tillage should also have a positive influence on soil organic matter if it increases crop production, as it appears to do at Indian Head.[36] Further analysis is required to determine if this influence may be offset by any extra use of herbicides or energy, which require C to manufacture.

VII. CONCLUSIONS

Results of two long-term (>30 years) experiments being conducted in the sub-humid region of the Canadian prairies showed that the degree to which C is sequestered in soil is a function of the amount of soil organic matter already present in the soil. It is very difficult to increase soil organic matter in a soil that is already high in organic matter but good management will reduce the rate of loss of soil organic C which is a positive occurrence in terms of C sequestration.

Factors that contribute to greater crop production per unit area (e.g., fertilizer use, legume green manure or hay crops in rotation, increased frequency of cropping) are the same ones that can be expected to enhance or maintain soil organic C. Frequent summer fallowing will tend to reduce organic C.

These experiments were conducted using conventional tillage management; it is of interest to see how such systems would respond under zero tillage which would likely reduce erosion and the rate of loss of soil organic matter. Second, organic matter has not been monitored on a regular basis in these two experiments; this needs to be done if we are to obtain a more complete picture regarding our ability to modify C sequestration by manipulating crop management.

ACKNOWLEDGMENTS

We wish to acknowledge the scientists at Indian Head and Melfort Research Stations who had the foresight to initiate these studies, and thank the many plotmen and technicians who have diligently conducted the associated field and laboratory work, kept the records, and analyzed the results throughout the years.

REFERENCES

1. Campbell, C.A., V.O. Biederbeck, R.P. Zentner, and G.P. Lafond, Effect of crop rotations and cultural practices on soil organic matter, microbial biomass and respiration in a thin Black Chernozem, *Can. J. Soil Soc.*, 71, 363, 1991a.
2. Campbell, C.A., K.E. Bowren, M. Schnitzer, R.P. Zentner, and L. Townley-Smith, Effect of crop rotations and fertilization on soil organic matter and some biochemical properties of a thick Black Chernozem, *Can. J. Soil Soc.*, 71, 377, 1991b.
3. Campbell C.A., G.P. Lafond, A.J. Leyshon, R.P. Zentner, and H.H. Janzen, Effect of cropping practices on the initial potential rate of N mineralization in a thin Black Chernozem, *Can. J. Soil Soc.*, 71, 43, 1991c.
4. Campbell, C.A., G.P. Lafond, R.P. Zentner, and V.O. Biederbeck, Influence of fertilizer and straw baling on soil organic matter in a thin Black Chernozem in western Canada, *Soil Biol. Biochem.*, 23, 443, 1991d.
5. Campbell, C.A., M. Schnitzer, G.P. Lafond, R.P. Zentner, and J.E. Knipfel, Thirty-year crop rotations and management practices effects on soil and amino nitrogen, *Soil Sci. Soc. Am. J.*, 55, 739, 1991e.
6. Campbell, C.A., A.P. Moulin, K.E. Bowren, H.H. Janzen, L. Townley-Smith, and V.O. Biederbeck, Effect of crop rotations on microbial biomass, specific respiratory activity and mineralizable nitrogen in a Black Chernozemic soil, *Can. J. Soil Sci.*, 72, 417, 1992.
7. Jenny, H., *The Soil Resource*, (Ecological Studies no. 37), Springer-Verlag, New York, 1980.
8. Campbell, C.A. and R.P. Zentner, Crop production and soil organic matter in long-term crop rotations in the semi-arid Northern Great Plains of Canada, in *Soil Organic Matter in Temperate Agroecosystems: Long-Term Experiments in North America*, Paul, E.A., Paustian, K., Elliott, E.T., and Cole, C.V., Eds., CRC Press, Boca Raton, FL, 1997, chap. 23.
9. Campbell, C.A., R.P. Zentner, H.H. Janzen, and K.E. Bowren, Crop rotation studies on the Canadian Prairies, Research Branch, Agriculture Canada, Ottawa, 1990.
10. Moss, H.C. and J.S. Clayton, Report on the soil survey of the Indian Head Experimental Farm, Saskatchewan Soil Surv., Dept. of Soil Sci., University of Saskatchewan, Saskatoon, 1941.
11. Mitchell, J., H.C. Moss, J.S. Clayton, and F.H. Edmunds, Soil Survey Rep. No. 12, University of Saskatchewan, Saskatoon, 1944.
12. Campbell, C.A., G.P. Lafond, V.O. Biederbeck, and G.E. Winkleman, Influence of legumes and fertilization on deep distribution of available phosphorus (Olsen-P) in a thin Black Chernozemic soil, *Can. J. Soil Sci.*, 73, 555, 1993a.
13. Zentner, R.P., C.A. Campbell, S.A. Brandt, K.E. Bowren, and E.D. Spratt, Economics of crop rotations in western Canada, in Wheat Production in Canada — A Review, A.E. Slinkard and D.B. Fowler, Eds., Canadian Wheat Production Symposium, Saskatoon, 3–5 March, 1986, 254.
14. Zentner, R.P., E.D., Spratt, H. Reisdorf, and C.A. Campbell, Effect of crop rotation and N and P fertilizer on yields of spring wheat grown on a Black Chernozemic clay, *Can. J. Plant Sci.*, 67, 965, 1987.
15. Zentner, R.P., K.E. Bowren, W. Edwards, and C.A. Campbell, Effect of crop rotations and fertilization on yields and quality of spring wheat grown on a Black Chernozem in North Central Saskatchewan, *Can. J. Plant Sci.*, 70, 383, 1990.
16. Saskatchewan Soil Testing Laboratory, Nutrient Requirement for Field crops in Saskatchewan, Mimeo, Saskatchewan Soil Testing Laboratory, University of Saskatchewan, Saskatoon, 1990.
17. Large, E.C., Growth stages in cereals — illustration of the Feekes scale, *Plant Pathol.*, 3, 128, 1954.
18. Atkinson, H.J., G.R. Giles, A.J. McLean, and J.R. Wright, Chemical methods of soil analysis, Department of Agriculture, Ottawa, Publ. 169, 1958.
19. Tessier, S. and H. Steppuhn, Quick mount soil core sampler for measuring bulk density, *Can. J. Soil Sci.*, 70, 115, 1990.
20. Selles, F. and R.P. Zentner, Spring wheat yield trends in long-term fertility trials, *Can. J. Plant Sci.*, 73, 83, 1993.
21. Campbell, C.A., G.P. Lafond, and R.P. Zentner, Yield trends of wheat — impact of fertilization and legume, in Soils and Crops Workshop, University of Saskatchewan, Saskatoon, Feb. 25–26, 1993b.
22. Freyman, S., C.J. Palmer, E.H. Hobbs, J.F. Dormaar, G.B. Schaalje, and J.R. Moyer, Yield trends on long-term dryland wheat rotations at Lethbridge, *Can. J. Plant Sci.*, 61, 609, 1981.
23. Greer, K.J., Evaluating the Soil Quality of Long-Term Crop Rotations at Indian Head, M.Sc. thesis, University of Saskatchewan, Saskatoon, 1989.
24. Pennock, D.J. and E. de Jong, The influence of slope curvature on soil erosion and deposition in hummocky terrain, *Soil Sci.*, 144, 209, 1987.

25. El-Harris, M.K., V.L. Cochran, L.F. Elliott, and D.F. Bezdicek, Effect of tillage, cropping and fertilizer management on soil nitrogen mineralization potential, *Soil Sci. Soc. Am. J.*, 47, 1157, 1983.
26. Janzen, H.H., Soil organic matter characteristics after long-term cropping to various spring wheat rotations, *Can. J. Soil Sci.*, 67, 845, 1987b.
27. Millar, H.C., F.B. Smith, and P.E. Brown, The rate of decomposition of various plant materials in soils, *Am. Soc. Agron. J.*, 28, 914, 1936.
28. Nuttall, W.F., K.E. Bowren, and C.A. Campbell, Crop residue management practices, and N and P fertilizer effects on crop response and on some physical and chemical properties of a Black Chernozem over 25 years in a continuous wheat rotation, *Can. J. Soil Sci.*, 66, 159, 1986.
29. Persson, J. and L. Mattsson, Soil C changes and size estimations of different organic C fractions in a Swedish long-term small plot experiment, *Swed. J. Agric. Res.*, 18, 9, 1987.
30. Biederbeck, V.O., C.A. Campbell, and R.P. Zentner, Effect of crop rotation and fertilization on some biological properties of a loam in southwestern Saskatchewan, *Can. J. Soil Sci.*, 64, 355, 1984.
31. Janzen, H.H., Effect of fertilizer on soil productivity in long-term spring wheat rotations, *Can. J. Soil Sci.*, 67, 165, 1987a.
32. Campbell, C.A., Soil organic matter, nitrogen and fertility, in *Soil Organic Matter Developments*, M. Schnitzer and S.U. Khan, Eds., Soil Science Vol. 8, Elsevier, Amsterdam, 1978.
33. Van Veen, J.A. and E.A. Paul, Organic carbon dynamics in grassland soils. I. Background information and computer simulation, *Can. J. Soil Sci.*, 61, 185, 1981.
34. Poyser, E.A., R.A. Hedlin, and A.O. Ridley, The effect of farm and green manures on the fertility of Blackearth-meadow clay soils, *Can. J. Soil Sci.*, 37, 48, 1957.
35. Campbell, C.A., V.O. Biederbeck, M. Schnitzer, F. Selles, and R.P. Zentner, Effect of 6 years of zero tillage and N fertilizer management on changes in soil quality of an orthic Brown Chernozem in southwestern Saskatchewan, *Soil Tillage Res.*, 14, 39, 1989.
36. Lafond, G.P., R.P. Zentner, R. Geremia, and D.A. Derksen, The effects of tillage systems on the economic performance of spring wheat, winter wheat, flax and field pea production in east-central Saskatchewan, *Can. J. Sci.*, 73, 47, 1993.
37. Campbell, C.A., A.P. Moulin, D. Curtin, G.P. Lafond, and L. Townley-Smith, Soil aggregation as influenced by cultural practices in Saskatchewan: 1. Black Chernozemic soils, *Can. J. Soil Sci.*, 73, 579, 1993c.
38. Campbell, C.A., G.P. Lafond, R.P. Zentner, and Y.W. Jame, Nitrate leaching in a Udic Haploboroll as influenced by fertilization and legumes, *J. Environ. Qual.*, 23, 195, 1994.

Chapter

Crop Production and Soil Organic Matter in Long-Term Crop Rotations in the Semi-Arid Northern Great Plains of Canada

C.A. Campbell and R.P. Zentner

CONTENTS

I. INTRODUCTION

Cultivated soils of the Canadian prairies are young (about 10,000 years) and occupy an area of about 30 million hectares. Swift Current is situated in the Brown soil zone (Aridic Haploboroll) in southwestern Saskatchewan. The native vegetation of this region was mainly xerophytic and mesophytic grasses and forbs. These soils, which were first cultivated for agriculture at the beginning of the 20th century, are

0-8493-2802-0/97/$0.00+$.50

generally quite fertile and require only moderate additions of N and P for rainfed agricultural production. The main constraint to production is limited and unpredictable available water. This has resulted in the widespread use of summer fallow as a hedge against drought, but this practice has also contributed significantly to soil erosion, loss of soil organic matter, saline seep, and a decline in the inherent fertility of the soil.[1,2]

It was in response to the need to stabilize agricultural production and to reduce soil degradation in this semi-arid region that the Canadian Government established a Research Station at Swift Current in 1920. Although various crop rotation studies have been conducted over the years, the most comprehensive is the ongoing study which was initiated in 1966.[3] This study has produced a plethora of scientific publications dealing with plant growth and development, water conservation and water use efficiency, nutrient uptake, release and losses, economics, energy, and changes in soil physical, chemical, and biological properties;[4] it is not our intention to summarize all of these findings here.

Our objective is to discuss the findings of the Swift Current long-term crop rotation experiment as they relate to the dynamics of soil organic matter, and to relate these changes to the feasibility of managing crops so as to sequester C in soil.

II. GEOGRAPHY

Swift Current is situated at 50°17′N, 107°48′W, on the Trans-Canada highway, about 220 km west of Regina (Figure 1).

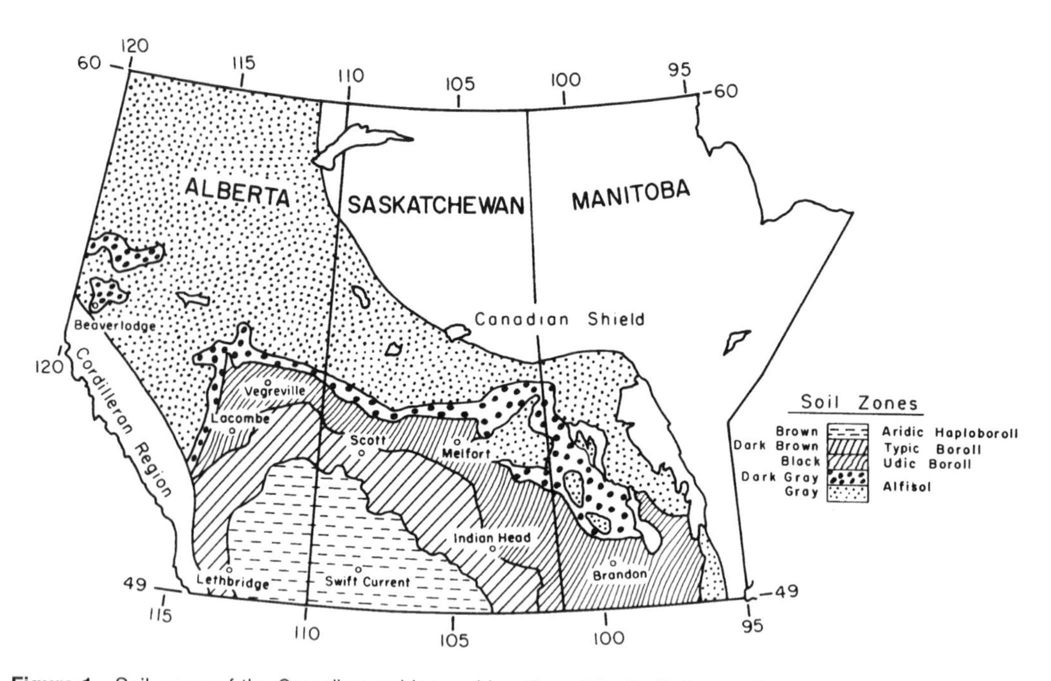

Figure 1 Soil zones of the Canadian prairies and location of the Swift Current long-term experiment.

The region is characterized by a continental climate with short, dry, hot summers and long, cold winters.[4] The warmest month is July and the coldest is January (Table 1). June is the month with greatest precipitation and July the month with highest potential evapotranspiration (ET_p). Temperatures fluctuate widely daily and seasonally. The mean annual temperature at Swift Current is 3.5°C, but the temperature difference between the warmest and coldest month is 33°C. Daily maximum-minimum temperature differences of 20 to 25°C are not unusual. About 50% of the annual precipitation falls from 1 May to 30 September, with about 30% falling as snow in winter. Snow is important because it provides soil moisture for spring-seeded crops, insulates the soil, and protects it against erosion. Potential evapotranspiration generally exceeds precipitation during the growing season, resulting in a large water deficit; the converse is true during winter months. At Swift Current the mean annual precipitation is 358 mm, ET_p is 729 mm, and moisture deficit 371 mm. Compounding the problems of soil degradation and limited water is the fact that Swift Current is one of the windiest locations in the Canadian prairies, with mean

Table 1 Mean Monthly Air Temperature, Precipitation, and Radiation and Long-Term Means for the Swift Current, Saskatchewan Site (1967 to 1991)

Period and Meteorological Characteristic	January	February	March	April	May	June	July	August	September	October	November	December	Mean[a]
						Temperature (°C)							
24-y mean													
Maximum	–8.1	–4.8	1.2	10.6	17.4	22.3	25.2	25.0	18.2	10.9	0.7	–5.7	9.4
Minimum	–17.4	–14.1	–8.4	–1.2	4.5	9.1	11.4	10.8	5.5	–0.4	–8.4	–15.1	–2.0
106-y mean													
Maximum	–8.3	–6.0	0.3	11.0	17.9	22.1	26.2	25.2	18.7	12.0	1.5	–4.8	9.7
Minimum	–18.5	–16.3	–10.1	–1.7	4.0	8.7	11.3	9.9	4.7	–0.8	–8.7	–14.5	–2.7
						Precipitation (mm)							
24-y mean	14.4	12.4	19.2	23.8	46.3	62.4	45.5	32.4	29.4	16.4	11.6	16.2	330
106-y mean	16.7	14.6	17.8	22.1	43.8	72.1	51.5	41.7	30.6	18.3	13.6	15.4	358
						Radiation (megajoules)							
24-y mean	4.8	8.4	13.1	17.7	21.2	23.2	23.7	19.2	13.0	8.6	5.2	3.7	171.6
27-y mean	4.9	8.5	13.6	17.9	21.3	23.1	24.1	19.4	13.4	8.9	5.2	3.8	164.4

Note: Values for radiation were only available for the period up to 1987.

[a] Mean temperatures and mean annual total precipitation and radiation.

annual wind speed of 22.9 km h^{-1} (10 m above ground level). This compares with Lethbridge, Alberta, at 20.4, Indian Head, Saskatchewan, at 15.8, and Melfort, Saskatchewan, at 15.4 km h^{-1}.[4] Frost-free days (>0°C) are equivalent to that at Lethbridge (117 days) being one of the highest on the Canadian prairies.

The soil at Swift Current is a Swinton loam to silt loam, an Orthic Brown Chernozem (Aridic Haploboroll) developed from aeolian deposits that overlie glacial till. The soil at the experimental site is situated on gently sloping land (<3%). It has an Ap horizon 0 to 8 cm thick with moderate, medium subangular, blocky primary structure that crushes to moderate, fine granular secondary structure. The Bm1 (8 to 24 cm) and Bm2 (24 to 48 cm) have a silt loam texture with strong, medium prismatic structure; the Ck (40 to 79 cm) has a silt loam texture with massive structure; and the 11C1 (79 to 104 cm) and 11C2 (104 to 129 cm) have clay loam texture with massive structure.[5]

III. DESCRIPTION OF EXPERIMENT

A. MANAGEMENT AND CROPPING

The site on which this experiment was conducted was broken out of native grassland in about 1903. Prior to the commencement of the experiment in 1966, the area was used for various cereal experiments, operated under a 2-year fallow-spring wheat (*Triticum aestivim* L.) system with weeds mainly controlled by conventional tillage management using a cultivator or rodweeder.

Details of the design and management of this experiment have been reported.[3,6–8] Therefore, only a review of aspects pertinent to the objective of this paper is presented here.

Twelve crop rotations were established on 81 0.04-ha plots in a randomized complete block design with 3 replicates; however, only 10 rotations are discussed here (Table 2). Two of the well-fertilized continuous wheat rotations were intended to be flexible rotations to include summer fallow based on annual assessment of soil water and weed infestations. Although during the first 12 years the criteria for summer fallowing these flexible rotations were met on several occasions, the action was not implemented; consequently, there were three similar continuous wheat systems (Table 2). In 1979, two of these rotations were changed to a spring wheat-grain lentil (*Lens culinaris* L.) rotation. All phases of each rotation (i.e., crops or fallow sequence in the rotation) were present every year and each rotation was cycled on its assigned plots.

Commercial farm equipment was used to perform cultural and tillage operations. Weed control was achieved by a combination of mechanical tillage (mainly cultivator and rod weeder) and herbicides (as required) using recommended methods and rates. The continuous cropping rotations required more herbicides for weed control than the other treatments.[7] The plots were seeded at the recommended rates (kg ha^{-1}) of 67 for spring wheat, 31 for flax, 63 for fall rye (*Secale cereale* L.), 67 for winter wheat, and 90 for lentil. Crops generally were seeded in early May, except for rye and winter wheat which generally were seeded during the 1st week of September. Recommended cultivars were used each year, but cultivars were changed over time as new ones became available.

Fertilizer N, as ammonium nitrate, was applied in accordance with rotation specifications (Table 2) at rates recommended by the Saskatchewan Advisory Council on Soils, based on soil NO_3-N (0- to 0.6-m) of the individual plots measured in fall. Ammonium phosphate was applied with the seed at an annual rate of 10 kg P ha^{-1}, to rotations that were designated to receive P (Table 2).

Crops were harvested at the full-ripe stage. Lentil was desiccated with herbicide to hasten ripening. On all other cropped plots, yield determinations were made by cutting a swath 5 m wide and 40 m long through the middle of the plot and harvesting the grain with a conventional combine. Smaller 2.32-m^2 areas also were harvested and used to determine straw/grain ratios and straw N concentrations. The straw/grain ratios were used together with combine grain yields to estimate combine straw yields. The straw was distributed on the plots by a paddle-type spreader attachment on the combine. The grain was weighed and its N concentration (Kjeldahl) determined.[9]

B. MONITORING SOIL ORGANIC MATTER

The pH of the 0 to 0.15 m of this soil is 6.0 in dilute $CaCl_2$ and the water held at –0.03 and –1.5 MPa are 230 and 100 g kg^{-1}, respectively.

Soil samples from the 0- to 0.15- and 0.15- to 0.3-m depths were removed with a soil probe from selected plots (Table 2) in spring 1976, in fall 1981, in 1984 (some in fall and some in spring 1985), and in fall 1990. In 1990 we sampled all rotations. The soil was air-dried, sieved (<2 mm), and ground to <1 mm with a Jr. Laboratory mill size 5, (Christy and Norris Ltd. Engineers, Chelmsford, England). Small pieces of crop residues passing through the 2-mm sieve were regarded as soil organic matter. In

Table 2 Crop Rotations and Treatments Sampled at Swift Current

Rotation[a]	Criteria	Average Fertilizer Applied to Rotation (1967–1990, kg ha^{-1} y^{-1}) N	P
F-(W)	N and P applied as required	3.2	6.4
F-W-(W)	N and P applied as required	11.2	6.4
(F)-W-(W)	P applied as required but no N applied except that in P fertilizer	3.9	6.4
F-W-W	N applied as required, no P applied	6.2	0
F-(Rye)-W[b]	N and P applied as required	10.4	6.0
CF-WW-WW	N and P applied as required	9.4	4.8
F-Flx-(W)	N and P applied as required	9.5	5.6
Cont (W)	N and P applied as required	29.7	9.6
Cont (W)[c]	(Fallow if less than 60 cm moist soil exists at seeding time: N and P applied as required)	27.5	9.6
Cont (W)[c]	(Fallow if grassy weeds become a problem: N and P applied as required)	29.2	9.6
(W)-Lent	N and P applied as required	20.4	9.1
Cont (W)[d]	P applied as required but no N applied except that in P fertilizer	9.0	9.6

[a] Selected plots, indicated in parentheses, were sampled for straw weight and straw N and P concentration at harvest. F = fallow; CF = chemical fallow; W = spring wheat; WW = winter wheat; Rye = fall rye; Flx = flax; Lent = grain lentil; Cont - continuous.

[b] After 1984, this rotation was changed to chemical fallow-winter wheat-winter wheat (spring wheat whenever winter wheat failed to survive the winter).

[c] During the first 12 years, the criteria necessary for summer fallowing in these two rotations were met on several occasions but the action was not implemented. In 1979, these two rotations were changed to the spring wheat-lentil rotation.

[d] In 1980 and 1982 nitrogen was inadvertently applied to this system at rates of 70 and 40 kg N ha^{-1}, respectively.

From Campbell, C.A. and Zentner, R.P., *Soil Sci. Soc. Am. J.*, 57, 1034, 1993. With permission.

the year of sampling, total soil N concentration (g kg^{-1}) was measured by Kjeldahl analysis,[9] inorganic C was measured as described by Allison,[10] and total C as described by McKeague.[11] For inorganic C the soil was treated with dilute HCl containing $SnCl_2$ catalyst. The CO_2 evolved was absorbed in NaOH, the resulting carbonate precipitated with 15% $BaCl_2$, and the nonreacted NaOH titrated with standard HCl. Total C was measured by mixing a weighed sample of soil with alundum R.R., placing this in a porcelain boat, covering the mixture with a thin layer of cupric oxide, and combusting the sample in a Lindberg combustion furnace at 925°C. Air, scrubbed free of CO_2, was used to flush the evolved CO_2 through NaOH and the analysis of the trapped CO_2 conducted as described for inorganic C. Organic C was calculated as the difference between total and inorganic C, but the latter varied from zero to trace amounts in this surface soil. Throughout the years, a standard soil sample was included in all chemical analyses as a control. The soil organic C and N concentrations were converted to weight basis using bulk densities of 1.22 and 1.30 Mg m^{-3} for the 0- to 0.15- and 0.15- to 0.30-m depths, respectively. The bulk densities were determined by taking two cores per plot using a Giddings soil probe with 0.05-m diameter cylinders. The values presented are the mean of 39 plots.[12]

C. NITRATE LEACHING

On 31 July 1990 a Giddings soil probe was used to remove soil samples by 0.3-m increments to a depth of 3 m from the fallow phase of the well-fertilized F-W-W, F-Flx-W, and CF-WW-WW rotations and from plots in the wheat phase of (W)-Lent and well-fertilized Cont (W). (See Table 1 for rotation identification.) Two cores were taken per plot in each of the three replicates and the samples analyzed separately. Subsamples were taken from each soil segment, a known volume isolated, dried at 105°C overnight for bulk density and soil water content (gravimetrically) determination; a separate subsample was used to determine NO_3-N.[13] Nitrate-N values were converted from gravimetric to volumetric basis using the measured bulk densities. Treatment had no effect on bulk density which averaged 1.29, 1.30, 1.50, 1.59, 1.64, 1.64, 1.67, 1.78, 1.73, and 1.68 Mg m^{-3} for the ten successive 0.3-m depth segments, respectively.

D. WEATHER CONDITIONS

Meteorological parameters of precipitation, solar radiation (Eppley), and maximum and minimum air temperatures were assumed to be the same at the site as those recorded at the meteorological station located 1 km away.

E. STATISTICAL ANALYSES

The various plant and soil characteristics were analyzed as a split-plot with the year as main plot and rotation phase as subplot.[14] These analyses were conducted on grain yield, grain N content (i.e., grain yield × N concentration), and soil organic C and N. For the plant variables, the analyses were done separately for wheat grown on fallow and wheat on stubble. Yield trends were also calculated as 5-year running means. A regression of straw yields on grain yields, using all data collected over the years, was conducted.

To assess the influence of rotation phase on soil organic matter C and N, samples taken in the five 3-year rotations in 1990 were analyzed as a split-plot with rotation as main plots and rotation phase as subplots.

To assess the effect of rotation on NO_3-N leaching in the well-fertilized F-W-W, F-Flx-W, and CF-WW-WW rotations, the quantities of soil water and NO_3-N in each 0.3-m segment to the 3-m depth, for samples taken in July 1990, were analyzed as a split-plot with rotation as main plots and depth as subplots. Standard errors ($S\bar{x}$) or LSD values were calculated for significant treatment effects in split-plot experiments.[15]

IV. RESULTS AND DISCUSSION

A. WEATHER CONDITIONS

The 24-year (1967 to 1990) mean monthly maximum/minimum temperatures were similar to the long-term (106-year) mean values for this area (Table 1). The 24-year mean annual precipitation was 330 mm, 28 mm less than the 106-year mean, and the 24-year mean growing season (1 May to 31 July) precipitation was 154 mm compared to the long-term mean of 167 mm. Severe droughts occurred in 1968, 1980, 1984, 1985, and 1988 (data not shown). Growing season precipitation was greater than average in 1970, 1976, 1977, 1981 to 1983, 1986, 1989, and 1990.

B. GRAIN YIELDS, N CONCENTRATIONS, AND N CONTENTS

The 24-year mean grain yields, N concentrations, and N contents of wheat grown on fallow and wheat grown on stubble were significantly influenced by the year × rotation-phase interaction; however, for simplicity we have only shown the 24-year means (Table 3). Wheat grown on fallow averaged 1880 kg ha^{-1} when P was nonlimiting; when P was limiting, grain yields averaged 10% less. Yields of wheat grown on stubble for the well-fertilized rotations containing fallow averaged 1380 kg ha^{-1} and were similar irrespective of the previous crop. Cont W (N + P) and F-W- (W) (+P) had similar yields which were 5% less than yields of stubble-wheat in well-fertilized 3-year rotations, while Cont W receiving little N in 24 years had average yields that were about 83.5% of the latter three rotations, or 61% of well-fertilized wheat grown on fallow. Yields were more variable on stubble (CV = 49%) than on fallow (CV = 36%).

There was little difference in the 24-year mean grain N concentrations of wheat grown on fallow irrespective of the N fertilizer regime (mean = 28.3 g kg^{-1}, Table 3). Grain N concentrations of wheat grown on stubble in well-fertilized F-W-W and F-Flx-W rotations were as high as those for wheat grown on fallow; N concentration was lowest for Cont W receiving minimal N fertilizer (24.7 g kg^{-1}) and was between 27.0 and 27.6 g kg^{-1} for the other four stubble-wheat rotations.

Grain N content (grain weight × N concentration) is mainly a function of grain weight; thus, these responses were similar to those described for grain weight (Table 3). The 24-year average N content for well-fertilized wheat grown on fallow was 52 kg ha^{-1} and, for well-fertilized wheat grown on stubble, 38 kg ha^{-1}. Failure to apply P to wheat grown on fallow significantly reduced grain N content, and failure to apply adequate N or P to stubble-wheat decreased N content, especially for Cont W.

Comparison of grain yields and grain N for the W-Lent (N+P) rotation with those in the corresponding Cont W (N+P) rotation, for the 12-year period 1979 to 1990, showed that there has been no significant effect of grain lentil on grain yields of wheat in the rotation (Table 3). However, the grain N concentration of wheat was increased by including lentil in the rotation, though the grain N contents were similar. Lentil grown in rotation with wheat produced about half as much grain dry matter as wheat, but its grain N concentration (almost 40 g kg^{-1}) was about 36% greater than that of well-fertilized Cont W.

Table 3 Mean Dry Weight, N Concentration and N Content of Grain and Straw for Crops Grown on Fallow and Stubble, and Amount of N Applied as Fertilizer and Exported in Grain for Rotations at Swift Current, SK During the Period 1967–1990

Rotation	Fertilizer		Dry Weight (kg ha^{-1})		N Concentration (g kg^{-1})		N Content (kg ha^{-1})		N Applied to or Exported in Grain from the Rotation in 24 years[a] (kg ha^{-1})	
	N	P	Straw	Grain	Straw	Grain	Straw	Grain	Applied	Export
F-(W)	Yes	Yes	3353	1881	4.9	28.6	16.4	53.8	84	646
F-(W)-W	Yes	Yes	ND	1896	ND	28.1	ND	53.3		
F-W-(W)	Yes	Yes	2406	1387	5.6	28.2	13.5	39.1	270	739
F-(W)-W	No[b]	Yes	ND	1865	ND	28.2	ND	52.6	93	703
F-W-(W)	No[b]	Yes	2117	1306	5.1	27.0	10.8	35.3		
F-(W)-W	Yes	No	ND	1697	ND	28.2	ND	47.9	150	661
F-W-(W)	Yes	No	ND	1257	ND	27.6	ND	34.7		
F-(Rye)-W[c]	Yes	Yes	ND	1749 (2118)	ND	22.5 (26.9)	ND	39.4 (56.6)	244	651
F-Rye-(W)	Yes	Yes	ND	1390	ND	27.1	ND	37.7		
F-(Flx)-W	Yes	Yes	ND	756	ND	39.7	ND	30.0	227	555
F-Flx-(W)	Yes	Yes	2320	1360	5.9	29.0	13.7	39.4		
Cont (W)	Yes	Yes	2260	1318	5.6	27.6	12.7	36.4	713	874
(W)-Lent[d]	Yes	Yes	1786	1208	6.4	29.8	11.4	36.0	586	803
W-(Lent)	Yes	Yes	1275	643	9.5	39.5	12.1	25.4		
Cont (W)	No[b]	Yes	1689	1153	4.1	24.7	6.9	28.5	216	684

Note: Unless indicated otherwise, the means are for the 24-year period; ND = not determined.

[a] For Cont W = grain N × 24; for F-W = grain N × 12; for F-W-W = Sum (grain N in fallow-wheat plus that in stubble-wheat) × 8. For W-Lent, the Cont W (N+P) data were used to represent the first 12 years.

[b] These treatments received minimal amounts of N as part of the P fertilizer; as well, Cont W inadvertently received other N in 1980 and 1982 (see Table 2).

[c] F-Rye-W for 18 years, changed to CF-WW-WW thereafter. The mean values in parentheses are for the 6 years of WW. F = fallow; CF = chemical fallow; W = spring wheat; WW = winter wheat; Rye = fall rye.

[d] Values are for the 12-year period (1979 to 1990) after Cont W (N+P) was changed to W-Lent. Cont = continuous; Lent = grain lentil.

From Campbell, C.A. and Zentner, R.P., *Soil Sci. Soc. Am. J.*, 57, 1034, 1993. With permission.

Nonetheless, due to its relatively low grain yield, the N exported in the lentil grain was only about two thirds of that exported in the grain of well-fertilized stubble wheat (25 vs. 38 kg N ha^{-1}).

Grain yield of flax grown on fallow in well-fertilized F-Flx-W rotation averaged 756 kg ha^{-1} and the N concentration of the grain was similar to that of grain lentil. The 24-year average N content of the flax grain was 30 kg ha^{-1}, 83% of that of well-fertilized stubble wheat.

In one of the well-fertilized, 3-year fallow-containing rotations, fall rye was grown for the first 18 years during which time its mean grain yield was 1749 kg ha^{-1}, its mean N concentration 22.5 g kg^{-1}, and its mean N content 39 kg ha^{-1} (Table 3). If we exclude 1 year when grasshoppers damaged the rye, the mean yield of rye was 1853 kg ha^{-1} (similar to spring wheat on fallow), but its N concentration was even lower than that of wheat grown on stubble. In this rotation, the yield of winter wheat grown on chemical fallow in the 6 years (1985 to 1990) was 2120 kg ha^{-1} and its N concentration 26.7 g kg^{-1} for a mean N export in grain of 57 kg ha^{-1}. However, this value should not be directly compared with the others collected over much longer periods.

The amount of grain produced per rotation per year was calculated by summing the yields for each rotation phase (including no yields for fallow) and dividing by the number of rotation phases in the rotation. For example, for Cont W the grain production would be the same as the unit area yield, while for F-W it would be one half the unit area yield. On this basis, production increased with cropping frequency and with fertilization (Figure 2). These results are similar to those reported for most other rotation experiments on the Canadian prairies except at Lethbridge where F-W produces as well as unfertilized Cont W and more than unfertilized F-W-W.[4] Substituting fall rye and winter wheat for spring wheat in the fertilized 3-year rotation had little influence on grain production per year, but substituting the low-yielding oilseed crop (flax) for spring wheat markedly decreased grain production per year.

Wheat yields tended to increase with time over the first 8 to 11 years of the experiment, but generally declined thereafter in stubble-wheat systems irrespective of the fertilizer treatment (Figure 3). Zentner et al.[16] showed that yields of stubble wheat were significantly ($P < 0.05$) correlated with stored soil moisture in the spring, but this was not so for wheat grown on fallow. They also showed that there were no significant time trends in wheat yields once they had accounted for the effects of stored soil moisture and growing season precipitation. Although the benefits of P fertilizer on F-W-(W) and N fertilizer on Cont W were apparent (Figure 3), there was no indication that fertilizer has so far caused an improvement in soil fertility nor that nonoptimal fertilization is leading to an impoverished soil. In this semi-arid environment, with its low nutrient export (harvest, leaching, and denitrification), we will likely need to wait a long time for this type of evidence to become apparent.

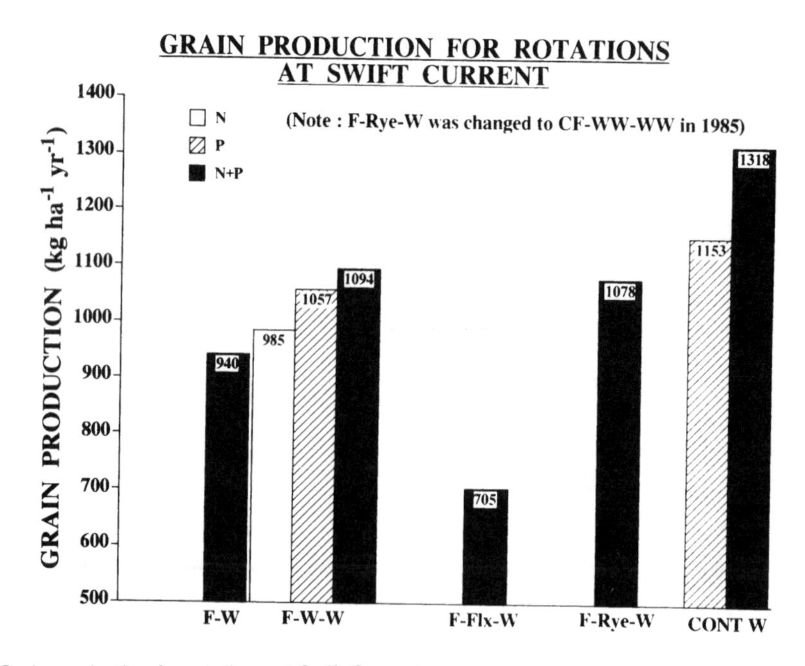

Figure 2 Grain production for rotations at Swift Current.

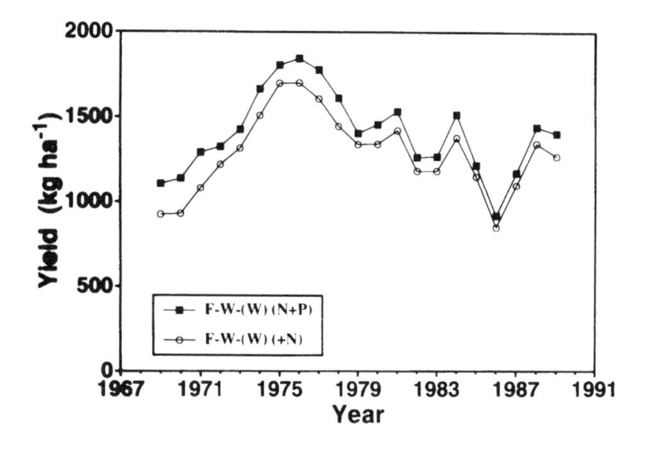

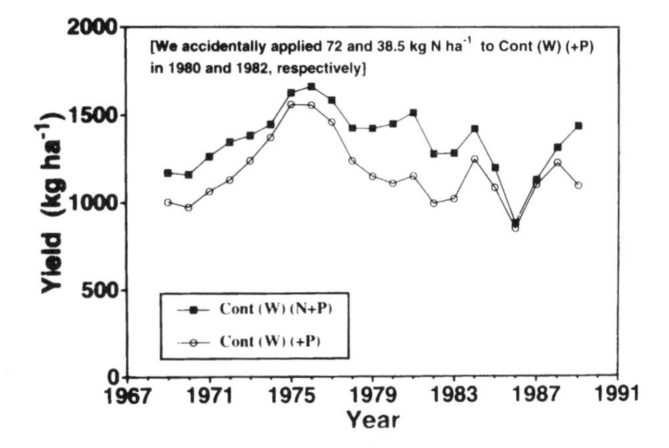

Figure 3 Yield trends of hard red spring wheat grown on stubble at Swift Current showing influence of fertilizer. (The points are 5-year running means.) In 1988 there was complete crop failure due to drought.

C. STRAW YIELDS, N CONCENTRATIONS, AND N CONTENTS

Straw samples were taken annually (1967 to 1990) only on the six selected plots (one fallow and five stubble seeded) shown in Tables 2 and 3. As well, straw samples were taken from both phases of the W-Lent rotation, starting in 1979 (Table 3). Straw yields for wheat grown on fallow in well-fertilized systems averaged 3353 kg ha^{-1}, 44% greater than that for well-fertilized wheat grown on stubble. Withholding N from wheat grown on stubble in the F-W-W rotation caused a 10% decrease in straw yields compared to well-fertilized stubble-wheat, but withholding N on Cont W was much more severe, decreasing straw yields by 27%. Straw yields of wheat grown in the W-Lent rotation were not significantly different from those of well-fertilized Cont W, but straw yields of lentil were only about two thirds of those of well-fertilized wheat grown on stubble.

A regression based on 24-year data, relating straw yields (SY) to grain yields (GY), both in kilograms per hectare (Figure 4), showed that

$$SY = 154 + 1.50\ GY\ (r^2 = 0.70^{***},\ n = 424) \qquad (1)$$

This relationship gives a slightly lower regression coefficient than was obtained after 18 years[4] (i.e., 1.50 vs. 1.61) and is similar to that obtained by Nuttall et al.[17] in a 25-year tillage × fertilizer study conducted on the thick Black Chernozem (Udic Boroll) at Melfort. The linear relationship also differs from a quadratic equation that was obtained in a 9-year snow management × fertilizer N study in which spring wheat was grown annually on zero tillage at Swift Current.[18] We believe that the difference between these two equations may be due to the much higher rates of N used in the zero tillage study; this would have resulted in greater straw/grain ratios at higher yields (i.e., higher moisture and N may result in luxury consumption of N and relatively greater production of vegetative matter compared to grain).

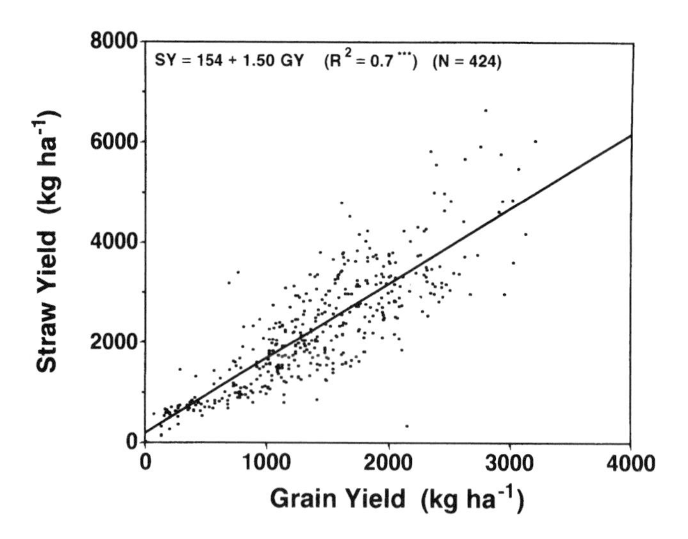

Figure 4 Relationship between straw yields and grain yields based on 24 years of data from the six special plots.

Straw N concentrations for wheat grown on fallow averaged about 5.0 g kg^{-1} and for well-fertilized wheat grown on stubble about 5.8 g kg^{-1} (lower straw yields than for fallow wheat) while for Cont W that received minimal N it averaged 4.1 g kg^{-1} (Table 3). These values are in the generally accepted range of about 5.0 g kg^{-1} N in cereal straw at maturity. When wheat was grown in rotation with grain lentil, its straw N concentration increased significantly (like the grain) averaging 6.4 g kg^{-1}. Nitrogen concentration of lentil straw was about 9.5 g kg^{-1} on average. This higher N concentration was not only because this crop is a legume, but also because its indeterminate growth habit (in contrast to wheat) results in its continued uptake of N long after wheat has stopped.

The N content of straw for wheat grown on fallow averaged 16.4 kg ha^{-1} while that for well-fertilized wheat grown on stubble averaged about 13.3 kg ha^{-1}; this value was almost halved when Cont W was grown with little N. Nitrogen content of lentil was only marginally greater than that for well-fertilized Cont W, due to the relatively low dry matter yields of lentil straw.

D. SOIL ORGANIC MATTER CHANGES

When statistically significant changes in soil organic matter occurred these were only in the top 0.15-m depth. Thus, only data for this depth are presented. Analyses of the five 3-year rotations in 1990 to determine the effect of rotation phase (within rotations) showed that these were not significant ($P > 0.05$).

In the 0- to 0.15-m segment, organic N and C were significantly affected ($P<0.05$) by rotation (Table 4). As reported by Biederbeck et al.,[6] the well-fertilized continuously cropped systems and the rotation containing fall cereals (winter wheat or fall rye) had the highest soil organic matter contents. The 3-year system containing the low-yielding flax crop (Table 3) had the lowest soil organic matter C and N, while fallow-spring wheat systems and the inadequately fertilized Cont (W) were intermediate (Table 4).

Measured organic N values were significantly different at different sampling times ($P<0.10$) but the interaction of year × rotation was not significant. Nonetheless, the trends in organic N for the rotation including fall cereals (F-Rye-W) appeared to differ from the other systems; therefore, we presented the data as shown in Figure 5. Trends in organic C did not differ significantly with year of sampling, nor was the year × rotation interaction significant ($P > 0.10$), but because the organic C trends were generally similar to the organic N trends we also presented these interactions (Figure 6).

Biederbeck et al.[6] estimated the initial organic N concentration in the upper 0- to 0.15-cm depth to be 1.78 g kg^{-1} (i.e., 3.26 t ha^{-1}). On the basis of this value it appears that, in the first 15 years, there may have been an increase in organic N in the well-fertilized continuously cropped systems and possibly in the F-Rye-W system, while the other fallow-containing systems remained essentially unchanged (Figure 5). The soil organic N content in the F-W-W (N+P), F-W-W (+P), and Cont (W) (+P) (not shown) behaved in a manner similar to F-W (N+P). Since 1981 all systems, except the one that included the fall-seeded cereals, have shown a steady decline in organic N; the F-Rye-W system has remained constant.

Table 4 Effect of Rotation on Soil Organic C and N Content of 0- to 0.15-m Depth at Swift Current

Rotation Sampled[a]	Fertilizer		Organic Component[b] (t ha^{-1})	
	N	P	C	N
F-(W)	Yes	Yes	30.9	3.18
F -W-(W)	No[c]	Yes	30.2	3.19
F-W-(W)	Yes	No	28.7	3.17
F-W-(W)	Yes	Yes	31.4	3.26
F-Flx-(W)	Yes	Yes	28.6	3.01
F-Rye-(W)[d]	Yes	Yes	33.0	3.46
Cont (W)	No[c]	Yes	31.4	3.23
Cont (W)	Yes	Yes	34.3	3.37
(W)-Lent[e]	Yes	Yes	35.2	3.58
LSD (P<0.05)			2.6	0.22

Note: Values averaged across years. Only data for 0 to 0.15 m showed consistent and significant effects; thus, only these are shown. The average organic C and N in the 0.15- to 0.3-m depth were 18.3 and 2.13 t ha^{-1}, respectively.

[a] The phase in parentheses was sampled. F = fallow; W = spring wheat; Flx = flax; Rye = fall rye; cont = continuous; Lent = grain lentil.

[b] Analysis was based on samples taken in 1976, 1981, 1984, and 1990. There were no significant year × rotation interactions but rotation was significant for both C and N.

[c] These treatments received minimal N with the P source and the Cont W inadvertently received other N in 1980 and 1982 (see Table 2).

[d] This rotation was F-Rye-W from 1967 to 1984, then it was changed to CF-WW-WW (spring wheat if WW fails).

[e] This was Cont W (N+P) from 1967 to 1978, then it was changed to W-Lent (N+P).

From Campbell, C.A. and Zentner, R.P., *Soil Sci. Soc. Am. J.*, 57, 1034, 1993. With permission.

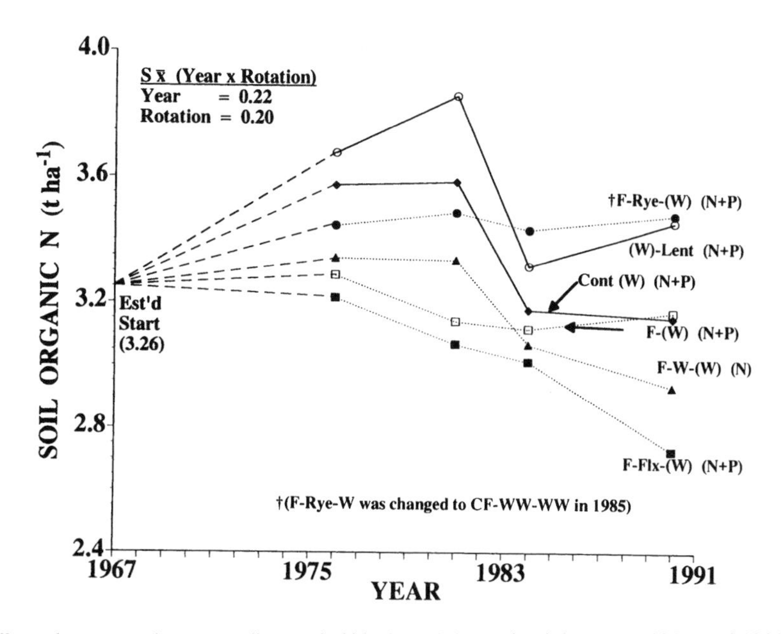

Figure 5 Effect of crop rotations on soil organic N in 0- to 0.15-m depth between 1967 and 1990. The rotation phase in parentheses was the one sampled. F = fallow; CF = chemical fallow; W = spring wheat; WW = winter wheat; Rye = fall rye; Flx = flax; Lent = grain lentil; Cont = continuous. Year was significant at P<0.10 and rotation at P<0.05; although year × rotation was not significant, the results are presented in this manner because the F-Rye-W system appeared to behave differently from the other systems. (From Campbell, C.A. and Zentner, R.P., *Soil Sci. Soc. Am. J.*, 57, 1034, 1993. With permission.)

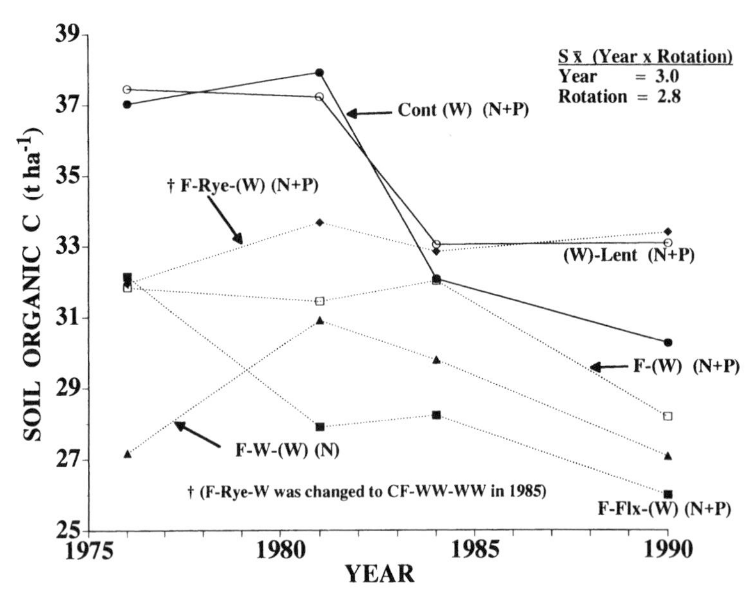

Figure 6 Effect of crop rotations on soil organic C in 0- to 0.15-m depth between 1967 and 1990. The rotation phase in parentheses was the one sampled. F = fallow; CF = chemical fallow; W = spring wheat; WW = winter wheat; Rye = fall rye; Flx = flax; Lent = grain lentil; Cont = continuous. Neither year nor year × rotation was significant ($P<0.10$); however, the data are presented in this manner to complement the data shown in Figure 5. (From Campbell, C.A. and Zentner, R.P., *Soil Sci. Soc. Am. J.*, 57, 1034, 1993. With permission.)

The decreasing trend in organic matter since 1981 coincides with a period when grain yields, and thus crop residue production, were very low in several years (Table 5). For example, up to 1981, grain yields for stubble crops exceeded 500 kg ha^{-1} in all years and 1000 kg ha^{-1} in 13 of the 15 years. However, in the 9 years thereafter, there were 3 years with yields of wheat grown on stubble being less than 700 kg ha^{-1}, including 1 year of complete crop failure and two consecutive years of low yields. These situations are very likely to result in soil erosion. Because crop residues are the main source of energy for soil microorganisms that build organic matter in soil, conditions favoring crop production would also favor maintenance or build-up of soil organic matter.[20,21] The relationship between soil organic matter C and crop residues as $r^2 = 0.67^*$ (data not shown).

The increased crop production associated with proper fertilization, more frequent cropping, and growing spring wheat compared to the lower-yielding flax (Table 3) was readily reflected in the maintenance of higher soil organic matter levels (Table 4). These results support those reported in the literature.[20–22] They can be explained in terms of the reduction in erosion accompanying reduced fallow frequency[21–23] and partly in terms of the difference in organic substrate (crop residue) inputs accruing from these systems (Figure 7).[20,21]

E. NITROGEN BALANCE

A true N balance for a grain crop system must consider inputs such as fertilizers, N fixation (symbiotic and nonsymbiotic), and dry and wet deposition, while outputs will include N lost in grain, by leaching, erosion, and gaseous means. For non-legume systems, the major source of N input in the semi-arid prairies will be fertilizers, and the main source of output will be N in harvested grain. To obtain a first approximation of how the various cropping systems in this study may be expected to influence the soil organic N, we calculated an apparent N deficit for each rotation, defined as N exported in the grain minus N applied as fertilizer. The calculated amount of N exported in grain and the amount of N applied as fertilizer for each rotation over the 24-year period (1967 to 1990) are shown in Table 3. A regression of soil organic N (average over the four sampling dates) vs. apparent N deficit (Figure 8) showed that soil organic matter N decreased as apparent N deficit increased. These results are similar to those reported by Rasmussen et al.[20] All systems, including those that were continuously cropped and fertilized annually according to soil test, showed an apparent N deficit. This implies that, if soil organic matter is to be maintained, not only must soil erosion be controlled, but sufficient N must be applied to the system to compensate for that removed in the grain. These results do not conform to those reported in the Lethbridge

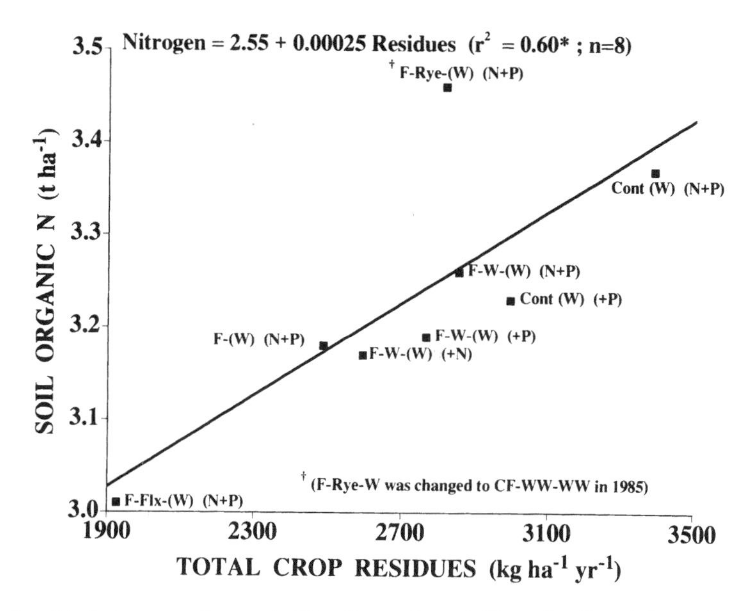

Figure 7 Relationship between soil organic N in the surface 0.15-m of soil and the estimated annual crop residue (including roots) produced by rotations in the period 1967 to 1990. (The W-Lent rotation was excluded because of inadequate data to estimate lentil residues.) The rotation phase in parentheses was the one sampled. F = fallow; CF = chemical fallow; W = spring wheat; WW = winter wheat; Rye = fall rye; Flx = flax; Lent = grain lentil; Cont = continuous. (From Campbell, C.A. and Zentner, R.P., *Soil Sci. Soc. Am. J.*, 57, 1034, 1993. With permission.)

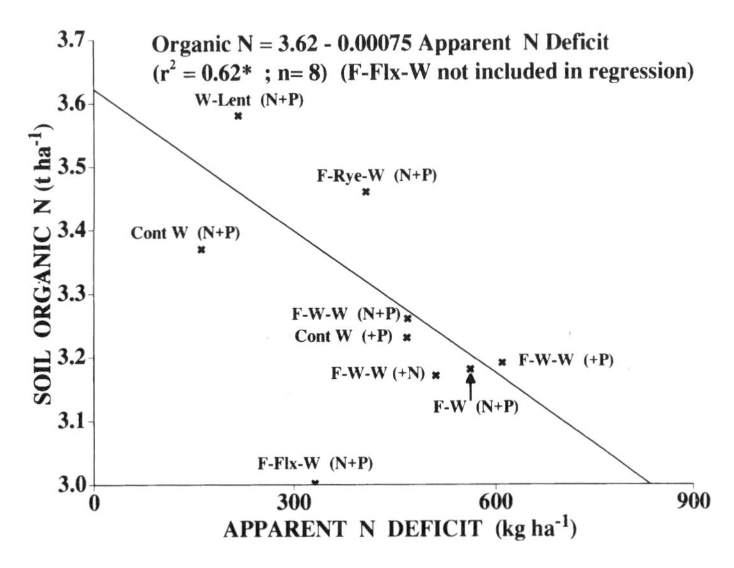

Figure 8 Relationship between soil organic N in the surface of 0.15 m of soil and apparent nitrogen deficit (i.e., nitrogen exported in grain minus nitrogen applied as fertilizer). F = fallow; CF = chemical fallow; W = spring wheat; WW = winter wheat; Rye = fall rye; Flx = flax; Lent = grain lentil; Cont = continuous. (From Campbell, C.A. and Zentner, R.P., *Soil Sci. Soc. Am. J.*, 57, 1034, 1993. With permission.)

long-term rotation experiment where yields of spring wheat grown without fertilizers for over 80 years have trended upward.[17] These authors speculate that there may be unknown inputs of N entering their rotation system.

Although the regression between soil organic N and crop residues (Figure 7) suggests an average relationship for the F-Flx-W rotation, Figure 8 suggests that N export from this rotation has been much greater than from grain export alone. Soil cores taken to 3-m depth in 1990 showed that there was a significantly greater amount of NO_3-N leached beyond the root zone under the F-Flx-W (N+P) compared to F-W-W (N+P), the difference being about 68 kg ha^{-1} (Figure 9). The latter amount is 27% of the 250 kg ha^{-1} estimated difference in soil organic N content (0- to 0.15-m depth) between these two rotations (Table 4). Although the flax is cut much closer to the ground than wheat, there should be minimal erosion occurring in these two rotations between the harvesting of the fallow crop and the seeding of the stubble crop. When flax is harvested, the straw tends to form "balls" which do not readily get "fixed" into the soil surface; thus, the straw can be blown from the plots by the frequent strong winds that are typical of the region. This contrasts with cereal straw which tends to remain in place on the plots. Further, the flax straw does not appear to decompose as readily as cereal straw (visual observation of farm foreman, Mr. Del Jensen). The net result may be greater physical loss of the N in the flax straw (compared to cereal straw) from the system. Similarly, we believe there was a loss of C from this F-Flx-W system, evidence of which is shown in the soil organic C data (Figure 6).

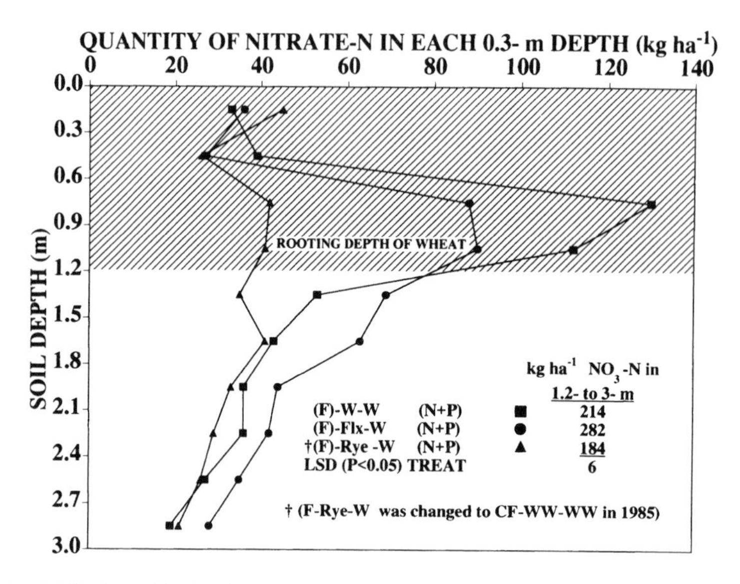

Figure 9 Nitrate distribution with depth in soil as influenced by flax or fall rye in rotation with spring wheat and summer fallow. Sampled 31 July, 1990. The rotation phase in parentheses was the one sampled. F = fallow; CF = chemical fallow; W = spring wheat; WW = winter wheat; Rye = fall rye; Flx = flax; Lent = grain lentil; Cont = continuous. (From Campbell, C.A. and Zentner, R.P., *Soil Sci. Soc. Am. J.*, 57, 1034, 1993. With permission.)

The 3-year rotation that contained the fall-seeded cereals had the highest organic N and C among the fallow-containing rotations (Table 4) and the organic matter levels remained constant throughout the study period (Figures 5 and 6). This suggests that the decreases after 1980 shown by the other rotations were not due to analytical error. The response of the rotation containing the fall-seeded cereals may be related to the efficiency with which fall-seeded crops use mineral N (taking up N till late fall and starting again early in spring leaves little soluble mineral N to be leached by snow-melt water or early spring rains). Thus, we found that the amount of NO_3-N that had been leached into the subsoil of this system during the first 16 years was less than under all other systems examined,[12] and this remained true up to 1990, 6 years after the system had been changed to chemical F-WW-WW (Figure 9). Further, the fall-seeded crops protect the soil against erosion because of the shorter fallow period (14 months compared to 21 for spring wheat). This would be especially advantageous after dry years such as 1980, 1984, 1985, and 1988 when little crop residue was produced (Table 5).

The wheat-lentil (N+P) rotation tended to have a higher soil organic N than Cont W (N+P), possibly because of the N fixed by the lentil, or due to the more efficient use of mineral N by the cereal-legume rotation, as evidenced by less NO_3-N leaching below the root zone of this system than under Cont W (N+P).[8] This may be due to greater synchrony of N uptake in W-Lent than in Cont W (N+P). If so, this augurs well for the use of W-Lent rotations for sustainable agriculture in areas where such systems can be employed without disease problems occurring.

Table 5 Grain Yields and Estimated Crop Residues (Straw and Roots) in Wheat Grown on Well-Fertilized Fallow (Fallow-Wheat [F-(W)]) and Stubble (Continuous Wheat [Cont (W)]) at Swift Current from 1967 to 1990

	Grain Yield (kg ha^{-1})		Estimated Crop Residues[a] (kg ha^{-1})	
Year	F-(W) (N+P)	Cont (W) (N+P)	F-(W) (N+P)	Cont (W) (N+P)
1967	1260	1285	3250	3310
1968	1370	580	3510	1620
1969	2010	1455	5040	3720
1970	1525	1300	3880	3340
1971	1870	1230	4700	3180
1972	1570	1240	3990	3200
1973	1230	1095	3180	2850
1974	2010	1865	5030	4690
1975	1640	1480	4160	3770
1976	2765	1560	6840	3960
1977	2835	2145	7010	5360
1978	1750	1275	4420	3280
1979	1995	1480	5000	3770
1980	1860	675	4680	1850
1981	2225	1550	5550	3940
1982	3020	2275	7450	5680
1983	2090	1595	5230	4050
1984	1140	290	2970	940
1985	765	690	2070	1890
1986	2615	2250	6480	5610
1987	1945	1165	4880	3030
1988	820	0	2210	300
1989	1970	1545	4950	3920
1990	2860	1605	7070	4070

[a] Straw yields = grain yields × 1.50 (see Figure 4). Root yields = straw yields × 0.59. Crop residues = straw + root yields.

From Campbell, C.A. and Zentner, R.P., *Soil Sci. Soc. Am. J.*, 57, 1034, 1993. With permission.

V. PERSPECTIVE

Unfortunately, organic C or N values were not taken at the initiation of the experiment; however, if one assumes that our estimated starting points are accurate, then it is readily apparent that, even with the addition of adequate fertilization to crops grown annually, we may only be able to maintain soil organic matter in equilibrium over a 25-year period. The reason is associated with the small yields usually obtained due to frequent and severe water deficits. As we demonstrated quite clearly, soil organic matter is directly related to crop production. Because we could not increase the level of soil organic C for any extended period even with the addition of fertilizer and because the manufacture of NH_3 fertilizer requires combustion of C which generates CO_2, it appears that, in the semi-arid regions of the Canadian prairies, most cropping systems that use frequent tillage for weed control and in which frequent use of fertilizers is commonplace will result in negative C sequestration.

The inclusion of lentil in a wheat-lentil rotation maintained soil organic matter at the same level as well-fertilized Cont W and had the advantage of requiring progressively less N fertilizer (Figure 10).[8] Thus, this lentil-containing rotation is one of the more promising systems for sequestering C in this region. Fortunately, in this semi-arid climate it is possible to use this system without experiencing disease problems such as infestation with aschochyta inoculum.

Unfortunately, the nature of the weather in this area dictates that summer fallow will likely remain an integral part of any cropping system.[4] As can be seen from our results, the only system that includes summer fallow which did not experience a loss of soil organic matter was the one that included a winter annual. This system was one of the most efficient users of N with minimal leaching losses; thus it appears

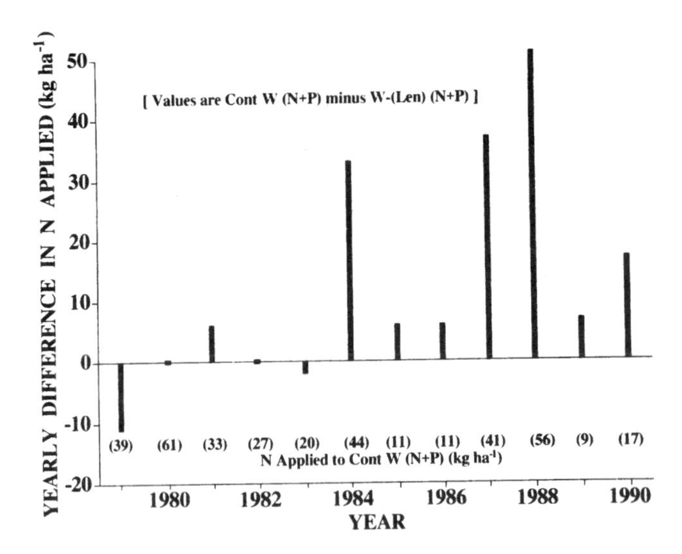

Figure 10 Trend in relative fertilizer-N requirements for W-(Len) (N + P). (The comparison is made with the lentil phase because its N requirements are dictated by the wheat phase.) (From Campbell, C.A. et al., *Can. J. Plant Sci.*, 72, 1091, 1992. With permission.)

to be promising for net C sequestration in this region (provided the economics of such a system is acceptable).

All the rotations used in this study were managed using conventional tillage which enhances organic matter decomposition and encourages erosion. However, if we employed zero tillage, combined with snow management to enhance stored soil water, cropped annually and fertilized diligently, it is possible that we might be able to increase soil organic C in reasonable time.[25] Ongoing studies at our Research Station will provide answers to these questions in the near future.

VI. CONCLUSIONS

This study showed that the amount of organic matter in the top 0.15-m depth of soil will vary depending on the amount and N content of the crop residues returned to the land. Thus, in the first 15 years of the experiment, when grain yields were above 1000 kg ha^{-1} in13 years and there were no complete crop failures, organic matter tended to increase in well-fertilized, annually cropped systems, and was maintained by the fallow-containing systems and by continuous wheat receiving inadequate N fertilizer. However, in later years when severe droughts in 3 years (one complete crop failure) markedly reduced crop residue input, all systems except the one containing the winter cereal lost organic matter. The advantage of the system containing the winter cereals was its ability to produce adequate amounts of crop residues in drought years, to protect the soil from erosion because of its much shorter fallow period compared to the spring-seeded crops (14 vs. 21 months). Soil organic matter was lowest under the fallow-flax-spring wheat system due to the low yields of flax and possibly due to the tendency of flax straw to be easily blown off the field after harvest. Carbon sequestration was generally greater when fallow frequency was reduced; however, the type of crop used to reduce fallow frequency has an important influence on the amount of C that may be sequestered, as shown by comparing the F-Flx-W system with the F-Rye-W system.

A nitrogen balance showed that even the annually cropped systems that were fertilized based on soil tests were not receiving sufficient fertilizer N to replace that being removed in the grain; under these circumstances it was difficult to increase soil organic matter. The amount of NO_3-N leached from three rotations examined confirmed the inefficient use of N by the low-yielding flax-containing rotation (more leaching) and the efficient use of N by the rotation containing the fall-seeded cereals. Although the wheat-lentil system has only been in place for 12 years, there was evidence that N fixation by the grain legume was tending to increase soil organic N compared to monoculture wheat systems.

All rotations were conventionally managed (mostly tillage used to control weeds) in this study and it appears that, with such systems in this low-yielding environment, C sequestration will be negative. It remains to be seen whether improved management systems that include reduced tillage, snow management, extended rotations, and proper fertilization will increase our ability to sequester C in this area.

ACKNOWLEDGMENTS

We wish to acknowledge the foresight of Dr. Wilf Ferguson and Mr. C. Hank Anderson who initiated this study and to thank the many scientists, plotmen, and technicians who have had the fortitude to diligently manage, analyze, and record the mountains of data that have emanated from this experiment.

REFERENCES

1. McGill, W.B., C.A. Campbell, J.F. Dormaar, E.A. Paul, and D.W. Anderson, Soil organic matter losses, in Agricultural Land: Our Disappearing Heritage. A Symposium, Proc. 18th Annu. Soil Sci. Workshop, Edmonton. Feb. 24–25, 1981, Alberta Soil and Feed Testing Laboratory, Edmonton, 1981, 72.
2. Campbell, C.A., E.A. Paul, and W.B. McGill, Effect of cultivation and cropping on the amounts and forms of soil N, in Proc. West. Can. Nitrogen Symp., Alberta Soil Sci. Workshop, Calgary, Jan. 20–21, 1976, Edmonton, 1976, 7.
3. Campbell, C.A., D.W.L. Read, R.P. Zentner, A.J. Leyshon, and W.S. Ferguson, First 12 years of a long-term crop rotation study in southwestern Saskatchewan — yield and quality of grain, *Can. J. Plant Sci.*, 63, 91, 1983.
4. Campbell, C.A., R.P. Zentner, H.H. Janzen, and K.E. Bowren, Crop Rotations on the Canadian Prairies, Publ. 1841/E, Canadian Government Publication Center, Ottawa, 1990.
5. Ayres, K.W., D.F. Acton, and J.G. Ellis, The Soils of the Swift Current Map Area 721 Saskatchewan, Extension Publ. 481, University of Saskatchewan, Saskatoon, 1985.
6. Biederbeck, V.O., C.A. Campbell, and R.P. Zentner, Effect of crop rotation and fertilization on some biological properties of a loam in southwestern Saskatchewan, *Can J. Soil Sci.*, 64, 355, 1984.
7. Zentner, R.P. and C.A. Campbell, Yield, grain quality and economics of some spring wheat cropping systems in the Brown soil zone — an 18-year study, *Can. J. Plant Sci.*, 68, 1, 1988.
8. Campbell, C.A., R.P. Zentner, F. Selles, V.O. Biederbeck, and A.J. Leyshon, Comparative effects of grain lentil-wheat and monoculture wheat on crop production, N economy and N fertility in a Brown Chernozem, *Can. J. Plant Sci.*, 72, 1091, 1992.
9. Atkinson, H.J., G.R. Giles, A.J. Mclean, and J.R. Wright, Chemical Methods of Soil Analysis, Publ. 169, Department of Agriculture, Ottawa, 1958.
10. Allison, L.E., Wet-combustion apparatus and procedure for organic and inorganic carbon in soils, *Soil Sci. Soc. Am. Proc.*, 24, 36, 1960.
11. McKeague, J.A., Manual on Soil Sampling and Methods of Analysis, 2nd ed., Canadian Soil Survey Committee, Canadian Society Soil Science, 1978.
12. Campbell, C.A., R. DeJong, and R.P. Zentner, Effect of cropping, summerfallow and fertilizer nitrogen on nitrate-nitrogen lost by leaching on a Brown Chernozemic loam, *Can J. Soil Sci.*, 64, 61, 1984.
13. Hamm, J.W., F.G. Radford, and E.H. Halstead, The simultaneous determination of nitrogen, phosphorus and potassium in sodium bicarbonate extracts of soils, in *Technicon Int. Congr. Advances in Automatic Analysis, Industrial Analysis 2*, M. Adelman et al. Eds., Futura, Mt. Kisco, NY, 1970, 65.
14. SAS Institute, SAS User's Guide: Statistics, Version 5, SAS Institute Cary, NC, 1985.
15. Little, T.M. and F.J. Hills, *Agricultural Experimentation — Design and Analysis*, John Wiley & Sons, New York, 1978.
16. Zentner, R.P., F.B. Dyck, K.R. Handford, C.A. Campbell, and F. Selles, Economics of flex-cropping in southwestern Saskatchewan, *Can. J. Plant Sci.*, 73, 749, 1993.
17. Nuttall, W.F., K.E. Bowren, and C.A. Campbell, Crop residue management practices, and N and P fertilizer effects on crop response and on some physical and chemical properties of a Black Chernozem over 25 years in a continuous wheat rotation, *Can. J. Soil Sci.*, 66, 159, 1986.
18. Campbell, C.A., F. Selles, R.P. Zentner, and B.G. McConkey, Available water and nitrogen effects on yield components and grain nitrogen of zero-till spring wheat, *Agron. J.*, 85, 114, 1993.
19. Campbell, C.A., D.R. Cameron, W. Nicholaichuk, and H.R. Davidson, Effects of fertilizer N and soil moisture on growth, N content and moisture use by spring wheat, *Can. J. Soil Sci.*, 57, 289, 1977.
20. Rasmussen, P.E., R.R. Allmaras, C.R. Rohde, and N.C. Roager, Jr., Crop residue influences on soil carbon and nitrogen in a wheat-fallow system, *Soil Sci. Soc. Am. J.*, 44, 596, 1980.
21. Campbell, C.A., K.E. Bowren, M. Schnitzer, R.P. Zentner, and L. Townley-Smith, Effect of crop rotations and fertilization on soil organic matter and some biochemical properties of a thick Black Chernozem, *Can J. Soil Sci.*, 71, 377, 1991.
22. Janzen, H.H., Effect of fertilizer on soil productivity in long-term spring wheat rotations, *Can J. Plant Sci.*, 67, 165, 1987a.
23. Janzen, H.H., Soil organic matter characteristics after long-term cropping to various spring wheat rotations, *Can. J. Soil Sci.*, 67, 845, 1987b.

24. Janzen, H.H., A.M. Johnston, J.M. Carefoot, and C.W. Lindwall, Soil organic matter dynamics in southern Alberta, in *Soil Organic Matter in Temperate Agroecosystems: Long-Term Experiments in North America*, Paul, E.A., Paustian, K., Elliott, E.T., and Cole, C.V., Eds., CRC Press, Boca Raton, FL, 1997, chap. 21.
25. Campbell, C.A. and R.P. Zentner, Soil organic matter as influenced by crop rotations and fertilization, *Soil Sci. Soc. Am. J.*, 57, 1034, 1993.

Chapter

A Conservation Tillage-Cropping Systems Study in the Northern Great Plains of the United States

A.L. Black and D.L. Tanaka

CONTENTS

I. INTRODUCTION

Soil erosion is recognized as a serious threat to sustainable agriculture and, in some areas of the world, to the production of food and fiber to meet human needs. In the semiarid northern Great Plains soil losses, predominantly by wind and to a lesser extent by water erosion, exceed 5 tons per acre annually on about 50% of the cropland. Cost-effective minimum-till and no-till conservation production systems, adaptable to the region, could potentially reduce soil losses by wind and water to negligible levels.

Soil erosion in the northern Great Plains is deemed to diminish soil productivity by also diminishing soil organic matter (OM) content.[1] Thus, the contribution of a unit increment of soil OM content to soil productivity can simultaneously be inferred to represent the loss of productivity associated with soil erosion when the extent of soil OM loss is known.[2] Quantity and management of crop residue is a major factor in controlling soil erosion. Left on the soil surface, crop residues reduce wind speed and subsequent wind erosion, acts as a barrier to water runoff, and absorb energy imparted by falling rain, thereby reducing water erosion. Simultaneously, residues improve water conservation by suppressing water evaporation rate and by retaining precipitation (including snow trapping) where it falls. High-level crop residue management systems therefore control soil erosion and increase soil water supplies, thus enhancing crop production and the quantity of residue produced to be returned to the soil to maintain and even enhance soil OM content over time. A given high-residue management tillage system can, therefore,

0-8493-2802-0/97/$0.00+$.50

become somewhat self-perpetuating and provide a positive influence on soil OM and soil productivity. In essence, the focal point of most recent dryland agricultural research in semiarid and subhumid regions is to utilize crop residue to save as much water as possible from precipitation and use it as efficiently as possible for crop and livestock production.

A conservation tillage-cropping systems study was established in 1984 on a 25-ha site to study long-term effects of conventional-, minimum-, and no-tillage residue management systems on water conservation, water and nitrogen-use efficiencies, and crop yields. The study was planned to allow long-term assessments of soil-C, -N, and -P concentration and cycling. A description of the field research site and average climatic information is shown in Table 1.

Table 1 Field Research Site Description

Item	Description
Legal description	NE $^1/_4$ Sec 18, Twn 138, R 81
Latitude	46°46′N
Longitude	100°55′W
Elevation	549 m
Soil type	Temvik-Wilton, Silt-Loam (fine-loamy, mixed, Typic Argiborolls)
Slope and aspect	2 to 4%, southeasterly
Average annual precipitation	402 mm
Average growing season precipitation	227 mm (May 1–September 1)
Average length of growing season	136 days
Average May, June, July, August temperature	12.5, 17.8, 21.0, 20.4°C
Mean annual temperature	5°C
Potential ET (Thornthwaite method)	600 mm (April–November 1)

II. PROJECT DESCRIPTION

The experimental variables in three replications include all combinations of two cropping sequences: spring wheat-fallow and spring wheat-winter wheat-sunflower (main blocks, 137.2 × 73.1 m); three conservation tillage treatments; conventional-tillage, minimum-tillage, and no-tillage (main plots, 45.7 × 73.1 m); three fertilizer nitrogen rates (0, 22, 45 kg N/ha for crop-fallow and 34, 67, 101 kg N/ha for continuous cropping sequence (subplots 45.7 × 24.4 m); and two cultivatars, standard (check) and current "best" (sub-subplots, 22.8 × 24.4 m).

The experimental area was uniformly cropped to sunflowers in 1983. Sunflower seed yield was 1680 kg/ha and all available soil water and nitrogen was used to a depth of 1.5 m. The minimum soil water depletion point after sunflower for the experimental site averaged 8.5% ± 0.5 by volume or 3.65 cm of total water per 30 cm-increment to a depth of 1.5 m. The soil will hold 22.0% water by volume or 9.53 cm of total water per 30 cm-increment to a depth of 1.5 m. These data serve as the base upon which available soil water and crop water use was calculated.

In the fall of 1983, we took soil samples (nine samples per main block) from the 0- to 7.6- to 7.6- to 15.2-, to 15.2- to 30.5-, 30.5- to 45.7-, and 45.7- to 60.9-cm depths and by 30.5 cm-increments from 60.9 to 152.4 cm for soil analysis to provide the base data for chemical and physical soil properties before treatment variables were initiated. Soil samples were obtained in this same manner for the same purpose in the spring of 1990. Sodium-bicarbonate soluble P ranged from 8 to 12 ppm in 1983. We applied a uniform broadcast application of 45 kg P/ha as concentrated superphosphate in October 1983. Soil test P in the 0 to 15.2 depth when sampled in the spring of 1984 ranged from 20 to 26 ppm.

In 1984, we initiated the fertilizer-N treatments and tillage treatments within main crop blocks. The crop-fallow sequence was established by planting spring wheat after sunflower in one block and summer fallowing in another block. For the spring wheat-winter wheat-sunflower sequence, spring wheat was planted in the crop blocks designated as spring wheat and winter wheat, and spring barley was planted in place of sunflower to avoid planting sunflower 2 years in a row in the same block.

By the 1985 crop season, all cropping sequences, tillage treatments, fertilizer-N rates, and cultivar variables were in place. The conventional tillage treatment consists of using the undercutter (81-cm sweeps) as needed between crops and using a tandem disk once just before planting. The minimum-tillage treatment consists of using the undercutter sweep once or twice between crops and using chemical weed control as may be needed thereafter. The no-tillage treatment consists of only herbicide spray operations as may be needed for weed control. Our tillage treatments are defined in terms of target crop residue levels after planting when the previous crop was spring wheat or winter wheat as follows: conventional tillage, 30% or less crop residue on the soil surface; minimum-tillage, 30 to 60% surface residue cover; and no-tillage, 60% or greater crop residue cover. Sixty percent crop residue cover is representative of a high residue management system or about 1680 kg/ha of small grain residue present on the soil surface. Therefore, we do not have a fixed number of tillage or spray operations but rather a system of tillage plus spray or spray-only operations as may be needed for weed control and for obtaining target residue maintenance levels. Our combine is equipped to spread both straw and chaff over the width of the header.

Fertilizer-N was broadcast as ammonium nitrate before spring planting of spring wheat or sunflowers and as a topdressing in late April each year for winter wheat. Fertilizer-N was not applied in the spring of 1991 or 1992 because of high residual soil NO_3-N levels caused by drought stress and low crop yields in 1988, 1989, and 1990. Soil samples were taken at 30-cm increments to a depth of 1.5 m before seeding and after harvest each year to determine soil water and soil N (NH_4 and NO_3).

III. DATA SETS AND MEASUREMENTS

The major measurements being taken each year (1985 to 1992) include (1) soil water at seeding, at harvest and at soil freeze-up; (2) crop residue quantity at harvest, in the spring, after planting, and periodically during fallow; (3) plant dry matter production, straw and grain yield; (4) N and P concentration and content of grain and straw; (5) soil physical properties such as soil aggregate stability, bulk density; (6) soil NO_3-N and NH_4-N by 30-cm increments to a depth of 1.5 or 2.0 m; (7) sodium-bicarbonate soluble P, periodically; (8) total organic C (1983 and 1990 samples) to 0.92 m; (9) plant population; (10) plant diseases and epidemiology; (11) minirhizotron root growth/density measurements; and (12) atmospheric environmental data (temperature, precipitation, humidity, wind, and solar radiation).

Most of the measurements and data sets have not been subjected to statistical analysis. However, tabular summaries of soil organic C and total N levels (1983, 1990) and the 8-year average of crop yields and crop residue production are presented. Soil organic C (1983 samples) was determined by a modified Walkley-Black method[5] and total N by microkjeldahl procedures[3,4] with slight modification. Soil organic C and total soil N (1990 samples) were determined by combustion in a C/N autoanalyzer.

Some chemical and physical properties of soil samples obtained by increment depth at the beginning of the study (Oct 1983) including total soil N and C are shown in Table 2. The cropping systems, tillage, and N-fertilizer treatments were initiated in 1984 and soil samples by increment depths were again taken in the spring of 1991 after 7 years of treatments (6 years with proper crops in place). Bulk density by depth increment and total soil N and C were determined and averaged for each treatment within the crop-fallow and annual cropping systems and reported in Tables 3 and 4, respectively. Each treatment value reported in Table 3 represents the average of three replications and two main blocks (spring wheat-fallow), and each treatment value in Table 4 represents the average of three replications and three main crop blocks (spring wheat-winter wheat-sunflower).

Table 2 Some Chemical and Physical Properties of Soil Samples Obtained October 1983 at the Beginning of the Study (Average of 30 Sites, 2 Sampling Sites/Main Block)

Sample Depth (cm)	pH	EC (mmhos)	Particle Size (%)			Total N (g/kg)	Organic C (g/kg)	C/N Ratio
			Sand	Silt	Clay			
0–7.6	6.4	0.34	28.5	51.4	20.1	1.98	21.4	10.8
7.6–5.2	6.3	0.21	28.1	51.6	20.3	1.91	20.5	10.7
15.2–30.4	6.7	0.18	26.2	52.0	21.8	1.41	14.1	10.0
30.4–45.7	7.2	0.21	30.4	47.1	22.5	1.05	9.3	8.9
45.7–60.9	7.6	0.29	38.2	37.8	24.0	0.76	6.9	9.1
60.9–91.4	7.9	0.35	42.5	31.9	25.6	—	—	—
91.4–121.9	8.0	0.45	45.2	30.0	24.8	—	—	—
121.9–152.4	8.1	0.44	46.9	30.5	22.6	—	—	—
152.4–182.9	8.0	0.45	47.5	30.4	22.1	—	—	—

Table 3 Bulk Density, Organic C, and Total Soil N in the Soil Profile after 6 Years of Cropping, Tillage, and N-Fertilizer Treatments in the Spring Wheat-Fallow Rotation

		Tillage Systems								
		Conventional Till			Minimum Till			No-Till		
N Treatment (kg N/ha)	Soil Depth (cm)	Bulk Density (g/cm³)	Soil C (g/kg)	Soil N (g/kg)	Bulk Density (g/cm³)	Soil C (g/kg)	Soil N (g/kg)	Bulk Density (g/cm³)	Soil C (g/kg)	Soil N (g/kg)
0	0–7.6	1.16	19.2	1.60	1.16	17.8	1.50	1.18	18.3	1.55
	7.6–15.2	1.33	19.1	1.60	1.36	17.4	1.50	1.36	17.3	1.45
	15.2–30.4	1.31	13.7	1.25	1.29	12.8	1.25	1.32	13.3	1.10
	30.4–60.9	1.42	9.2	0.80	1.42	8.6	0.80	1.42	8.6	0.75
	60.0–91.2	1.42	4.3	0.40	1.42	4.1	0.35	1.42	4.4	0.45
20	0–7.6	1.14	20.0	1.70	1.11	17.9	1.50	1.20	17.7	1.55
	7.6–15.2	1.32	18.5	1.55	1.35	17.6	1.50	1.37	17.2	1.45
	15.2–30.4	1.34	14.1	1.25	1.29	14.5	1.25	1.28	11.8	1.05
	30.4–60.9	1.42	8.6	0.70	1.42	8.0	0.75	1.42	8.5	0.75
	60.9–91.2	1.42	4.7	0.40	1.42	4.7	0.45	1.42	4.0	0.40
40	0–7.6	1.15	19.9	1.65	1.14	19.3	1.55	1.20	17.8	1.50
	7.6–15.2	1.34	20.4	1.65	1.35	19.0	1.55	1.36	16.6	1.40
	15.2–30.4	1.29	15.6	1.25	1.28	13.2	1.10	1.32	12.5	1.10
	30.4–60.9	1.42	9.2	0.80	1.42	8.8	0.75	1.42	7.1	0.65
	60.9–91.2	1.42	6.4	0.55	1.42	4.5	0.45	1.42	3.8	0.35

Table 4 Bulk Density, Organic C, and Total Soil N in the Soil Profile after 6 Years of Cropping, Tillage, and N-Fertilizer Treatments in the Spring Wheat-Winter Wheat Sunflower Annual Cropping System

		Tillage Systems								
		Conventional Till			Minimum Till			No-Till		
N Fertilizer Added (kg N/ha)	Soil Depth (cm)	Bulk Density (g/cm³)	Soil C (g/kg)	Soil N (g/kg)	Bulk Density (g/cm³)	Soil C (g/kg)	Soil N (g/kg)	Bulk Density (g/cm³)	Soil C (g/kg)	Soil N (g/kg)
34	0–7.6	1.14	21.2	1.80	1.09	21.1	1.83	1.07	24.5	2.00
	7.6–15.2	1.37	19.7	1.67	1.33	19.2	1.73	1.31	19.9	1.70
	15.2–30.4	1.29	15.0	1.33	1.28	15.2	1.33	1.32	15.8	1.43
	30.4–60.9	1.42	8.3	0.77	1.42	8.3	0.77	1.42	7.8	0.77
	60.0–91.2	1.42	5.4	0.50	1.42	4.7	0.43	1.42	3.7	0.40
67	0–7.6	1.12	20.3	1.80	1.08	23.2	1.90	1.13	23.6	2.00
	7.6–15.2	1.38	18.1	1.63	1.36	20.0	1.76	1.36	19.6	1.73
	15.2–30.4	1.33	13.1	1.23	1.31	16.0	1.46	1.33	14.9	1.33
	30.4–60.9	1.42	6.3	0.67	1.42	9.6	0.96	1.42	7.5	0.73
	60.9–91.2	1.42	3.8	0.37	1.42	5.3	0.60	1.42	4.0	0.43
101	0–7.6	1.16	21.3	1.90	1.12	22.7	1.96	1.09	24.9	2.13
	7.6–15.2	1.39	18.9	1.73	1.33	19.3	1.73	1.31	20.8	1.80
	15.2–30.4	1.29	13.8	1.23	1.31	16.3	1.37	1.27	17.1	1.47
	30.4–60.9	1.42	7.6	0.77	1.42	9.6	0.87	1.42	9.0	0.87
	60.9–91.2	1.42	4.0	0.40	1.42	4.8	0.56	1.42	4.7	0.50

IV. RESULTS AND DISCUSSION

After 7 years in the crop-fallow system, tillage and N-fertilizer treatments did not have much influence on bulk density, total soil N, or soil C (Table 3). However, in the annual cropping system, minimum- and no-till systems appear to be reducing bulk density and increasing both soil C and soil N compared to conventional tillage (Table 4). These results could be anticipated based on the much higher levels of crop residue being produced annually in minimum- or no-till continuous cropping system compared to the crop-fallow system.

The 8-year average annual grain yield for the spring wheat-fallow (Table 5) and for the spring wheat-winter wheat-sunflower (Tables 6, 7, and 8) cropping systems show that tillage and N-treatment effects on the grain yield were essentially negligible in the spring wheat-fallow system, but were increased in the annual cropping system. The annual cropping system with minimum- or no-till produced 1.5- to nearly 2.0-fold more grain annually than crop-fallow.

Table 5 Eight-Year Average Annual Spring Wheat Grain Yields in the Spring Wheat-Fallow Rotation as Influenced by Tillage System, N-Fertilizer Level, and Cultivar (1985–1992)

Cultivar	Rate of N Added (kg/ha)	Tillage System (kg/ha)			
		Conventional Till	Minimum Till	No-Till	Average (N-Rate)
Butte 86	0	1220	1180	1140	1180
	22	1200	1150	1150	1170
	45	1180	1190	1140	1170
	Average	1200	1170	1140	1170
Stoa	0	1190	1200	1170	1190
	22	1180	1170	1180	1180
	45	1140	1160	1190	1160
	Average	1170	1180	1180	1180
	Average (tillage)	1190	1180	1160	

Table 6 Eight-Year Average Annual Spring Wheat Grain Yields in the Spring Wheat-Winter-Wheat-Sunflower Cropping System as Influenced by Tillage System, N-Fertilizer Level, and Cultivar (1985–1992)

Cultivar	Rate of N Added (kg/ha)	Tillage System (kg/ha)			
		Conventional Till	Minimum Till	No-Till	Average (N-Rate)
Butte 86	34	1320	1420	1410	1380
	67	1400	1510	1530	1480
	101	1470	1590	1690	580
	Average	1400	1510	1540	1480
Stoa	34	1090	1280	1230	1200
	67	1260	1370	1280	1300
	101	1300	1480	1570	1450
	Average	1220	1380	1360	1320
	Average (tillage)	1310	1450	1450	

Amount of crop residue produced and returned to the land is the major source of organic C input back into the soil. The 8-year annual crop residue (straw plus chaff) production for the spring wheat-fallow is shown in Table 9 and for the spring wheat-winter wheat-sunflower rotation in Tables 10, 11, and 12, respectively. Averaged across all three crops in the annual cropping system and comparing the same tillage and N treatment in the crop-fallow sequence, crop residue production in annual cropping was increased 790 to 990 kg/ha/y with conventional till, 1000 to 1300 kg/ha/y with minimum tillage, and 1050 to 1580 kg/ha/y with no-tillage for the low-and high-N treatments, respectively.

Table 7 Eight-Year Average Annual Winter Wheat Grain Yields in the Spring Wheat-Winter-Wheat-Sunflower Cropping System as Influenced by Tillage System, N-Fertilizer Level, and Cultivar (1985–1992)

Cultivar	Rate of N Added (kg/ha)	Tillage System (kg/ha)			
		Conventional Till	Minimum Till	No-Till	Average (N-Rate)
Roughrider	34	1650	1840	1900	1800
	67	1690	1920	1970	1860
	101	1740	1940	2080	1920
	Average	1690	1900	1980	1860
Norstar	34	1710	1800	1940	1820
	67	1630	1840	2060	1840
	101	1740	1910	2000	1880
	Average	1690	1850	2000	1850
	Average (tillage)	1690	1880	1990	

Table 8 Eight-Year Average Annual Sunflower Seed Yields in the Spring Wheat-Winter Wheat-Sunflower Cropping System as Influenced by Tillage System, N-Fertilizer Level, and Cultivar (1985–1992)

Cultivar	Rate of N Added (kg/ha)	Tillage System (kg/ha)			
		Conventional Till	Minimum Till	No-Till	Average (N-Rate)
Early Maturity	34	1240	1320	1290	1280
	67	1330	1460	1360	1380
	101	1370	1430	1610	1470
	Average	1310	1400	1420	1380
Med-Maturity	34	1220	1300	1260	1260
	67	1310	1430	1340	1360
	101	1370	1470	1420	1420
	Average	1300	1400	1340	1350
	Average (tillage)	1310	1400	1380	

V. CONCLUSIONS

In order to maintain and enhance soil productivity in the Great Plains, soil erosion by wind and water must be controlled.

The use of high-residue minimum- or no-till systems controls soil erosion and improves soil water storage and precipitation use efficiency, more for annual cropping systems than for crop-fallow. Total grain and crop residue production was increased 54 to 88% each year in the annual cropping system when compared to conventional spring wheat-fallow.

The soil C and N changes that are beginning to appear reveal the positive influences of high-residue management systems on water conservation and its efficient use in annual cropping systems that produce higher grain and crop residue yields than conventional crop-fallow. In this scenario, the additional water conserved and its efficient use in producing additional yield and crop residue is the driving variable controlling the opportunity to improve soil OM content in a northern Great Plains agroecosystem.

Table 9 Eight-Year Average Annual Spring Wheat Residue Production in the Spring Wheat-Fallow Cropping System as Influenced by Tillage System, N-Fertilizer Level, and Cultivar (1985–1992)

Cultivar	Rate of N Added (kg/ha)	Tillage System (kg/ha)			
		Conventional Till	Minimum Till	No-Till	Average (N-Rate)
Butte 86	0	1760	1710	1720	1730
	22	1730	1680	1750	1720
	45	1890	1800	1840	1840
	Average	1790	1730	1770	1760
Stoa	0	1780	1810	1770	1790
	22	1720	1780	1720	1740
	45	1760	1790	1740	1760
	Average	1750	1790	1740	1760
	Average (tillage)	1770	1760	1760	

Table 10 Eight-Year Average Annual Spring Wheat Residue Production in the Spring Wheat-Winter Wheat-Sunflower Cropping System as Influenced by Tillage System, N-Fertilizer Level, and Cultivar (1985–1992)

Cultivar	Rate of N Added (kg/ha)	Tillage System (kg/ha)			
		Conventional Till	Minimum Till	No-Till	Average (N-Rate)
Butte 86	34	2230	2440	2460	2380
	67	2370	2540	2840	2580
	101	2450	2710	3000	2720
	Average	2350	2560	2770	2560
Stoa	34	1940	2220	2240	2130
	67	2160	2400	2530	2360
	101	2260	2550	2840	2550
	Average	2120	2390	2540	2350
	Average (tillage)	2240	2480	2660	

Table 11 Eight-Year Average Annual Winter Wheat Residue Production in the Spring Wheat-Winter Wheat-Sunflower Cropping System as Influenced by Tillage System, N-Fertilizer Level, and Cultivar (1985–1992)

Cultivar	Rate of N Added (kg/ha)	Tillage System (kg/ha)			
		Conventional Till	Minimum Till	No-Till	Average (N-Rate)
Roughrider	34	2610	2930	2970	2840
	67	2760	3170	3360	3100
	101	2980	3330	3740	3350
	Average	2780	3140	3360	3100
Norstar	34	2870	3070	3100	3010
	67	2870	3310	3800	3330
	101	3130	3600	3790	3510
	Average	2960	3330	3560	3280
	Average (tillage)	2870	3240	3460	

Table 12 Eight-Year Average Annual Sunflower Residue Production in the Spring Wheat-Winter Wheat-Sunflower Cropping System as Influenced by Tillage System, N-Fertilizer Level, and Cultivar (1985–1992)

Cultivar	Rate of N Added (kg/ha)	Tillage System (kg/ha)			
		Conventional Till	Minimum Till	No-Till	Average (N-Rate)
Early Maturity	34	2730	2940	2920	2860
	67	2900	3150	3060	3040
	101	2880	3180	3220	3090
	Average	2840	3090	3030	3000
Med-Maturity	34	2970	3020	3070	3020
	67	3060	3180	3360	3200
	101	3190	3210	3620	3340
	Average	3070	3140	3350	3190
	Average (tillage)	2960	3120	3190	

REFERENCES

1. Bauer, A. and A.L. Black, Soil carbon, nitrogen, and bulk density comparisons in two cropland tillage systems after 25 years and in virgin grassland, *Soil Sci. Soc. Am. J.*, 45, 1166, 1981.
2. Bauer, A. and A.L. Black, A quantification of the effect of soil organic matter content on soil productivity, *Soil Sci. Soc. Am. J.*, 58, 185, 1994.
3. Bremmer, J.M., total nitrogen in *Methods of Soil Analysis, Part 2*, C.A. Black, et al., Eds., Vol. 9, American Society of Agronomy, Madison, WI, 1965, 1149.
4. Nelson, D.W. and L.E. Sommers, A simple digestion procedure for estimation of total nitrogen in soils and sediments, *J. Environ. Qual.*, 1, 423, 1972.
5. Peech, M., L.T. Alexander, L.A. Dean, and J.F. Reed, Methods of Soil Analysis for Soil-Fertility Investigations, Circ. No. 757, USDA, U.S. Government Printing Office, Washington, DC, 1947.

Chapter 25

Soil Organic Matter Changes over Two Decades of Winter Wheat-Fallow Cropping in Western Nebraska

D.J. Lyon, C.A. Monz, R.E. Brown, and A.K. Metherell

CONTENTS

I. INTRODUCTION

In 1969 and 1970, Fenster and Peterson[1] initiated two studies to compare the effects of no-tillage (chemical), stubble-mulch, and plow (bare fallow) systems of fallow on winter wheat (*Triticum aestiuum* L. emend. Thell.) grain yield, grain protein, residue retention, soil NO_3-N accumulation, and soil water accumulations during fallow. The studies were established on the High Plains Agricultural Laboratory located 8.3 km north of Sidney, NB. The first study site was established in 1969 on land that had been farmed from 1920 until 1957, then seeded to crested wheatgrass [*Agropyron cristatum* (L.) Gaertn.] for 10 years prior to being broken out of sod with a moldboard plow in 1967. The site has been variously referred to in the literature as the reseeded sod, previously cultivated, Alliance silt loam, and Alliance site.

The second study site was situated on land that had remained in native grasses until 1970 when it was broken out of sod with a moldboard plow. The predominant grasses on the site prior to plowing were western wheatgrass (*Agropyron smithii* Rydb.), needleandthread (*Stipa comata* Trin. & Rupr.), blue grama [*Bouteloua gracilis* (H.B.K.) Lag. *ex* Steud.], sideoats grama [*Bouteloua curtipendula* (Michx.) Torr.], and sand dropseed [*Sporobolus cryptandrus* (Torr.) Gray]. The site has been variously referred to in the literature as the native sod, Duroc loam, and Duroc site. The native sod site has attracted more interest by researchers over the last 2 decades than has the reseeded sod site, probably because the native sod site had never been cropped prior to the establishment of plots in 1970 and because native grass was maintained in each replication. It is for these reasons that the remainder of this chapter focuses on the research findings from the native sod site over the 2 decades of the 1970s and 1980s.

The authors would be remiss not to mention the contribution of the researchers who have worked very diligently to understand the influences of different tillage practices on the various aspects of crop

0-8493-2802-0/97/$0.00+$.50

production, soil chemistry, and soil microbiology. It is especially important to recognize the efforts of Charles R. Fenster, who initiated and maintained the research sites for many years. His commitment to these sites has allowed many other researchers the opportunity to hypothesize and study the influence of tillage and crops on the soils of the semi-arid High Plains. As of 1990, work conducted at either the reseeded sod or the native sod sites had resulted in 26 journal articles (Table 1) and numerous abstracts, proceedings, theses, popular publications, and congressional reports. The authors have tried to bring some of the information gathered by many researchers over the past 2 decades into one report. We hope we have done justice to their efforts.

Table 1 Publications Developed, at Least in Part, from Work Conducted at the Long-Term Tillage Plots Located at the High Plains Agricultural Laboratory near Sidney, NE between 1970 and 1990.

Broder, M.W., J.W. Doran, G.A. Peterson, and C.R. Fenster, Fallow tillage influence on spring populations of soil nitrifiers, denitrifiers, and available nitrogen, *Soil Sci. Soc. Am. J.*, 48, 1060, 1984.

Cambardella, C.A., and E.T. Elliott, Particulate soil organic-matter changes across a grassland cultivation sequence, *Soil Sci. Soc. Am. J.*, 56, 777, 1992.

Dickey, E.C., C.R. Fenster, J.M. Laflen, and R.H. Mickelson, Effects of tillage on soil erosion in a wheat-fallow rotation, Trans. of the ASAE, 26(3), 814, 1983.

Doran, J.W., Soil microbial and biochemical changes associated with reduced tillage, *Soil Sci. Soc. Am. J.*, 44, 765, 1980.

Doran, J.W. and L.N. Mielke, A rapid, low-cost method for determination of soil bulk density at several depths, *Soil Sci. Soc. Am. J.*, 48, 717, 1984.

Doran, J.W. , Microbial biomass and mineralizable nitrogen distributions in no-tillage and plowed soils, *Biol. Fertil. Soils*, 5, 68, 1987.

Doran, J.W., L.N. Mielke, and J.F. Power, Tillage/residue management interactions with the soil environment, organic matter, and nutrient cycling, *INTECOL Bull.*, 15, 33, 1987.

Elliott, E.T., Aggregate structure and C, N, and P in native and cultivated soils, *Soil Sci. Soc. Am. J.*, 50, 627, 1986.

Fenster, C.R., Conservation tillage in the northern Great Plains, *J. Soil Water Cons.*, 37, 42, 1977.

Fenster, C.R. and G.A. Wicks, Minimum tillage fallow systems for reducing wind erosion, Transactions ASAE, 2(5), 906, 1982.

Fenster, C.R. and G.A. Peterson, Effects of no-tillage fallow as compared to conventional tillage in a wheat-fallow system, Nebr. Exp. Stn. RB 289, 1979.

Fenster, C.R. and G.A. Wicks, Fallow systems for winter wheat in western Nebraska, *Agron. J.*, 74, 9, 1982.

Follett, R.F. and G.A. Peterson, Surface soil nutrient distribution as affected by wheat-fallow tillage systems, *Soil Sci. Soc. Am. J.*, 52, 141, 1988.

Follett, R.F. and D.S. Schimel, Effect of tillage practices on microbial biomass dynamics, *Soil Sci. Soc. Am. J.*, 53, 1091, 1989.

Hooker, M.L., G.A. Peterson, and D.H. Sander, Ammonia N losses from Simulated Plowing of Natural Sods, *Soil Sci. Soc. Am. Proc.*, 37, 247, 1973.

Lamb, J.A., G.A. Peterson, and C.R. Fenster, Wheat fallow tillage systems effect on a newly cultivated grassland soils's N budget, *Soil Sci. Soc. Am. J.*, 49, 352, 1985.

Lamb, J.A., G.A. Peterson, and C.R. Fenster, Fallow nitrate accumulation in a wheat-fallow rotation as affected by tillage system, *Soil Sci. Soc. Am. J.*, 49, 1441, 1985.

Lamb, J.A., J.W. Doran, and G.A. Peterson, Nonsymbiotic dinitrogen fixation in no-till and conventional wheat-fallow systems, *Soil Sci. Soc. Am. J.*, 51, 356, 1987.

Linn, D.M. and J.W. Doran, Aerobic and anaerobic microbial populations in no-till and plowed soils, *Soil Sci. Soc. Am. J.*, 48, 794, 1984.

Mielke, L.N., W.W. Wilhelm, K.A. Richards, and C.R. Fenster, Soil physical characteristics of reduced tillage in a wheat-fallow system, *Trans. Am. Soc. Agric. Eng.*, 27(6), 1724, 1984.

Mielke, L.N., J.W. Doran, and K.A. Richards, Physical environment near the surface of plowed and no-tilled soils, *Soil Till. Res.*, 7, 355, 1986.

Power, J.F., W.W. Wilhelm, and J.W. Doran, Recovery of fertilizer nitrogen by wheat as affected by fallow method, *Soil Sci. Soc. Am. J.*, 50, 1499, 1986.

Siddoway, F.H. and C.R. Fenster, Soil Conservation: Western Great Plains. Dryland Agriculture, Agronomy Monograph No. 23. ASA, 1983.

Tracy, P.W., D.G. Westfall, E.T. Elliott, and G.A. Peterson, Tillage influence on soil sulfur characteristics in winter wheat-summer fallow systems, *Soil Sci. Soc. Am. J.*, 54, 1630, 1990.

Varvel, G.E., J.L. Havlin, and T.A. Peterson, Nitrogen placement evaluation for winter wheat in three fallow tillage systems, *Soil Sci. Soc. Am. J.*, 58, 219, 1988.

Wilhelm, W.W., L.N. Mielke, and C.R. Fenster, Root development of winter wheat as related to tillage in western Nebraska, *Agron. J.*, 74, 85, 1982.

II. SITE DESCRIPTION AND MANAGEMENT

The soil type at the native sod site is a Duroc loam, a fine silty, mixed, mesic Pachic Haplustoll. Its parent material is mixed loess and alluvium. Soil texture within the native sod treatment is about 40% sand, 35% silt, and 25% clay at the 0- to 15-cm depth. The slope is approximately 0.1%.

The climate at the site is characterized as continental, semi-arid with warm summers and cold winters. Mean annual maximum temperature is 16.7°C and mean annual minimum temperature is 0.2°C (1970 to 1989). Annual precipitation averages 380 mm (1970 to 1989) and occurs predominantly (75%) from April through August. The weather data were collected approximately 1.5 km from the experimental site and, therefore, weather patterns at the site may have varied at times from these data.

The native sod site is divided into two major sections: one section is seeded to winter wheat and the other is left fallow each year. The west section (also referred to as "site C") has a winter wheat crop harvested in every odd-numbered year, and the east section (also referred to as "site D") has a winter wheat crop harvested in every even-numbered year. The fallow management treatments in each major block are plowing (bare fallow), stubble-mulch, and no-tillage (chemical fallow). A sod treatment (native mixed prairie species) has been maintained in each replicate to serve as a control.

For the plowing treatment, a moldboard plow is used to a depth of approximately 15 cm in the spring, followed by two or three operations with a field cultivator and then one or two operations with the rotary rodweeder (Table 2). Stubble-mulch fallow treatments are performed with 90- to 150-cm sweeps, two to four times a year to a maximum depth of approximately 10 cm, followed by one or two operations with a rotary rodweeder. Initial tillage operations are performed at the deepest depth, with subsequent operations conducted at decreasing depths to provide a firm mellow seedbed. The no-tillage treatment relies on nonresidual herbicides such as glyphosate [*N*-(phosphonomethyl)glycine], paraquat (1,1′-dimethyl-4,4′-bipyridinium ion), 2,4-D [(2,4-dichlorophenoxy)acetic acid], and dicamba (3,6-dichloro-2-methoxybenzoic acid) to control weeds. Herbicides are applied as needed. Seeding is the only operation in the no-tillage treatment that causes soil disruption.

All non-sod plots are seeded in September with a drill equipped with large coulters, slot openers for the seed, and press wheels spaced 30 cm apart. For the period 1971 to 1983, "Centurk" winter wheat was used. Beginning with the September seeding of 1984, winter wheat varieties have varied. No fertilizer has been applied to date. Proso millet (*Panicum miliaceum* L.) was planted at site C in 1971, due to the very dry soil conditions in the fall of 1970 after plowing under the sod in the spring of 1970.

Winter wheat is harvested in July with a combine. From 1971 to 1983, yield data were taken by harvesting the entire plot. Beginning with the 1984 harvest, yield data were collected from a 1.8 × 15.2 m portion of each plot. Residue data are collected by clipping standing stubble and gathering all surface residue from 3 m^2 in each plot and weighing. Grain protein is determined by standard Kjeldahl techniques. Soil samples have been collected and numerous soil measurements made at irregular intervals and various incremental depths over the 2 decades of this study. Soil nitrogen has been determined by various modifications of Kjeldahl digestion, and soil carbon has been measured by the wet oxidation methods of Snyder and Trofymow[2] and the Smith-Weldon modification[3] of the Walkley-Black procedure.

Experimental design for each section ("C" and "D") is a randomized complete block with three replications. Experimental plots are 8.5 m wide by 45.7 m long.

Table 2 Management History at Sidney, NE from 1970 to 1989

			Fallow Tillage System (month)			
Field Equipment	**Depth (cm)**	**Operations (# y^{-1})**	**Plow**	**Stubble-Mulch**	**No-Till**	**Sod**
Plow	15	1	April–May	—	—	—
Sweep	7–10	2–4	—	May–August	—	—
Field cultivator	5–7.5	2–3	April–July	—	—	—
Rotary rodweeder	2–5	1–2	July–September	July–September	—	—
Sprayer	—	As needed	—	—	April–September	—
Drill	4–10	1	September	September	September	—

III. RESULTS AND DISCUSSION

The High Plains region is characterized by variable weather conditions and severe weather has had an influence on these sites. Winterkill was experienced in 1977 and 1979, a late hail in 1980 caught the no-till wheat at a susceptible stage and greatly reduced grain yields, hail destroyed all the wheat in 1988, and extreme drought combined with heavy downy brome (*Bromus tectorum* L.) populations destroyed the 1989 wheat crop. The sites were flooded twice since their initiation. On June 8, 1974, a 4.4-cm rainfall event at the site resulted in the plots being flooded. In July of 1981, a heavy rain to the west of the site resulted in a great amount of water moving down the watershed and flooding the site under as much as 25 cm of water. These are the realities of research and crop production in the Panhandle of Nebraska.

Fenster and Peterson[1] found no-tillage fallow to have positive effects overall, with increased water storage, improved erosion protection, and increased yields when weed control was adequate. Downy brome populations have been very high and interfered with winter wheat production in the no-till treatments on several occasions.

Surface levels of C (Figure 1, Table 3), N (Figure 2, Table 4), and water were found to be higher in no-till.[4] Broder et al.[5] reported soil organic C at significantly higher levels with reduced tillage than with plowing. Nitrifier populations were found to be 22 to 56% and 16 to 35% lower, respectively, in no-till and stubble-mulch fallow than in plow treatments. Denitrifiers tended to increase with a decrease in tillage. The results of studies conducted by Lamb et al.[6] suggest that switching from plowing to minimum or no-till systems may result in a more N-conserving system (Figure 2, Table 4).

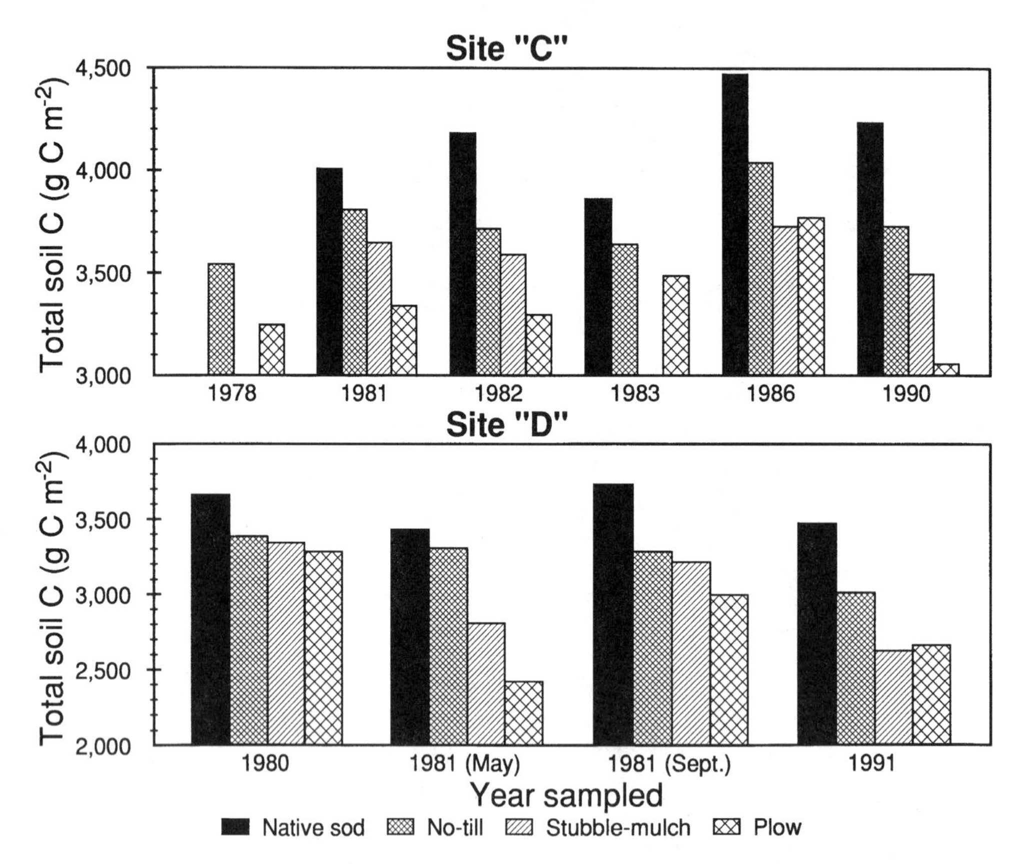

Figure 1 Fallow tillage system effects on soil C dynamics at Sidney, NE from 1970 to 1990. Prestudy C level reported at 4132 g C m^{-2} for both site "C" and "D". (From Fenster, C.R. and G.A. Peterson, Effects of No-Tillage Fallow as Compared to Conventional Tillage in a Wheat-Fallow System, Nebr. Exp. Stn. RB 289, Lincoln, 1979.)

Mielke et al.[7] described no-till soils as having a greater bulk density (Table 5) and water content than plowed soils. No-till soils generally had more pore space filled with water and more water retained by organic materials, but plowed soils were able to accept more water before biological activity became limited by aeration.

Decreased soil stability with cultivation was found to be paralleled by a reduction of organic matter concentrations and an increase in microaggregate formation.[9] Microaggregates were found to have less C, N, and P and lower specific N mineralization rates than macroaggregates. It was suggested that organic

Table 3 Total Soil Carbon for the 0- to 10- and 10- to 20-cm Depth Increments at Sidney, NE 1970-1990

	Fallow Tillage System							
	0- to 10-cm Soil Depth				10- to 20-cm Soil Depth			
Year Sampled	Sod	No-Till	Stubble-Mulch	Plow	Sod	No-Till	Stubble-Mulch	Plow
				g C m^{-2}				
1970	2225	2225	2225	2225	1907	1907	1907	1907
				Site C				
1982[6]	2483	2308	2068	1751	1699	1409	1522	1544
1986[10]	2369	2226	2084	1827	1834	1811	1643	1943
				Site D				
1981[6]	2065	2049	1664	1135	1369	1259	1145	1289

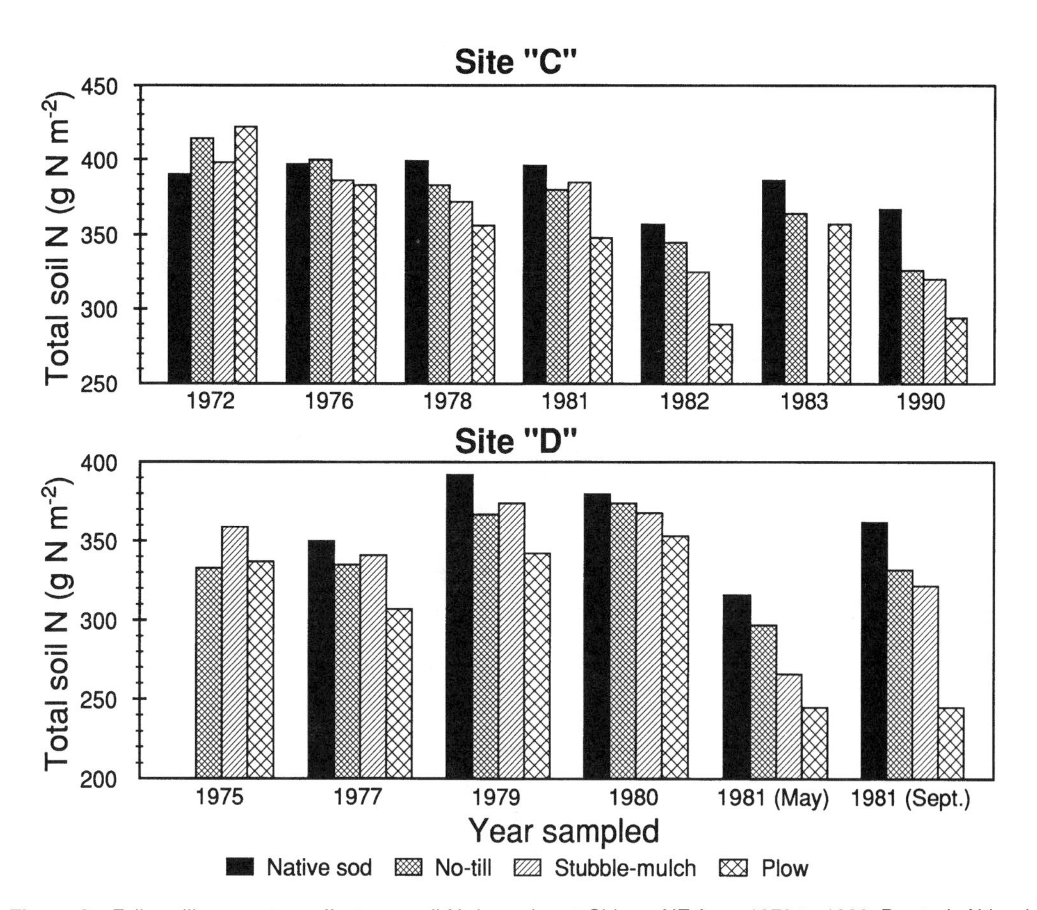

Figure 2 Fallow tillage system effects on soil N dynamics at Sidney, NE from 1970 to 1990. Prestudy N level reported at 344 g N^{-2} for both site "C" and "D". (From Fenster, C.R. and G.A. Peterson, Effects of No-Tillage Fallow as Compared to Conventional Tillage in a Wheat-Fallow System, Nebr. Exp. Stn. RB 289, Lincoln, 1979.)

matter binding microaggregates into macroaggregates is the primary source of nutrients released when organic matter is lost on cultivation.

Doran[9] found soil microbial biomass levels to be highly correlated with total organic C and N content in surface soils, both of which are largely influenced by tillage and cropping management practices. The stratification of crop residues, organic matter, and soil organisms within the profile of no-till is suggested as a major mechanism for immobilization of N near the soil surface as compared to conventional tillage

Table 4 Total Soil Nitrogen for the 0- to 10- and 10- to 20-cm Depth Increments at Sidney, NE 1970-1990

	Fallow Tillage System							
	0- to 10-cm Soil Depth				10- to 20-cm Soil Depth			
Year Sampled	Sod	No-Till	Stubble-Mulch	Plow	Sod	No-Till	Stubble-Mulch	Plow
				$g\ N\ m^{-2}$				
1970	180	180	180	180	165	165	165	165
				Site C				
1982[6]	222	207	189	147	135	138	136	143
				Site D				
1981[6]	176	183	148	129	140	114	118	116

Table 5 Soil Bulk Density at Sidney, NE in 1980 and 1982

	Fallow Tillage System ($Mg\ m^{-3}$)											
	0- to 7.5-cm Soil Depth				7.5- to 15-cm Soil Depth				15- to 30-cm Soil Depth			
Year Sampled	Sod	No-Till	Stubble-Mulch	Plow	Sod	No-Till	Stubble-Mulch	Plow	Sod	No-Till	Stubble-Mulch	Plow
						Site C and D						
1980[5]	—	1.03	1.09	1.20	—	1.29	1.35	1.33	—	1.36	1.37	1.40
						Site D						
1980[5]	0.99	1.07	1.10	1.20	1.36	1.35	1.34	1.38	1.35	1.34	1.33	1.44
1982[7]	—	1.02	—	1.18	—	1.32	—	1.34	—	1.37	—	1.35

(Table 4). Doran's findings were reiterated by Follett and Peterson,[10] who found no-till resulted in the maintenance of higher soil organic matter compared to plow tillage, especially in the surface 5 cm. Follett and Schimel[11] also reported that C available for microbial growth declined with increasing tillage intensity. Increased tillage intensity decreased the soil's capacity to immobilize and conserve mineral N. Soil P (Elliott, unpublished) and S[12] measurements collected in 1984 and 1985 also suggest a decrease in total P and S with increasing tillage intensity (Table 6).

Table 6 Total Soil Phosphorus and Sulfur in the Surface 0 to 15 cm at Sidney, NE in 1984 and 1985

	Fallow Tillage System			
Year Sampled	Sod	No-Till	Stubble-Mulch	Plow
		g P m^{-2}		
1984	90.7	90.8	88.4	83.2
1985	94.9	95.5	95.3	93.5
		g S m^{-2}		
1984[12]	52.1	47.3	49.1	41.1

Doran (unpublished) states that total C and N contents of surface soil with plowing declined by 35 to 40% over the first 11 years of this study. No-till resulted in the least loss of C and N with time. Two mechanisms responsible for the change in surface soil C and N levels are redistribution with depth in soil and biological oxidation. Losses of surface soil C and N are apparently greatest during the first 8 to 10 years after a sod is cultivated and the rate of loss appears to decline with time thereafter. Bulk densities increase with time and increasing degrees of tillage accompany declines in soil organic matter. Fallow tillage management changed the relative predominance and distributions of microbial populations in soil, especially in the top 2.5 cm.

The trends described by Doran (unpublished) after the first 11 years of the study appear to hold true for the next 9 years of the study (Figures 1 and 2). Total C and N levels in the surface 20 cm of soil, expressed as a percentage of the native sod treatment, continued to decrease with increasing tillage intensity, with the relative ranking of total C and N maintenance being sod > no-till > stubble-mulch > plow. A new equilibrium had not been reached for any of the fallow tillage systems even after 20 years of fallow tillage.

Despite the apparent detrimental effects of increased tillage on soil quality factors such as C and N level, the effect of fallow tillage on grain yields (Figure 3) and crop residue weights (Figure 4) is not evident. The relative ranking of grain yields with fallow tillage has not been consistent over time (Figure 3). No one fallow system appears to be superior to the others over the 2 decades of this study. In fact, in the majority of years there is no difference in grain yield as a result of the fallow tillage system. Occasionally, one system does produce superior grain yield, but this is usually the result of a specific weather event, e.g., winterkill being worst in the plow treatments in 1977 or hail detrimentally affecting no-till treatments to a greater extent in 1980. The inability to consistently control downy brome in the no-till plots has created some problems with interpreting the yield data over time. The problem of controlling winter annual grasses in a no-till winter wheat-fallow system greatly limits its adoption on a wide scale in the Central High Plains.

IV. CONCLUDING REMARKS

As C and N continue to be lost from the soil as a result of tillage and crop production, differences in grain yield between treatments may become more apparent. Nitrogen available to the crop through mineralization of organic matter will decrease as organic matter is lost through oxidation as a result of tillage. Farmers in the Nebraska Panhandle began to notice a benefit from N application in winter wheat approximately 50 years after breaking their land out of native sod. It will be interesting to follow the progress of this study over the next 20 years to see if or when consistent grain yield differences between fallow tillage systems become evident. It is clear from the work conducted in the first 20 years that soil quality factors are being differentially affected by the different fallow tillage systems. Soil C and N can be conserved by minimizing tillage operations, particularly inversion tillage, within a cropping system.

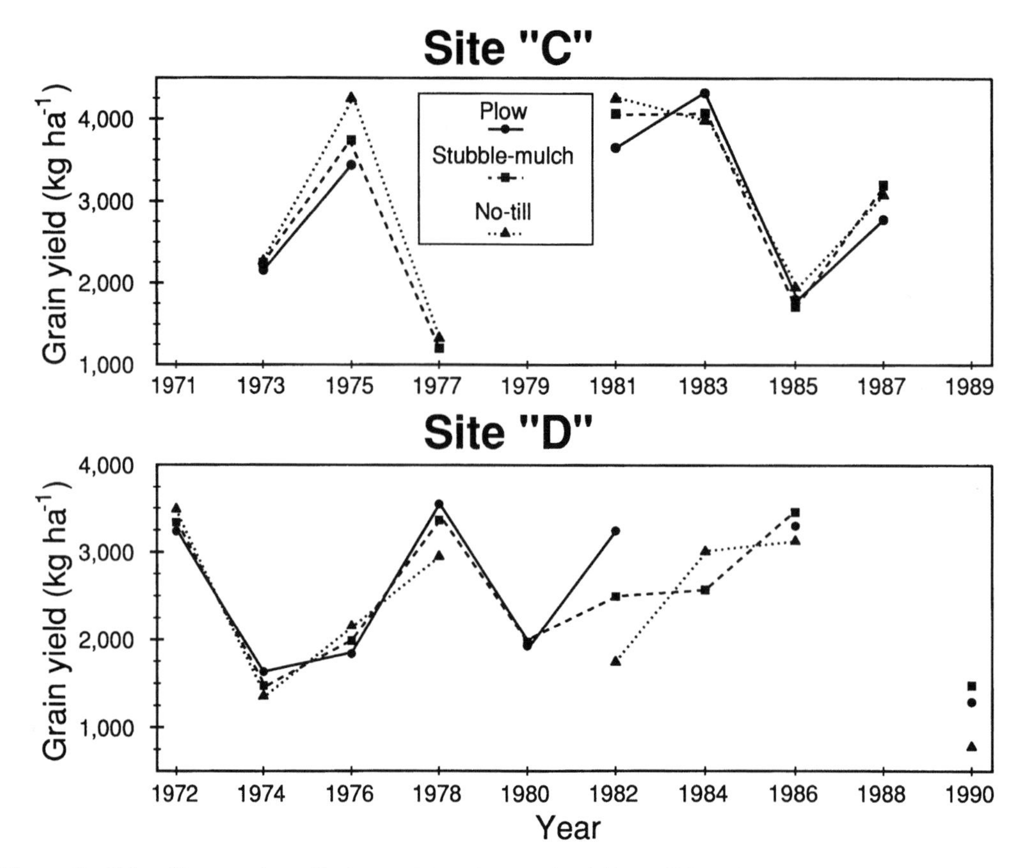

Figure 3 Fallow tillage system effects on winter wheat grain yields at Sidney, NE from 1971 to 1990.

The sampling procedures used to collect the data presented in this chapter has varied over time. Differences in depth of sampling alone have made comparisons over time difficult to make. Add to this differences in time of year sampled, sampling in-crop vs. in-fallow, and different analytical procedures for determining C and N levels and you have a situation that stifles interpretation of results. Perhaps simulation modeling of existing data could help with interpretation.[13]

The long-term tillage sites located at the High Plains Agricultural Laboratory near Sidney, NE have provided an excellent opportunity to study the effects of fallow tillage and winter wheat crop production on the physical, chemical, and biological parameters of a Central High Plains native prairie soil over time. The site will continue to provide researchers with the opportunity to study soil organic matter dynamics, as well as other changes occurring to the soil as a result of a very common winter wheat-fallow cropping system.

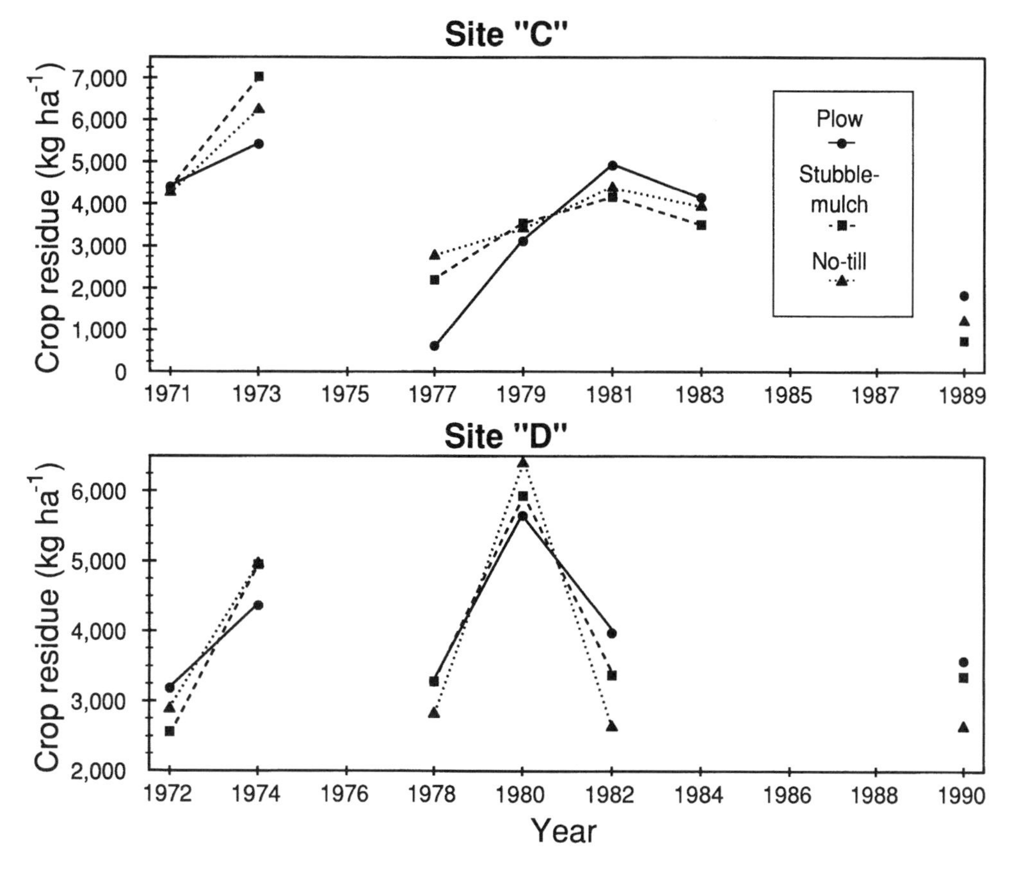

Figure 4 Fallow tillage system effects on winter wheat residue weights after grain harvest at Sidney, NE from 1971 to 1990. Residue weights reported for 1971 are from proso millet (*Panicum miliaceum* L.).

REFERENCES

1. Fenster, C.R. and G.A. Peterson, Effects of no-tillage fallow as compared to conventional tillage in a wheat-fallow system, Nebr. Exp. Stn. RB 289, 1979.
2. Snyder, J.D. and J.A. Trofymow, A rapid accurate wet oxidation diffusion procedure for determining organic and inorganic carbon in plant and soil samples, *Commun. Soil Sci. Plant Anal.*, 15:587, 1984.
3. Smith, H.W. and M.D. Weldon, A comparison of some methods for the determination of soil organic matter, *Soil Sci. Soc. Am. Proc.*, 5, 177, 1984.
4. Doran, J.W., Soil microbial and biochemical changes associated with reduced tillage, *Soil Sci. Soc. Am. J.*, 44, 765, 1980.
5. Broder, M.W., J.W. Doran, G.A. Peterson, and C.R. Fenster, Fallow tillage influence on spring populations of soil nitrifiers, denitrifiers, and available nitrogen, *Soil Sci. Soc. Am. J.*, 48, 1060, 1984.
6. Lamb, J.A., G.A. Peterson, and C.R. Fenster, Wheat fallow tillage systems effect on a newly cultivated grassland soil's N budget, *Soil Sci. Soc. Am. J.*, 49, 352, 1985.
7. Mielke, L.N., J.W. Doran, and K.A. Richards, Physical environment near the surface of plowed and no-tilled soils, *Soil Till. Res.*, 7, 355, 1986.
8. Elliott, E.T., Aggregate structure and C, N, and P in native and cultivated soils, *Soil Sci. Soc. Am. J.*, 50, 627, 1986.
9. Doran, J.W., Microbial biomass and mineralizable nitrogen distributions in no-tillage and plowed soils, *Biol. Fertil. Soils*, 5, 68, 1987.
10. Follett, R.F. and G.A. Peterson, Surface soil nutrient distribution as affected by wheat-fallow tillage systems, *Soil Sci. Soc. Am. J.*, 52, 141, 1988.
11. Follett, R.F. and D.S. Schimel, Effect of tillage practices on microbial biomass dynamics, *Soil Sci. Soc. Am. J.*, 53, 1091, 1989.
12. Tracy, P.W., D.G. Westfall, E.T. Elliott, and G.A. Peterson, Tillage influence on soil sulfur characteristics in winter wheat-summer fallow systems, *Soil Sci. Soc. Am. J.*, 54, 1630, 1990.
13. Metherell, A.K., C.A. Cambardella, W.J. Parton, G.A. Peterson, L.A. Harding, and C.V. Cole, Simulation of soil organic matter dynamics in dryland wheat-fallow cropping systems, in *Soil Management and Greenhouse Effect*, R. Lal, J. Kimble, E. Levine, and B.A. Stewart, Eds., Lewis Publishers, Boca Raton, FL, 1995, 259.

Chapter 26

Soil Carbon and Nitrogen Change In Long-Term Agricultural Experiments at Pendleton, Oregon

P.E. Rasmussen and R.W. Smiley

CONTENTS

I. INTRODUCTION

The Pendleton Agricultural Research Center has several ongoing long-term experiments. The earliest was started in 1931, the latest in 1982. The Residue Management and Tillage-Fertility experiments are among the oldest replicated research experiments in the western U.S.[1] All have a documented history of crop variety, tillage, date of seeding, and grain yield. The studies are representative of most of the cropping systems in the Pacific Northwest intermountain cereal region that receives less than 450 mm precipitation. All research activities on the long-term experiments are presently monitored by an oversight committee consisting of four members from Oregon and one each from Washington and Idaho.

The Pendleton Agricultural Research Center was established in 1929 as a branch station of Oregon State University. It is presently known as the Columbia Basin Agricultural Research Center, and is administered by the OSU Agricultural Experiment Station. The Columbia Plateau Conservation Research Center, administered by USDA-ARS, is immediately adjacent. Research facilities are shared jointly by the staff of both agencies.

II. SITE DESCRIPTION

The center is located 15 km northeast of Pendleton, in the northeastern corner of Oregon. It lies within the Columbia Plateau physiographic province between the Cascade and Rocky mountains. The elevation is 455 m above sea level. The climate is semi-arid but partially influenced by maritime winds from the Pacific Ocean. Winters are cool and wet and summers hot and dry. Average temperature is 10.2°C, but ranges from –0.6 in January to 21.2 in July. Precipitation occurs primarily during the winter, in direct contrast to climatic patterns in the midwest and eastern U.S. Annual precipitation is 420 mm, about 70% of which occurs between September 1 and March 31. Winter precipitation falls mainly as rain with limited duration of snow cover in most years.

Soils consist of loess overlying basalt and are young geologically. Soils are coarse silty mixed mesic Typic Haploxerolls (Walla Walla silt loam). The upper 30 cm of soil contains about 18% clay and 70% silt.

III. EXPERIMENT DESIGN

Six sets of research plots are designated as long-term experiments (Table 1). Site features, agronomic practices, and soil quality measurements are described below, along with details on experimental design and treatment histories for three of the studies.

Table 1 Designated Long-Term Research Sites at Pendleton

Year Initiated	Symbol	Experiment Name	Treatment Variables
1931	GP	Grass pasture	None
1931	CW	Continuous cereal	None
1931	CR	Residue management	Nitrogen, manure, burning
1940	TP	Tillage fertility	Tillage, fertility
1963	WP	Wheat/pea	Tillage
1982	SF	No-till wheat	Nitrogen

A. GRASS PASTURE (GP)

This site contains no experimental variables, but has been maintained since 1931. It approximates near-virgin grassland and serves as a baseline for evaluating changes in the other systems. It is periodically reseeded with introduced-grass selections, occasionally fertilized, and infrequently irrigated. The dominant grass species is tall fescue *(Festuca arundinacea* Scheeber) with lesser amounts of bulbous bluegrass (*Poa bulbosa* L.), green foxtail (*Setina viridis* (L.) P. Beauv.), and yellow foxtail [*S. pumila* (Poiret) Roemer & Schultes]. This site received limited grazing from 1931 to 1985. It has not been grazed since, but vegetation is clipped once or twice during summer growth.

B. CONTINUOUS CEREAL (CW)

This experiment was established in 1931 and cropped annually to winter wheat (*Triticum aestivum* L.) from 1931 to 1982. The site was modified in 1982 to accommodate winter wheat, spring wheat, and spring barley, each grown every year in the same location. The site is conventionally tilled (moldboard plowed) 20 to 22 cm deep, and receives both chemical and mechanical weed control. The original experiment consisted of eight plots, each 11.6 × 80.4 m, with no replication. The eight plots received different rates of N fertilization (0 to 168 kg ha^{-1} y^{-1}) from 1943 to 1951. No fertilizer was applied from 1952 to 1959, and 90 kg N ha^{-1} y^{-1} has been uniformly applied to the entire area since 1960. The experiment has periodically received P and S fertilization since 1982. The different N rates created a pH (1:2 soil:0.01 $CaCl_2$) difference that currently ranges from 5.2 for low N to 4.6 for high N. This site currently serves as a cereal monoculture baseline for comparing changes in crop rotation systems.

C. RESIDUE MANAGEMENT (CR)

This is the most comprehensive of the long-term experiments (Table 2). It was established in 1931 and has had only two major revisions (1967, 1979). The rotation is winter wheat/fallow with conventional tillage (moldboard plow). The experimental design is an ordered block consisting of nine treatments (ten originally) and two replications. The experiment contains duplicate sets of experiments that are offset by 1 year so that data can be obtained annually. Plot size is 11.6 × 40.2 m. Present treatments by number are (2) spring burning-45 kg N ha^{-1}, (3) spring burning-90 kg N ha^{-1}, (4) no burning-45 kg N ha^{-1}, (5) no burning-90 kg N ha^{-1}, (6) fall burning, no N, (7) spring burning, no N, (8) no burning-22.4 t ha manure, (9) no burning-2.24 t pea vines, and (10) no burning-no N. Organic amendments are applied in the spring of the fallow year (late March-early April). All plots are plowed 22 cm deep after the organic amendments have been applied and spring burns implemented. Weeds are controlled by tillage during the fallow phase and with herbicides during the crop phase. Nitrogen fertilizer is applied 5 to 15 days prior to seeding of wheat.

Table 2 Treatment History of the Residue Management Experiment

Treatment No.	Organic N Addition	1931–1966		1967–1978		1979 +	
		RTa	Nb	RT	N	RT	N
1	—	—	—	—	—	—	—
2	—	FD	0	NB	45	SB	45
3	—	SD	0	NB	90	SB	90
4	—	NB	34	NB	45	NB	45
5	—	NB	34	NB	90	NB	90
6	—	FB	0	FB	0	FB	0
7	—	SB	0	SB	0	SB	0
8	MN[c]	NB	0	NB	0	NB	0
9	PV[d]	NB	0	NB	0	NB	0
10	—	NB	0	NB	0	NB	0

Note: Organic residues applied in late March of fallow year and N fertilizer in early October of crop year.

[a] RT = residue treatment: FD = fall disk, SD = spring disk, NB = no burn, FB = fall burn, SB = spring burn.

[b] N = nitrogen rate (kg $ha^{-1}crop^{-1}$).

[c] MN = manure = 22.4 t $ha^{-1}crop^{-1}$ wet wt; 0.508 DM, 1482 kg C, 111 kg N ha^{-1} per crop.

[d] PV = pea vines = 2.24 t $ha^{-1}crop^{-1}$, 0.897 DM, 793 kg C, 34 kg N ha^{-1} per crop.

The C and N content of the upper 60-cm of soil has been determined about every 10 years (1931, 1941, 1951, 1964, 1976, and 1986).[2–4] Straw yield, grain, and straw N content and the nutrient content of organic amendments have been determined since 1977.[4] Straw yield and nutrient uptake from 1931 to 1976 has been estimated by utilizing variety-trial data coupled with periodic measurements in this experiment.

D. TILLAGE-FERTILITY (TF)

This experiment was established in 1940 and had major revisions in 1952, 1962, and 1988. The rotation is winter wheat/fallow. This experiment has only one set of plots; thus, yield is obtained only in odd years. The experimental design is a randomized block split-plot, with three replications. Main plots consist of three primary tillage systems (moldboard plow, offset disk, and subsurface sweep) and subplots of six fertility levels (currently, rates from 0 to 180 kg N ha^{-1} in 45-kg increments, with one duplication, Table 3). Individual plot size is 5.5 × 40.2 m. Primary tillage is performed in April. Secondary tillage operations are the same for all treatments. All plots are smoothed 10 to 15 cm deep with a field cultivator and harrow following primary tillage. They are then rodweeded four to five times between April and October to control weeds and maintain seed zone moisture. Nitrogen fertilizer is applied about October 1 and winter wheat seeded about October 10. Nitrogen was broadcast as NH_4NO_3 from 1963 to 1987 and applied thereafter as urea-ammonium nitrate shanked 15 cm deep with 25-cm band spacing. The experiment relies on both mechanical and chemical weed control, but the stubble mulch treatments (DI and SW) have occasionally received extra chemical treatment when grassy weeds have been a problem.

The replicates differ in depth to a root-restricting layer by virtue of landscape position. Replicate 1 (208 ± 9 cm deep) is located on a north-facing 3% back slope, replicate 2 (132 ± 25 cm) on east-west facing foot slopes of 0 to 2%, and replicate 3 (111 ± 9 cm deep) on an east-facing 2% back slope. Medium-tall soft white winter wheat was grown from 1940 to 1962 and semi-dwarf soft white winter

Table 3 Treatment History of the Tillage-Fertility Experiment

Main Plots		Tillage	Residue Cover
Symbol	Tillage	Depth (cm)	at Seeding (%)
MP	Plow	22	7
DI	Disk	15	34
SW	Sweep	15	43

Subplots		Nitrogen Rate[a] (kg ha^{-1} crop^{-1})			
No.	Symbol	1941–1952	1953–1962	1963–1988	1989+
1	45–	0	0	45	0
2	45+	11	34	45	45
3	90–	0	0	90	90
4	90+	11	34	90	90
5	135+	11	34	135	135
6	180+	11	34	180	180

[a] Nitrogen applied as ammonium sulfate from 1941 to 1962, ammonium nitrate from 1963 to 1988, and urea-ammonium nitrate since 1989. Nitrogen broadcast from 1941 to 1988 and banded 15 cm deep with 25-cm row spacing since 1989. Nitrogen applied 7 to 14 days prior to seeding. Treatments with a minus symbol have no history of sulfur application.

wheat since. Straw yield and grain and straw N content have been determined since 1977. Soil N, C, and pH were determined in 1984 in 7.5-cm increments to a depth of 30 cm.[5,6]

E. WHEAT/PEA (WP)

This experiment consists of four tillage treatments in a winter wheat/pea rotation (Table 4). It was established in 1963 and modified in 1972, 1976, and 1989. The experimental design is a randomized block with four replications and duplicated treatment sets within each replication. Offset duplicate replications allow yearly data collection for both wheat and peas. Individual plot size is 7.3 × 36.6 m. Tillage intensity ranges from maximal inversion to noninversion. The current tillage treatments are (1) MAXTILL (fall rototill for peas, fall disk/chisel for wheat), (2) FALLPLOW (fall plow for both peas and wheat), (3) SPRINGPLOW (spring plow for peas, fall plow for wheat), and (4) MINTILL (skewtread for peas, subsurface sweep for wheat). Depth of tillage is 20 to 22 cm for moldboard plow, 15 cm for disk, 30 cm for chisel, and 10 cm for sweep. Semidwarf soft white winter wheat is seeded after October 10 whenever soil moisture is sufficient for germination and early crop growth. Peas are seeded in late March or early April and harvested in late June or early July. The type of peas grown was changed from fresh-green processing to dry-edible seed in 1989. From 1963 to 1988, wheat received from 45 to 90 kg N ha^{-1} broadcast as NH_4NO_3 prior to seeding. After 1988, each wheat plot receives 22 kg N ha^{-1} as ammonium phosphate-sulfate (16-9-0-14S). One-half of the plot (lengthwise) then receives 90 kg N ha^{-1} as urea-ammonium nitrate and the other half none. The side receiving extra N reverses in each succeeding wheat crop, so that the average N application for wheat is now 66 kg ha^{-1} y^{-1}. Peas receive 20 kg N and 45 kg S ha^{-1} as broadcast $(NH_4)_2SO_4$ every second or third crop. The east half of the experiment received 2.00 t ha^{-1} of lime in 1976. A 7.32 × 7.32 m area on the western edge of certain plots was fumigated in the early 1980s.

F. NO-TILL WHEAT (SF)

This experiment was established in 1982 and modified in 1983 and 1988. The initial crop rotation was winter wheat/spring wheat, but this was changed in 1988 to winter wheat/fallow. The experimental design

Table 4 Treatment Design of the Wheat/Pea Experiment

Treatment		Primary Tillage	
No.	Identification	Wheat Stubble	Pea Vines
1	Maxtill	Rototill (fall)	Disk/chisel (fall)
2	Falltill	Plow (fall)	Plow (fall)
3	Springtill	Plow (spring)	Plow (fall)
4	Mintill	Skewtread (fall)	Sweep (summer)

consists of ten treatments and four replications. Treatments consist of five N rates (0, 56, 112, 112, and 168 kg ha^{-1}), in duplicate. The design included a residue burn vs. no-burn variable from 1983 to 1988. The burn variable was converted to a date-of-seeding variable in 1989. Plot size is 2.4 × 30.5 m. Tillage is strictly no-till, with only chemical weed control and no tillage other than for seeding and stubble flailing. It is a recent addition and was implemented to evaluate N fertilizer effects on crop yield and soil quality under no-till cropping.

IV. CHANGES IN SOIL CARBON AND NITROGEN

A. GENERAL TRENDS

Carbon inputs, soil C and N, biomass C and N, and pH of several different cropping systems are shown in Table 5. In general, annual cropping favors greater soil C and N retention than does wheat/fallow rotation. The total C/N ratio in soil is generally higher and biomass C/N ratio lower in cultivated soil than in pasture soil.[7] Soil C and N are responsive to residue input.[8] Fertilizer N has generally improved soil C and N levels.[5]

Table 5 Average C Input, Soil C and N, Microbial Biomass C and N, and pH in the Long-Term Experiments in 1987

Experiment (treatment)	Organic-C Input ($t\ ha^{-1}\ y^{-1}$)	Soil ($g\ kg^{-1}$) C	Soil ($g\ kg^{-1}$) N	Microbial Biomass ($mg\ kg^{-1}$) C	Microbial Biomass ($mg\ kg^{-1}$) N	pH
GP	(2.51)	22.2	2.00	870	96	7.4
CW	2.25	15.0	1.27	410	51	5.4
WP(2)	2.29	15.5	1.21	430	65	5.8
CR(8)	2.41	15.2	1.05	350	62	6.7
CR(5)	1.68	11.2	0.85	230	35	5.6
CR(10)	1.13	12.0	0.80	260	31	6.1
CR(6)	0.40	10.5	0.72	210	22	6.1

Adapted from Collins, H.P., Rasmussen, P.E., and Douglas, C.L., Jr., *Soil Sci. Soc. Am. J.*, 56, 783, 1992.

B. RESIDUE MANAGEMENT EXPERIMENT

There is a continuing steady linear decline in OM in the top 30 cm of soil with time for all treatments except the manure application (Figure 1). The rate of change is related to residue input, with about 5.5 t ha^{-1} required to prevent continuing decline in C or N.[8] Soil N changes closely reflect those of soil C.[4] Wheat yield trends with time (Figure 2) mimic the soil organic matter changes illustrated in Figure 1. Unfertilized treatments show a nearly 40% loss in productivity after 60 years of cultivation. Changes in both yield and soil organic C and N content correlate reasonably well with predicted change by the Century model.[9]

There has also been a slow decline in C and N in the 30- to 60-cm zone independent of treatment,[4] which is probably the result of a change in oxidation equilibrium resulting from the conversion from virgin grassland to a cultivated system. Carbon input from root material is generally much less under cultivation than under virgin grassland.

C. TILLAGE-FERTILITY EXPERIMENT

Nitrogen fertilization has had a positive linear influence on soil C and N (Figure 3). Soil C levels for treatments with no S application fit the regression line less precisely than do the N values. This suggests that an S deficiency may exist, which has affected C/S transformations more than C/N transformations.

A major concern in assessing long-term changes in C and N in soil over time is the effect of acid-forming fertilizer.[6] Ammonium-N based fertilizers have increased soil N (Figure 3) while simultaneously decreasing soil pH (Figure 4). Increasing acidity appears to be retarding N mineralization and apparently slowing microbial N turnover. Lime addition in soil incubation studies increased NO_3-N mineralization, especially in annual-crop systems.[10] The degree of stability of this N fraction is presently unknown — it may be more labile than the native fraction and could mineralize rapidly if soil is limed to or above the original pH.

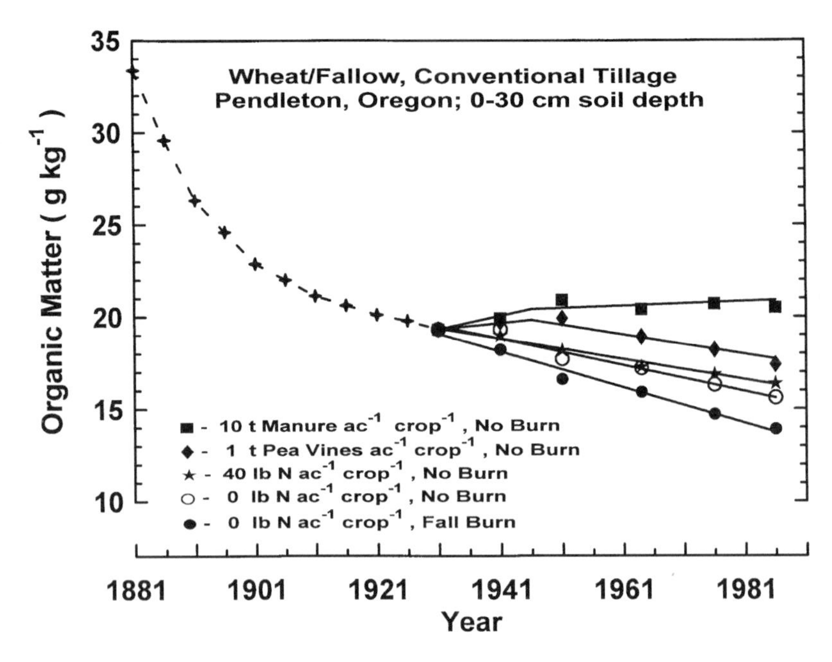

Figure 1 Treatment effects on organic matter in the 0- to 30-cm soil zone of the residue management experiment, 1931 to 1986. (Adapted from Rasmussen, P.E., Collins, H.P., and Smiley, R.W., Stn. Bull. 675, USDA-ARS and Oregon State Univ. Agri. Expt. Stn., Corvallis, OR, 1989.)

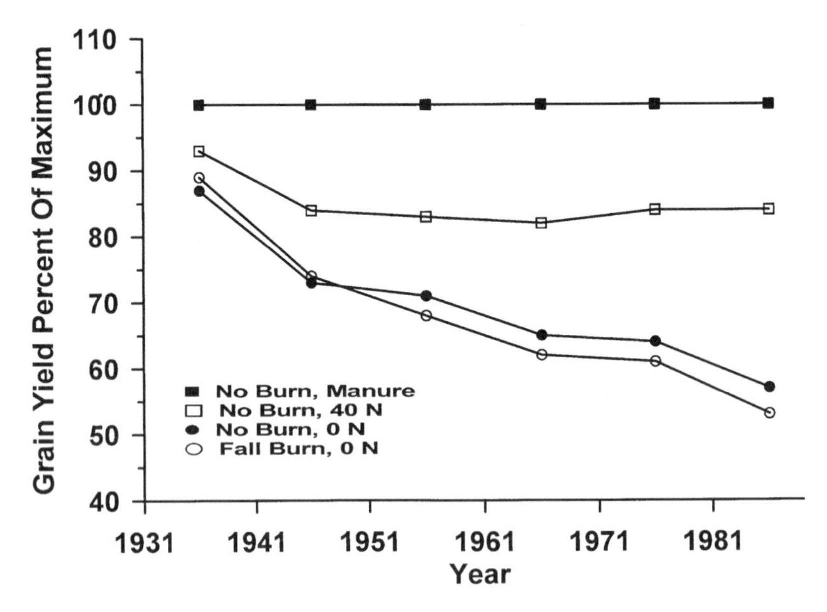

Figure 2 Treatment effects on wheat yield trends in the residue management experiment, 1931 to 1986. (Adapted from Rasmussen, P.E., Collins, H.P., and Smiley, R.W., Stn. Bull. 675, USDA-ARS and Oregon State Univ. Agri. Expt. Stn., Corvallis, OR, 1989.)

V. RELATED RESEARCH ON LONG-TERM EXPERIMENTS

Related research on the long-term experiments includes studies of silica movement,[11] biological and enzyme activity,[12,13] and physical properties.[14,15] The influence of stubble burning on wheat yield and soil properties is being examined periodically.[16,17] The influence of tillage, soil depth, and precipitation on crop yield and N utilization was determined for the TP study.[18] Studies of soil quality are presently

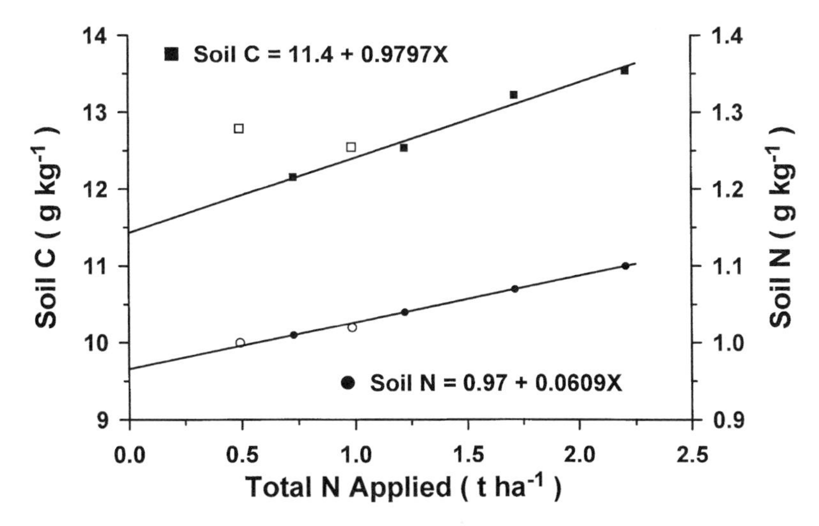

Figure 3 The effect of N fertilizer on C and N in the 0- to 22.5-cm soil zone of the tillage-fertility experiment, 1984. Solid symbols for treatments with S fertilization history; open symbols for treatments with no S application. (Adapted from Rasmussen, P.E. and Rohde, C.R., *Soil Sci. Soc. Am. J.*, 52, 1114, 1988.)

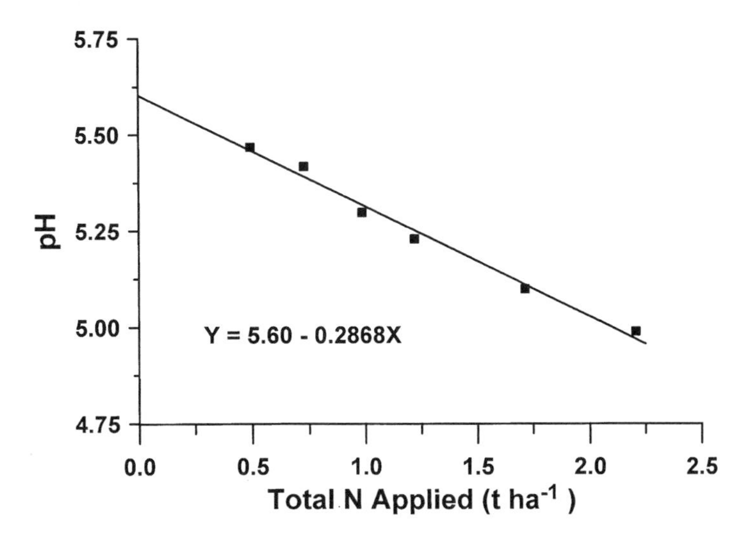

Figure 4 The effect of ammonium-based N fertilizer on soil pH (1:2 soil:0.01 M $CaCl_2$) in the 0- to 22.5-cm zone of the tillage-fertility experiment, 1984. (Adapted from Rasmussen, P.E. and Rohde, C.R., *Soil Sci. Soc. Am. J.*, 53, 119, 1989.)

in progress. Nutrient mineralization potentials and N release studies are being conducted. Sulfur and phosphorus adequacy and soil acidity-liming relations will be addressed in the near future, with some treatment modifications planned for the late 1990s.

VI. SUMMARY

Organic C and N in semi-arid soils appear quite sensitive to level of residue return. Increasing N application has increased N and C retention in soil organic matter, primarily because of increased residue production. Soil cropped annually has higher C and N content than soil cropped in wheat-fallow rotation. Soil under stubble mulch tillage contains more C and N than soil that is plowed. Ammonium-based N fertilizers are progessively acidifying the upper 8 cm of soil and may be influencing microbial diversity N transformations and soil quality.

None of the above responses would have been identified by 1- to 3-year duration experiments. Well-managed long-term experiments appear essential to properly evaluate crop and soil management practices on soil quality and sustainable agroecosystems.

REFERENCES

1. Mitchell, C.C., Westerman, R.L., Brown, J.R., and Peck, T.R., Overview of long-term agronomic research, *Agron. J.*, 83, 24, 1991.
2. Oveson, M.M., Conservation of soil nitrogen in a wheat-summer fallow farming practice, *Agron. J.*, 58, 444, 1966.
3. Rasmussen, P.E., Allmaras, R.R., Rohde, C.R., and Roager, N.C., Jr., Crop residue influences on soil carbon and nitrogen in a wheat-fallow system, *Soil Sci. Soc. Am. J.*, 44, 596, 1980.
4. Rasmussen, P.E. and Parton, W.J., Long-term effects of residue management in wheat/fallow. I. Inputs, yield, and soil organic matter, *Soil Sci. Soc. Am. J.*, 58, 523, 1994.
5. Rasmussen, P.E. and Rohde, C.R., Long-term tillage and nitrogen fertilization effects on organic nitrogen and carbon in a semi-arid soil, *Soil Sci. Soc. Am. J.*, 52, 1114, 1988.
6. Rasmussen, P.E. and Rohde, C.R., Soil acidification by NH_4-N fertilization with conventional and stubble mulch tillage in a wheat/fallow system, *Soil Sci. Soc. Am. J.*, 53, 119, 1989.
7. Collins, H.P., Rasmussen, P.E., and Douglas, C.L., Jr., Crop rotation and residue management effects on soil carbon and microbial dynamics, *Soil Sci. Soc. Am. J.*, 56, 783, 1992.
8. Rasmussen, P.E. and Collins, H.P., Long-term impacts of tillage, fertilizer, and crop residue on soil organic matter in temperate semi-arid regions, *Adv. Agron.*, 45, 93, 1991.
9. Parton, W.J. and Rasmussen, P.E., Long-term effects of residue management in wheat/fallow. II. Century model simulations, *Soil Sci. Soc. Am. J.*, 58, 530, 1994.
10. Rasmussen, P.E., Allmaras, R.R., and Douglas, C.L. Jr., Nitrogen fertilizer effects on soil acidity and N mineralization, *Agron. Abstr.*, Am. Soc. Agron., Madison, WI, 1991, 298.
11. Douglas, C.L., Jr., Allmaras, R.R., and Roager, N.C., Jr., Silicic acid and oxidizable carbon movement in a Walla Walla silt loam as related to long-term management, *Soil Sci. Soc. Am. J.*, 48, 156, 1984.
12. Castellano, S.D. and Dick, R.P., Distribution of sulfur fractions in soil as influenced by management of organic residues. *Soil Sci. Soc. Am. J.*, 52, 1403, 1988.
13. Dick, R.P., Rasmussen, P.E., and Kerle, E.A., Influence of long-term management on soil enzyme activities in relation to soil chemical properties of a wheat-fallow system, *Biol. Fert. Soils*, 6, 159, 1988.
14. Pikul, J.L. Jr. and R.R. Allmaras, Physical and chemical properties of a Haploxeroll after fifty years of residue management, *Soil Sci. Soc. Am. J.*, 50, 214, 1986.
15. Zuzel, J.F., Pikul, J.L., Jr., and Rasmussen, P.E., Tillage and fertilizer effects on water infiltration, *Soil Sci. Soc. Am. J.*, 54, 205, 1990.
16. Rasmussen, P.E., Collins, H.P., and Smiley, R.W., Long-term management effects on soil productivity and crop yield in semi-arid regions of Eastern Oregon, Stn. Bull. 675, USDA-ARS and Oregon State Univ. Agric. Expt. Stn., Corvallis, OR, 1989.
17. Rasmussen, P.E. and Rohde, C.R., Stubble burning effects on winter wheat yield and nitrogen utilization, *Agron. J.*, 80, 940, 1988.
18. Rasmussen, P.E. and Rohde, C.R., Tillage, soil depth, and precipitation effects on wheat response to nitrogen, *Soil Sci. Soc. Am. J.*, 55, 121, 1991.

Chapter

Long-Term Tillage and Crop Residue Management Study at Akron, Colorado

A.D. Halvorson, M.F. Vigil, G.A. Peterson, and E.T. Elliott

CONTENTS

I. INTRODUCTION

Following the dust bowl years of the 1930s, summer fallow became a mechanism to conserve soil water for future crop production. Intensive tillage was used for weed control during the fallow period. Precipitation storage efficiency during the fallow period was relatively low with conventional tillage practices. Stubble mulch tillage practices were developed during the 1960s that improved precipitation storage efficiency by maintaining more crop residue on the soil surface during the fallow period. During the late 1980s, herbicides became available that could be used to control weed growth during the fallow period rather than using mechanical tillage. Use of chemicals for weed control and reduction in mechanical tillage during the fallow period had potential to increase precipitation storage efficiency. The long-term tillage study reported in this paper was initiated to evaluate the effects of various combinations of tillage and herbicides for weed control during the fallow period in a winter wheat-fallow farming system. This chapter reports comparative data for the conventional-till, reduced-till, and no-till systems from this study.

II. SITE DESCRIPTION

The site is at the Central Great Plains Research Station (40°9′N, 103°9′W, 1384 m a.m.s.l) located 6.4 km east of Akron, CO on Highway 34 (NW 1/4 of SE 1/4 of Sec. 12, T2N, R52W, Washington County). The climate is continental, characterized by warm summers (mean daily temperature in July is 23°C) and cold winters (mean temperature in January is –3.9°C), with large diurnal temperature differences associated with the high altitude and predominantly clear skies. Mean annual precipitation

0-8493-2802-0/97/$0.00+$.50

is 420 mm with about 70% occurring between April and August. Year to year variability is high with annual totals varying between 250 and 680 mm, since records began at the station (1908).

The soil is a Weld silt loam (Aridic Paleustoll) with a mean surface soil texture of about 30% sand, 40% silt, and 30% clay. The site has been under cultivation since 1907. From 1907 to 1954, most of the area was used to study numerous crop rotations.[1] During this time, most of the crop residues were removed from the site. In 1955, the entire area was cropped to grain sorghum. From 1956 to 1966 the area was uniformly cropped to winter wheat in a crop-fallow rotation. In 1967, a study was initiated on the area by Wally Greb and Darryl Smika (USDA-ARS) to evaluate the use of herbicides and tillage for weed control during fallow. There were two sets of plots established in separate blocks so that there was a cropped and fallowed plot for each treatment each year. Eight tillage and crop rotation treatments in all were established. Data from three of the tillage treatments are reported in this paper. Tillage treatments are

- *Conventional (stubble-mulch) till fallow.* Tillage is performed with sweeps and rodweeder operations during the 19-month fallow period as needed. Average number of operations is six, which incorporates about 55 to 65% of the previous crop residues by wheat planting time.
- *Reduced till fallow.* A combination of tillage, residual herbicide (Atrazine), and contact herbicides (Paraquat + 2,4-D) are used. When the residual herbicide no longer controls weeds (average 11 months after application), a sweep plow or rodweeder tillage is used as needed for weed control. Generally, two or three tillage operations, which incorporate 30 to 35% of the residue, are needed to control weeds.
- *No-till fallow.* Residual herbicide (atrazine) and contact herbicides (Paraquat + 2,4-D) are used. When residual herbicides no longer control weeds (average 11 months after application), contact herbicides are used as needed for weed control. Average estimated residue loss is 20% from natural causes (wind, etc.).

Details of the tillage operations and treatments are provided by Smika.[2] Figure 1 shows the physical layout of the plots in 1991, as modified in 1989. Five of the previous reduced till treatments were converted to other tillage and crop rotation systems in 1989. Plot numbers 102, 205, 304, and 407 correspond to the conventional till (CT) plots; 108, 206, 303, and 401 to the reduced till (RT) plots; and 105, 207, 301, and 403 to the no-till (NT) plots. These plot numbers correspond to the north set of plots (top of Figure 1). The south set of plots have the same plot numbers (plus 400) for each of the treatments mentioned above. In the data tables, only one set of numbers is used to identify the treatments in the north and south sets of plots. The north plots are 11 × 30 m in size and the south plots 7.3 × 30 m.

A native grass site is located adjacent to the tillage plot area. This area could be sampled as a reference site to document changes in soil quality caused by long-term crop production practices.

III. RESULTS

A. GRAIN AND STRAW YIELDS

Semi-dwarf wheats were grown during the time period reported here. "Vona" winter wheat was grown from 1979 to 1988 and "TAM 107" from 1989 to 1991. A hoe-type drill with a 30-cm row spacing was used to plant the wheat until 1989, then a disk drill with a 18-cm row spacing was used through 1991. Grain yields for each of the treatments are reported in Table 1 for 1979 through 1991 for the north and south sets of plots. Grain yields were determined using field and plot combines to harvest a minimum of 45 m^2 from each plot. Grain yields were corrected to 12% moisture content. Grain and straw yields were not measured in 1985 because the crop was harvested as forage in early June and removed from the plot area because of a heavy infestation of jointed goatgrass (*Aegilops cyclindrica* Host.). Grain yields were generally not significantly affected by tillage system at the 0.05 probability level in any year. Grain yields varied from year to year depending on amount and distribution of precipitation during growing season. The average grain yield over all years was 2954, 2844, and 2788 kg/ha for the CT, RT, and NT tillage treatments, respectively, excluding 1985. The nearly identical yields between tillage systems is probably not surprising when one considers the fact that, after a fallow period, the root zone is at field capacity regardless of tillage system used.

Straw yields are reported in Table 2 for 1979 through 1991 for the north and south sets of plots. Straw yields were determined by harvesting a small area (1 to 2 m^2) of the plot at grain harvest for total biomass production (straw + grain), then subtracting the grain yield as determined by combine to get estimated straw yield. In 1985, the reported 560 kg/ha of straw stubble was only a visual estimate after the wheat crop was harvested as a forage. In 1988, straw yield was not measured. Tillage system did not have a significant affect on straw yield in any of the years reported, except for 1979. In 1979, straw yields of the CT system were significantly greater than those for the RT and NT systems. Straw yields, like grain yields, varied greatly from year to year depending on amount and distribution of precipitation

LONG TERM TILLAGE (AKRON) YEAR 1991

101	CC	201	CFW	301	NT	401	RT
102	CT	202	PLOW	302	CFW	402	WC
103	FWC	203	FWC	303	RT	403	NT
104	WCF	204	WCF	304	CT	404	PLOW
105	NT	205	CT	305	FWC	405	CFW
106	PLOW	206	RT	306	CC	406	WCF
107	CFW	207	NT	307	PLOW	407	CT
108	RT	208	CC	308	WCF	408	CC

501	CC	701	NT
502	CT	702	CFW
503	FCW	703	RT
504	WCF	704	CT
505	NT	705	FCW
506	PLOW	706	CC
507	CFW	707	PLOW
508	RT	708	WCF
601	CFW	801	RT
602	PLOW	802	FWC
603	FWC	803	NT
604	WCF	804	PLOW
605	CT	805	CFW
606	RT	806	WCF
607	NT	807	CT
608	CC	808	CC

TREATMENTS

CT=CONVENTIONAL TILLAGE ONLY
RT=RESIDUAL HERBICIDE, TILLAGE
NT=NO TILL, HERBICIDES ONLY
PLOW=PLOW, CONVENTIONAL TILLAGE
FWC=FALLOW-WHEAT-CORN
CFW=CORN-FALLOW-WHEAT
WCF=WHEAT-CORN-FALLOW
CC=CONTINUOUS CORN

NORTH
WEST **EAST**
SOUTH

Figure 1 Plot diagram of the long-term tillage study at Akron, CO.

during the growing season. The average straw yields over all years were 7373, 7269, and 7206 kg/ha for the CT, RT, and NT tillage systems, respectively, excluding 1985 and 1988.

Straw-to-grain ratios (straw yield/grain yield) for each year are reported in Table 3. Since straw yields were not measured in 1988, we were not able to calculate a straw-to-grain ratio for 1988. Straw-to-grain ratios varied from year to year, but were not significantly affected by tillage system in any given year. Straw-to-grain ratios averaged 2.57, 2.69, and 2.75 for the CT, RT, and NT systems, respectively.

B. SOIL ANALYSES

Elliott et al.[3] analyzed soil samples collected in 1982 from the 0- to 2.5, 2.5- to 5, 5- to 10-, and 10- to 20-cm depths from the CT and NT treatments (Table 4). Bulk densities tended to be greater for NT than for CT at all depths. Total inorganic and organic C, total N, total P, total inorganic P, and total organic P on the same samples used for bulk density are reported in Table 5. Little, if any, inorganic carbon was found in the top 20 cm of soil at this site. Most of the carbon was in the organic form. When averaging the north and south sites, the general trends were for the NT to have a higher organic carbon level in the 0- to 2.5-cm depth than CT. At deeper depths, soil organic carbon was similar for both tillage treatments. Total soil N tended to be higher for NT than CT in the 0- to 2.5-cm depth and nearly identical at deeper soil depths. Inorganic P was greater for NT than CT in the 0-2.5 cm depth. At other depths, inorganic and organic P levels were similar between tillage treatments.

Table 1 Winter Wheat Grain Yields (12% Moisture Content) for the North and South Rotations of the CT, RT, and NT Treatments

		Grain Yield at North Site (kg ha^{-1})							Grain Yield at South Site (kg ha^{-1})						
	Plot	**1991**	**1989**	**1987**	**1985**	**1983**	**1981**	**1979**	**1990**	**1988**	**1986**	**1984**	**1982**	**1980**
CT	102	4154	1925	3147	n.d.[a]	1256	3194	4270	2720	2479	2920	3014	2806	3341
	205	4482	2534	3722	n.d.	1156	3561	4223	1892	1951	4223	3140	2960	4029
	304	3222	1446	3294	n.d.	1931	3194	4043	1685	1964	4069	2619	3040	3495
	407	3369	1892	2826	n.d.	1537	2472	3421	1994	2907	4664	3341	3274	2993
RT	108	4194	2311	3267	n.d.	1363	2873	3949	1840	1470	3434	2913	2699	3869
	206	5567	2094	3581	n.d.	1229	3561	3882	1746	2045	3461	2913	2786	3976
	303	3592	2174	3040	n.d.	1330	3167	3568	2096	2125	4430	2927	2225	3181
	401	2814	1595	2853	n.d.	1290	2780	3221	1449	1944	4063	2753	3154	3708
NT	105	4412	2519	4223	n.d.	1163	3354	3221	2064	2158	4363	2699	3949	3495
	207	3913	2938	3769	n.d.	1617	2606	3622	2228	2172	2626	2967	3020	3815
	301	3924	1834	2826	n.d.	1624	2125	3936	2406	2118	2125	1644	2760	3815
	403	3226	1843	2693	n.d.	1229	2432	4330	1821	2459	2011	2132	2245	3341
Mean		3906	2092	3270	n.d.	1394	2943	3807	1995	2149	3532	2755	2910	3588
CT		3807	1949	3247	n.d.	1470	3105	3989	2073	2325	3969	3029	3020	3465
RT		4042	2044	3186	n.d.	1303	3095	3655	1783	1896	3847	2877	2716	3683
NT		3869	2283	3378	n.d.	1408	2629	3777	2130	2227	2781	2360	2993	3617
$LSD_{.05}$		n.s.	n.s.	n.s.	n.d.	n.s.	n.s.	n.s.	n.s.	n.s.	n.s.	n.s.	n.s.	n.s.
C.V. (%)		13.8	15.3	10.9	n.d.	16.6	13.5	12.5	16.9	15.1	28.0	12.6	18.5	8.3

[a] Not harvested, due to jointed goatgrass infestation.

Table 2 Estimated Straw Quantities Remaining on Each Plot after Harvest for Each Crop Year for the North and South Rotations of the CT, RT, and NT Treatments

		Straw Yield at North Site (kg ha^{-1})							Straw Yield at South Site (kg ha^{-1})					
	Plot	1991	1989	1987	1985	1983	1981	1979	1990	1988	1986	1984	1982	1980
CT	102	6256	1932	7076	560[a]	6031	9451	7282	9953	n.d.[b]	13222	14455	5820	5234
	205	7141	2894	7039	560	7115	10571	6816	7996	n.d.	14060	11445	7130	7551
	304	3560	1429	8180	560	6582	8706	6994	4728	n.d.	14641	2470	9016	7172
	407	6345	2311	6320	560	5749	5706	5756	7140	n.d.	12527	5221	9671	7718
RT	108	4581	3930	6955	560	6252	8284	6517	5754	n.d.	11340	11960	6147	7279
	206	5517	1074	8653	560	7288	9453	6582	7703	n.d.	13514	14661	6885	6522
	303	4667	1339	8751	560	5466	10594	5725	8514	n.d.	10704	4630	5901	5757
	401	4499	1069	7556	560	7144	9122	5032	5847	n.d.	12854	14403	7130	6294
NT	105	10058	2946	9575	560	10305	10034	6078	6668	n.d.	12570	9054	10081	7688
	207	5474	1723	8186	560	5996	13766	6563	8586	n.d.	9625	12294	5737	5900
	301	6980	2677	9993	560	4842	9781	6229	8389	n.d.	4799	3730	10163	8021
	403	5653	2522	4648	560	4993	7981	5464	7478	n.d.	5708	5505	6393	6204
Mean		5894	2154	7744	560	6480	9454	6253	7396	n.d.	11297	9152	7506	6778
CT		5825	2142	7154	560	6369	8608	6712	7454	n.d.	13612	8398	7909	6919
RT		4816	1853	7979	560	6537	9363	5964	6955	n.d.	12103	11414	6516	6463
NT		7041	2467	8100	560	6534	10391	6084	7780	n.d.	8175	7646	8093	6953
$LSD_{.05}$		n.s.	n.s.	n.s.	n.d.	n.s.	n.s.	534	n.s.	n.d.	n.s.	n.s.	n.s.	n.s.
C.V. (%)		27.2	43.6	16.3	n.d.	25.5	15.6	4.9	24.4	n.d.	20.5	30.0	25.5	17.4

[a] Estimated standing stubble, most of biomass removed.
[b] Not determined.

Table 3 Straw-to-Grain Ratios for the CT, RT, and NT Treatments from the North and South Rotations

		Straw/Grain Ratio at North Site							Straw/Grain Ratio at South Site					
	Plot	1991	1989	1987	1985	1983	1981	1979	1990	1988	1986	1984	1982	1980
CT	102	1.51	1.00	2.25	n.d.[a]	4.80	2.96	1.71	3.66	n.d.	4.53	4.80	2.07	1.57
	205	1.59	1.14	1.89	n.d.	6.16	2.97	1.61	4.23	n.d.	3.33	3.64	2.41	1.87
	304	1.10	0.99	2.48	n.d.	3.41	2.73	1.73	2.81	n.d.	3.60	0.94	2.97	2.05
	407	1.88	1.22	2.24	n.d.	3.74	2.31	1.68	3.58	n.d.	2.69	1.56	2.95	2.58
RT	108	1.09	1.70	2.13	n.d.	4.59	2.88	1.65	3.13	n.d.	3.30	4.11	2.28	1.88
	206	0.99	0.51	2.42	n.d.	5.93	2.65	1.70	4.41	n.d.	3.90	5.03	2.47	1.64
	303	1.30	0.62	2.88	n.d.	4.11	3.34	1.60	4.06	n.d.	2.42	1.58	2.65	1.81
	401	1.60	0.67	2.65	n.d.	5.54	3.28	1.56	4.04	n.d.	3.16	5.23	2.26	1.70
NT	105	2.28	1.17	2.27	n.d.	8.86	2.99	1.89	3.23	n.d.	2.88	3.35	2.55	2.20
	207	1.40	0.59	2.17	n.d.	3.71	5.28	1.81	3.85	n.d.	3.67	4.14	1.90	1.55
	301	1.78	1.46	3.54	n.d.	2.98	4.60	1.58	3.49	n.d.	2.26	2.27	3.68	2.10
	403	1.75	1.37	1.73	n.d.	4.06	3.28	1.26	4.11	n.d.	2.84	2.58	2.85	1.86
Mean		1.52	1.04	2.39	n.d.	4.82	3.27	1.65	3.72	n.d.	3.21	3.27	2.59	1.90
CT		1.52	1.09	2.21	n.d.	4.53	2.74	1.68	3.57	n.d.	3.54	2.74	2.60	2.02
RT		1.25	0.87	2.52	n.d.	5.04	3.04	1.63	3.91	n.d.	3.20	3.99	2.42	1.76
NT		1.80	1.15	2.43	n.d.	4.90	4.04	1.64	3.67	n.d.	2.91	3.09	2.75	1.93
$LSD_{.05}$		n.s.	n.s.	n.s.	n.d.	n.s.	n.s.	n.s.	n.s.	n.d.	n.s.	n.s.	n.s.	n.s.
C.V. (%)		20.3	38.7	16.7	n.d.	35.1	21.7	9.6	11.1	n.d.	18.1	32.4	15.2	16.7

[a] n.d. = not determined.

Table 4 Soil Bulk Density Levels for the CT and NT Treatments from the North and South Rotations

Tillage	Depth (cm)	North Site Soil Bulk Density (g cm^{-3})			Average	South Site Soil Bulk Density (g cm^{-3})			Average
		9-13-82	10-4-82	5-9-83		9-13-82	10-4-82	5-9-83	
CT	0–2.5	0.99	0.97	1.24	1.07	0.95	1.02	1.27	1.08
	2.5–5	1.05	1.11	1.24	1.13	1.08	1.08	1.41	1.19
	5–10	1.11	1.38	1.51	1.33	1.31	1.16	1.44	1.30
	10–20	1.35	1.44	1.54	1.44	1.36	1.44	1.58	1.46
NT	0–2.5	0.97	1.18	1.27	1.14	1.14	1.00	1.15	1.10
	2.5–5	1.29	1.41	1.47	1.39	1.35	1.35	1.30	1.33
	5–10	1.45	1.50	1.47	1.47	1.41	1.54	1.54	1.50
	10–20	1.40	1.53	1.59	1.51	1.43	1.48	1.57	1.49

Note: Each soil bulk density value is the average of three replicated plots for each date.

Table 5 Total Soil Inorganic and Organic Carbon, Total P, Total Inorganic P, and Total Organic P in 1982

Tillage	Depth (cm)	Inorganic C (g kg⁻¹)	Organic C (g kg⁻¹)	Total N (g kg⁻¹)	Total P (mg kg⁻¹)	Inorganic P (mg kg⁻¹)	Organic P (mg kg⁻¹)
				North Site			
CT	0–2.5	0.024 (0.024)[a]	8.81 (0.3)	0.95 (0.04)	349 (24.4)	233 (16.2)	117 (29.2)
	2.5–10	0.000 (0)	8.14 (0.54)	0.93 (0.05)	368 (24.3)	243 (16.9)	125 (7.8)
	5–10	0.000 (0)	7.97 (0.41)	0.93 (0.08)	367 (24.0)	238 (18.6)	129 (6.1)
	10–20	0.000 (0)	6.76 (0.36)	0.76 (0.03)	359 (27.3)	234 (16.6)	125 (10.8)
NT	0–2.5	0.000 (0)	9.85 (0.89)	1.01 (0.09)	368 (25.0)	250 (16.9)	118 (9.5)
	2.5–10	0.000 (0)	8.38 (0.45)	0.89 (0.05)	361 (23.2)	244 (16.5)	118 (7.0)
	5–10	0.015 (0.015)	6.96 (0.55)	0.79 (0.06)	354 (28.2)	235 (21.9)	119 (7.1)
	10–20	0.005 (0.005)	6.33 (0.55)	0.72 (0.04)	350 (26.4)	234 (21.9)	116 (5.2)
				South Site			
CT	0–2.5	0.000 (0)	7.44 (0.38)	0.88 (0.13)	345 (4.3)	239 (3.3)	105 (1.3)
	2.5–10	0.000 (0)	7.19 (0.30)	0.78 (0.03)	345 (1.5)	234 (3.3)	110 (1.9)
	5–10	0.007 (0.007)	6.49 (0.12)	0.68 (0.06)	335 (1.5)	229 (1.5)	106 (1.7)
	10–20	0.000 (0)	5.57 (0.19)	0.82 (0.09)	329 (4.0)	217 (4.8)	112 (1.3)
NT	0–2.5	0.042 (0.042)	8.60 (0.33)	0.95 (0.03)	370 (32.9)	248 (12.0)	122 (9.5)
	2.5–10	0.000 (0)	7.43 (0.47)	0.81 (0.04)	359 (13.2)	248 (15.6)	111 (3.4)
	5–10	0.003 (0.003)	6.31 (0.55)	0.74 (0.03)	353 (13.8)	240 (11.7)	113 (2.2)
	10–20	0.003 (0.003)	5.86 (0.22)	0.74 (0.07)	342 (11.9)	226 (10.7)	116 (1.2)

[a] Each number is an average of three plots. Numbers in parentheses are standard errors.

Table 6 Total Soil Organic Carbon, Nitrogen, and Phosphorus for the South Site in 1982 with Soil Depth

Tillage	Depth (cm)	Total C (g kg⁻¹)	Total N (g kg⁻¹)	Inorganic P (mg kg⁻¹)	Organic P (mg kg⁻¹)
CT	0–20	6.19 (0.03)[a]	0.75 (0.018)	249 (13.1)	127 (5.6)
	20–40	5.25 (0.37)	0.70 (0.041)	460 (4.0)	131 (6.8)
	40–60	2.76 (0.40)	0.37 (0.057)	433 (10.3)	79 (7.5)
	60–90	1.66 (0.17)	0.25 (0.002)	448 (14.1)	42 (6.0)
	90–120	1.45 (0.16)	0.21 (0.006)	456 (3.1)	38 (2.9)
	120–150	1.25 (0.02)	0.19 (0.024)	466 (4.7)	32 (4.6)
	150–180	1.27 (0.26)	0.17 (0.009)	440 (7.7)	35 (4.5)
	180–210	0.97 (0.24)	0.13 (0.017)	344 (34.1)	17 (3.8)
	210–240	0.76 (0.22)	0.10 (0.013)	307 (51.0)	20 (2.3)
	240–270	0.49 (0.13)	0.10 (0.004)	274 (58.5)	20 (1.4)
	270–300	0.49 (0.12)	0.09 (0.003)	242 (50.9)	27 (4.8)
NT	0–20	6.70 (0.12)	0.75 (0.017)	250 (6.5)	120 (3.3)
	20–40	5.00 (0.10)	0.70 (0.016)	417 (76.4)	123 (3.7)
	40–60	3.44 (0.88)	0.44 (0.103)	459 (10.8)	76 (18.6)
	60–90	1.80 (0.19)	0.28 (0.048)	449 (5.0)	61 (9.8)
	90–120	1.32 (0.11)	0.22 (0.041)	450 (5.2)	34 (7.2)
	120–150	0.99 (0.28)	0.19 (0.034)	462 (3.2)	36 (5.2)
	150–180	0.99 (0.06)	0.17 (0.021)	444 (4.4)	28 (0.6)
	180–210	0.75 (0.05)	0.11 (0.014)	361 (31.1)	26 (1.8)
	210–240	0.38 (0.09)	0.09 (0.012)	265 (36.3)	36 (15.1)
	240–270	0.32 (0.06)	0.09 (0.011)	303 (31.9)	23 (8.5)
	270–300	0.24 (0.07)	0.09 (0.008)	305 (21.6)	33 (5.4)

[a] Each number is an average of three plots. Numbers in parentheses are standard errors.

Total soil organic C, N, inorganic P, and organic P was also determined on the south site for depths up to 2.8 m (Table 6). The trends were for the total soil organic carbon, total soil N, and total soil organic P to decrease with increasing depth in the soil profile for both tillage treatments. Total inorganic P tended to be higher in the 30- to 195-cm depth than at shallower or deeper soil depth for both tillage treatments.

Soil was sampled for the 0- to 20-cm soil depth in the CT, RT, and NT treatments of both sites in April 1989 (Halvorson and Vigil, unpublished). The trend was for the NT treatment to have a higher total soil C and N than the CT and RT tillage systems at the 0- to 5-cm soil depth (Table 7). There was little difference in total soil C and N between tillage treatments below 5 cm.

Soil NO_3-N levels, to 180-cm depth, for each of the CT, RT, and NT treatments are shown in Table 8 for the north and south rotation. The north rotation tended to have a higher level of residual soil NO_3-N than the south rotation for all tillage systems. When averaged over north and south rotations, the soil profile (0- to 180-cm depth) contained 167, 197, and 173 kg N/ha for the CT, RT, and NT tillage systems, respectively. Little or no difference in soil NO_3-N existed between tillage systems.

Table 7 Total Soil Carbon and Nitrogen on April 1989 Samples by Carlo Erba N-C-S Analyzer

Site	Tillage	Depth (cm)	Total C (g kg^{-1})	Total N (g kg^{-1})
North	CT	0–5	9.52 (0.72)*	0.93 (0.07)
		5–10	8.41 (0.65)	0.79 (0.07)
		10–20	7.26 (0.46)	0.72 (0.04)
	NT	0–5	10.17 (0.87)	0.98 (0.08)
		5–10	7.79 (0.75)	0.77 (0.07)
		10–20	7.14 (0.45)	0.70 (0.03)
	RT	0–5	9.74 (1.08)	0.91 (0.09)
		5–10	7.79 (0.80)	0.75 (0.06)
		10–20	6.78 (0.46)	0.66 (0.03)
South	CT	0–5	9.01 (0.41)	0.72 (0.01)
		5–10	7.18 (0.15)	0.62 (0.02)
		10–20	6.13 (0.08)	0.61 (0.01)
	NT	0–5	10.10 (0.20)	0.99 (0.16)
		5–10	7.35 (0.24)	0.60 (0.02)
		10–20	6.47 (0.21)	0.58 (0.01)
	RT	0–5	9.25 (0.77)	0.73 (0.05)
		5–10	7.19 (0.40)	0.61 (0.02)
		10–20	6.29 (0.21)	0.56 (0.01)

[a] Each number is an average of three plots. Numbers in parentheses are standard errors.

Table 8 Soil NO_3-N Levels in the Root Zone of Winter Wheat at Planting Time for the North Rotation (September 19, 1990) and South Rotation (September 6, 1989)

Tillage	Plot	Soil Depth (cm)						Profile
		0–15	15–30	30–60	60–90	90–120	120–180	Total
				Soil NO_3-N (kg ha^{-1})				
North Site								
CT	102	12.5	10.1	50.0	53.5	32.5	39.6	198.2
	205	38.6	10.4	22.0	19.7	11.0	38.4	140.1
	304	26.7	8.8	24.3	38.5	20.5	22.9	141.7
	407	15.6	18.1	116.2	52.6	30.2	43.9	276.6
RT	108	15.3	16.0	52.3	69.3	35.1	39.1	227.0
	206	15.9	24.3	68.8	80.7	46.9	40.6	277.2
	303	59.6	18.2	69.6	32.2	21.1	33.3	233.9
	401	22.5	22.3	52.0	38.3	43.6	29.1	207.8
NT	105	41.2	22.4	42.0	52.1	24.4	32.9	215.1
	207	120.9	21.3	42.8	54.6	27.5	57.7	324.8
	301	9.3	14.0	48.2	41.5	29.3	28.8	171.2
	403	13.5	27.6	73.8	30.2	15.7	26.4	187.2
South Site								
CT	102	22.6	26.4	65.9	16.7	10.7	31.9	174.1
	205	18.4	14.4	32.4	15.7	3.6	16.1	100.6
	304	34.7	18.7	27.1	24.8	14.0	57.2	176.4
	407	36.8	14.8	27.3	12.7	11.0	23.1	125.6
RT	108	4.4	6.2	27.5	9.4	4.4	8.7	60.6
	206	58.2	28.6	58.0	26.8	8.8	77.3	257.5
	303	23.4	23.9	46.1	24.9	14.5	16.3	149.1
	401	25.5	18.2	50.5	26.9	9.9	32.7	163.8
NT	105	21.2	16.0	35.6	25.1	11.7	18.4	128.0
	207	28.6	14.2	37.1	3.6	9.8	19.3	112.7
	301	17.1	15.1	36.8	36.5	8.1	8.2	122.0
	403	37.1	14.1	24.2	22.3	8.7	17.3	123.8

REFERENCES

1. Brandon, J.F. and Mathews, O.R., Dry Land Rotation and Tillage Experiments at the Akron (Colorado) Field Station, USDA Circular no. 700, Washington, DC, 1944.
2. Smika, D.E., Fallow management practices for wheat production in the Central Great Plains, *Agron. J.*, 82, 319, 1990.
3. Elliott, E.T., Horton, K., Moore, J.C., Coleman, D.C., and Cole, C.V., Mineralization dynamics in fallow dryland wheat plots, Colorado, *Plant Soil*, 76, 149, 1984.
4. Schepers J.S., Francis, D.D., and Thompson, M.T., Simultaneous determination of total C, total N, and ^{15}N on soil and plant material, *Commun. Soil Sci. Plant Anal.*, 20, 949, 1989.

Chapter

Management of Dryland Agroecosystems in the Central Great Plains of Colorado

G.A. Peterson and D.G. Westfall

CONTENTS

I. INTRODUCTION

Decades of farming with the wheat (*Triticum aestivum* L.)-fallow system in the Great Plains have allowed large losses in soil C content. Haas et al.[1] report C losses of over 50% after 30 to 40 years of cultivation in a wheat-fallow rotation, which is typical for much of the Great Plains. This agroecosystem has stabilized short-term production and sustainability,[2] but evidence for its negative influence on soil and environmental quality continues to build. Cultivation practices often have left the soil bare for much of the summer fallow period and with little protection from wind and water erosion during early growth stages of the grain crop. Even without fallow periods, maximum tillage has encouraged soil organic matter loss and its attendant negatives such as decreased water infiltration and lower fertility status.[1,3,4] Haas et al.[1] have documented that systems containing fallow decreased soil organic matter content more rapidly than continuous cropping.

Peterson and Westfall,[5] in a review paper, present evidence that shows summer fallowing in semiarid regions has actually allowed movement of large quantities of nitrate-N to soil layers beneath the root zones of the commonly grown dryland crops. Hundreds of kilograms of nitrate-N now reside in the vadose zone of our Central Plains soils. Continued practice of fallow, incorporating practices that improve soil water storage, such as reduced tillage, may move this nitrate into groundwater. The nitrate load in the vadose zone today is mainly the result of mineralized N from the breaking of the prairies. With the use of N fertilizers and improved fallow practices we could negatively impact ground water even more rapidly. Further improvements in fallow techniques within the wheat-summer fallow system will enhance the possibility for ground water degradation.

0-8493-2802-0/97/$0.00+$.50

The use of monoculture winter wheat also has resulted in large increases in winter annual grassy weed populations [Jointed goatgrass (*Aegilops cylindrica Host*), Downy brome (*Bromus tectorum* L.), etc.], especially when conservation tillage practices are adopted for erosion control. In natural ecosystems, diversity of plant life is always present and proper competition controls problems with invading plants. Our continuous use of monoculture wheat has allowed the grassy weed problem to build up to a point that threatens sustainability in the Central Great Plains.

Halvorson and Black[6] have shown that fallowing has created "saline seeps" in some areas with subsequent degradation of the soil. Ironically, the better the water conservation techniques during fallow, the greater the problem with seep formation. They concluded that more-intensive cropping with less fallow time was essential to prevent further seep formation and loss of productivity.

The wheat-fallow system has contributed greatly to the degradation of the Central Great Plains' agroecosystem, even though it has provided temporary economic stability to the area. Large quantities of C and N have been removed from the landscape or have been redistributed in ways that are equivalent to losses for a given farm. Our research addresses this problem from two directions. We are testing hypotheses regarding (1) increased C inputs into the soil system via intensified crop rotations and (2) decreasing C losses by decreasing tillage, thus allowing smaller perturbations in C addition, increased soil aggregate formation, and stabilization, both of which would decrease rates of C oxidation.

Increasing C additions via intensified crop rotations is only possible if water conservation is maximized. Smika and Wicks[7] provide some of the most clear-cut evidence that water conservation techniques allow us to store more water. They showed that changing from moldboard plow primary tillage to no-till brought water storage to 80% of field capacity by the spring of the fallow year following a wheat crop. This amounted to an extra 15 cm of stored water by spring. During the summer the soil under no-till reached field capacity well before wheat seeding and further useful storage of rainfall could not occur. In fact, no-till wheat fallow would promote nitrate-N loss by leaching or potential "saline seep" formation. The solution is to take advantage of the extra water and plant a spring crop in the rotation.

We think that conversion from conventionally tilled wheat-fallow systems to extended rotations with less tillage and fewer summer fallow periods will greatly impact soil C. There is a possibility that soil C levels, at least in the immediate soil surface, can be increased by these techniques, thereby reversing the net loss of C to the atmosphere as carbon dioxide gas. Blevins et al.[8] report that conversion from tilled to no-till environments has decreased organic matter decomposition rate with a concomitant increase in soil organic matter level in the surface 5 cm of soil. Lamb et al.[4] provide evidence that C conservation will be accompanied by proportional conservation of soil N.

The overall objective of our research is to identify dryland crop and soil management systems which will maximize water use efficiency of the total annual precipitation. A complete discussion of the philosophy and experimental approach has been published by Peterson et al.[9]

Specific objectives are to

1. Determine if cropping sequences with fewer or shorter summer fallow periods are feasible.
2. Quantify the relationship of climate (precipitation and evaporative demand), soil type, and cropping sequences that involve fewer or shorter fallow periods.
3. Quantify the effects of long-term use of no-till management systems on soil structural stability, microorganism and faunal populations of the soil, soil organic matter, and soil N and P contents, all in conjunction with various crop sequences, climates, and soils.
4. Identify cropping or management systems that will minimize soil erosion by crop residue maintenance.
5. Develop a database across climatic zones that will allow economic assessment of entire management systems.

II. MATERIALS AND METHODS

Interactions of climate, soils, and cropping systems are being studied in an experimental array that was established in the fall of 1985. The field sites are located at Sterling (40.37°N, 103.13°W), Stratton (39.18°N, 102.26°W), and Walsh (37.23°N, 102.17°W), CO. They represent a potential evapotranspiration (PET) gradient from north to south: long-term (growing season) open pan evaporation averages of 1000, 1400, and 1900 mm/growing season at Sterling (low ET), Stratton (medium ET), and Walsh (high ET), respectively[10] (Figure 1). All sites have long-term precipitation averages of approximately 400 to 450 mm. Elevations are 1341 (4400 ft), 1335 (4380 ft), and 1134 (3720 ft) above sea level at Sterling, Stratton, and Walsh, respectively.

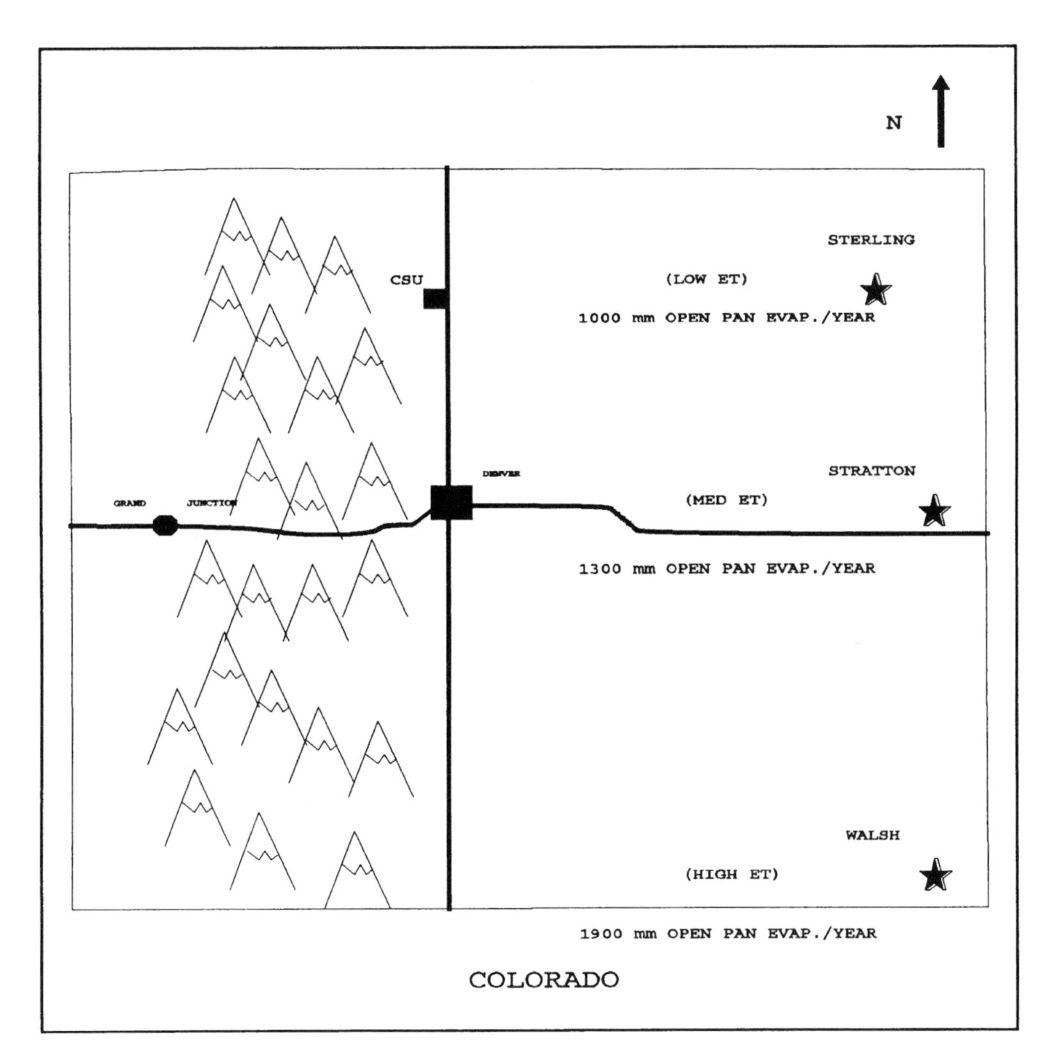

Figure 1 Study site locations.

Each site was selected so that it represents a catenary sequence of soils common to the geographic area. All fields chosen had been cultivated for more than 50 years (see soil characteristics in Table 1). The cropping system during this time had been dryland winter wheat fallow with some inclusion of grain sorghum (*Sorghum bicolor L.*) at Walsh and corn (*Zea mays L.*) at Sterling. Cropping system treatments were placed on the soil sequence at each site (Figure 2) and are identified in Table 2. All systems are managed with no-till techniques. Residual herbicides such as atrazine are used in rotation phases following wheat and preceding corn or sorghum and for the corn and sorghum crops. Weed control during fallow is primarily accomplished with non-residual materials such as glyphosate. Specific herbicide programs and costs are reported annually in technical bulletins.[10,11]

The experimental design is a split block including location (Sterling, Stratton, and Walsh), slope position (summit, sideslope, and toeslope), and cropping system variables.[9] The cropping system treatment plots are 6 m wide by 225 to 325 m long and are replicated twice. All phases of all rotations are present every year.

To illustrate the treatment differences, consider the following: the wheat-fallow system (least intensive) has two crops and approximately 480 days of summer fallow in two cycles of the rotation (4 years), while wheat-corn-millet fallow (more intensive) has three crops and approximately 270 days of summer fallow in a 4-year period. Summer fallow months include May, June, July, August, and September. Cropping system greatly influences the ratio of crop to non-crop periods, which in turn alters surface soil water conditions and ultimately decomposition rates of residue inputs.

Table 1 Soil Classification and Soil Organic C and N Contents (0- to 5-cm Depth) in 1985 at the Three Experimental Sites

Slope Position	Classification	Surface[a] Texture	Organic C ($g\ kg^{-1}$)	Organic N ($g\ kg^{-1}$)
	Sterling			
Summit	Fine, montmorillonitic, mesic Aridic Paleustoll	L	9.80	0.96
Sideslope	Fine, montmorillonitic, mesic Aridic Paleustoll	L	9.69	0.93
Toeslope	Fine, montmorillonitic, mesic Pachic Argiustoll	L	14.59	1.37
	Stratton			
Summit	Fine-silty, mixed, mesic Aridic Argiustoll	CL	10.58	1.08
Sideslope	Fine-silty, mixed (calcareous), Ustic Torriorthent	L	10.44	1.03
Toeslope	Fine-silty, mixed, mesic, Pachic Argiustoll	L	19.21	1.87
	Walsh			
Summit	Fine, montmorillonitic, mesic, Ustollic Haplargid	SL	2.73	0.30
Sideslope	Fine, montmorillonitic, mesic, Ustollic Haplargid	SL	3.04	0.33
Toeslope	Fine, montmorillonitic, mesic, Ustollic Paleargid	SCL	8.49	0.88

[a] L = loam, CL = clay loam, LS = loamy sand, SL = sandy loam, SCL = sandy clay loam.

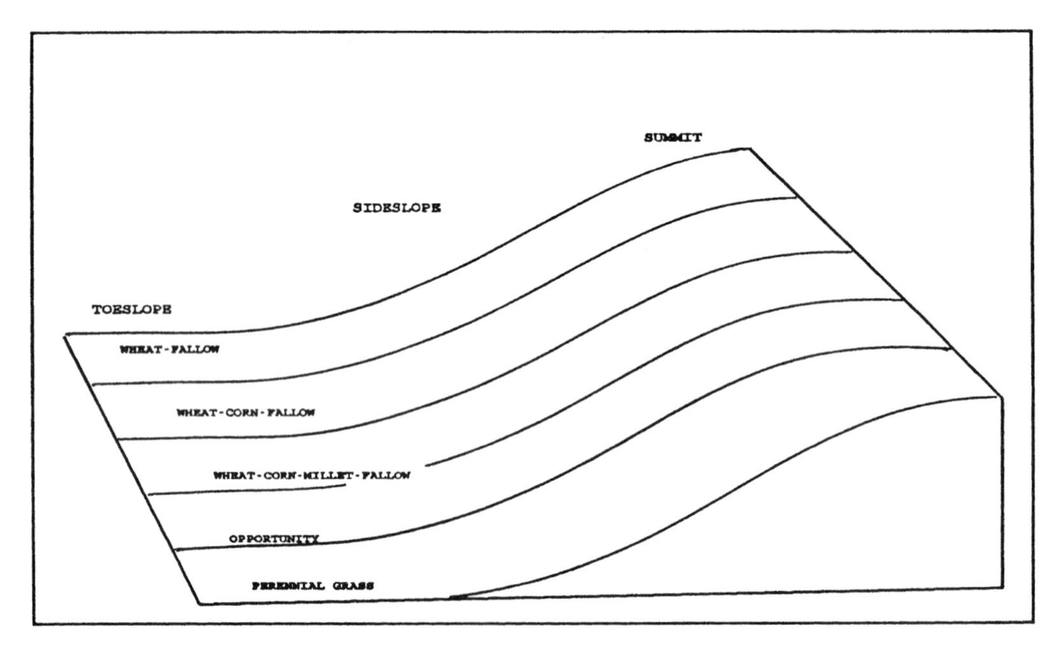

Figure 2 Experimental layout on soil catena.

Sorghum replaces corn in WCMF at Walsh because it is better adapted to higher PET in those areas.[10] Wheat (hard red winter wheat), corn, sorghum, and proso millet (*Panicum milacium L.*) are planted in mid-September, mid-May, early June, and mid-June, respectively, at all sites. A forage sorghum was inserted into the 4-year rotation at Walsh in place of proso millet in 1991 because of poor success with

Table 2 Management Systems for Each Site in 1990

Site		Rotations
Sterling	(1)	Wheat-fallow
	(2)	Wheat-corn-fallow
	(3)	Wheat-corn-millet-fallow
	(4)	Opportunity cropping[a]
	(5)	Perennial grass
Stratton	(1)	Wheat-fallow
	(2)	Wheat-corn-fallow
	(3)	Wheat-corn-millet-fallow
	(4)	Opportunity cropping[a]
	(5)	Perennial grass
Walsh	(1)	Wheat-fallow
	(2)	Wheat-sorghum-fallow
	(3)	Wheat-sorghum-hay-fallow
	(4)	Continuous sorghum
	(5)	Opportunity cropping[a]
	(6)	Perennial grass

[a] Opportunity cropping is to be as continuous as possible without fallow, but not monoculture.

Opportunity Cropping History

	Site		
Year	**Sterling**	**Stratton**	**Walsh**
1985	Wheat	Fallow	Sorghum
1986	Wheat	Wheat	Sorghum
1987	Corn	Sorghum	Millet
1988	Corn	Sorghum	Sudex
1989	Attempted hay millet	Attempted hay millet	Sorghum
1990	Wheat	Wheat	Attempted sunflower
1991	Corn	Corn	Wheat

proso in this climate. The forage was grown in 1991 and 1992, but was found to be of low economic value and therefore was changed to a 2nd year of grain sorghum. For the forseeable future the 4-year rotation at Walsh will be WSSF.

The continuous grass (CG) system was established at each site in the spring of 1986 with a mixture of six perennial species: crested wheatgrass (*Agropyron cristatum* Fisch. ex Link), western wheatgrass [*Agropyron smithii* (Rydb.) Love], blue grama [*Bouteloua gracilis* (H.B.K.) ag ex Steund.], sideoats grama [*Bouteloua curtipendula* (Michx.) Torr.], little bluestem [*Schizachyrium scoparium* (Mich.) Nash], and buffalograss [(*Buchloe dactyloides* (Nutt.) Engelm]. The CG treatment is clipped once each year, and, beginning in summer 1990, the grass was removed. Prior to 1990 all grass was mowed but was not removed. The CG treatment has received no N or P fertilizer.

Nitrogen fertilizer is applied in accordance with the NO_3-N content of the soil profile (0 to 180 cm) and expected yield on each soil position at each site. Therefore, N rate changes by year, crop grown, and soil position. Nitrogen fertilizer is applied at planting for wheat and proso millet, while N applications to corn and sorghum are sidedressed when the corn or sorghum is 30 to 50 cm tall.

Phosphorus is applied as 10-34-0, at planting near the seed, on one half of each plot over all soils. The rate of P is determined by the lowest soil test on the catena, which is usually found on the sideslope position. This rate has been 9.5 kg/ha of P at each site each year. Zinc (1 kg/ha) is banded annually near the seed at corn planting at Sterling and Stratton to correct a deficiency identified by soil test.

Grain, stover, and total plant biomass are measured annually. Nitrogen and P content of both grain and stover are determined on each crop. Plant residue remaining on the field prior to planting and harvesting of each crop is measured with a quadrat technique.

Soil samples are taken to a depth of 180 cm, or a restrictive layer, each year prior to planting a crop. These samples are collected in 30-cm increments for nitrate analyses. Prior to wheat planting in a given rotation, the surface soil samples are segmented into 0- to 2.5-, 2.5- to 5.0-, 5- to 10-, 10- to 20-, and 20- to 40-cm increments for total C and N analyses. Increments below 40 cm are in 30 cm units.

Soil water is measured with the neutron scatter technique. Aluminum access tubes were installed, two per soil position, in each treatment at each site in 1988. These tubes are not removed for any field operation and remain in the exact positions year to year. Precautions are taken to prevent soil compaction around each tube. By not moving the tubes over years we get the best possible estimates of soil water use in each rotation. Soil water measurements are made on all soils and rotations at the beginning and end of fallow periods. Soil water also is measured biweekly throughout the summer in each crop on the summit and toeslope soil positions.

III. RESULTS AND DISCUSSION

Total biomass, stover and grain yields are reported by year in the appendix section (see Part IV of this volume). These values represent yields for a particular crop in a particular year as affected by climate, soil, and cropping system. They do not allow easy comparison of the cropping system effect on total productivity. Annualization of production for each rotation is a means of evaluating the negative impact of the fallow year when no production occurs. Annualized crop stover and total biomass values for each cropping system are summarized for 1988 to 1991 data in Table 3.

Cropping systems with fewer months of summer fallow increased annualized total biomass production in most situations. Wheat fallow has been extremely productive at the medium ET site (Stratton) on summit and sideslope soils and therefore the net improvement of more crops and less summer fallow is lessened. The 3-year rotation produced 30% more total biomass than the 2-year wheat-fallow rotation. When averaged over all soils and ET environments, the 4-year rotation produced 33% more as compared to wheat fallow. Interestingly, the increases in crop stover, compared to wheat fallow, were only 15 and 19% for the 3- and 4-year rotations, respectively. This indicates that the water savings are being translated into grain production and not just unsalable crop residues and provides a potential economic gain for the producer. On the other hand C inputs into the soil system are still increased, but not proportional to grain production.

Crop stover becomes crop residue upon harvest of the crop and the longevity of that residue on the soil surface is of interest in terms of soil water conservation; both water infiltration properties and evaporation potential are affected. Residue data collected at wheat planting in fall 1991 (Table 4) illustrate the differences in residue quantity as related to climate, soil, and cropping system. Comparison of the total amount present at harvest of a crop with amount remaining after fallow provide an assessment of decomposition rates. Residue remaining in the 3- and 4-year rotations compared to wheat-fallow was about 62% greater when averaged over sites and soils. This greater amount of residue lessens the chances of soil erosion after wheat planting and should enhance water infiltration rates. The primary reason for the differences in residue loss attributable to cropping system is probably related to length of fallow period. Residue loss rates per month of fallow were 295, 220, and 185 kg ha^{-1} mo^{-1} for the 2-, 3-, and 4-year rotations, respectively. The 2-year rotation has more summer months during the fallow which may account for the apparently higher loss rate. We also observe large winter losses of residues, which are likely related to wind weathering and removal from the field.

Water use efficiency (WUE) for grain production was increased by 30 to 100% compared to wheat fallow for the 3- and 4-year rotations, respectively[12] (Figure 3). This is evidence that extending rotation length and decreasing summer fallow improves water conservation in the Central Great Plains.

Soil water storage during fallow in the 2-year rotation is generally greater than in other rotations with shorter summer fallow periods. The most critical water storage period in the 3-and 4-year rotations is the summer, fall, and winter after wheat harvest just prior to corn or sorghum planting. Weed control is critical during this period. The stored water will not be enough to grow the entire summer grain crop, but must be large enough to provide a buffer for extended periods without rain.

No-till management measureably increased soil C and N contents of surface soils (0 to 10 cm) after only 4 years of practice. More intense cropping systems such as wheat-corn-millet fallow maintained C and N levels at higher levels than wheat fallow.[13] At deeper soil levels there was C and N loss in some

Table 3 Annualized Stover and Total Biomass Yields as Affected by site (ET Zone), Soil, and Rotation (1988–1991)

		Soil Position (kg/ha)							
		Summit		Sideslope		Toeslope		Mean (kg/ha)	
Site	**Rotation**[a]	**Stover**	**Total**	**Stover**	**Total**	**Stover**	**Total**	**Stover**	**Total**
Low ET[b]	W-F	1720	2530	2570	3100	1980	2920	1920	2850
	W-C-F	2390	4002	2570	4490	3150	5620	2700	4704
	W-C-M-F	2150	3740	2760	4690	3330	5650	2750	4693
Medium ET[c]	W-F	2970	4254	2950	4130	4100	6030	3340	4805
	W-C-F	2420	3690	2600	4070	4460	7030	3160	4930
	W-C-M-F	2900	4540	2820	4420	4420	7000	3380	5320
High ET[d]	W-F	900	1550	1470	2370	1180	2020	1180	1980
	W-S-F	980	1860	1600	2850	2030	4043	1540	2918
	W-S-M-F	1150	2023	1610	2560	2020	3890	1590	2824

[a] WF = wheat fallow; WC(S)F = wheat-corn (sorghum) fallow; WC(S)MF = wheat-corn (sorghum)-proso millet.
[b] Sterling.
[c] Stratton.
[d] Walsh.

Table 4 Residue Amounts Present at the Beginning of Fallow in 1990 and Ending Amounts of Residue Present at Wheat Planting in September 1991

			Soil Position (kg/ha)							
		Months of	Summit		Sideslope		Toeslope		Mean (kg/ha)	
Site	Rotation[a]	Fallow	Beginning	End	Beginning	End	Beginning	End	Beginning	End
Low ET[b]	W-F	14	4610	1620	5260	1340	7740	2840	5870	1930
	W-C-F	10	6010	3480	6190	4410	8330	4840	6840	4240
	W-C-M-F	11	6650	4100	—	4810	8900	5460	7780	4790
Medium ET[c]	W-F	14	9400	3260	9490	2280	9920	4100	9600	3210
	W-C-F	10	4900	2770	4900	3670	8100	4810	5970	3750
	W-C-M-F	11	5230	2760	5140	2930	6730	4290	5700	3330
High ET[d]	W-F	14	2080	410	2990	480	3300	1350	2790	750
	W-S-F	10	2540	700	2580	1170	4890	2760	3340	1540
	W-S-M-F	11	2470	470	2640	1500	5320	2480	3480	1480

[a] WF = wheat fallow; WC(S)F = wheat-corn (sorghum) fallow; WC(S)MF = wheat-corn (sorghum)-proso millet.
[b] Sterling.
[c] Stratton.
[d] Walsh.

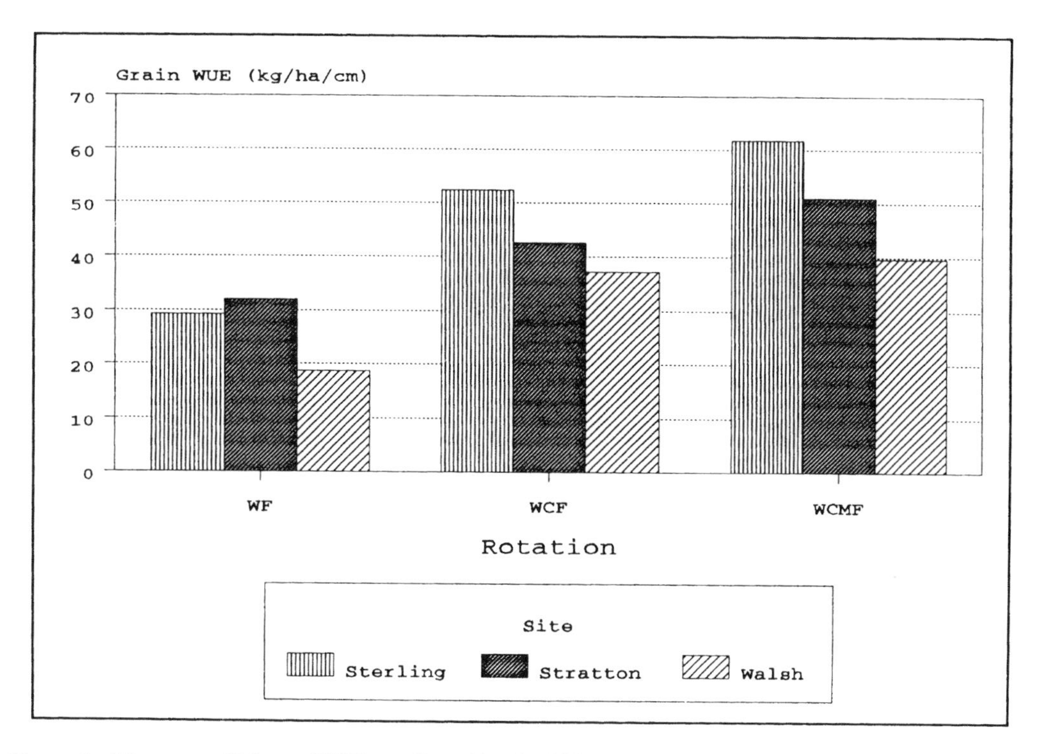

Figure 3 Water use efficiency (WUE) as affected by site (ET environment) and cropping system 1988 to 1990.

cases. The net effects are not yet known. One would not expect large changes in just a few years of time, given the relatively small amounts of C being returned in residues annually. In fact it is surprising that Wood et al.[13] were able to detect any significant changes. Reasons for the increased C and N contents are related to amount of crop stover material being returned to the soil surface and the lack of disturbance with no-till practices. Wood et al.[14] state that plant residues were added in sufficient quantities in both the 2- and 4-year rotations to maintain soil C at present levels on summit soils, but were added at rates in the 4-year rotation that increased soil C levels on sideslope and toeslope positions.

IV. CONCLUSION

Extended cropping systems have increased total biomass production and harvestable product by over 30% compared to wheat fallow. This was accompanied by greater residue accumulations on the soil surface, 60% compared to wheat fallow. These benefits were accomplished via improvements in WUE, which are associated with water capture and its retention in the soil profile plus using that water in summer crops prior to the hotter, dryer, summer fallow period. Soil C and N levels in the surface soils have tended to increase after only 4 years of no-till intensive crop rotation practices.

REFERENCES

1. Haas, H.J., Evans, C.E., and Miles, E.F., Nitrogen and Carbon Changes in Great Plains Soils as Influenced by Cropping and Soil Treatments, USDA Tech. Bull. 1164. U.S. Government Printing Office, Washington, DC, 1957.
2. Haas, H.J., Willis, W.O., and Bond, J.J., Summer fallow in the western United States, USDA-ARS Conservation Research Report No. 17, U.S. Government Printing Office, Washington DC, 1974.
3. Jenny, H., Soil Fertility Losses under Missouri Conditions, Missouri Agric. Exp. Stn. Bull. 324, University of Missouri, Columbia, MO, 1933.
4. Lamb, J.A., Peterson, G.A., and Fenster, C.R., Wheat-fallow tillage systems' effect on a newly cultivated grassland soil's nitrogen budget, *Soil Sci. Soc. Am. J.*, 49, 352, 1985.
5. Peterson, G.A. and Westfall, D.G., Dryland cropping systems to enhance water quality, in Proceedings of the Non-point Water Quality Symposium, Sponsored by the Colorado Chapter of the Soil and Water Conservation Society, Colorado Springs, CO, March 22–23, 93, 1990.

6. Halvorson, A.D. and Black, A.L., Saline-seep development in dryland soils of northeastern Montana, *J. Soil Water Cons.*, 29, 77, 1974.
7. Smika, D.E. and Wicks, G.A., Soil water storage during fallow in the Central Great Plains as influenced by tillage and herbicide treatments, *Soil Sci. Soc. Am. Proc.*, 32, 591, 1968.
8. Blevins, R.L., Smith, M.S., and Thomas, G.W., Changes in soil properties under no-tillage, in *Notill Agriculture*, Phillips, R.E. and Phillips, S.H., Eds., Van Nostrand Reinhold, New York, 1984, 190.
9. Peterson, G.A., Westfall, D.G., and Cole, C.V., Agroecosystem approach to soil and crop management research, *Soil Sci. Soc. Am. J.*, 57, 1354, 1993.
10. Peterson, G.A., Westfall, D.G., Wood, C.W., and Ross, S., Crop and Soil Management in Dryland Agroecosystems, Tech. Bull. LTB88-6. Agric. Exp. Stn., Colorado State University, Fort Collins, 1988.
11. Peterson, G.A., Westfall, D.G., Sherrod, L., Kolberg, R., and Rouppet, B., Sustainable Dryland Agroecosystem Management, Tech. Bull. TB93-4. Agric. Exp. Stn., Colorado State University, Fort Collins, 1993.
12. McGee, E.A., Peterson, G.A., and Westfall, D.G., Water-use efficiency of dryland no-till cropping systems in the west central Great Plains, in *Agronomy Abstracts*, American Society of Agronomists, Madison, WI, 1991, 153.
13. Wood, C.W., Peterson, G.A., Westfall, D.G., Cole, C.V., and Willis, W.O., Nitrogen balance and biomass production of newly established no-till dryland agroecosystems, *Agron. J.*, 83, 519, 1991.
14. Wood, C.W., Westfall, D.G., Peterson, G.A., and Burke, I.C., Impacts of cropping intensity on carbon and nitrogen mineralization under no-till dryland agroecosystems, *Agron. J.*, 82, 1115, 1990.

Chapter

Management Effects on Soil Organic Carbon and Nitrogen in the East-Central Great Plains of Kansas

J.L. Havlin and D.E. Kissel

CONTENTS

I. INTRODUCTION

The "steady-state" level of soil organic matter (SOM) is determined by numerous soil properties which are influenced by cropping practices such as residue management, crop rotation, and many others.[1–3] Numerous long-term studies have documented the decline in SOM with tillage.[4–7] Compared to native grassland, total soil N decreased 3, 8, and 19% and organic C decreased 4, 14, and 16% with no-tillage (NT), stubble-mulch, and conventional tillage (CT) treatments, respectively, after 12 years of wheat-fallow.[8] Similar findings have been reported by Bauer and Black[9] and Blevins et al.[10]

Several long-term studies also have demonstrated the beneficial effect of crop rotation, especially with legumes, on SOM content and on crop productivity.[6–7] These studies generally showed that nonlegume grain crops in rotation with forage legumes increased organic C and N after several decades of cropping, compared to continuous grain crops produced without manure or fertilizer N inputs. Hargrove and Frye[11] reported that NT grain sorghum/clover (*Sorghum bicolor* L. /*Trifolium incarnatum* L. Moench) rotations resulted in greater SOM accumulation than wheat/soybean rotations. Larson et al.[12] reported that soil organic C was linearly related to the quantity of residue added and that 5 Mg ha^{-1} of corn (*Zea mays* L.) or alfalfa (*Medicago sativa* L.) residue were needed to maintain initial organic C content. Larson et al.[13] showed that SOM accumulation rates were similar for alfalfa or wheat straw applied in equal amounts.

Few studies have reported on the combined effects of crop rotation and tillage in grain/soybean rotations on SOM. On two Ohio soils, long-term corn/soybean (C/B) rotations were evaluated with three

0-8493-2802-0/97/$0.00+$.50

levels of tillage.[14,15] After 19 years, organic C and N (0- to 7.5-cm depth) were 1.5 and 1.3 times greater, respectively, in the NT treatments compared to CT at both locations. Compared to the C/B rotation, continuous corn (C/C) resulted in higher organic C and N levels, especially in the NT treatments. Differences were not as great under CT. A corn-oats-meadow rotation also was included in the studies and resulted in even higher organic C and N levels than C/C under NT; however, similar levels were reported for CT. Crop rotation effects on organic C and N were related to the total quantity of residue produced under each rotation.

Maintaining, or even increasing, SOM at the highest levels attainable under economic crop production requires reduced tillage systems and crop rotations that maximize the amount of residue produced and returned to the soil surface. Thus, the objective of this study was to quantify the long-term effects of tillage and crop rotation on soil organic C and N.

II. SITE DESCRIPTION

A long-term tillage/crop rotation study was initiated in eastern Kansas (Riley Co.) on a Muir silt loam soil (fine-silty, mixed, mesic Cumulic Haplustoll) in 1975 (Table 1). The climate is characterized by the mid-continental location of the site (39°07′N, 96°37′W), with hot summers and cold winters. Mean annual temperature is 12.8°C and mean annual precipitation is 835 mm, occurring mainly from April through October.

Table 1 Soil Characteristics and Tillage Management at the Riley Co. Location

pH	5.5
CEC (cmol kg^{-1})	20
Clay (g kg^{-1})	264
Fall tillage[a]	Chisel plow
Spring tillage[a]	Disc

[a] Tillage practices applied to CT treatments only.

Grain sorghum (S/S) and soybean (B/B) were grown continuously and in a sorghum-soybean rotation (S/B). Tillage variables included NT and CT treatments, where crop residues were left on the soil surface in the NT treatments and crops were planted directly into shredded residue. Preplant herbicides were used for weed control in NT. Tillage management used for the CT treatment is shown in Table 1. The CT plots were cultivated once after planting to control weeds at both sites. Sixty-five kg N ha^{-1} as NH_4NO_3 was broadcast and incorporated with a disk prior to planting on all CT plots each year. Fertilizers were not incorporated in the NT plots.

Treatments were arranged in a split plot experimental design with four replications, although main plots were not replicated. Main plots were crop rotation and subplots were tillage. Plots were 6 × 18 m and row spacing was 76 cm.

III. SOIL SAMPLING AND ANALYSIS

Soil in tillage-rotation plots was sampled prior to spring tillage and planting in April, 1986, at 0- to 2.5-, 2.5- to 7.5-, 7.5- to 15-, and 15- to 30-cm depths. Six cores (2.5-cm diameter) were taken randomly from the row middles in each plot and composited. In the NT plots, all surface residues were removed before the soil sampler was inserted. Samples were forced-air dried at 25°C and sieved through a 1-mm screen. All visible residue particles were separated from each sample. Duplicate subsamples were analyzed for total C by dry combustion using a LECO carbon analyzer[16] and for total N by salicylic-sulfuric acid digestion[17] followed by analysis on a Technicon Auto Analyzer[18]. Since soil pH was <7.0 in all samples, total C was assumed to be equivalent to organic C. Ammonium and NO_3-N also were determined by extraction with 2 *M* KCL and colorimetric analysis similar to that used for total N analysis.[19] Organic N was calculated as the difference between total N and NH_4-N + NO_3-N. Data were interpreted using analysis of variance and regression techniques.[20]

IV. EFFECTS OF ROTATION AND TILLAGE ON SOIL C AND N

Averaged over rotation, tillage, and soil depths, organic C and N content was 12.4 and 1.09 g kg^{-1}, respectively. Organic C and N decreased with soil depth in all rotation/ tillage treatments on the Muir soil (Table 2). Compared to CT, maintaining crop residues on the soil surface (NT treatment) resulted in higher organic C and N at the 0- to 2.5-cm depth, but not at the lower depths. The increase in organic C and N with NT also varied with rotation. The NT treatment resulted in higher organic C and N with all rotations (0- to 2.5-cm depth). Compared to CT, the NT treatment was higher in organic C by 25, 49, and 40% and organic N by 16, 37, and 40% for the B/B, S/B, and S/S rotations, respectively.

Table 2 Crop Rotation and Tillage Effect on Soil Organic C and N

Tillage[a]	Soil Depth (cm)	Organic C[b] (g kg^{-1})			Organic N (g kg^{-1})		
		B/B	S/B	S/S	B/B	S/B	S/S
	0–2.5	11.4	12.5	14.1	1.0	1.14	1.20
	2.5–7.5	11.3	12.3	13.6	1.04	1.14	1.18
CT	7.5–15	11.0	11.6	12.6	1.00	0.99	1.13
	15–30	9.6	10.4	11.0	0.90	0.88	0.99
	0–2.5	14.3	18.6	19.8	1.26	1.56	1.68
	2.5–7.5	11.3	13.0	12.5	1.02	1.14	1.12
NT	7.5–15	11.0	11.9	11.6	0.99	1.01	1.03
	15–30	10.0	10.3	10.6	0.94	0.88	0.96

ANOVA	df	Organic C (P > F)	Organic N (P > F)
Tillage (T)	1	0.01	0.03
Rotation (R)	2	0.01	0.04
Depth (D)	3	<0.01	<0.01
T × R	2	0.19	0.34
T × D	3	<0.01	0.01
R × D	6	0.28	0.28
CV %		6.3	6.2

[a] CT = conventional tillage; NT = no tillage.

[b] B/B = continuous soybeans; S/B = sorghum/soybeans; S/S = continuous sorghum.

Increasing the frequency of sorghum in the rotation resulted in higher soil organic C and N contents in the Muir soils, especially at the 0- to 2.5-cm depth (Table 2). Compared to B/B and S/B, S/S rotations resulted in 10 and 24% more organic C under CT, respectively, and 30 and 39% more under NT, respectively (0- to 2.5-cm depth). Similarly, surface soil organic N was 5 and 10% higher under CT and 24 and 33% higher under NT for the S/B and S/S rotations, respectively. Compared to CT B/B, organic C and N was 7 and 6% higher with CT S/S, respectively, but did not change with the CT S/B rotation. Similar differences in rotation effects on organic C and N were observed between soils of various clay contents in studies in Ohio.[14–15]

The higher soil organic C and N contents associated with NT compared to CT and in rotations with sorghum compared to B/B were directly related to the quantity of residue produced and left on the soil surface after harvest (Figure 1). The quantity of residue produced was estimated from grain yield data from the Muir soils[21] by multiplying average annual grain production for each rotation (Table 3) by a residue weight to grain weight ratio of 1.0 and 1.5 for sorghum and soybean, respectively.[12] The soil organic C and N mass (Mg ha^{-1} 0 to 15cm) were calculated from soil bulk density measured in each tillage treatment (BD = 1.40 g cm^{-3} in NT; BD = 1.36 g cm^{-3} in CT). Bulk densities were similar between rotations. Organic C and N was greater under NT than under CT at each level of residue produced with the three rotations. The rate of increase in organic C and N with increasing residue level was greater for NT.

Soil organic N decreased with increasing frequency of soybean in the rotation because soybean generally depletes soil N. In the midwest US, average symbiotic N_2 fixation in soybean does not exceed 50% of total crop N.[22,23] Total N removed in a 2 Mg ha^{-1} soybean crop (grain + residue) would be approximately 160 kg N ha^{-1}.[24] Assuming that grain N represents two-thirds total crop N and 50% of total N is symbiotic N, soil N removed by the grain is 52 kg N ha^{-1} y^{-1} and symbiotic N in the residue

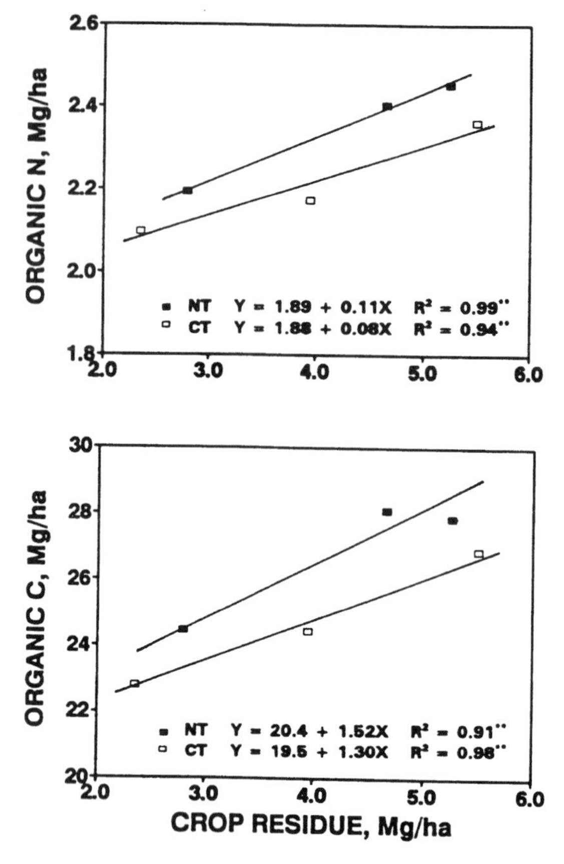

Figure 1 Effect of crop residue mass and tillage on soil organic C and N (0-15 cm depth). The three data points, left to right on each line, represent continuous soybean, sorghum/soybean, and continuous sorghum rotations. NT = no tillage; CT = conventional tillage.

Table 3 Mean Soybean and Sorghum Grain Yield from 1975 to 1986 at the Riley Co. Site

Tillage System	Rotation (kg ha^{-1})		
	B/B	S/B	S/S
No-tillage			
Sorghum	—	5856	5263
Soybean	1855	2296	—
Conventional tillage			
Sorghum	—	4936	5531
Soybean	1616	1943	—

returned to the soil is 27 kg N ha^{-1} y^{-1}. Thus, each soybean crop potentially depletes soil N by 25 kg N ha^{-1} y^{-1}. These data show that increasing the quantity of residue returned to the soil by increasing yields or increasing the frequency of high-residue-producing crops in the rotation will increase soil organic N greater than including soybean in the rotation.

In summary, both tillage and rotation had significant effects on soil organic C and N accumulation, primarily at the surface (0 to 2.5 cm). Additional increases in organic C and N at the surface and at depths below 2.5 cm are expected as these studies continue. Tillage/rotation effects were greater in the Ohio studies after 19 years;[14,15] the present studies were conducted over only 10 to 12 years. Results from these studies show that SOM, and perhaps soil productivity, can be strongly influenced by maintaining surface residues through reduced tillage and increasing the quantity of residues with appropriate rotations.

V. SUMMARY

Sustaining or increasing soil productivity may depend on soil and crop management practices which maintain or increase SOM. This study was conducted to determine the effects of tillage and crop rotation on soil organic carbon (C) and nitrogen (N). Soils were sampled from conventional (CT) and no-tillage (NT) treatments applied to continuous sorghum (S/S), continuous soybean (B/B), and sorghum-soybean (S/B) rotations. Organic C and N generally decreased with depth and treatment effects were observed mostly in the 0- to 2.5-cm depth increment. Compared to CT, NT had greater organic C and N contents. Compared to B/B, S/B and S/S increased organic C 10 and 24% under CT, respectively, and 30 and 38% under NT, respectively (0- to 2.5-cm depth).

Similar treatment effects occurred with organic N. Increases in organic C and N with NT compared to CT and with sorghum rotations compared to B/B were directly related to the quantity of residue produced and left on the soil surface after harvest (S/S > S/B > B/B). Crop management systems that include rotations with high-residue-producing crops and maintenance of surface residue cover with reduced or no-tillage result in greater soil organic C and N, which may improve soil productivity.

REFERENCES

1. Tate, R.L., *Soil Organic Matter: Biological and Ecological Effects*, John Wiley & Sons, New York, 1978.
2. Stevenson, F.J., *Cycles of Soil: Carbon, Nitrogen, Phosphorus, Sulfur, and Micronutrients*, John Wiley & Sons, New York, 1986.
3. Cole, C.V., Williams, J., Schaffer, M., and Hanson, J., Nutrient and organic matter dynamics as components of agricultural production systems, in *Soil Fertility and Organic Matter as Critical Components of Production Systems*, Follett, R.F. et al. Eds., ASA Spec. Pub. 19, SSSA, Madison, WI, 1987, 147.
4. Haas, H.J., Evans, C.E., and Miles, E.F., Nitrogen and Carbon Changes in Great Plains Soils as Influenced by Cropping and Soil Treatments, USDA Tech. Bull. 1164, U.S. Government Printing Office, Washington, DC, 1957.
5. Tiessen, H., Stewart, J.W.B., and Bettany, J.R., Cultivation effects on the amounts and concentration of carbon, nitrogen, and phosphorus in grassland soils, *Agron. J.*, 74, 831, 1982.
6. Odell, R.T., Melsted, S.W., and Walker, W.M., Changes in organic carbon and nitrogen of Morrow plot soils under different treatments, 1904–1973, *Soil Sci.*, 137, 160, 1984.
7. Johnston, A.E., Soil organic matter, effects on soils and crops, *Soil Use Manage.*, 2, 97, 1986.
8. Lamb, J.A., Peterson, G.A., and Fenster, C.R., Wheat fallow tillage systems' effect on a newly cultivated grassland soils' nitrogen budget, *Soil Sci. Soc. Am. J.*, 49, 352, 1985.
9. Bauer, A. and Black, A.L., Soil carbon, nitrogen, and bulk density comparisons in two cropland tillage systems after 25 years and in virgin grassland, *Soil Sci. Soc. Am. J.*, 45, 1166, 1981.
10. Blevins, R.L., Thomas, G.W., Smith, M.S., Frye, W.W., and Cornelius, P.L., Changes in soil properties after 10 years of continuous non-tilled and conventionally-tilled corn, *Soil Tillage Res.*, 3, 135, 1983.
11. Hargrove, W.L. and Frye, W.W., The need for legume cover crops in conservation tillage production, in *The Role of Legumes in Conservation Tillage Systems*, Power, J.F., Ed., Soil Conservation Society of America, Ankeny, IA, 1987, 1.
12. Larson, W.E., Holt, R.F., and Carlson, C.W., Residues for soil conservation, in *Crop Residue Management Systems*, Oschwald, W.R., Ed., ASA Spec. Pub. 31, ASA, CSSA, SSSA, Madison, WI, 1978, 1.
13. Larson, W.E., Clapp, C.E., Pierre, W.H., and Morachan, Y.B., Effects of increasing amounts of organic residues on continuous corn. II. Organic carbon, nitrogen, phosphorus, and sulfur, *Agron. J.*, 64, 204, 1972.
14. Dick, W.A., Van Doren, D.M., Jr., Triplett, G.B., Jr., and Henry, J.E., Influence of Long-Term Tillage and Rotation Combinations on Crop Yields and Selected Soil Parameters. I. Results Obtained for a Mollic Ochraqualf Soil, Research Bulletin No. 1180 of The Ohio State University/The Ohio Agricultural Research and Development Center, Wooster, 1986.
15. Dick, W.A., Van Doren, D.M., Jr., Triplett, G.B., Jr., and Henry, J.E., Influence of Long-Term Tillage and Rotation Combinations on Crop Yields and Selected Soil Parameters. II. Results Obtained for a Typic Fragiudalf Soil, Research Bulletin No. 1181 of The Ohio State University/The Ohio Agricultural Research and Development Center, Wooster, 1986.
16. Tabatabai, M.A. and Bremner, J.M., Use of the Leco automatic 70-second carbon analyzer for total carbon analysis in soils, *Soil Sci. Soc. Am. Proc.*, 34, 608, 1970.
17. Bremner, J.M. and Mulvaney, C.S., Salicylic acid-thiosulfate modification of Kjeldahl method to include nitrate and nitrite, in Miller, R.H. and Keeney, D.R., Eds., Methods of soil analysis, Part 2, 2nd ed, *Agronomy*, 9, 621, 1982.
18. Technicon Industrial Systems, Industrial/Simultaneous Determination of Nitrogen and/or Phosphorus in BD Acid Digests, Industrial method no. 334-74W/Bt, Tarrytown, NY, 1977.
19. Technicon Industrial Systems, Nitrate and Nitrite in Soil Extracts, Industrial method no. 487-77A, Tarrytown, NY, 1977.
20. SAS Institute, SAS User's Guide: Statistics, SAS Institute, Inc., Cary, NC, 1982.

21. Peterson, D.E., Crop Production as Affected by Cropping Sequence and Method of Seedbed Preparation in Conservation Tillage, M.S. thesis, Kansas State University, Manhattan, 1983.
22. LaRue, T.A. and Patterson, T.G., How much nitrogen do legumes fix?, *Adv. Agron.* 34, 15, 1981.
23. Heichel, G.H., Legumes as a source of nitrogen in conservation tillage systems, in *The Role of Legumes in Conservation Tillage Systems*, Power, J.F., Ed., Soil Conservation Society of America, Ankeny, IA, 1987, 29.
24. Mengel, D.B., Segars, W., and Rehm, G.W., Soil fertility and liming, in Wilcox, J.R., Ed., Soybeans: improvement, production, and uses, *Agronomy*, 16, 461, 1987.

Chapter 30

Management of Dry-Farmed Southern Great Plains Soils for Sustained Productivity

Ordie R. Jones, Bobby A. Stewart, and Paul W. Unger

CONTENTS

I. INTRODUCTION

When native grassland is cultivated for crop production, soil organic carbon (SOC) and nitrogen (N) concentrations usually decline with time. The rate of decline is influenced by tillage practices, crops grown, crop rotations, soil characteristics, management of crop residues, fertilizer or manure applications, and climate.[1–9]

Effects of various management practices on SOC concentration have been evaluated in several long-term studies in the southern Great Plains.[10–16] Studies show that all types of tillage result in decreased SOC concentration compared to native grass. As the intensity and frequency of tillage are increased, the rate of SOC decline increases. However, more frequent cropping decreases the rate of SOC decline because more organic material is added to the soil. Cropping to small grains results in slower declines in SOC concentrations than cropping to row crops. In only a few studies, however, did an increase in

SOC concentration occur when tillage or cropping systems were changed, even if green manure or legume crops were involved in the rotation. The rate of SOC decline can be reduced by reducing tillage intensity and frequency or changing cropping rotations, but increases in SOC rarely occur in the highly oxidative environment of the southern Great Plains unless tillage is stopped, grass is planted, or large quantities of manure or residues are added periodically.

Sustainable dryland farming systems in the southern Great Plains will have lower SOC contents than native grass because of C and N mineralization. However, sustainable systems should approach a SOC content equilibrium, the level of which can be affected by management system. At Bushland, TX, Johnson et al.[6] showed that SOC contents were still declining after 39 years of cropping but that different tillage and cropping management resulted in differential rates of SOC decline. After 24 years (1943 to 1966) in a wheat-fallow cropping sequence on a Pullman clay loam, sweep-tilled delayed fallow contained 11.8 g kg^{-1} SOC, while soil on a one-way plowed clean fallow treatment contained 9.3 g kg^{-1} SOC (Figure 1). With the delayed sweep treatment, tillage was delayed until the spring following wheat harvest, some 10 months after harvest. This allowed considerable weed growth and maintenance of surface residues, resulting in higher SOC content. Because soil water used by weeds was replenished during the last 6 months of fallow, major effects of delayed tillage on wheat yield were not experienced. While we certainly do not recommend letting weeds grow during fallow, the research demonstrates that reduced tillage systems have the potential of maintaining SOC content in the southern Great Plains soils. This potential might be enhanced by addition of manure or fertilizer.

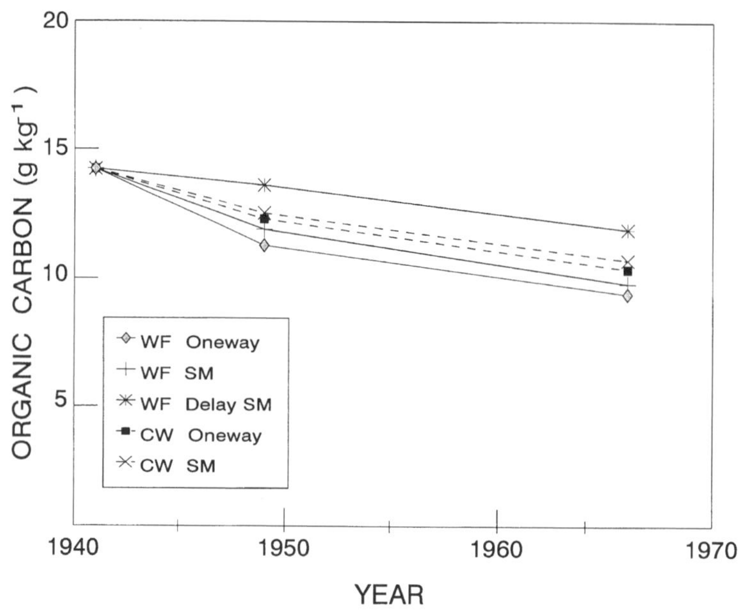

Figure 1 Organic carbon concentration in the surface 15 cm of soil for tillage plots at Bushland, TX, as related to cropping system and tillage method. (Adapted from Johnson, W.C., Van Doren, C.E., and Burnett, E., Summer Fallow in the Western United States USDA Agric. Res. Serv. Conserv. Res. Rep. No. 17, U.S. Government Printing Office, Washington, DC, 86, 1974.)

Several cropping and tillage experiments at the USDA Conservation and Production Research Laboratory, Bushland, TX, offer the potential to determine the effects of reduced- and no-tillage management on SOC and N dynamics in the Southern Great Plains (Table 1). We describe two of these experiments and present current findings.

II. SOILS

Soil type has a large effect on SOC concentration. This variability was shown by Mathers et al.,[17] who sampled seven southern Great Plains soils by soil horizon, obtaining samples from both cultivated fields and adjacent native grass sites (Figure 2). Eck et al.[18] sampled six southern Great Plains soils, some of which were native grass and some cultivated (Table 2). The data from native grass sites show the potential that exists for C sequestration on various soils in the southern Great Plains.

Table 1 Climatic Data for USDA Conservation and Production Research Laboratory, Bushland, TX

Month	Precipitation[a] (mm)	Pan Evaporation[b] (mm)	Temperature[c] Minimum	Temperature[c] Maximum	Wind Velocity[d] (km h^{-1})
January	12	—	–7	8	9.4
February	13	—	–5	11	10.0
March	19	—	–1	16	11.8
April	26	257	4	21	11.7
May	68	283	9	25	11.3
June	75	333	15	30	10.1
July	64	362	17	32	9.5
August	71	298	16	31	8.4
September	50	229	12	27	7.3
October	41	—	5	22	8.8
November	19	—	–1	14	9.2
December	15	—	–6	9	9.6

Note: Elevation = 1167 m; location = 35°10′N latitude, 102°3′W longitude frost-free growing season = 185 days.

[a] 53 yr average (1939–1991).
[b] Class A pan, 20 yr average (1972–1991).
[c] 20 year average (1972–1991).
[d] 53 cm height, 20 yr average (1972–1991).

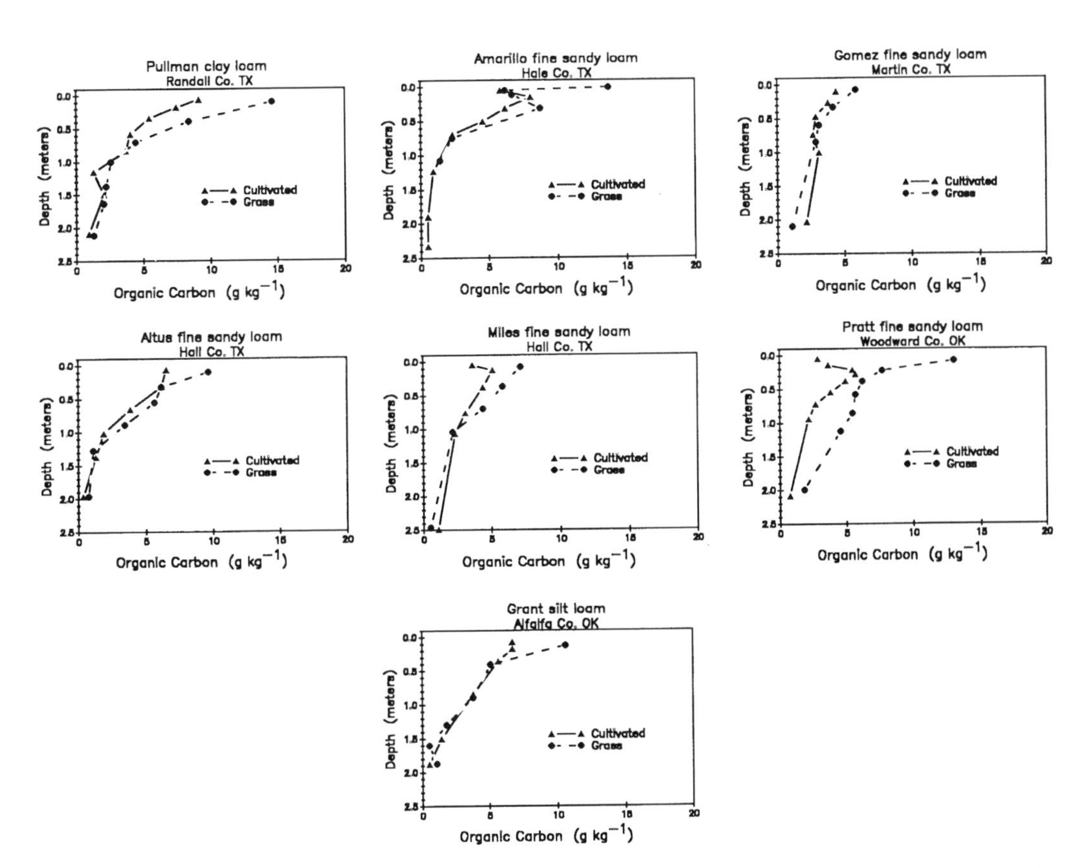

Figure 2 Organic carbon concentration of cultivated and native grass profiles for seven southern Great Plains soils. Organic matter was determined by Walkley and Black procedure. (Adapted from Mathers, A.C., Gardner, H.R., Lotspeich, F.B., Taylor, H.M., Laase, G.R., and Daniell, R.E., Some Morphological, Physical, Chemical, and Mineralogical Properties of Seven Southern Great Plains Soils, USDA Rep. ARS 41-85, Washington, DC, 1963.)

Table 2 Total Nitrogen and Organic Carbon Concentrations and pH of Soil Horizons for Six Southern Great Plains Soils

	Hale, TX Amarillo Fine Sandy Loam					Cimarron, OK Dalhart Fine Sandy Loam			
Depth (m)	Horizon	Total N (g kg^{-1})	pH	Organic C (g kg^{-1})	Depth (m)	Horizon	Total N (g kg^{-1})	pH	Organic C (g kg^{-1})
0–0.07	Ala	0.73	6.1	5.2	0–0.10	Alp	0.24	7.5	2.7
0.07–0.18	Alb	0.66	5.9	5.0	0.10–0.20	A12	0.32	7.3	2.8
0.18–0.36	B21ta	0.76	6.0	5.6	0.20–0.73	B21	0.74	6.0	4.9
0.36–0.51	B21tb	0.58	6.0	4.9	0.33–0.58	B22	0.65	6.5	3.8
0.51–0.76	B22ta	0.33	6.3	2.0	0.58–0.91	B31ca	0.36	7.9	1.7
0.76–1.02	B222tb	0.29	6.3	1.6	0.91–1.37	B32ca	0.31	7.9	0.9
	Hall, TX Miles Fine Sandy Loam					Chickasha, OK Reinach Loam			
Depth (m)	Horizon	Total N (g kg^{-1})	pH	Organic C (g kg^{-1})	Depth (m)	Horizon	Total N (g kg^{-1})	pH	Organic C (g kg^{-1})
0–0.09	Ala	0.69	7.1	6.7	0–0.28	Ap	0.47	7.2	6.1
0.09–0.18	Alb	0.69	7.1	4.7	0.28–0.53	A/C	0.48	7.1	4.2
0.18–0.41	B21t	0.69	6.7	4.4	0.53–0.76	C1	0.66	7.8	4.2
0.41–0.61	B22t	0.64	6.7	5.3	0.76–0.97	C2	0.37	7.7	3.5
0.61–0.81	B23t	0.56	6.5	4.5	0.97–1.39	C3	0.18	7.9	2.3
0.81–1.06	B3	0.31	6.5	2.3					
	Hardeman, TX Abilene Silty Clay Loam					Alfalfa, OK Grand Silt Loam			
Depth (m)	Horizon	Total N (g kg^{-1})	pH	Organic C (g kg^{-1})	Depth (m)	Horizon	Total N (g kg^{-1})	pH	Organic C (g kg^{-1})
0–0.18	Ap	0.70	7.4	8.2	0–0.13	Ala	14.1	6.5	16.6
0.18–0.30	B1	0.64	7.6	7.5	0.13–0.25	Alb	0.86	6.6	10.6
0.30–0.63	B2	0.45	7.5	5.5	0.25–0.41	A3a	0.66	6.4	8.4
0.63–1.01	B2ca	0.32	7.9	3.5	0.41–0.56	A3a	0.55	6.5	7.4
1.01–1.27	Cca	0.25	8.0	3.0	0.56–1.21	B2t	0.42	6.3	6.0
1.27–1.52	C1	0.25	8.1	2.7	1.21–1.37	Cca	0.38	6.5	4.8

Note: Total N was determined by the Kjeldahl method and organic carbon was determined with the Walkley and Black procedure.

Adapted from Eck, H.V., Ford, R.H., and Fanning, C.D., Productivities of Horizons of Seven Benchmark Soils of the Southern Great Plains, USDA Agric. Res. Serv. Conserv. Res. Rep. No. 11, Washington, DC, 1967.

Pullman clay loam (fine mixed, thermic Torrertic Paleustoll) is the dominant fine-textured soil on the southern High Plains, occupying all or part of 21 counties.[19] All cultivated land at the USDA Conservation and Production Research Laboratory, Bushland, TX, is located on this soil. The soil is slowly permeable with a silty clay loam AP horizon, underlain with silty clay and clay horizons (Table 3). Clay mineralogy is mixed, with montmorillonite predominating. Depth to the calcic horizon ranges from 1.5 to 1.8 m on 0 to 1% slopes and from 0.6 to 1.2 m on 1 to 3 percent slopes. The calcic zone does not limit rooting with most crops. Winter wheat commonly roots to 1.8 m and sunflower roots have been observed at 2.7-m depth. About 195 mm of plant-available water can be stored to a 1.2-m depth. Fertility is high and dryland crops grown with fallow are not usually fertilized. With annual cropping, the addition of nitrogen fertilizer is often required.

III. CONSERVATION TILLAGE ON GRADED TERRACES

The objectives of the experiment are to determine the effects of no-tillage management of a dryland wheat-sorghum-fallow (WSF) sequence on runoff, soil erosion, nutrient and pesticide loss, infiltration rate, soil water storage, soil physical and chemical characteristics, and crop physiological development and yield. The no-tillage (NT) management system is compared with stubble-mulch (SM) tillage, a proven conservation tillage system that is well adapted for dryland crop production in the Great Plains.

A. DESCRIPTION

Tillage treatments are described in Table 4. The NT treatment was started in 1981 on one half of paired, field size (2 to 5 ha) graded-terraced watersheds that were constructed in 1949[20] and had been farmed in a dryland WSF sequence with SM tillage since that time. The three initial watersheds were instrumented with 0.76 m "H" flumes in 1958 (one in each phase of WSF sequence) and with Chickasha sediment samplers in 1978. The three no-tillage, WSF watersheds (one half of each of the three initial watersheds) were instrumented with H flumes and sediment samplers in 1984. Additionally, a seventh watershed, cropped in a NT continuous wheat sequence, was instrumented in 1984. Thus, there are seven watersheds, six of which are in WSF with paired NT and SM management. One pair of watersheds is in each phase of WSF annually. In a WSF sequence, one crop of wheat and one crop of sorghum are grown in a 36-month period, with an 11-month fallow between each crop. The 3-year sequence is called WSF, even though it has two fallow periods, because during the 1st calendar year, a wheat crop is harvested; during the 2nd year, sorghum is harvested; and during the 3rd year, no crop is harvested.

Each watershed is farmed as a field unit; thus tillage treatments and runoff and erosion data are not replicated. However, for other data (soil or plant samples, yields, infiltration with rainfall simulator etc.), multiple samples or observations are obtained. The experimental arrangement provides the opportunity to examine the effects of imposing NT management on a long-term SM tilled soil in relation to SOC and N dynamics since none of the watersheds has ever been fertilized. Additional comparisons can be made with samples from an adjacent native grass site that has never been cultivated.

B. RESULTS AND DISCUSSION

1. Climatic Data, Grain, and Residue Yields

Long-term average and ranges in crop yields, seasonal precipitation and runoff, soil water content at planting, evapotranspiration, and seasonal water-use-efficiency for SM watersheds are shown in Table 5. These data show the low yields and extreme variabilities that can occur with dryland farming in the southern Great Plains. Wheat grain yields for the 1982 to 1991 period averaged 1.41 Mg ha^{-1} with NT and 1.47 Mg ha^{-1} with SM. Sorghum yields averaged 3.43 Mg ha^{-1} with NT and 3.07 Mg ha^{-1} with SM. Sorghum yields were greater on NT because of higher soil water content at planting resulting from reduced evaporation. Estimates of crop residues remaining after harvest were calculated using equations developed by Howell,[21] which resulted in average harvest index (HI) of 0.35 for wheat and 0.37 for sorghum. Estimated residues remaining on the soil surface after harvest averaged 2.0 Mg ha^{-1} for wheat and 3.4 Mg ha^{-1} for sorghum.

2. Runoff, Erosion, and Infiltration

Long-term monthly mean precipitation and storm runoff values from SM watersheds are shown in Figure 3. Most runoff from the WSF sequence occurred from the row crop (sorghum) and during fallow after sorghum with much lower runoff from wheat. Interestingly, NT resulted in a 57% increase in runoff (Table 6) compared to SM primarily because tillage destroys soil crusts and roughens the soil. NT has

Table 3 Characteristics of Pullman Clay Loam, Bushland, TX

Horizon	Depth (cm)	Sand (%)	Silt (%)	Clay (%)	Texture	Organic C (g kg^{-1})	pH	Bulk Density (Mg m^3)	Water Content at Potential of (MPa)		Plant Available Water	
									–0.03	–1.5	%	m^{3}m^{-3}
Ap	0–15	17.0	53.0	30.0	Silty clay loam	2.06	6.70	1.26	25.0	16.0	9.0	0.113
B21t	15–40	13.0	38.8	48.2	Clay	1.29	6.77	1.48	28.8	18.5	10.3	0.152
B22t	40–73	13.0	40.0	47.0	Silty clay	0.95	7.24	1.60	26.7	18.1	8.6	0.138
B23t	73–112	15.0	40.8	44.2	Silty clay	0.76	7.58	1.58	24.8	15.7	9.1	0.144
B24t	112–147	19.3	37.2	43.5	Clay	0.39	7.65	1.65	24.4	16.7	7.7	0.127
Depth weighted mean		15.5	40.7	43.9		1.03	7.29	1.55	25.8	17.0	8.8	0.137

From Unger, P.W. and Pringle, F.B., Pullman Soils: Distribution, Importance, Variability, and Management, Texas Agric. Exp. Stn. Bull. No. B-1372, College Station, 1981.

Table 4 Tillage and Herbicide Treatments Imposed on a 36-Month Dryland Winter Wheat-Sorghum Fallow (WSF) Sequence, Bushland, TX, 1981–1991

WSF Phase	Month[a]	Description of Tillage Treatments
		No-Tillage
Fallow after wheat	1–11	Immediately after wheat harvest, apply 3.4 kg ha^{-1} (AI) atrazine + 0.84 kg ha^{-1} (AI) 2,4-D
Sorghum crop	12–16	At or a few days before sorghum planting, apply 0.3 kg ha^{-1} glyphosate (low volume) + surfactant; apply 1.7 kg ha^{-1} propazine (pre-emergence)
Fallow after sorghum	17–27	Early March, apply 23 to 35 g ha^{-1} (AI) chlorsulfuron + surfactant to sorghum stubble
Wheat crop	28–36	At or a few days before wheat planting, apply 0.3 kg ha^{-1} glyphosate (low volume) + surfactant; glyphosate is also used for weeds and grasses that may escape chlorsulfuron or atrazine during 11-month fallow periods; A 0.6 kg ha^{-1} rate is necessary if weeds are large or drought stressed
		Conventional (Stubble-Mulch Tillage with 5-ft "V" Blades)
Fallow after wheat	1–11	Sweep 3–4 times during summer and fall and once next spring
Sorghum crop	12–16	Sweep and plant sorghum; apply 1.7 kg ha^{-1} propazine (pre-emergence)
Fallow after sorghum	17–27	Sweep 4–5 times during spring and summer
Wheat crop	28–36	Sweep and plant wheat; apply 0.56 kg ha^{-1} 2,4-D for mustard control in March

[a] Month 1 is July, when wheat is harvested and fallow begins.

Table 5 Long-Term (34-y) Variable Means for Winter Wheat and Grain Sorghum Grown on Graded Terraced Watersheds Cropped in a Wheat-Sorghum-Fallow Sequence with Stubble Mulch Tillage, Bushland, TX, 1958–1991

Variable	34-y Mean	Range
	Winter Wheat	
Annual precipitation	468 mm	234–801 mm
Growing season precipitation	260 mm	134–456 mm
Growing season runoff	11 mm	0–80 mm
Available soil water at planting (0- to 1.8-m depth)	157 mm	66–238 mm
Wheat grain yield	1.20 Mg ha^{-1}	0–2.82 Mg ha^{-1}
Wheat straw yield (estimated)	1.98 Mg ha^{-1}	0.68–4.47 Mg ha^{-1}
Seasonal ET	332 mm	195–509 mm
Seasonal water use efficiency (grain)	0.36 kg m^{-3}	0–0.92 kg m^{-3}
Median planting date	October 1	September 18–November 20
Median harvest date	June 22	June 8–July 6
	Grain sorghum	
Growing season precipitation	253 mm	48–506 mm
Growing season runoff	25 mm	0–130 mm
Available soil water at planting (0- to 1.8-m depth)	158 mm	66–251 mm
Sorghum grain yield	2.3 Mg ha^{-1}	0–5.23 Mg ha^{-1}
Sorghum stover yield (estimated)	3.4 Mg ha^{-1}	1.1–6.2 Mg ha^{-1}
Seasonal ET	303 mm	157–457 Mg ha^{-1}
Seasonal water use efficiency (grain)	0.74 kg m^{-3}	0–1.45 kg m^{-3}
Median planting date	June 12	May 27–July 2
Median harvest date	October 27	September 29–November 27

its greatest effect on infiltration during fallow after sorghum because sorghum residue does not provide effective ground cover to prevent crusting. Infiltration data from rainfall simulator runs verify much lower infiltration rates with NT. No-till, however, resulted in 53% less soil erosion than SM. Although NT had more runoff, field observations revealed the smooth crusted surface was resistant to erosion and water moved across the NT field as a sheet, whereas with SM, water tended to channelize, resulting in rill erosion and more soil loss. Both systems, however, had erosion losses well under the soil loss tolerance of 11.2 Mg ha^{-1} y^{-1} established for the Pullman soil.[22]

3. Water Conservation

NT consistently stores substantially more water in the soil during fallow than SM, particularly with wheat stubble on the surface (Figure 4). Since runoff is greater from NT, we conclude that the greater water conservation with NT results from decreased evaporation resulting from (1) residue and soil crust effects of NT on vapor movement and (2) increased evaporation on SM resulting from drying of the tillage zone after each tillage operation.

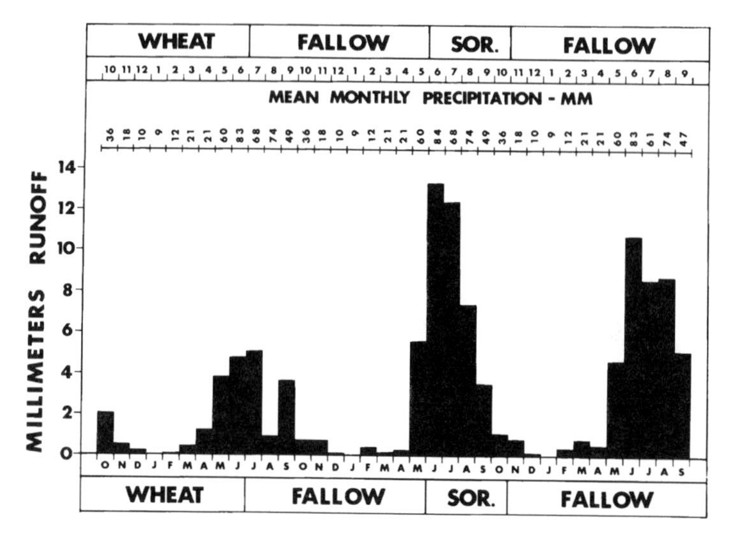

Figure 3 Average monthly precipitation and runoff from graded terraces cropped in a 3-year wheat-sorghum-fallow sequence with stubble-mulch tillage. Bushland, Texas 1958 to 1989.

Table 6 Tillage Effects on Annual (Calendar Year) Runoff and Erosion from Graded Terraced Watersheds Cropped in a Wheat-Sorghum-Fallow Sequence, 1984-1991, Bushland, TX

Phase of Sequence	Tillage	Average Runoff (mm y^{-1})	Average Soil Loss (Mg ha^{-1} y^{-1})
Wheat	NT	28	0.315
Wheat	SM	11	0.206
Sorghum	NT	41	0.432
Sorghum	SM	36	1.641
Fallow	NT	61	0.752
Fallow	SM	34	1.364

Note: Precipitation averaged 513 mm yr^{-1} compared to 465 mm yr^{-1} long-term average.

C. ORGANIC CARBON AND NITROGEN

Unger[9] sampled soils on NT and SM watersheds incrementally to a 20-cm depth 8 years after NT was initiated (Table 7). Average SOC concentration to a 10-cm depth (tillage zone) was 16.3 and 15.8 g kg^{-1} for NT and SM, respectively, indicating a trend toward OC gains in NT treatment compared to SM. The increase occurred in the top 2 cm, as would be expected. The same trend was exhibited with nitrate-N and inorganic P, slightly higher concentrations in the profile, with most of the difference occurring in

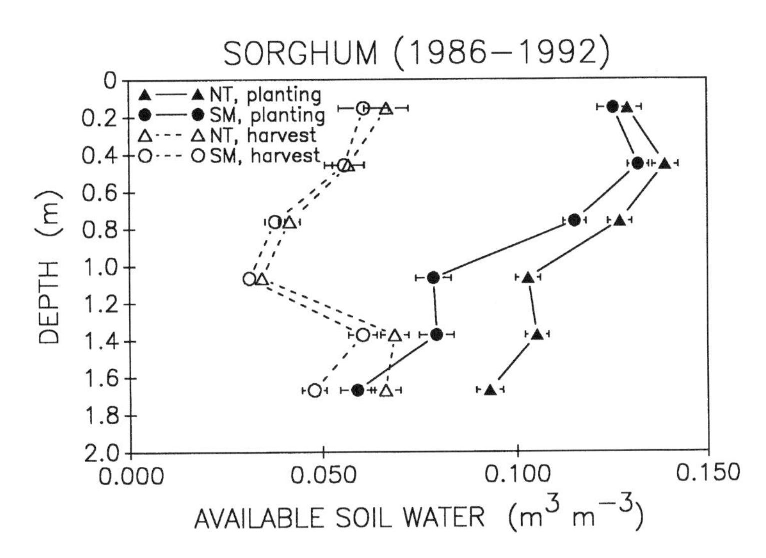

Figure 4 Tillage effects on soil water content at sorghum planting and harvest; 11 months of fallow (after wheat) preceded sorghum planting; Bushland, Texas.

the top 2 cm. These results were expected since with NT residues accumulate and decompose on or near the surface.

Eck and Jones[23] sampled NT and SM watersheds periodically for nitrate-N to a 1.8-m depth from 1985 to 1989. Nitrate-N showed an overall decrease for the study period with increases during fallow periods and decreases during cropping. Decreases were more drastic with sorghum than with wheat due to higher grain yields and N uptake by sorghum. The overall decrease over the period indicates that more N was removed than was mineralized from both tillage treatments. Nitrogen levels in the root zone were reduced enough so that N deficiency was observed on NT sorghum in 1989. Stubble-mulch watersheds almost always contained more nitrate-N than NT watersheds to a 1.8-m depth. Sampling to a 6-m depth in 1990 revealed that the NT and SM watersheds contained 65 and 105 kg ha^{-1} nitrate-N in the 1.8-m root zone and 511 and 470 kg ha^{-1} nitrate-N in the 6-m profile. The high levels of nitrate-N in these unfertilized soil profiles are a result of mineralized nitrate-N accumulated in excess of plant needs during 30+ years of SM tillage on WSF. It appears that the additional water conserved with NT had leached nitrate-N deeper into the soil profile, below the root zone for wheat or sorghum (Figure 5). Differential leaching rather than differential N mineralization may be responsible for lower nitrate-N accumulations in the root zone with NT.

IV. CROPPING AND TILLAGE SYSTEMS FOR MINIBENCHES

The objective is to develop cropping and tillage systems for minibenches[24] that conserve and use precipitation efficiently. The hypothesis is that a combination of the soil and water conservation practices of reduced or NT combined with land leveling will result in more effective use of water and soil resources. This study also provides the opportunity to evaluate tillage effects on soil physical and chemical characteristics for several cropping systems.

A. DESCRIPTION

The study was established in 1982 on a 16-ha site that had been in continuous dryland wheat or wheat fallow for at least 30 years. The soil is Pullman clay loam with a 1 to 1.5% slope. Narrow bench terraces (minibenches), 9.1 × 158 m, are constructed on the contour with berms around the perimeter to retain all precipitation on the plot. Check (nonlevel) plots are also on the contour, are the same size, have berms on the sides of the plots, but the ends are open so runoff can occur. Treatments are combinations of leveling — level (L) and nonlevel (NL); tillage — NT and SM; and cropping system — continuous wheat (CW), continuous sorghum (CS), wheat-fallow (WF), and wheat-sorghum-fallow (WSF). Treatments are arranged in a randomized block design with three replications. All phases of all rotations are present every year. For NT treatments, a combination of herbicides is used for weed control during both the cropped and fallow periods. Stubble-mulch tillage with large (1.5-m) V blades is used to control

Table 7 Tillage Method Effects on Soil Organic Carbon (SOC), Total N, Nitrate- N, and P Concentrations with Depth in Pullman Clay Loam, Bushland, TX, 1988

	Tillage Method[a]			
	SM		NT-81	
Depth (cm)	Mean	CI[b]	Mean	CI
		SOC (g kg^{-1})		
0–1	9.6	0.6	12.5	2.8
1–2	10.0	0.8	11.5	2.6
2–4	9.7	0.9	9.7	1.3
4–6	9.2	0.6	8.6	0.5
6–8	8.9	0.3	8.5	0.7
8–10	8.5	0.6	8.5	0.8
10–15	7.8	0.5	8.3	0.3
15–20	7.1	0.5	7.5	0.3
Prot. LSD[c]	0.7	—	1.6	—
Average 0–20	8.3	—	8.7	—
		Total N (g kg^{-1})		
0–1	1.11	0.09	1.23	0.14
1–2	1.08	0.08	1.21	0.16
2–4	1.11	0.13	1.07	0.10
4–6	1.14	0.15	0.96	0.05
6–8	1.11	0.13	0.95	0.15
8–10	1.07	0.15	0.99	0.10
10–15	1.00	0.12	0.97	0.06
15–20	0.94	0.10	0.96	0.05
Prot. LSD	0.13	—	0.11	—
Average 0–20	1.03	—	1.00	—
		Nitrate N (mg kg^{-1})		
0–1	29.1	7.1	34.0	15.7
1–2	21.6	7.7	25.5	10.4
2–4	25.6	6.1	29.2	10.3
4–6	23.4	7.9	31.3	6.4
6–8	22.1	8.2	27.5	0.2
8–10	16.8	3.7	21.1	1.8
10–15	12.3	1.1	16.1	6.4
15–20	8.9	0.8	10.0	1.5
Prot. LSD	6.3	—	8.7	—
Average 0–20	16.6	—	20.5	—
		Phosphorus (mg kg^{-1})		
0–1	12.3	3.3	21.3	4.3
1–2	14.2	2.5	20.7	5.3
2–4	13.5	1.4	14.5	5.2
4–6	9.8	2.6	11.5	2.3
6–8	10.0	2.8	9.0	1.7
8–10	7.1	2.5	6.9	1.3
10–15	5.0	1.3	4.5	1.9
15–20	3.2	0.6	3.0	1.6
Prot. LSD	2.4	—	3.5	—
Average 0–20	7.4	—	8.1	—
		pH		
0–1	6.14	0.37	6.02	0.31
1–2	6.15	0.29	6.10	0.24
2–4	6.13	0.30	6.10	0.21
4–6	6.15	0.29	6.09	0.17
6–8	6.15	0.24	6.12	0.15
8–10	6.15	0.27	6.13	0.13

Table 7 (continued) Tillage Method Effects on Soil Organic Carbon (SOC), Total N, Nitrate- N, and P Concentrations with Depth in Pullman Clay Loam, Bushland, TX, 1988

	Tillage Method[a]			
	SM		NT-81	
Depth (cm)	**Mean**	**CI[b]**	**Mean**	**CI**
10–15	6.20	0.26	6.20	0.10
15–20	6.32	0.23	6.29	0.09
Prot. LSD	NS[e]	—	0.20	—
Average 0–20	6.20	—	6.17	—

[a] Tillage methods are SM — stubble mulch tillage, NT-81 — no-tillage since 1981.
[b] CI — confidence interval = standard error of the mean (t = 0.05)
[c] Protected least significant difference at the P = 0.05 level.
[d] $NaHCO_3$ extractable.
[e] Not significant at P = 0.05.
From Unger, P.W., *Agron J.*, 83, 186, 1991. With permission.

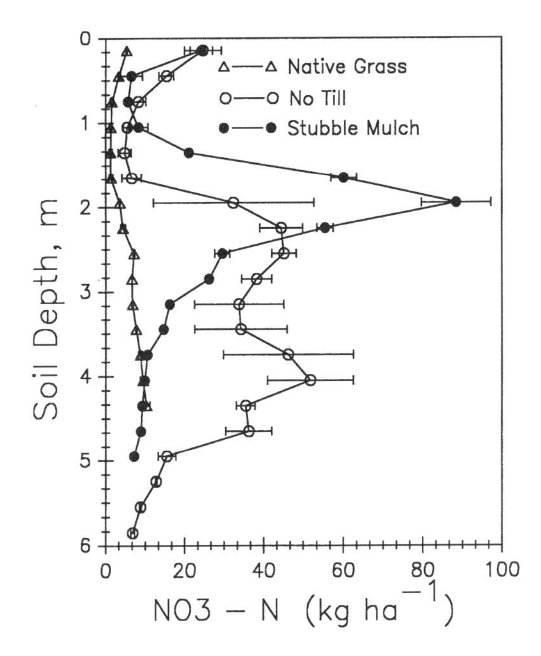

Figure 5 Depth profile of soil NO_3-N in unfertilized WSF graded-terraced notill and stubble mulch watersheds after 8 years of NT and 41 years of SM. A nearby virgin sod site was also sampled. Error bars represent one standard deviation. (From Eck, H.V. and Jones, O.R., *Agron. J.*, 84, 660, 1992.

weeds during fallow or noncrop periods and prepare seedbeds on all SM treatments. Stubble-mulch sorghum plots have herbicides applied (atrazine or propazine) for growing season weed control. To correct N deficiencies with some treatments, 56 kg ha^{-1} N is applied at planting of all crops (beginning in 1988). There are two 10- × 15-m check strips that are not fertilized in each plot; thus, fertilized and nonfertilized areas are available for sampling.

B. RESULTS AND DISCUSSION

1. Chemical Analyses

Unger et al.[25] sampled the plots at establishment in 1982 to determine the effects of land leveling on soil texture, organic carbon content, and aggregate stability. Mean SOC content differences for the 0- to 10-cm depth between nonleveled (18.0 g kg^{-1}) and leveled (17.5 g kg^{-1}) plots were not significant, but the sampling location on the leveled bench was significant with the cut and fill areas averaging 16.8 and

18.1 g kg^{-1}, respectively. The maximum depth of cut ranged from 3 to 10 cm and averaged 6.7 cm for the 34 minibenches that were leveled.

Leveled plots were sampled in three increments to a 30-cm depth in 1989 for chemical analyses (Table 8). For all cropping systems except WF, NT plots had significantly higher or tended toward higher SOC content in the 0- to 7.5-cm depth than on SM plots, indicating that NT is accumulating SOC. Wheat systems (CW and WF) had higher SOC means than WSF and CS systems, even though CS had the highest annual grain and forage yields of any system. Neither tillage nor cropping system had a significant effect on total N content. Available P appeared to be lower on CS, but the difference was not significant. Plots at the top of the slope had lower SOC and P contents than plots at lower slope positions. For the 7.5-to 15- and 15- to 30-cm depths, there were no apparent effects of cropping or tillage system on SOC or P content, but total N content tended to be higher for wheat systems and was significantly higher at the 15- to 30-cm depth. Detailed sampling with smaller depth increments near the surface is needed to delineate where changes in chemical properties are occurring in the soil profile. Also, non-leveled plots need to be included in the sampling plan.

2. Soil Nitrates

Eck and Jones[23] reported crop yields, nitrate accumulation during fallow and non-crop periods, and soil nitrate levels for leveled plots (unfertilized areas) for the 1986 to 1989 period. Yields with NT and SM were similar, except for CS where N deficiency was encountered and SM had higher yield. Average nitrate-N accumulation (kg ha^{-1}) from crop harvest until crop planting for the surface 1.2 m were CW NT, +20; CW SM, +37; CS NT, +28; CS SM, +24; WSF NT, +34; WSF SM, +52; WF NT, +57; and WF SM, +60. Tillage significantly affected N accumulation only on the WSF sequence. Nitrogen mineralization is not supplying sufficient N for annual cropping. Sorghum with a 3.4 Mg ha^{-1} yield requires in excess of 50 kg ha^{-1} N, about twice the N that is being mineralized during the non-crop period with CS. With similar nitrification rates for CS NT and CS SM, no reason is apparent for explaining the severe N deficiency observed on CS NT. Possible explanations might be differential leaching or increased volatilization of N from the surface with NT. With WSF, nitrate-N moved deeper in the profile under NT than under SM, indicating that differences in the root zone may have resulted from differential leaching rather than from differential nitrification where fallow was involved. Additional investigation of the N dynamics of NT and SM tillage systems is needed.

3. Grain Production

Jones et al.[26] reported tillage and crop sequence effects on grain yields and total production for level plots from 1984 to 1987. Tillage system had little effect on grain yield in any sequence. Crop sequence, however, had a fourfold effect on annualized grain production. Continuous sorghum had the greatest production with 3.45 Mg ha^{-1} y^{-1}. Annualized production of other sequences were CW, 1.08; WSF, 1.88; and WF, 0.78 Mg ha^{-1} y^{-1}. Systems with sorghum had much greater production than systems with wheat because of favorable summer rainfall distribution relative to plant needs. Beginning in 1988, sorghum yields on CS have been reduced due to N deficiency on unfertilized strips, with the yield reduction being greatest for NT. In 1989, adding 56 kg ha^{-1} N increased sorghum yields 18% (from 3820 to 4520 kg ha^{-1}) on CS SM and 96% (from 2700 to 5280 kg ha^{-1}) on CS NT.[23]

V. SUMMARY

Soil OC and N concentrations decline when native grasslands are converted to croplands. Soil OC loss is influenced by such factors as crops grown, crop rotations, crop residue management practices, climate, soil characteristics, fertilizer and manure applications, and tillage practices. In general, rates of OC loss (1) increase as tillage intensities increase, (2) are greater for cropping systems involving fallow than with continuous (annual) cropping, and (3) are greater for row crops than for small grain crops. Cropping systems and tillage methods are being evaluated in the semiarid southern Great Plains with respect to their potential, among other factors, for retarding the rates of soil OC loss. These systems are (1) conservation tillage on graded terraces and (2) cropping and tillage systems for minibenches.

For the graded terrace study, soil SOC to the 10-cm depth (tillage zone) was 16.3 g kg^{-1} with NT and 15.8 g kg^{-1} with SM tillage. The greatest difference due to tillage was in the upper 2 cm of soil where SOC was higher with NT than with SM. Organic carbon concentrations are greater in NT soils than in SM soils because NT retains crop residues at the surface where they decompose slower. In contrast, SM tillage partially incorporates residues with soil, thereby mixing them throughout the tillage

Table 8 Tillage and Cropping System Effects on Soil Organic Carbon (SOC), Total Nitrogen, and Available Phosphorus

		SOC (g kg^{-1})			Total N (g kg^{-1})			Available Phosphorus (mg kg^{-1})		
Sequence	**Tillage**	**0–7.5cm**	**7.5–15cm**	**15–30cm**	**0–7.5cm**	**7.5–15cm**	**15–30cm**	**0–7.5cm**	**7.5–15cm**	**15–30cm**
CS	NT	9.31	6.41	4.67	1.08	0.83	0.65	8.28	3.21	2.85
CS	SM	8.74	6.18	4.45	0.97	0.76	0.60	7.56	3.94	5.01
CW	NT	10.00	6.53	4.94	1.09	0.87	0.71	9.01	3.94	4.29
CW	SM	9.13	7.14	5.14	1.07	0.81	0.71	10.88	4.29	4.29
WSF-S	NT	9.48	6.57	5.26	1.03	0.78	0.61	10.12	4.65	4.29
WSF-S	SM	8.87	5.98	4.28	0.99	0.79	0.57	10.16	4.29	5.00
WSF-W	NT	9.09	6.19	4.15	0.99	0.75	0.61	10.53	4.65	4.65
WSF-W	SM	8.78	6.37	4.75	1.04	0.76	0.67	9.03	5.01	4.29
WF	NT	9.70	6.37	5.01	1.00	0.79	0.65	9.79	3.21	3.92
WF	SM	9.50	6.88	5.10	1.13	0.82	0.69	9.76	3.93	5.37
LSD (P = 0.05)		1.05	NS	NS	0.13	NS	NS	NS	NS	NS

Note: Only level plots were sampled, Bushland, TX.

From Eck, H.V., unpublished data, 1989.

layer. In this study, water conservation has been greater with NT, which has resulted in deeper leaching of nitrate-N than where SM tillage is used.

Organic matter concentrations were or tended to be significantly greater to the 7.5-cm depth on NT plots than SM tillage plots, indicating that SOC is accumulating on the NT plots of the minibench study. Although annual grain and forage yields have been greatest with continuous sorghum, SOC concentrations have been greater with continuous wheat and wheat-fallow cropping systems. Nitrate-N has moved deeper in the soil on NT than on SM tillage plots, probably due to greater soil water contents on NT plots.

REFERENCES

1. Finnell, H.H., Raw Organic Matter Accumulations under Various Systems of Culture, Oklahoma Agric. Exp. Stn. Bull., No. 216, Oklahoma Agric. and Mech. Coll., Stillwater, 1933.
2. Johnson, W.C., Stubble-Mulch Farming on Wheatlands of the Southern High Plains, USDA Circular No. 860, Washington, DC, 1950.
3. Harper, H.J., Sixty-Five Years of Continuous Wheat on Reddish Prairie Soil in Central Oklahoma, Oklahoma Agric. Exp. Stn. Bull., No. B-531, Oklahoma State University, Stillwater, 1959.
4. Hobbs, J.A. and Brown, P.L., Effects of Cropping and Management on Nitrogen and Organic Carbon Contents of a Western Kansas Soil, Kansas State Univ. of Agric. and Appl. Sci. Tech. Bull., No. 144, Manhattan, 1965.
5. Unger, P.W., Soil organic matter and nitrogen changes during 24 years of dryland wheat tillage and cropping practices, *Soil Sci. Soc. Am. Proc.*, 32, 427, 1968.
6. Johnson, W.C., Van Doren, C.E., and Burnett, E., Summer fallow in the Southern Great Plains, in Summer Fallow in the Western United States, USDA Agric. Res. Serv. Conserv. Res. Rep. No. 17, U S. Government Printing Office, Washington, D.C, 86, 1974.
7. Mathers, A.C. and Stewart, B.A., Manure effects on crop yields and soil properties, *Trans. Am. Soc. Agric. Eng.*, 27, 1022, 1984.
8. Unger, P.W., Surface soil physical properties after 36 years of cropping to winter wheat, *Soil Sci. Soc. Am. J.*, 46, 796, 1982.
9. Unger, P.W., Organic matter, nutrient, and pH distribution in no- and conventional-tillage semiarid soils, *Agron. J.*, 83, 186, 1991.
10. Chang, C.W., Effect of long-time cropping on soil properties in Northeastern New Mexico, *Soil Sci.*, 69, 359, 1950.
11. Locke, L.F. and Mathews, O.R., Relation of Cultural Practices to Winter Wheat Production, Southern Great Plains Field Station, Woodward, Oklahoma, USDA Circular No. 917, Washington, D.C, 1953.
12. Haas, H.J., Evans, C.E., and Miles, E.F., Nitrogen and Carbon Changes in Great Plains Soils as Influenced by Cropping and Soil Treatments, USDA Tech. Bull. No. 1164, Washington, DC, 1957.
13. Zingg, A.W. and Whitfield, C.J., A summary of Research Experience with Stubble-Mulch Farming in the Western States, USDA Tech. Bull. No. 1166, Washington, DC, 1957.
14. Unger, P.W., Dryland Winter Wheat and Grain Sorghum Cropping Systems — Northern High Plains of Texas, Texas Agric. Exp. Stn. Bull. No. B-1126, College Station, 1972.
15. Johnson, W.C. and Davis, R.G., Research on Stubble-Mulch Farming of Winter Wheat, USDA Agric. Res. Serv. Conserv. Res. Rep. No. 16, Washington, DC, 1972.
16. Webb, B.B., Tucker, B.B., and Westerman, R.L., The Magruder Plots: Taming the Prairies through Research, Oklahoma Agric. Exp. Stn. Bull. No. B-750, Oklahoma State University, Stillwater, 1980.
17. Mathers, A.C., Gardner, H.R., Lotspeich, F.B., Taylor, H.M., Laase, G.R., and Daniell, R.E., Some Morphological, Physical, Chemical, and Mineralogical Properties of Seven Southern Great Plains Soils, USDA Rep. ARS 41-85, Washington, DC, 1963.
18. Eck, H.V., Ford, R.H., and Fanning, C.D., Productivities of Horizons of Seven Benchmark Soils of the Southern Great Plains, USDA Agric. Res. Serv. Conserv. Res. Rep. No. 11, Washington, DC, 1967.
19. Unger, P.W. and Pringle, F.B., Pullman Soils: Distribution, Importance, Variability, and Management, Texas Agric. Exp. Stn. Bull. No. B-1372, College Station, 1981.
20. Hauser, V.L., Van Doren, C.E., and Robins, J.S., A comparison of level and graded terraces in the Southern High Plains, *Trans. Am. Soc. Agric. Eng.*, 5, 75, 1962.
21. Howell, T.A., Grain, dry matter relationships for winter wheat and grain sorghum — Southern High Plains, *Agron. J.*, 82, 914, 1990.
22. Wischmeier, W.H. and Smith, D.D., Predicting Rainfall Erosion Losses, A Guide to Conservation Planning, USDA Handbook No. 537, Washington, DC, 1978.
23. Eck, H.V. and Jones, O.R., Soil N status as affected by tillage, crops, and crop sequences, *Agron. J.*, 84, 660, 1992.
24. Jones, O.R., Land forming effects on dryland sorghum production in the Southern Great Plains, *Soil Sci. Soc. Am. J.*, 45, 606, 1981.
25. Unger, P.W., Fulton, L.J., and Jones, O.R., Land-leveling effects on soil texture, organic matter content, and aggregate stability, *J. Soil Water Conserv.*, 45, 412, 1990.

26. Jones, O.R., Wiese, A.F., and Johnson, G.L., Tillage and crop sequence effects on dryland crop production, in *Challenges in Dryland Agriculture — A Global Perspective, Proc. Int. Conf. on Dryland Farming*, Unger, P.W., Sneed, T.V., Jordan, W.R., and Jensen, R., Eds., Texas Agric. Exp. Stn., College Station, 1990, 826.

Index

Index*

*In this index, (**D**) indicates information on disk.

D

E

F

G

H

I

J

K

L

M

N

O

P

R

S

T

U

W

Y

Z